The Ideal Theory and Arithmetic of Rings, Monoids, and Semigroups

AF795874

CONTEMPORARY MATHEMATICS

836

The Ideal Theory and Arithmetic of Rings, Monoids, and Semigroups

2nd International Joint Meeting of the AMS–UMI
Special Session on The Ideal Theory
and Arithmetic of Rings, Monoids, and Semigroups
July 23–26, 2024
Palermo, Italy

Scott T. Chapman
Editor

AMERICAN MATHEMATICAL SOCIETY

EDITORIAL COMMITTEE

Michael Loss, Chair

John Etnyre Angela Gibney Alex Kontorovich

2020 *Mathematics Subject Classification.* Primary 13A05, 13A15, 13B22, 20M14, 05E40, 20M12, 11B75, 11R04, 11R27, 20M13.

Library of Congress Cataloging-in-Publication Data

Names: International Joint Meeting of the AMS-UMI Special Session on the Ideal Theory and Arithmetic of Rings, Monoids, and Semigroups (2nd : 2024 : Palermo, Italy) author | Chapman, Scott T. editor

Title: The ideal theory and arithmetic of rings, monoids, and semigroups / Scott T. Chapman, editor.

Description: Providence, Rhode Island : American Mathematical Society, [2026] | Series: Contemporary mathematics, 0271-4132 ; volume 836 | Includes bibliographical references.

Identifiers: LCCN 2026004990 | ISBN 9781470480455 paperback | ISBN 9781470485443 ebook

Subjects: LCSH: Commutative algebra | Commutative rings | Algebraic number theory | Group theory | AMS: Commutative algebra – General commutative ring theory – Divisibility; factorizations | Commutative algebra – General commutative ring theory – Ideals; multiplicative ideal theory | Commutative algebra – Ring extensions and related topics – Integral closure of rings and ideals | Number theory – Sequences and sets – Other combinatorial number theory | Number theory – Algebraic number theory: global fields – Algebraic numbers; rings of algebraic integers | Number theory – Algebraic number theory: global fields – Units and factorization | Group theory and generalizations – Semigroups – Ideal theory | Group theory and generalizations – Semigroups – Arithmetic theory of monoids | Group theory and generalizations – Semigroups – Commutative semigroups | Combinatorics – Algebraic combinatorics – Combinatorial aspects of commutative algebra | LCGFT: Conference papers and proceedings

Classification: LCC QA251.3 .I593 2026

LC record available at https://lccn.loc.gov/2026004990

DOI: https://doi.org/10.1090/conm/836

Color graphic policy. Any graphics created in color will be rendered in grayscale for the printed version unless color printing is authorized by the Publisher. In general, color graphics will appear in color in the online version.

Copying and reprinting. Individual readers of this publication, and nonprofit libraries acting for them, are permitted to make fair use of the material, such as to copy select pages for use in teaching or research. Permission is granted to quote brief passages from this publication in reviews, provided the customary acknowledgment of the source is given.

Republication, systematic copying, or multiple reproduction of any material in this publication is permitted only under license from the American Mathematical Society. Requests for permission to reuse portions of AMS publication content are handled by the Copyright Clearance Center. For more information, please visit `www.ams.org/publications/pubpermissions`.

Send requests for translation rights and licensed reprints to `reprint-permission@ams.org`.

© 2026 by the American Mathematical Society. All rights reserved.
The American Mathematical Society retains all rights
except those granted to the United States Government.
Printed in the United States of America.

∞ The paper used in this book is acid-free and falls within the guidelines
established to ensure permanence and durability.
Visit the AMS home page at `https://www.ams.org/`

10 9 8 7 6 5 4 3 2 1 31 30 29 28 27 26

Contents

Preface

As the title reflects, this volume is a collection of papers which explores the ideal theory, arithmetic, and factorization properties of various algebraic objects. Its topics are quite broad, as the systems covered include, but are not limited to, general rings and integral domains, algebraic number rings, power series rings, numerical semigroups, Puiseux monoids, modules, and Rees algebras. The original idea for this volume emanated from a Special Session with the same title which took place from July 23 to July 26, 2024 at the 2nd International Joint Meeting of the Unione Matematica Italiana (UMI) and the American Mathematical Society (AMS), held at the Università degli Studi di Palermo. This Special Session was co-organized by Marco D'Anna of the Dipartimento di Matematica e Informatica at the Università di Catania and myself. The session included twenty-one talks by speakers from ten different countries.

There is a strong history of Italian participation in algebra of this type. There was a somewhat similar Special Session (Commutative Rings and Integer-Valued Polynomials) at the 1st International Joint Meeting of the UMI and AMS in Pisa in 2002. That meeting was preceeded a week earlier by the Algebra Conference – Venezia 2002, which was organized by COFIN 2000, a joint research group of six Italian universities. There have been numerous workshops held at the Incontro INdAM, in Cortona (in particular, 2004, 2006, and 2018) which have touched on similar topics.

Less than two months following the session in Palermo (September 11–13), the Università di Catania hosted a conference titled *Advances in Commutative Ring Theory*. One of the purposes of the conference was to honor the retirement of Professor Alan Loper from the Department of Mathematics at the Ohio State University in Columbus, Ohio. This meeting featured seventeen talks by speakers from five different countries. As many of the topics covered overlapped with the Palermo meeting, I invited several of the speakers from Catania to contribute to this volume, and happily, several accepted my offer. I thank the organizers of Catania meeting, Carmelo Finocchiaro, Lorenzo Guerrieri, Giulio Peruginelli, Dario Spirito, Francesca Tartarone for their help in completing this collection of papers.

The final result is that this volume contains fourteen articles contributed by thirty-eight coauthors from at least seven different countires. Why does the popularity of the topics covered in these pages endure over the years? Over the past three decades, the theory and structure of the ideals and arithmetic of structures related to rings, integral domains, and monoids have taken several important steps forward. Many applications of this theory, with particular attention to the multiplicative monoids of integral domains and their combinatorial or numerical applications to ring theory, have appeared in high-quality journals throughout the mathematical literature. It is the aim of this volume to review recent developments in this area by bringing together researchers from different areas of algebra under the umbrella of commutative monoids, semigroups, and rings. We hope that you agree that these meetings and the papers contained within this book were not only a success, but reflect the interests and trends that this research area will follow for decades to come.

Scott Chapman
San Antonio, Texas
August 1, 2025

Contemporary Mathematics
Volume **836**, 2026
https://doi.org/10.1090/conm/836/16701

On long-term problems in multiplicative ideal theory and factorization theory

Alfred Geroldinger, Hwankoo Kim, and K. Alan Loper

ABSTRACT. In this survey article we discuss key open problems which could serve as a guidance for further research directions of multiplicative ideal theory and factorization theory.

1. Introduction

The origins of multiplicative ideal theory can be traced back to the beginnings of commutative algebra in the first half of the twentieth century. We refer to articles of Halter-Koch [**86**, **90**] for a discussion of the early development. The monographs of Gilmer [**79**] and Larsen and McCarthy [**109**], which appeared in the 1970s, strongly influenced the area. Building on the work of Lorenzen and Aubert, Halter-Koch showed (in his book [**86**], published 1998) that large parts of multiplicative ideal theory can be derived in the purely multiplicative setting of commutative monoids. This viewpoint also turned out to be crucial for the further development of factorization theory.

Factorization theory has its origin in algebraic number theory. Starting points were results (in the 1960s) of Carlitz and of Narkiewicz, who studied the asymptotic behavior of arithmetic counting functions in orders of algebraic number fields. From the 1990s on, factorization theory expanded its scope into commutative ring theory and since then, multiplicative ideal theory and factorization theory developed in parallel and stimulated each other. At present, both areas are in the midst of a stormy development. This is documented by recent research monographs of Chabert, Elliott, Fontana et al., Geroldinger et al., Grynkiewicz, Halter-Koch, Knebusch et al., Wang and Kim ([**21**, **40**, **55**, **65**, **84**, **90**, **105**, **106**, **140**]), by a steady flow of conferences devoted to these fields and the associated conference proceedings (for some of the last ten years, see [**8**, **19**, **22**, **28**, **47**, **53**, **54**]).

2020 *Mathematics Subject Classification.* Primary 13A05, 13A15; Secondary 13F05, 13F15, 13D22, 13D30, 20M12, 20M13.

Key words and phrases. Multiplicative ideal theory, factorization theory, Prüfer domains, Gorenstein homological algebra, Mori monoids, Krull monoids.

The first author was supported by the Austrian Science Fund FWF, Project P36852-N.

The second author was supported in part by the Basic Science Research Program through the National Research Foundation of Korea (NRF), funded by the Ministry of Education (2021R1I1A3047469).

©2026 American Mathematical Society

The above mentioned monographs contain an abundance of open problems (in addition, we would like to refer to the survey [**20**] by Cahen et al. on unsolved questions). Thus we made no attempt to give a comprehensive overview on open problems, but we tried to offer brief discussions of some research directions which we hope will stimulate further research in multiplicative ideal and factorization theory.

2. Preliminaries

Let $\mathbb{N}$ denote the set of positive integers and set $\mathbb{N}_0 = \mathbb{N} \cup \{0\}$. For integers $a, b \in \mathbb{Z}$, let $[a, b] = \{x \in \mathbb{Z} \colon a \le x \le b\}$ denote the discrete interval between a and b. For subsets A, B of an additive abelian group G, let $A + B = \{a + b \colon a \in A, b \in B\} \subseteq G$ denote their sumset. If S is any set, then $\mathcal{P}(S)$ denotes the system of all subsets of S and $\mathcal{P}_{\text{fin}}(S)$ the system of all finite, nonempty subsets of S.

All semigroups are commutative and have an identity element. If not stated otherwise, we use multiplicative notation. Let S be a semigroup. We denote by $S^\times$ the group of invertible elements, and we say that S is *reduced* if $S^\times = \{1_S\}$. Then $S_{\text{red}} = S/S^\times = \{aS^\times : a \in S\}$ denotes the associated reduced semigroup of S. We say that S is *cancellative* if $a, b, c \in S$ and $ab = ac$ implies that $b = c$. By a *monoid*, we mean a cancellative semigroup. Let M be a monoid. An element $a \in M$ is called an *atom* (or *irreducible*) if $a \notin M^\times$ and an equation $a = bc$, with $b, c \in M$, implies that $b \in M^\times$ or $c \in M^\times$. We denote by $\mathcal{A}(M)$ the set of atoms and by $\mathsf{q}(M)$ the quotient group of M.

Next we introduce a key concept of multiplicative ideal theory, namely ideal systems or, equivalently, star-operations ($*$-operations). The term star-operation was popularized by Gilmer, following the term *Strich-operation* introduced by Krull. Following Aubert, Halter-Koch used the term *ideal system*. These concepts were generalized to semistar operations and then to abstract closure operations by Elliott. Ideal systems were generalized to weak ideal systems and then to module systems by Halter-Koch [**88, 90**]. In the present survey we use both the terms star-operation and ideal system.

Let M be a monoid. An *ideal system* on M is a map $r \colon \mathcal{P}(M) \to \mathcal{P}(M)$, defined by $X \mapsto X_r := r(X)$ for all $X \subseteq M$, such that the following conditions are satisfied for all subsets $X, Y \subseteq M$ and all $c \in M$:

- $X \subseteq X_r$.
- $X \subseteq Y_r$ implies $X_r \subseteq Y_r$.
- $cM \subseteq \{c\}_r$.
- $cX_r = (cX)_r$.

Let r be an ideal system on M. A subset $I \subseteq M$ is called an r*-ideal* if $I_r = I$. We denote by $\mathcal{I}_r(M)$ the set of all nonempty r-ideals, and we define r-multiplication by setting $I \cdot_r J = (IJ)_r$ for all $I, J \in \mathcal{I}_r(M)$. Then $\mathcal{I}_r(M)$ together with r-multiplication is a reduced semigroup with identity element M, and it is a multiplicative lattice. Conversely, by [**39**], for every multiplicative lattice there are a commutative semigroup H and a weak ideal system r on H such that L is isomorphic to the lattice of r-ideals of H.

A subset $X \subseteq \mathsf{q}(M)$ is called (M-)*fractional* if $cX \subseteq M$ for some $c \in M$. We define

$$X_r := c^{-1}(cX)_r \,,$$

and note that the definition does not depend on the choice of c. A subset $I \subseteq \mathsf{q}(M)$ is a *fractional r-ideal* if $I = X_r$ for some fractional subset $X \subseteq \mathsf{q}(M)$. We denote by $\mathcal{F}_r(M)$ the set of fractional r-ideals. For $I, J \in \mathcal{F}_r(M)$, the set IJ is M-fractional, and we call $I \cdot_r J := (IJ)_r \in \mathcal{F}_r(M)$ the r-product of I and J. A fractional r-ideal $J \in \mathcal{F}_r(M)$ is called *r-invertible* if there exists a fractional r-ideal J' such that $J \cdot_r J' = M$. $(\mathcal{F}_r(M), \cdot_r)$ is a semigroup containing $\mathcal{I}_r(M)$, $\mathcal{F}_r(M)^\times$ is the group of r-invertible fractional r-ideals, and $\mathcal{I}_r^*(M) = \mathcal{F}_r^\times(M) \cap \mathcal{I}_r(M)$ the monoid of r-invertible r-ideals of M with r-multiplication.

The map $\delta\colon \mathsf{q}(M) \to \mathcal{F}_r(M)^\times$, defined by $\delta(a) = aM$, is a group homomorphism with kernel $\ker(\delta) = M^\times$. Its cokernel

$$\mathcal{C}_r(M) = \mathcal{F}_r(M)^\times / \delta(\mathsf{q}(M))$$

is called the *r-class group* of M. For subsets $X, Y \subseteq \mathsf{q}(M)$, we denote by $X_s = XM$, by $(X \colon Y) = \{x \in \mathsf{q}(M) \colon xY \subseteq X\}$, by $X^{-1} = (M \colon X)$, by $X_v = (X^{-1})^{-1}$ and by $X_t = \bigcup_{E \subseteq X, |E| < \infty} E_v$. We will use the s-system, the v-system, the t-system, and the w-system. The v-system was introduced by Krull and v-ideals are also called divisorial ideals. The w-system was introduced by R. McCasland and F. Wang (see Section 7).

All rings are commutative and have an identity element. For a commutative ring D, we denote by $D^\times$ its group of units and by $D^\bullet$ its monoid of regular elements. Let D be a domain with quotient field K. We denote by $\mathcal{A}(D)$ the set of atoms, by $\mathsf{q}(D)$ the quotient field of D, by $\overline{D}$ the integral closure of D, and by $\widehat{D} = \widehat{D^\bullet} \cup \{0\}$ the complete integral closure of D.

An *ideal system* on D is a map $r\colon \mathcal{P}(D) \to \mathcal{P}(D)$ such that the following conditions are satisfied for all subsets $X, Y \subseteq D$ and all $c \in D$:

- $X \cup \{0\} \subseteq X_r$.
- $X \subseteq Y_r$ implies $X_r \subseteq Y_r$.
- $cM \subseteq \{c\}_r$.
- $cX_r = (cX)_r$.

If $r\colon \mathcal{P}(D) \to \mathcal{P}(D)$ is an ideal system on D, then $r^\bullet\colon \mathcal{P}(D^\bullet) \to \mathcal{P}(D^\bullet)$, defined by $r^\bullet(X) = r(X) \setminus \{0\}$ for every $X \subseteq D$, is an ideal system on $D^\bullet$. Conversely, if $r\colon \mathcal{P}(D^\bullet) \to \mathcal{P}(D^\bullet)$ is an ideal system on $D^\bullet$, then $r^\circ\colon \mathcal{P}(D) \to \mathcal{P}(D)$, defined by $r^\circ(X) = r(X) \cup \{0\}$ for every $X \subseteq D$, is an ideal system on D. Thus, in the sequel we will not distinguish between $r, r^\bullet$, and r° and simply speak of r-ideals of D and of $D^\bullet$. In particular, $\mathcal{I}_r(D)$, $\mathcal{F}_r(D)$, $\mathcal{I}_r^*(D)$, and $\mathcal{C}_r(D) = \mathcal{C}_r(D^\bullet)$ have their obvious meaning.

The d-system $d\colon \mathcal{P}(D) \to \mathcal{P}(D)$, defined by $X_d = \langle X \rangle$ for every $X \subseteq D$, is the system of classical ring ideals of D. Whenever we speak of ideals of D, we mean d-ideals and in all notation introduced above we omit the d-suffix. In particular, $\mathcal{I}^*(D) = \mathcal{I}_d^*(D)$ is the monoid of invertible ideals of D and $\mathcal{C}_d(D) = \mathrm{Pic}(D)$ is the Picard group of D.

A monoid (resp. a domain) is said to be

- *Mori* (resp. *v-Noetherian*) if it satisfies the ascending chain condition on v-ideals.
- *strong Mori* (resp. *w-Noetherian*) if it satisfies the ascending chain condition on w-ideals.
- *Krull* if it is completely integrally closed and Mori.

Every v-ideal is a t-ideal and if the monoid is Mori, then $v = t$. Krull monoids and Noetherian domains are strong Mori, and every strong Mori monoid is Mori, because every t-ideal is a w-ideal.

3. Intersecting valuation domains

Let D be a domain with quotient field K and suppose that $D \subsetneq K$. In general, there are a lot of domains between D and K. For example, if D is a Noetherian domain with Krull dimension greater than one, then there are uncountably many domains between D and K.

A valuation domain V with quotient field L is, in a sense, close to being all of L. For example, every domain between V and L is again a valuation domain and is a localization of V. And if V has finite Krull dimension then there are only finitely many such intermediate domains.

A theorem of Krull should make the study of valuation domains worthwhile.

THEOREM 3.1. *Every integrally closed domain D (not a field) can be expressed as the intersection of all the valuation domains between D and its quotient field.*

It seems reasonable then that valuation domains should play a central role in the study of integrally closed domains. They are the indivisible building blocks from which everything is made. Some work has been done in this direction but not a lot. A principal difficulty is that given an integrally closed domain D it is usually hard to determine what the valuation domains are. And, given a collection of valuation domains it is hard to discern what the structure of the intersection is. A valuation overring V of D is local, and its maximal ideal contracts to a prime ideal of D. But, often, the maximal ideals of uncountably many valuation overrings all contract to the same prime ideal of D.

Clearly, many efforts are made to build new examples of domains as overrings of well-understood domains. This is generally done by starting with a domain that is well-understood and adding in some fractions to obtain an overring. Instead of this bottom-up approach we explore here a top-down approach of intersecting valuation domains.

PROBLEM 1.

(1) Given an integrally closed domain what can we say about the collection of its valuation overrings?
(2) Given a collection of valuation domains what can we say about the domain that results from intersecting them?

These are both difficult questions about which little is known. In this note we focus on one facet of the second one. As we observed above, valuation domains are local. One could then hope that, given a domain, we could obtain valuation overrings by localizing at prime ideals. This rarely succeeds. But sometimes it does. We say that a domain is a Prüfer domain if the localization at any prime ideal is a valuation domain. So, if given a Prüfer domain, it is easy to write the domain as an intersection of valuation domains. The representation is simply the intersection of the localizations at prime ideals. The other direction is a fascinating question.

PROBLEM 2. Given a collection of valuation domains, what are necessary and sufficient conditions that we can put on the collection that would make the intersection a Prüfer domain?

The following theorem gives partial answers to Problem 2.

THEOREM 3.2. *Let $\{V_i \mid i \in \Omega\}$ be a collection of valuation domains that all have the same quotient field. In each of the following cases the intersection of the valuation domains is a Prüfer domain.*

(1) *Ω is finite.* [**115**]
(2) *There is a monic non-constant polynomial $f \in K[X]$, whose coefficients lie in each V_i and which has no root in the residue field of any V_i.* [**80**]
(3) *There is a field L, whose cardinality is strictly larger than the cardinality of Ω, such that $L \subseteq V_i$ for each $i \in \Omega$.* [**116**]

The common theme of these three properties (1) - (3) seems to be that the collection of valuation domains needs to be, in some sense, thin or sparse. This observation, in turn, might lead one to seek a topological condition. We digress to describe a construction of Krull which shows that looking at the standard topologies on a collection of valuation domains is not sufficient to determine when the intersection is a Prüfer domain.

We start by giving a quick demonstration that Prüfer domains have properties that make them easy to work with.

Let K be a field and let $D = K[X, Y]$ be the polynomial ring in X and Y. Let $J = (X, Y)$ and $I = (X^2, Y^2)$ be ideals of D. It is easy to see that $J^3 = JI$. It seems intuitively clear then that $J^2 = I$ should be true. But it is not. On the other hand, suppose that T is a Prüfer domain and let $d, r \in T$ be nonzero. Let $B = (d, r)$ and $H = (d^2, r^2)$ be ideals of T. As above, $BH = B^3$. is clear. Now, however, in the context of a Prüfer domain, we can cancel B and conclude that $B^2 = H$ is true.

Krull's idea was to define certain closure operations on the collection of fractional ideals of a domain. These closure operations are known as star operations. Of particular interest are the star operations he identified as ab (arithmetisch brauchbar = arithmetically useful). Let D be an integrally closed domain and let $*$ be an ab star operation. Then when I, J, H are finitely generated ideals such that $(IH)^* \subseteq (JH)^*$ then we can conclude that $I^* \subseteq J^*$.

Before pressing on, we recap what we have discussed so far and where we want to go with it.

- Every integrally closed domain is the intersection of its valuation overrings.
- Prüfer domains are precisely the domains for which any localization at a prime ideal is a valuation domain.
- We would like to know what conditions on a collection of valuation rings with a common quotient field characterize those collections for which the intersection is a Prüfer domain.
- We have some results which seem to indicate that the property (or properties) that work will be topological in nature.

There are two well-known topologies which can be placed on a collection of valuation domains (for a sample of contributions in this direction see [**33, 37, 51, 52, 83**]). These are known as

(1) the Zariski topology, and

(2) the constructible or patch or ultrafilter topology.

Some amazing work of Krull demonstrates clearly that neither of these topologies alone can provide a solution to our problem.

Let V be a valuation domain. For a polynomial $f \in V[X]$, let $c(f)$ be the ideal of V generated by the coefficients of f. Let $N = \{f \in V[X] \mid c(f) = V\}$ and finally, let $V(X) = N^{-1}V[X]$. $V(X)$ is known as the *Nagata extension* of V. $V(X)$ is a valuation domain which is almost identical to V. The only significant difference between V and $V(X)$ is that $V(X)$ has more units.

Let D be an integrally closed domain (not a field) with quotient field K. Krull showed how to use ab star operations on D to build what he called Kronecker function domains. [**56**] There is always a unique minimal Kronecker function domain which we label as $\mathrm{Kr}(D)$ which has the following properties.

(1) $\mathrm{Kr}(D)$ is an overring of $D[X]$.
(2) $\mathrm{Kr}(D) \cap K = D$.
(3) $\mathrm{Kr}(D)$ is a Prüfer domain.
(4) The localizations $\mathrm{Kr}(D)_{\mathfrak{p}}$ where $\mathfrak{p}$ runs over the prime ideals of $\mathrm{Kr}(D)$ are precisely the valuation domains $V(X)$, where V runs over the valuation overrings of D.

The key importance of $\mathrm{Kr}(D)$ for us is that when D is an integrally closed domain the space of valuation overrings of D and the space of valuation overrings of $\mathrm{Kr}(D)$ are homeomorphic for both of the topologies noted above. This is problematic because, while D can be any integrally closed domain, $\mathrm{Kr}(D)$ is always a Prüfer domain. Hence, there is no hope of characterizing when an intersection of valuation domains is a Prüfer domain using just information about the topology of the collection of valuation domains.

4. Classification of integral domains between $\mathbb{Z}[X]$ and $\mathbb{Q}[X]$

We survey recent developments in the classification of domains that lie strictly between the polynomial ring $\mathbb{Z}[X]$ and $\mathbb{Q}[X]$. We focus on two major studies:

[**111**] which classifies *integrally closed domains* (including Prüfer and Dedekind domains) via order-theoretic arguments, extending classical valuation-theoretic methods.

[**119**] which classifies *Dedekind domains* using the concept of *generalized rings of integer-valued polynomials*, broadening earlier results on factorization and class groups.

Both works build on classical ideas from valuation theory, factorization theory, and the arithmetic of polynomial rings, offering new insights into how certain local constructions (valuation overrings) glue together to form global integrally closed domains between $\mathbb{Z}[X]$ and $\mathbb{Q}[X]$.

The primary motivation of [**111**] is to give an order-theoretic classification of integrally closed domains situated between $\mathbb{Z}_{(p)}[X]$ and $\mathbb{Q}[X]$, where p is a prime number and $\mathbb{Z}_{(p)}$ denotes the localization of $\mathbb{Z}$ at $p\mathbb{Z}$. In particular, [**111**] focuses on the structure of valuation overrings and extends classical results of MacLane on constructing valuations via *key polynomials*.

A representative theorem from [**111**] is:

THEOREM 4.1 ([**111**]). *Let $p \in \mathbb{Z}$ be a fixed prime, and let D be a domain such that*

$$\mathbb{Z}_{(p)}[X] \subseteq D \subseteq \mathbb{Q}[X].$$

Then D is Prüfer if and only if D is integrally closed and all the p-unitary valuation overrings of D are limit.

Constructive aspects hinge on the notion of *key polynomials*, which allow for systematic building of valuation overrings, ultimately classifying the possible integrally closed domains in a step-by-step manner.

Having explored the order-theoretic classification of integrally closed domains, we now shift our focus to Dedekind domains. To facilitate this discussion, we first introduce the relevant notation.

Let $\mathbb{P}$ be the set of all prime numbers. For a fixed $p \in \mathbb{P}$, we adopt the following notation:

- $\mathbb{Z}_p$ denotes the ring of p-adic integers.
- $\mathbb{Q}_p$ denotes the field of p-adic integers.
- $\mathbb{C}_p$ denotes the completion of the algebraic closure of the field $\mathbb{Q}_p$.
- $\overline{\mathbb{Z}}_p$ denotes the absolute integral closure of $\mathbb{Z}_p$.
- $\overline{\overline{\mathbb{Z}}} = \prod_{p\in\mathbb{P}} \overline{\mathbb{Z}}_p$.

A *generalized ring of integer-valued polynomials* is defined, for a subset $\underline{E} \subset \overline{\overline{\mathbb{Z}}}$, as

$$\mathrm{Int}_{\mathbb{Q}}(\underline{E}, \overline{\overline{\mathbb{Z}}}) = \{ f \in \mathbb{Q}[X] \mid f(\alpha) \in \overline{\overline{\mathbb{Z}}} \text{ for all } \alpha \in \underline{E}\}.$$

This idea generalizes the classical ring of integer-valued polynomials.

Let $\underline{E} = \prod_p E_p$ be a subset of $\overline{\overline{\mathbb{Z}}}$. Following [**118**, Definition 2.11], we say that $\underline{E}$ is *polynomially factorizable* if, for every $g \in \mathbb{Z}[X]$ and every $\alpha = (\alpha_p) \in \underline{E}$, there exist $n, d \in \mathbb{Z}$ with $n, d \geq 1$ such that $g(\alpha)^n/d$ is a unit of $\overline{\overline{\mathbb{Z}}}$.

In [**118**], the author classifies *Dedekind domains with finite residue fields of prime characteristic* that lie between $\mathbb{Z}[X]$ and $\mathbb{Q}[X]$. These domains are precisely realized as generalized rings of integer-valued polynomials. The work is inspired by earlier studies on factorization theory, integer-valued polynomials, and class group computations.

A key observation is that if W is a one-dimensional valuation overring of such a Dedekind domain with $\mathbb{Q} \subset W$, then $W = \mathbb{Q}[X]_{(q)}$ for some irreducible polynomial $q \in \mathbb{Q}[X]$, implying that its residue field has infinite cardinality. In [**119**], the author extends this classification by relaxing the condition on the cardinality of residue fields of prime characteristic.

Recall that an element $\alpha \in \mathbb{C}_p$ has *bounded ramification* if the extension $\mathbb{Q}_p \subseteq \mathbb{Q}_p(\alpha)$ has finite ramification. Denote by $\mathbb{C}_p^{br}$ the set of all elements of $\mathbb{C}_p$ with bounded ramification, and let $\mathbb{O}_p$ be the unique local ring of $\mathbb{C}_p$. Define $\mathcal{O} = \prod_{p\in\mathbb{P}} \mathbb{O}_p \subset \prod_{p\in\mathbb{P}} \mathbb{C}_p$.

THEOREM 4.2 ([**119**, Theorem 3.4]). *Let R be a Dedekind domain such that $\mathbb{Z}[X] \subset R \subseteq \mathbb{Q}[X]$. Then R is equal to $\mathrm{Int}_{\mathbb{Q}}(\underline{E}, \mathcal{O})$ for some polynomially factorizable subset $\underline{E} = \prod_{p\in\mathbb{P}} E_p \subset \mathcal{O}$, where for each prime p, the set $E_p \subset \mathbb{O}_p \cap \mathbb{C}_p^{br}$ consists of finitely many transcendental elements over $\mathbb{Q}$.*

This result completes the classification of Dedekind domains between $\mathbb{Z}[X]$ and $\mathbb{Q}[X]$ and provides a powerful characterization of Dedekind domains in this setting, linking local valuation data and integer-valued polynomial constraints to global structural properties such as the class group. The result is obtained by characterizing discrete valuation rings (DVRs) W of the field $\mathbb{Q}(X)$ that are residually algebraic over $\mathbb{Z}_{(p)}$ for some prime p (see [**119**, Corollary 2.28]).

When the extension of residue fields has infinite degree, such a domain is given by

$$\{\phi \in \mathbb{Q}(X) \mid v_p(\phi(\alpha)) \geq 0\}$$

for some $\alpha \in \mathbb{C}_p$, which is transcendental over $\mathbb{Q}_p$. The case where the residue field extension has finite degree was addressed in [**117**].

This result also generalizes findings by Loper, Werner, and others concerning Prüfer domains and valuation-theoretic constructions.

Although [**111**] and [**118, 119**] offer significant advances in classifying integrally closed domains between $\mathbb{Z}[X]$ and $\mathbb{Q}[X]$, the broader landscape of such domains is still only partially explored. In particular, unifying *generalized integer-valued polynomial* viewpoints with *valuation-theoretic* frameworks remains an active area of research, as do possible applications to:

- *Krull or (strong) Mori domains*: Extending classification results to settings with additional constraints (e.g., finite residue fields, specific v-class groups, etc.).
- *Factorial domains*: Investigating when generalized integer-valued polynomial rings are factorial.

Recall that an integral domain D is called an *almost Dedekind domain* (resp., a *t-almost Dedekind domain*) if $D_{\mathfrak{m}}$ is a DVR for all maximal ideals (resp., maximal t-ideals) $\mathfrak{m}$ of D.

PROBLEM 3. Classify all integral domains of a specified type (e.g., (t-)almost Dedekind domains and Krull domains satisfying additional conditions) and determine their placement between $\mathbb{Z}[X]$ and $\mathbb{Q}[X]$. Furthermore, investigate the extent to which such domains can be realized as:

- Generalized rings of integer-valued polynomials, building on [**118, 119**], or
- Valuation overrings constructed via MacLane-type key polynomial techniques, as in [**111**].

5. Generalizing Kronecker function domains

In a previous section we discussed a construction of Krull known as the Kronecker function domain. We start with an integrally closed domain D with quotient field K and build the domain $\mathrm{Kr}(D)$ which satisfies

(1) $\mathrm{Kr}(D)$ lies between $D[X]$ and $K(X)$.
(2) $\mathrm{Kr}(D)$ is a Prüfer domain.
(3) The prime ideals of $\mathrm{Kr}(D)$ correspond precisely with the valuation overrings of D.

There are two features of this that might strike one as annoyances that one would wish to avoid. We start as above with an integrally closed domain D with quotient field K.

- The actual construction is built using the ab star operations. It is cumbersome. One could wish for a simpler construction.
- One could wonder whether the base ring D is essential. Is it possible to build a Kronecker function domain using just K—without D?

In 2003 Halter-Koch [**87**] gave what he called an axiomatic treatment of what he called Kronecker domains which gives us the construction we were looking for above.

Let K be a field which has valuation subrings. Let X be an indeterminate. We say that a ring T is a *Kronecker domain* provided

(1) $T \subseteq K(X)$.
(2) X is a unit in T.
(3) For every nonzero polynomial $f \in K[X]$, we have $f(0) \in fT$.

If these conditions are satisfied, then we have also

- T is a Prüfer domain.
- If $\mathfrak{p}$ is a prime ideal of T then $T_{\mathfrak{p}}$ has the form $V(X)$ for some valuation domain V contained in K.
- The conventional Kronecker function domains defined for subrings D of K are included as Kronecker domains.

Halter-Koch's notion of Kronecker domain has been known since 2003 and very little has been done in that time to study them. Actually, Halter-Koch himself extended what is known by himself by demonstrating that they can be built with a variant of the conventional construction ([**90**, Theorem 6.5.5]). Many questions could be posed. We will content ourselves with one.

PROBLEM 4. Fabbri and Heubo-Kwegna [**43**] give explicit details for the construction of a non-traditional Kronecker domain in a very specialized setting. Can their results be generalized to a much wider setting?

6. Coherent domains

Coherent domains were first defined by Chase [**30**] in 1960. A domain D is *coherent* if it satisfies one of the following three equivalent conditions (note there are many more equivalent ones).

(a) The intersection of any two finitely generated ideals of D is finitely generated.
(b) Any finitely generated ideal of D is finitely presented.
(c) The direct product of any family of flat D-modules is flat.

We focus on the first of these conditions. It is clear from this condition that any Noetherian domain is coherent. It is not as simple, but still straightforward to prove that any Prüfer domain is coherent. When one is considering all domains there are other natural examples of coherent domains. In the realm of one-dimensional domains things are much more restrictive. A one-dimensional coherent domain which is integrally closed is a Prüfer domain. The integral closure of a one-dimensional Noetherian domain is a Dedekind domain which is both Noetherian and Prüfer. The class of one-dimensional coherent domains which are neither Noetherian nor Prüfer is vanishingly small. We are aware of one such example in the literature. This example is not integrally closed but has Prüfer integral closure.

This state of affairs led Vasconcelos to ask the following.

PROBLEM 5. Is the integral closure of a one-dimensional coherent domain a Prüfer domain? [**81**]

One might hope to make progress on this question by considering many examples of one-dimensional coherent domains that are neither integrally closed nor Noetherian. But again, such examples scarcely exist.

Instead of looking at this question from a forward direction (examine the integral closure of a one-dimensional coherent domain) maybe one can make progress by looking at the question from a backwards perspective.

- Let D be a one-dimensional integrally closed domain which is not Prüfer. Is D in any sense close to being coherent?

A recent result shows that if such a domain D is local it is, in a sense anti-coherent.

THEOREM 6.1. *Let D be a one-dimensional integrally closed local domain which is not a Prüfer domain. Let I, J be two principal ideals of D with no containment relations. Then $I \cap J$ is not finitely generated.* [**85**]

This seems to be evidence toward a positive answer to Vasconcelos' question. How could a coherent domain - where the intersection of two finitely generated ideals is always finitely generated - have as its integral closure a domain where the intersection of two principal ideals is almost never finitely generated?

PROBLEM 6. Let K be a field. Let X, Y be indeterminates. Consider the valuation domain $V = K(X)[Y]_{(Y)}$. Let M be the maximal ideal of V. Let $D = K + M$. Then D is a one-dimensional local domain which is integrally closed and not Prüfer. Can we prove that D is not the integral closure of a coherent domain?

7. Theory of w-modules

Originally formulated by McCasland and Wang [**141**], [**112**], *w-module theory* refines classical module and ideal theories in commutative algebra by focusing on a specific star operation, the w-operation. The concept of w-ideals coincides with the F_∞-operation previously considered by J. R. Hedstrom and E. G. Houston [**92**], and the w-ideals coincide with the semidivisorial ideals of S. Glaz and W. Vasconcelos [**82**].

The framework of w-ideals generalizes earlier results on t-ideals and divisorial ideals while providing new insights in non-Noetherian settings. The core motivation for w-module theory arose from studying the structure of commutative rings and the interplay of their star operations with modules and ideals.

DEFINITION 7.1. Let R be a commutative ring, possibly with zero divisors.

(1) A nonzero ideal J of R is called a *Glaz–Vasconcelos ideal* (GV-ideal) if J is finitely generated and the natural homomorphism

$$\varphi\colon R \longrightarrow \mathrm{Hom}_R(J, R)$$

is an isomorphism. The set $\mathrm{GV}(R)$ of all GV-ideals forms a multiplicative system of ideals in R.

(2) An R-module M is called *GV-torsion-free* if, for any $J \in \mathrm{GV}(R)$, the condition $Jx = 0$ (for some $x \in M$) implies $x = 0$. On the other hand, M is called *GV-torsion* if, for every nonzero element $x \in M$, there exists $J \in \mathrm{GV}(R)$ such that $Jx = 0$. It is evident that the pair (GV-torsion modules, GV-torsion-free modules) defines a hereditary torsion theory, first introduced in [**104**, p. 1618].

(3) For a GV-torsion-free R-module M, define

$$M_w = \{x \in E(M) \mid \text{there exists } J \in \mathrm{GV}(R) \text{ such that } Jx \subseteq M\},$$

where $E(M)$ is the injective envelope of M. The module M_w is called the *w-envelope* of M.

(4) A GV-torsion-free R-module M is called a *w-module* if $\mathrm{Ext}^1_R(R/J, M) = 0$ for every $J \in \mathrm{GV}(R)$; equivalently, if $M = M_w$. In particular, an ideal I of R is called a *w-ideal* if $I = I_w$.

(5) An R-module M is said to be *w-flat* if, for every maximal w-ideal $\mathfrak{m}$ of R, the localization $M_{\mathfrak{m}}$ is flat over $R_{\mathfrak{m}}$.

The w-operation builds a bridge towards homological algebra. The resulting w-module theory can be used to study certain "high-dimensional" domains from a module-theoretic perspective. In fact, the w-module theory is a type of locally homological method.

Every flat R-module is w-flat, and if D is a Prüfer domain, w-flat D-modules coincide with flat D-modules. This shows that w-module theory specializes to classical notions in more restricted settings.

It is well known that Noetherian rings play an important role in commutative algebra and algebraic geometry. Using w-operations, strong Mori domains were introduced in [**141**]. It was previously claimed ([**59**, Corollary IX.5.5]) that for a domain D with quotient field K, the injective module $E(K/D)$ is Σ-injective if and only if D is a Mori domain. However, this statement is *incorrect*: one must replace "Mori" with "strong Mori" (see [**58**, Theorem 9]). In recent years, many classical theorems about Noetherian rings have been generalized to strong Mori domains.

In parallel with w-module theory, the study of *Krull domains* has been pivotal in commutative algebra via the lens of star operations. Indeed, a Krull domain D can be characterized by the following conditions.

(i) For every height-one prime ideal $\mathfrak{p}$, the localization $D_{\mathfrak{p}}$ is a discrete valuation domain.

(ii) $D = \bigcap D_{\mathfrak{p}}$, the intersection taken over all height-one prime ideals $\mathfrak{p}$.

(iii) Every nonzero non-unit in D is contained in only finitely many height-one prime ideals.

Analogous to how Dedekind domains can be characterized by the invertibility of nonzero ideals, Krull domains can be characterized as follows ([**103**, Theorem 3.6], [**140**, Theorem 7.9.3]):

$$D \text{ is Krull} \iff \text{every nonzero ideal of } D \text{ is } t\text{- (or } w\text{-) invertible.}$$

This implies that a Krull domain can essentially be considered a t- (or w-) Dedekind domain. Furthermore, it suggests that Krull domains can be studied in a manner similar to Dedekind domains, utilizing star-operator methods.

PROBLEM 7. Prove or disprove the assertion that

> *Every result in classical commutative algebra has a corresponding analogue in w-module theory.*

Finding an accurate analogue for this may be challenging and could require the development of new concepts and techniques. For instance, in a Dedekind domain, any finitely generated module M decomposes as a direct sum of a projective module and a torsion module. A central question in w-module theory (and more broadly in star-operation settings) is whether such classical decompositions have analogues in Krull domains. For Krull domains, what is the corresponding formulation for finitely generated (w-finitely generated) modules?

Recent extensions explore w-factor rings. For a domain D, the following conditions are equivalent.

(a) D is Krull.
(b) The w-factor ring $(D/I)_w$ is a zero-dimensional semilocal principal ideal ring for every proper w-ideal I of D, [**26**, Theorem 4.5].
(c) $(D/I)_w$ is a quasi-Frobenius ring [**149**, Theorem 9].

These results highlight a strong parallel between Dedekind and Krull domains when analyzed through w-operations and w-ideal structures. This raises the question of whether PvMDs can be characterized by their w-factor rings, where an integral domain D is called a *Prüfer v-multiplication domain* (*PvMD* for short) if every nonzero finitely generated ideal of D is t-invertible (or w-invertible).

PROBLEM 8. Investigate whether the PvMDs D are precisely those domains for which their w-factor rings $(D/I)_w$ are FC-rings for every proper w-ideal I of D, where a commutative ring R is an *FC-ring* if it is coherent and self-FP-injective.

8. Homological versus elementary methods

Every ring considered in this section is an integral domain.

In 1937 Krull defined a regular local ring to be a Noetherian local ring D with maximal ideal $\mathfrak{m}$ for which the minimum number of elements in a generating set for $\mathfrak{m}$ is the same as the height of $\mathfrak{m}$ [**108**]. This definition is completely elementary.

In 1956 Serre [**134**] proved that a Noetherian local ring is regular if and only if it has finite global dimension. This homological characterization of regular local rings was pursued because it led to an easy proof that a localization of a regular local ring at a prime ideal was still regular, which had proven to be quite elusive using non-homological methods.

This event is often cited as validating the use of homological methods in the study of ring theory. In fact, homological methods have become quite dominant. We propose several settings in which the elementary methods might regain prominence.

QUESTION I. Does there now exist a non-homological proof that any localization of a regular local ring is still regular?

Answer: There exist informal sites on the internet that indicate that the answer is yes. These sites indicate that the proof involves careful use of regular sequences. This is not surprising since it is well-known that an ideal generated by a regular sequence has finite projective dimension. But we were unable to find a published proof. So we get the following.

PROBLEM 9. Can one prove non-homologically that any localization of a regular local ring is still regular?

We push this a little farther. Ratliff [**127**] defined a set $\{d_0, d_1, \dots, d_n\}$ of elements in a local domain $(D, \mathfrak{m})$ to be analytically independent if, whenever $F(d_0, \dots, d_n) = 0$ for some homogeneous polynomial $F(X_0, \dots, X_n) \in D[X_0, \dots, X_n]$, then all coefficients of F are in $\mathfrak{m}$. This generalizes the notion of a regular sequence. Say that the analytically independent dimension is the maximal size of an analytically independent set. Is it possible that the notion of finite projective dimension could be replaced by the notion of analytically independent dimension?

PROBLEM 10. An ideal generated by a regular sequence has finite projective dimension. A regular sequence is analytically independent. Is it true that an ideal generated by an analytically independent set has finite projective dimension?

QUESTION II. Can we use homological terminology to define classes of rings that are not being studied at all and might be profitably studied with elementary techniques?

Answer: Perhaps. Recall that a Dedekind domain is a domain for which every nonzero ideal is invertible. The more general notion of a Prüfer domain is a domain for which every finitely generated ideal is invertible. We could similarly say that a domain has pseudo-finite global dimension provided there is an upper bound on the projective dimensions of all finitely generated modules. Perhaps the union of an ascending tower of regular local rings could provide nontrivial examples of such rings. Examples of such towers are given in recent papers concerning the so-called Shannon extensions [**93**].

PROBLEM 11. Can the property of having pseudo-finite global dimension be defined in elementary terms?

PROBLEM 12. Do Shannon extensions have pseudo-finite global dimension?

9. Gorenstein multiplicative ideal theory

Gorenstein homological algebra arose as an extension of classical homological algebra to better accommodate *non-regular* contexts. Traditional projective, injective, and flat modules often fail to capture subtler behaviors in more general rings and modules, especially those beyond Noetherian or regular situations. Gorenstein dimensions (and related Gorenstein notions) were introduced to address these shortcomings, building on Auslander and Bridger's foundational work on Gorenstein projective modules [**7**]. Their investigations were initially guided by syzygy theory over Noetherian rings and the concept of complete intersections.

Subsequent expansions by Enochs, Jenda, and Torrecillas introduced a range of new tools—such as relative homological dimensions [**41, 42**]. Classical homological algebra heavily emphasizes projective, injective, and flat modules. The Gorenstein versions of these notions "relax" certain regularity assumptions.

Throughout this section, let R denote a commutative ring, possibly with zero divisors.

DEFINITION 9.1. Let M be an R-module.

(1) M is said to be *Gorenstein projective* if there exists an exact sequence of projective R-modules
$$\cdots \longrightarrow P_1 \longrightarrow P_0 \longrightarrow P^0 \longrightarrow P^1 \longrightarrow \cdots$$
such that $M \cong \ker(P^0 \to P^1)$ and $\operatorname{Hom}_R(-, Q)$ preserves exactness of this sequence for every projective R-module Q.

(2) M is said to be *Gorenstein injective* if there exists an exact sequence of injective R-modules
$$\cdots \longrightarrow I_1 \longrightarrow I_0 \longrightarrow I^0 \longrightarrow I^1 \longrightarrow \cdots$$

such that $M \cong \ker(I^0 \to I^1)$ and $\mathrm{Hom}_R(Q, -)$ preserves exactness of this sequence for every injective R-module Q.

(3) M is said to be *Gorenstein flat* if there exists an exact sequence of flat R-modules

$$\cdots \longrightarrow F_1 \longrightarrow F_0 \longrightarrow F^0 \longrightarrow F^1 \longrightarrow \cdots$$

such that $M \cong \ker(F^0 \to F^1)$ and exactness is preserved under the functor $- \otimes_R E$ for every injective R-module E.

(4) The *Gorenstein projective dimension* of M, denoted $\mathrm{Gpd}_R(M)$, is defined as the smallest integer n (if such an n exists) such that there exists an exact sequence of Gorenstein projective modules:

$$0 \longrightarrow G_n \longrightarrow G_{n-1} \longrightarrow \cdots \longrightarrow G_1 \longrightarrow G_0 \longrightarrow M \longrightarrow 0,$$

where each G_i is a Gorenstein projective module. If no such finite n exists, we set $\mathrm{Gpd}_R(M) = \infty$.

EXAMPLE 9.2.

(1) If R is a *Gorenstein ring*, then every finitely generated R-module has finite Gorenstein projective dimension. Over a regular local ring, the Gorenstein projective dimension agrees with the usual projective dimension.

(2) Let $R = K[X]/(X^2)$ for a field K. This is a Frobenius algebra, so R is both Gorenstein projective and Gorenstein injective as an R-module.

DEFINITION 9.3. An integral domain D is called a

- *Gorenstein Dedekind domain* (briefly, G-Dedekind domain) if *every ideal of D is G-projective.*
- *Gorenstein Prüfer domain* (briefly, G-Prüfer domain) if D is coherent and every submodule of a flat D-module is Gorenstein flat.

The notion of a *Gorenstein Krull domain* (G-Krull domain for short) is also introduced in [**124**]. Examples of G-Prüfer and G-Krull domains are provided in [**145**].

Holm's metatheorem [**94**] proclaims that

Every result in classical homological algebra has a Gorenstein counterpart.

This sweeping statement has been largely corroborated by numerous analogies and parallels between classical and Gorenstein homological results. Yet there remain instances in which Gorenstein analogies are only partially true or require stricter hypotheses.

For example, in the classical setting, a domain D is Prüfer if and only if it is locally a valuation domain ([**140**, Corollary 3.7.22]). In the Gorenstein setting, while every G-Prüfer domain is locally G-Prüfer [**146**, Corollary 11], the converse *fails* in general [**148**, Example 4.1]. Hence, the direct classical analogy does not hold perfectly in Gorenstein contexts.

Classically, D is Prüfer if and only if every overring of D is flat [**109**, Theorem 6.10]. Analogously, it has been shown that if R is G-Prüfer, then every overring of R is Gorenstein flat [**146**, Theorem 6]. However, it remains *open* whether the converse also holds. Namely, is a domain D G-Prüfer if every overring of D is Gorenstein flat? This is a natural Gorenstein-homological parallel to a well-known Prüfer criterion, but it awaits resolution.

A key feature of Dedekind domains D is that *every nonzero ideal is invertible*, which is equivalent to demanding that D/I be quasi-Frobenius for each nonzero ideal I [**149**, Theorem 13]. Similarly, in the Gorenstein Dedekind setting, one has:

- An integral domain D is G-Dedekind if and only if, for every nonzero element $a \in D$, the factor ring $D/(a)$ is quasi-Frobenius [**140**, Theorem 11.7.7]. This implies that every nonzero ideal of D is a v-ideal. In fact, D is a G-Dedekind domain if and only if it is a Noetherian divisorial domain.
- In general, a G-Dedekind domain is one-dimensional and Noetherian [**140**, Theorem 11.7.7 and Corollary 11.7.8], yet can exhibit different behaviors regarding *integral closure*, reminiscent of how Gorenstein rings relax some classical regularity or integrality conditions.

Recent connections show that factor rings of the form $(D/xD)_w$ being quasi-Frobenius for all nonzero x in D characterizes G-Krull domains [**147**, Theorem 2.7]. Such *Gorenstein-enabled* statements suggest deep analogies between multiplicative ideal theory and Gorenstein homological algebra, but also underscore certain exceptions where classical and Gorenstein theories diverge.

PROBLEM 13. Develop explicit characterizations of Gorenstein homological properties—with an emphasis on G-Dedekind and G-Prüfer domains—through both ideal-theoretic and ring-theoretic lenses. In particular:

- Clarify how the v-operation and $*$-invertibility interact in the Gorenstein setting, and determine whether the v-operation fully captures Gorenstein analogues of classical invertibility.
- Investigate if and how integral closure behaves differently for G-Dedekind domains compared to classical Dedekind domains.
- Resolve open questions about overring flatness: under what conditions does "every overring is Gorenstein flat" imply that the base domain is G-Prüfer?
- Investigate whether $\overline{D}$ is a Prüfer domain when D is G-Prüfer.

10. Cotorsion theory

Cotorsion theory is a branch of homological algebra that uses the Ext functor to organize module categories (or abelian categories) into two complementary classes of objects, generalizing concepts such as injectivity, flatness, and their relative versions. By drawing on ideas from torsion theories and projective/injective dimensions, cotorsion theory provides a powerful framework for decomposing modules and understanding the structure of extension groups.

Throughout this section, let R denote a commutative ring, possibly with zero divisors.

DEFINITION 10.1. A *cotorsion pair* $(\mathcal{A}, \mathcal{B})$ in the category of R-modules (or in any abelian category) consists of two classes of modules $\mathcal{A}$ and $\mathcal{B}$ such that:

(1)
$$\mathcal{A} = \{ M \mid \operatorname{Ext}^1_R(M, B) = 0 \text{ for all } B \in \mathcal{B} \}$$
and
$$\mathcal{B} = \{ N \mid \operatorname{Ext}^1_R(A, N) = 0 \text{ for all } A \in \mathcal{A} \}.$$

(2)
- $\mathcal{A}$ is closed under direct sums, summands, and extensions.
- $\mathcal{B}$ is closed under direct products, summands, and extensions.

The classes $\mathcal{A}$ and $\mathcal{B}$ are often called *orthogonal complements* of each other with respect to $\mathrm{Ext}^1_R(-,-)$.

EXAMPLE 10.2.

(1) Let $\mathcal{M}$ be the category of all R-modules, $\mathscr{P}$ the class of all projective modules, and $\mathscr{I}$ the class of all injective modules. Then

$$(\mathscr{P}, \mathcal{M}) \quad \text{and} \quad (\mathcal{M}, \mathscr{I})$$

are both cotorsion pairs in $\mathcal{M}$.

(2) Over any ring D, the cotorsion pair

$$(\mathcal{A}, \mathcal{B}) = (\mathrm{Flat}, \mathrm{Cotorsion})$$

satisfies:

$$\mathcal{A} = \{\text{flat } D\text{-modules}\}, \quad \mathcal{B} = \{N \mid \mathrm{Ext}^1_R(F, N) = 0 \text{ for all flat } F\}.$$

Cotorsion pairs underlie many useful constructions in homological algebra, including relative homological dimensions, abelian model categories, and exact categories. By focusing on how $\mathrm{Ext}^1_R(-,-)$ vanishes, one can obtain powerful decomposition theorems for modules.

A variant of cotorsion theory, called *Tor-torsion theory*, shifts attention from Ext to Tor and hence emphasizes torsion properties and tensor products.

DEFINITION 10.3. Let $\mathcal{A}$ and $\mathcal{B}$ be two classes of modules. The pair $(\mathcal{A}, \mathcal{B})$ is called a *Tor-torsion pair* (or *Tor-torsion theory*) if

$$\mathcal{A} = \mathcal{B}^{\top} \quad \text{and} \quad \mathcal{B} = \mathcal{A}^{\top},$$

where

$$\mathcal{A}^{\top} = \{B \in \mathcal{M} \mid \mathrm{Tor}^R_1(A, B) = 0 \text{ for every } A \in \mathcal{A}\}.$$

EXAMPLE 10.4. Let $\mathscr{F}$ be the class of all flat R-modules. Then $(\mathscr{F}, \mathcal{M})$ is a (classical) Tor-torsion pair in the category of R-modules. In a Tor-torsion context, we focus on R-modules M with $\mathrm{Tor}^R_1(F, M) = 0$ rather than on $\mathrm{Ext}^1_R(F, M) = 0$.

Cotorsion theory (and its Tor-torsion counterpart) can effectively describe structural aspects of rings that are central in multiplicative ideal theory. For instance:

(1) An integral domain D is Dedekind if and only if the global dimension of D is at most 1. Equivalently, every nonzero ideal of D is projective (i.e., invertible).
(2) An integral domain D is Krull if and only if the global *weak w-projective dimension* $w.w.gl.\dim(D)$ is at most 1; see [**142**, Theorem 4.3].

These homological characterizations blend naturally with cotorsion-theoretic viewpoints: the condition $gl.\dim(R) \leq 1$ can be recast via orthogonality requirements on R-modules, while the w-projective setting relies on extension functors in the w-module theory sense. Indeed, building robust (Tor-)cotorsion frameworks broadens the scope of such equivalences, reinforcing links between ring-theoretic properties (e.g., invertibility of ideals) and vanishing results in homological algebra.

Although there are many ways to characterize Prüfer domains, cotorsion viewpoints yield elegant homological statements. First, we recall additional module-theoretic definitions.

DEFINITION 10.5. Let D be a domain with quotient field K, and let M be an D-module.

(1) M is called a *cotorsion module* if

$$\operatorname{Ext}_D^1(F, M) = 0 \quad \text{for every flat } D\text{-module } F.$$

(2) M is called a *Warfield cotorsion module* if

$$\operatorname{Tor}_1^D(M, F) = 0 \quad \text{for every torsion-free } D\text{-module } F.$$

(3) M is called a *Lee cotorsion module* if

$$\operatorname{Ext}_D^1(K/D, M) = 0.$$

THEOREM 10.6 ([**140**, Theorem 12.9.12]). *For a domain D, the following conditions are equivalent.*

(a) *D is a Prüfer domain.*
(b) *Every cotorsion D-module is Warfield cotorsion.*
(c) *Every divisible D-module is FP-injective.*
(d) *Every h-divisible D-module is FP-injective.*
(e) *Every Lee cotorsion D-module is FP-injective.*

Thus, a domain is Prüfer precisely when several natural classes of modules (cotorsion, divisible, etc.) coincide homologically or admit uniform homological dimensions. This result interlinks classical ideals-based characterizations (e.g., localizing at height-one primes or checking invertibility of ideals) with module-theoretic vanishing conditions.

The interplay between cotorsion theory and w-theory highlights how star operations can refine homological statements, particularly for non-Noetherian rings or those with specialized divisorial properties. For example, focusing on w-flatness and w-cotorsion can unify certain results in multiplicative ideal theory with relative homological vanishing conditions. In this direction, the cases of PvMDs and Krull domains have been studied in [**122, 123**].

PROBLEM 14. Develop explicit *cotorsion* and *Tor-torsion* characterizations of rings and integral domains that arise in multiplicative ideal theory. In particular, investigate how homological invariants such as

$$\operatorname{Ext}_R^1(-,-), \quad \operatorname{Tor}_1^R(-,-), \quad \text{global dimensions, etc.}$$

can be used to classify such rings. Examples of interest include:

- *Prüfer domains*: Extend Theorem 10.6 by identifying further cotorsion-theoretic criteria.
- *Krull domains*: Relate the w-projective dimension viewpoint to (Tor-)cotorsion approaches.
- *Other star-operation domains*: Characterize specialized classes (e.g., Gorenstein domains, G-Prüfer, PvMDs, etc.) via novel cotorsion or Tor-cotorsion properties.

11. Factorization theory in Mori monoids

Factorization theory has spread into various subfields of algebra. However, in this survey we restrict ourselves to factorization theory in Mori and Krull monoids and domains. We discuss neither factorization theory in non-commutative ring theory [**11**, **16**, **137**], nor combinatorial factorization theory with its overlap to additive combinatorics, [**65**], nor factorization theory in non-BF-monoids [**35**], nor factorizations into distinguished building blocks other than irreducibles [**32**].

We start with a lemma highlighting the abundance of Mori and Krull monoids in commutative ring theory. For more on the ideal theory of Mori monoids we refer to [**90**, Chapter 5] and for more of this flavor in the setting of v-Marot and t-Marot rings see [**14**, **27**]. For Mori monoids with a combinatorial flavor see [**44**]. Moreover, we refer to a survey of Facchini [**45**] for the role of Krull monoids in module theory.

LEMMA 11.1.

(1) *A domain D is Mori if and only if $D^\bullet$ is a Mori monoid if and only if $\mathcal{I}_v^*(D)$ is a Mori monoid.*
(2) *If D is a v-Marot ring, then D is a Mori ring if and only if $D^\bullet$ is a Mori monoid.*
(3) *Let D be a domain and let r be an ideal system on D with $\mathcal{I}_r(D) \subseteq \mathcal{I}(D)$. Then D is a Krull domain if and only if $D^\bullet$ is a Krull monoid if and only if $\mathcal{I}_r^*(D)$ is a Krull monoid.*
(4) *If D is a Noetherian domain, then its complete integral closure $\widehat{D}$ is Krull and $\mathcal{I}^*(D)$ is a Mori monoid.*
(5) *Let M be a Mori monoid. If M is seminormal or if the conductor $(M\colon \widehat{M})$ is nonempty, then the complete integral closure $\widehat{M}$ is Krull. Furthermore, the complete integral closure of a strong Mori domain is Krull.*

PROOF. For (1) and (2) we refer to [**72**] and for (3) to [**71**, Theorem 4.5]. The first claim in (4) is the Theorem of Mori-Nagata and the second claim follows from [**67**, Example 2.1].

(5) For a seminormal M the claim follows from [**90**, Theorem 5.2.8] and if the conductor is nonempty, then it follows from [**66**, Theorem 2.3.5]. If M is a strong Mori domain, then the claim follows from [**140**, Theorems 7.7.14 and 7.10.21]. □

We start the discussion of Arithmetic Finiteness Properties. Let M be a monoid. We say that M is

(1) *atomic* if every element $a \in M \setminus M^\times$ has a factorization into atoms.
(2) *factorial* if one of the following two equivalent conditions hold.
 (a) M is atomic and factorization into atoms is unique up to the order and up to associates.
 (b) M is Krull and each divisorial ideal of M is principal.

The uniqueness statement in Condition 2.(a) means that whenever $a = u_1 \ldots u_k = v_1 \ldots v_\ell$, with $k, \ell \in \mathbb{N}$ and $u_1, \ldots, u_k, v_1, \ldots, v_\ell \in \mathcal{A}(M)$, then $k = \ell$ and there is a permutation $\sigma\colon [1,k] \to [1,k]$ such that $u_i M^\times = v_{\sigma(i)} M^\times$ for all $i \in [1,k]$.

Mori monoids are atomic. Indeed, suppose that M is Mori and assume to the contrary that $\Omega = \{aM\colon a \in M \setminus M^\times \text{ is not a product of atoms}\}$ is nonempty. Since principal ideals are divisorial, Ω has a maximal element $aM \in \Omega$. Then there are $b, c \in M \setminus M^\times$ such that $a = bc$, $aM \subsetneq bM$, and $aM \subsetneq cM$. Thus, b and c are products of atoms and so is $a = bc$, a contradiction. The overall goal is to describe the non-uniqueness of factorizations in Mori monoids which are either not Krull or Krull with nontrivial v-class group. This is done by arithmetic invariants which are set in context with algebraic parameters of the respective class of Mori monoids.

We introduce length sets and some parameters controlling their structure. If $a = u_1 \ldots u_k$, with $k \in \mathbb{N}$ and $u_1, \ldots, u_k \in \mathcal{A}(M)$, then the number of factors k is called a factorization length of a and the set $\mathsf{L}_M(a) = \mathsf{L}(a)$ of all factorization lengths of a is called the *length set* of a. Note that $1 \in \mathsf{L}(a)$ if and only if a is an atom if and only if $\mathsf{L}(a) = \{1\}$. It is a convenient convention to set $\mathsf{L}(a) = \{0\}$ if a is invertible. We denote by

$$\mathcal{L}(M) = \{\mathsf{L}(a) \colon a \in M\}$$

the *system of length sets* of M. The monoid M is said to be *half-factorial* if $|\mathsf{L}(a)| = 1$ for all $a \in M$. Suppose that M is not half-factorial. Then there exists some $a \in M$ with factorizations of two different lengths, say

$$a = u_1 \ldots u_k = v_1 \ldots v_\ell \,,$$

where $k, \ell \in \mathbb{N}$ with $k < \ell$ and atoms $u_1, \ldots, u_k, v_1, \ldots, v_\ell$. Then, for every $N \in \mathbb{N}$, we have

$$a^N = (u_1 \ldots u_k)^i (v_1 \ldots v_\ell)^{N-i} \quad \text{for all } i \in [0, N] \,,$$

whence $|\mathsf{L}(a^N)| > N$. Thus, length sets may become arbitrarily large. For an element $a \in M$, let $\omega_M(a) = \omega(a)$ be the smallest $N \in \mathbb{N}_0 \cup \{\infty\}$ for which the following holds:

- If $m \in \mathbb{N}$ and $v_1, \ldots, v_m \in M$ with $a \mid v_1 \ldots v_m$, then there exists a subset $\Omega \subseteq [1, m]$ such that $|\Omega| \le N$ and $a \mid \prod_{\nu \in \Omega} v_\nu$.

For an atom $a \in M$, let $\mathsf{t}(a)$ be the smallest $N \in \mathbb{N}_0 \cup \{\infty\}$ for which the following holds:

- If $m \in \mathbb{N}$ and $v_1, \ldots, v_m \in \mathcal{A}(M)$ are such that $a \mid v_1 \ldots v_m$, but a divides no proper subproduct of $v_1 \ldots v_m$, then there exist $\ell \in \mathbb{N}$ and $u_2, \ldots, u_\ell \in \mathcal{A}(M)$ such that $v_1 \ldots v_m = a u_2 \cdot \ldots \cdot u_\ell$ and $\max\{\ell, m\} \le N$.

We say that M is *locally tame* if $\mathsf{t}(a) < \infty$ for all $a \in \mathcal{A}(M)$. By definition, $\omega(a) = 1$ if and only if a is a prime element. If a is an atom but not prime, then $\omega(a) \le \mathsf{t}(a)$.

The next lemma gathers some basic arithmetic properties of Mori monoids.

LEMMA 11.2. *Let M be a Mori monoid.*

(1) *All length sets are finite and nonempty* ([**66**, Theorem 2.2.9]).
(2) $\omega(a) < \infty$ *for all* $a \in M$ ([**67**, Theorem 4.2]).

For a finite, nonempty set $L = \{m_1, \ldots, m_k\} \subset \mathbb{N}$ with $k \in \mathbb{N}$ and $m_1 < \ldots < m_k$, we call

- $\Delta(L) = \{m_{i+1} - m_i \colon i \in [1, k-1]\} \subseteq \mathbb{N}$ the *set of distances* of L, and
- $\rho(L) = m_k / m_1$ the *elasticity* of L.

For convenience, we set $\rho(\{0\}) = 1$. Let M be a Mori monoid and let $k \in \mathbb{N}$. Then

- $\Delta(M) = \bigcup_{L \in \mathcal{L}(H)} \Delta(L) \subseteq \mathbb{N}$ is the *set of distances* of M,
- $\rho(M) = \sup\{\rho(L) \colon L \in \mathcal{L}(H)\} \in \mathbb{R}_{\ge 1} \cup \{\infty\}$ is the *elasticity* of M, and
- $\mathcal{U}_k(M) = \bigcup_{k \in L, L \in \mathcal{L}(M)} L \subseteq \mathbb{N}$ is the *union of length sets* containing k.

Thus, by definition,

$$M \text{ is half-factorial} \quad \text{if and only if} \quad \Delta(M) = \emptyset \quad \text{if and only if} \quad \rho(M) = 1 \,,$$

and, by a straightforward calculation, we get

$$\rho(M) = \lim_{k \to \infty} \frac{\sup \mathcal{U}_k(M)}{k} \,.$$

We say that M satisfies the *Structure Theorem for Unions* if there are $k^* \ge 1$ and $N \ge 0$ such that, for all $k \ge k^*$, $\mathcal{U}_k(M)$ has the form

$$\mathcal{U}_k(M) = y_k + \left(L' \cup L^* \cup L''\right) \subset y_k + d\mathbb{Z},$$

where $y_k \in \mathbb{Z}$, L^* is a nonempty (finite or infinite) arithmetic progression with difference d, $\min L^* = 0$, $L' \subset [-N, -1]$, and $L'' \subset \sup L^* + [1, N]$, with the convention that $L'' = \emptyset$ if L^* is infinite (if M is half-factorial, then $\Delta(M) = \emptyset$, $d = 0$, $L' = L'' = \emptyset$, $L^* = \{0\}$, and $\mathcal{U}_k(M) = y_k + \{0\} = \{k\}$ for all $k \ge 1$).

We say that M satisfies the **Main Arithmetic Finiteness Properties** if

(a) M is locally tame.
(b) The set of distances is finite.
(c) The Structure Theorem for Unions holds.
(d) The Structure Theorem for Length Sets holds (it says that all length sets are almost arithmetic multiprogressions with global bounds for all parameters; for the technical definition see [**66**]).

Consider two simple examples of Mori domains, Cohen-Kaplansky domains and one-dimensional local Mori domains. Their arithmetic is extremely simple and it stimulates research to which extent the arithmetic of more general domains fulfills some resp. all of their arithmetic finiteness properties. Let D be a domain and let r be an ideal system on D such that $\mathcal{I}_r(D) \subseteq \mathcal{I}(D)$. Then D is called a *Cohen-Kaplansky domain* if it satisfies one of the following equivalent conditions ([**73**, Proposition 4.5])

(a) D is atomic and has only finitely many atoms up to associates.
(b) The monoid of principal ideals is finitely generated.
(c) The monoid $\mathcal{I}_r^*(D)$ of r-invertible r-ideals is finitely generated.
(d) The semigroup $\mathcal{I}_r(D)$ of all nonzero r-ideals is finitely generated.

This class of domains was introduced by Anderson and Mott [**4**] (see also [**31**, **36**]). Every Cohen-Kaplansky domain is one-dimensional, semilocal, Noetherian with nonzero conductor.

THEOREM 11.3.

(1) *If D is a one-dimensional local Mori domain, then D satisfies the Main Arithmetic Finiteness Properties. Moreover, M has finite elasticity if and only if the conductor is nonzero and the natural map* $\operatorname{spec}(\widehat{D}) \to \operatorname{spec}(D)$ *is injective.*
(2) *Let M be a monoid such that M_{red} is finitely generated (e.g., the monoid of nonzero elements of a Cohen-Kaplansky domain). Then M has finite elasticity and satisfies the Main Arithmetic Finiteness Properties.*

PROOF.

(1) The statement on the elasticity follows from [**2**, Theorem 2.12] and from [**66**, 2.9.2+2.10.7]. The other statements follow from [**63**, Theorem 4.1] and from [**74**, Theorem 3.9].

(2) This follows from [**66**, 3.1.4+4.4.11] and from [**139**]. □

12. Some long-term arithmetic problems

12.1. Half-factoriality. In 1960 Carlitz proved that a ring of integers in an algebraic number field is half-factorial if and only if its class group has at most two elements. Ever since then half-factoriality is a key property studied in factorization theory and an abundance of results has been achieved (we refer to the surveys [**29**, **34**] for some early work and to [**38**] for a recent characterization of half-factoriality). We can pick only one recent highlight and one of our favorite problems.

After Carlitz's result, the question which (non-principal) orders in number fields are half-factorial received a lot of attention. Many special cases were handled (including orders in quadratic number fields and seminormal orders), but only recently Rago [**125**] characterized half-factoriality for arbitrary orders in number fields.

In 1966 Claborn proved that every abelian group is isomorphic to the class group of a Dedekind domain. His result stimulated much further research. There is a lot of work on the realization of various types of abelian groups as class groups of Dedekind domains with wanted additional properties. We refer to some recent results and the literature cited there ([**25**, **118**, **120**]). The following problem was first posted by Steffan [**138**] in 1986. It has an affirmative answer for all Warfield groups [**62**] but is open in general.

PROBLEM 15. Is every abelian group isomorphic to the class group of a half-factorial Dedekind domain?

12.2. Finiteness of arithmetic invariants. A key direction of research is to characterize (in a distinguished class of monoids) the finiteness of one resp. of all the arithmetic properties as they hold true in Theorem 11.3. We discuss a sample of results and then we give a more precise formulation of some problems. We start with a result by Kainrath [**101**].

THEOREM 12.1. *Let D be a domain that is finitely generated as a $\mathbb{Z}$-algebra. Then the elasticity $\rho(D)$ is finite if and only if $\mathcal{C}_v(\overline{D})$ and $\overline{D}/D$ are both finite and the natural map* $\operatorname{spec}(\overline{D}) \to \operatorname{spec}(D)$ *is injective.*

We continue with finiteness results for Krull monoids which highlight the significance of their v-class group for the arithmetic.

THEOREM 12.2. *Let M be a Krull monoid and let $G_0 \subseteq \mathcal{C}_v(M)$ denote the set of classes containing prime divisors.*

(1) *If G_0 is finite, then the Main Arithmetic Finiteness Properties hold and $\rho(M) < \infty$.*
(2) *Suppose that $\mathcal{C}_v(M)$ is infinite cyclic, say $\mathcal{C}_v(M) = \mathbb{Z}$. Then $\rho(M) < \infty$ if and only if $\Delta(M)$ is finite if and only if all $\mathcal{U}_k(M)$'s are finite and the Structure Theorem for Unions holds if and only if either $G_0 \cap \mathbb{N}$ or $G_0 \cap (-\mathbb{N})$ is finite.*

PROOF. The results of (1) can be found in [**66**] and (2) follows from [**64**, Theorem 4.2]. □

More is known than what is given by the above results. Among others, the assumption that G_0 is finite can be generalized to the assumption that the Davenport constant of G_0 is finite. If $\mathcal{C}_v(M)$ is finitely generated, then there is a characterization (in geometric terms of G_0) when $\rho(M)$ is finite ([**84**]). In particular, we do know that, in the above setting, length sets are completely controlled by the set G_0

(see the next subsection), but we are far away from a characterization of when one of the Main Arithmetic Finiteness Properties (a) - (d) hold true in general.

PROBLEM 16. Let M be a Krull monoid and let $G_0 \subseteq \mathcal{C}_v(M)$ denote the set of classes containing prime divisors. Characterize (in terms of G_0) when one resp. all of the Main Arithmetic Finiteness Properties hold.

In the local case Theorem 11.3 provides a satisfactory answer for one-dimensional domains, but only little is known for higher-dimensional local domains. Apart from the local case and apart from Krull monoids, C-monoids are the most studied class of Mori monoids with respect to the arithmetic, and they include a class of higher-dimensional Mori domains (Theorem 12.4.2). We recall their definition.

Let F be a factorial monoid and let $M \subseteq F$ be a submonoid. Two elements $y, y' \in F$ are called M-equivalent (we write $y \sim y'$) if $y^{-1}M \cap F = y'^{-1}M \cap F$ or, in other words,

$$\text{if for all } x \in F, \text{ we have } xy \in M \quad \text{if and only if} \quad xy' \in M\,.$$

M-equivalence defines a congruence relation on F and for $y \in F$, let $[y]_M^F = [y]$ denote the congruence class of y. Then

$$\mathcal{C}(M,F) = \{[y] \mid y \in F\} \quad \text{and} \quad \mathcal{C}^*(M,F) = \{[y] \mid y \in (F \setminus F^\times) \cup \{1\}\}$$

are commutative semigroups with identity element $[1]$. $\mathcal{C}(M,F)$ is the *class semigroup* of M in F and the subsemigroup $\mathcal{C}^*(M,F) \subseteq \mathcal{C}(M,F)$ is the *reduced class semigroup* of M in F.

A monoid M is a C-*monoid* if it is a submonoid of some factorial monoid F such that $M^\times = M \cap F^\times$ and the reduced class semigroup is finite, and a domain D is a C-domain if $D^\bullet$ is a C-monoid. The next theorem gathers the main algebraic and arithmetic properties of C-monoids (see [**66**]).

THEOREM 12.3. *Let M be a C-monoid.*

(1) *M is Mori, $(M : \widehat{M}) \neq \emptyset$, and $\mathcal{C}_v(\widehat{M})$ is finite.*
(2) *M satisfies the Main Arithmetic Finiteness Properties.*

This result gives rise to the following two problems.

PROBLEM 17. Let M be a C-monoid. By Theorem 12.3.(2), we know that arithmetic invariants are finite but the present methods just provide abstract finiteness results. Thus the goal is to derive the precise values (or at least good upper bounds) for arithmetic invariants in terms of suitable algebraic invariants (say in terms of the class semigroup). Much is known is the special case when M is Krull (see the next subsection). But in the general case of C-monoids only first steps are done so far (e.g., [**70**]).

PROBLEM 18. Let M be a Mori monoid with $(M : \widehat{M}) \neq \emptyset$ such that $\mathcal{C}_v(\widehat{M})$ is finite. Characterize when M is a C-monoid. The cases when M is seminormal or strong Mori are of special interest ([**76**]).

There is a simple characterization of C-monoids within the class of finitely generated monoids and a partial answer for Mori domains. But the general case is open.

THEOREM 12.4.

(1) *Let M be a monoid such that M_{red} is finitely generated. Then M is Mori with $(M:\widehat{M}) \neq \emptyset$, and M is a C-monoid if and only if $\mathcal{C}_v(\widehat{M})$ is finite.*

(2) *Let D be Mori domain with $(D:\widehat{D}) \neq \{0\}$ and with $\mathcal{C}_v(\widehat{D})$ finite. If $\widehat{D}/(D:\widehat{D})$ is finite, then D is a C-domain. The converse holds if D is semilocal, Noetherian but not local.*

PROOF. For Part (1), we refer to [**68**, Proposition 4.8]. The first statement in Part (2) follows from [**66**] and the converse statement is proved by Reinhart in [**128**, Corollary 4.5]. □

Every generalized Cohen-Kaplansky domain (this is an atomic domain that has only finitely many atoms, up to associates, that are not prime) is a C-domain [**73**, Proposition 4.11]. Part 2 of Theorem 12.4 shows that, in particular, orders in algebraic number fields are C-domains. A classical result of Furtwängler characterizes the ideals that can be realized as conductor ideals of an order [**89**, Theorem 2.12.10]. This was generalized by Reinhart to a more general setting of Noetherian domains [**129**], and his result gives rise to the following problem.

PROBLEM 19. Let H be a Krull monoid with finite v-class group. Characterize the ideals $\mathfrak{f} \subseteq H$ for which there is a C-monoid $M \subseteq H$ with $\widehat{M} = H$ and with conductor $\mathfrak{f} = (M:\widehat{M})$.

C-domains have been generalized to weakly C-domains which cover a larger class of Mori domains, as given in Theorem 12.4(2). They also satisfy most of the Main Arithmetic Finiteness Properties, but it is open whether the Structure Theorem for Length Sets holds true for them.

PROBLEM 20. Let M be a weakly C-monoid in the sense of [**102**]. Prove or disprove that the Structure Theorem for Length Sets holds (for a first step see [**69**]).

Based on a result of Nagata, Auslander and Buchsbaum proved that every regular local Noetherian domain is factorial. The following problem opens the door for a wide open area where only first steps have been done so far (e.g., [**144**]).

PROBLEM 21. Study the interplay of homological properties (as discussed in the previous sections) and arithmetic finiteness properties of a domain.

12.3. The transfer Krull property. A monoid homomorphism $\theta\colon M \to B$, between atomic monoids M and B, is called a *transfer homomorphism* if the following two properties hold.

(T 1) $B = \theta(M)B^{\times}$ and $\theta^{-1}(B^{\times}) = M^{\times}$.

(T 2) If $u \in M$, b, $c \in B$ and $\theta(u) = bc$, then there exist v, $w \in M$ such that $u = vw$, $\theta(v) \in bB^{\times}$ and $\theta(w) \in cB^{\times}$.

It is easy to check that $\mathsf{L}_M(a) = \mathsf{L}_B(\theta(a))$ for all $a \in M$ and $\mathcal{L}(M) = \mathcal{L}(B)$. In particular, we have $\rho(M) = \rho(B)$ and $\Delta(M) = \Delta(B)$.

Let M be a Krull monoid and let $G_0 \subseteq \mathcal{C}_v(M)$ denote the set of classes containing prime divisors. Then there is a transfer homomorphism $\theta\colon M \to \mathcal{B}(G_0)$, where $\mathcal{B}(G_0)$ is the monoid of zero-sum sequences with terms from G_0. This allows to study arithmetic invariants of M in the combinatorial monoid $\mathcal{B}(G_0)$, which is studied with methods from additive combinatorics. The interplay is most strong if every class contains a prime divisor (so, if $G_0 = \mathcal{C}_v(M)$) and $\mathcal{C}_v(M)$ is finite, say $\mathcal{C}_v(M) \cong C_{n_1} \oplus \ldots \oplus C_{n_r}$ with $1 < n_1 \mid \ldots \mid n_r$. We refer to [**65, 133**] which demonstrates that arithmetic invariants of M can be studied in terms of the group invariants $(n_1, \ldots, n_r)$.

Along the way, it was observed that there are monoids and domains, that are not Krull, but nevertheless they allow transfer homomorphisms to monoids of zero-sum sequences. Inspired by the success of the powerful methods from additive combinatorics, intense investigations have been started to figure out under which conditions we do have transfer homomorphisms to monoids of zero-sum sequences. Deep results revealed that this may happen even for non-commutative rings ([**137**]). Here we restrict our discussion to monoids (so, to commutative cancellative semigroups) and just mention two results. A one-dimensional local Mori domain D is transfer Krull if and only if D is half-factorial (whence in the trivial case only; see [**75**, Theorem 5.5]). Recent characterizations for orders in Dedekind domains are given in [**126**]. For a survey and more, see [**14, 15, 77**].

PROBLEM 22. Take a distinguished class of Mori monoids and characterize (in terms of algebraic invariants) when they are transfer Krull.

Module theory offers a rich source of not necessarily cancellative semigroups of modules that are transfer Krull (for such an example, see [**12**]). We formulate the following problem.

PROBLEM 23. Let R be a not necessarily commutative ring and let $\mathcal{C}$ denote a class of left R-modules, that is closed under direct sums, direct summands, and isomorphisms. Let $\mathcal{V}(\mathcal{C})$ be the semigroup of isomorphism classes of modules from $\mathcal{C}$, with operation induced by the direct sum. If every module from $\mathcal{C}$ has a semilocal endomorphism ring, then $\mathcal{V}(\mathcal{C})$ is Krull (this was proved by Facchini; for examples, see [**46**]). For a given class $\mathcal{C}$, for which $\mathcal{V}(\mathcal{C})$ is Krull, study classes $\mathcal{C}'$ of modules having a transfer homomorphism $\theta\colon \mathcal{V}(\mathcal{C}') \to \mathcal{V}(\mathcal{C})$. For a simple (artificial) example, of how this strategy could work, see [**77**, Example 5.4].

12.4. What can a family of length sets look like? In contrast to Theorem 12.2 there is the following result by Kainrath (see [**66**, Theorem 7.4.1]).

THEOREM 12.5. *Let M be a Krull monoid and let $G_0 \subseteq \mathcal{C}_v(M)$ denote the set of classes containing prime divisors. If G_0 contains an infinite abelian group, then every finite, nonempty subset of $\mathbb{N}_{\ge 2}$ occurs as a length set of M.*

The same result is true for large classes of integer-valued polynomials over Krull domains and of weakly Krull Mori domains (for a sample, see [**24, 48, 49**]). Surprisingly in almost all settings studied so far, monoids either satisfy the Structure Theorem for Length Sets or every finite, nonempty subset of $\mathbb{N}_{\ge 2}$ occurs as a length set. This dichotomy gave rise to the following problem. Let $\mathcal{P}_{\text{fin}}(\mathbb{N}_0)$ denote the system of all finite, nonempty subsets of $\mathbb{N}_0$.

PROBLEM 24. Which families $\mathcal{L} \subseteq \mathcal{P}_{\text{fin}}(\mathbb{N}_0)$ can be realized as the system of length sets of a Mori monoid?

Let $\mathcal{L} \subseteq \mathcal{P}_{\text{fin}}(\mathbb{N}_0)$ be a family of finite, nonempty subsets. If $\mathcal{L}$ is the system of length sets of a Mori monoid, which is not a group, then the following properties hold.

(i) $\{0\}, \{1\} \in \mathcal{L}$ and all other sets of $\mathcal{L}$ lie in $\mathbb{N}_{\ge 2}$.
(ii) For every $k \in \mathbb{N}_0$ there is $L \in \mathcal{L}$ with $k \in L$.

(iii) If $L_1, L_2 \in \mathcal{L}$, then there is $L \in \mathcal{L}$ with $L_1 + L_2 \subseteq L$.

If $\mathcal{L}$ is a family satisfying the stronger condition that for each two $L_1, L_2 \in \mathcal{L}$ the sumset $L_1 + L_2 \in \mathcal{L}$, then there is a Dedekind domain D such that $\mathcal{L}(D) = \mathcal{L}$ [**78**]. But the general case is open.

The system $\mathcal{L}(M)$ gathers all length sets of a given monoid M and Theorems 12.2 and 12.5 describe all the phenomena that might occur. But, in addition, one might ask after the typical structure of a length set when we pick a random element. Let M be an analytic monoid. This is a Krull monoid with finite v-class group and with a norm function that allows to establish a theory of L-functions [**100**]. Then almost all (in the sense of Dirichlet density) length sets are intervals [**66**, Theorem 9.4.11]. Other concepts of density include the Zariski density (say for $\mathbb{Q}$-algebras). There is much heuristic evidence (in a variety of settings) that *most* (in what sense ever) length sets are much simpler compared with the possible structure of all length sets popping up in $\mathcal{L}(M)$. Thus, we present the following problem.

PROBLEM 25. Consider a distinguished class of Mori monoids which allows us to establish a natural concept of density. Describe the structure of length sets of subsets of elements having high density.

We end with a short discussion on semigroups of ideals that are not necessarily cancellative. The algebraic and arithmetic structure of ideal semigroups has found wide attention in the literature (for a sample of recent contributions, see [**3, 18, 50, 60, 71, 73, 98, 99, 110**]). In general, these semigroups need not be cancellative (indeed, the semigroup of all nonzero ideals of a domain is cancellative if and only if the domain is almost Dedekind [**86**, Theorem 23.2]). However, semigroups of ideals are unit-cancellative in a variety of significant cases. To discuss this situation, let M be a commutative semigroup with identity. We say that M is unit-cancellative if $a, u \in M$ and $au = a$ implies that $u \in H^\times$. Factorization theory in unit-cancellative monoids can be developed along the same lines as in the cancellative setting. In particular, atoms and length sets are defined as it is done for monoids in Section 2.

Let D be a domain and let r be an ideal system on D. Suppose that D satisfies the r-Krull Intersection Theorem, which means that

$$\bigcap_{n \in \mathbb{N}_0} (I^n)_r = \{0\} \quad \text{for all } I \in \mathcal{I}_r(D) \setminus \{D\}.$$

Now let $I, J \in \mathcal{I}_r(D)$ with $(IJ)_r = I$. This implies that $(IJ^n)_r = I$ for all $n \in \mathbb{N}$, whence $\{0\} \neq I \subseteq \cap_{n \in \mathbb{N}} (J^n)_r$. Since D satisfies the r-Krull Intersection Theorem, we infer that $J = D$, whence $\mathcal{I}_r(D)$ is unit-cancellative.

PROBLEM 26. Let D be a domain and let r be an ideal system on D such that $\mathcal{I}_r(D)$ is unit-cancellative. Study the system of length sets of $\mathcal{I}_r(D)$.

13. Appendix: Non-commutative rings

Multiplicative ideal theory and factorization theory spread from commutative ring theory to noncommutative ring theory in the first half of the 20th century. Although this is beyond the scope of our present article, we would be remiss not even to mention this development.

A starting point of ideal theory in the non-commutative setting was the pioneering work of Asano [**5**], followed by work of Asano, Murato, and Harada [**6, 91**], which led to a theory of Dedekind-like rings (including Asano prime rings and Dedekind prime rings, see [**113**]). From the 1970s on, non-commutative Krull rings have been introduced [**23, 97, 114**], and it was Wauters who introduced non-commutative Krull monoids in order to study non-commutative semigroup algebras [**61, 96, 143**]. We refer to some recent articles, with a focus on surveys [**1, 40, 130–132**].

The early development of factorization theory in non-commutative rings is discussed in a survey by Smertnig [**136**]. We briefly address two current research directions.

As it is the case in commutative ring theory, a topic of research are the questions which atomic prime rings have the finite factorization property (meaning that each element has only finitely many essentially different factorizations) resp. the bounded factorization property (meaning that all length sets are finite and nonempty; see [**16, 17**]). Whereas it is quite easy to verify that a commutative Noetherian domain has bounded factorizations (see Lemma 11.2), it is still an open problem whether every Noetherian prime ring has bounded factorizations.

A further key problem asks which Noetherian prime rings allow a transfer homomorphism to a commutative (cancellative) monoid or even to a commutative Krull monoid. In his seminal paper [**135**], Smertnig characterized maximal orders in central simple algebras over algebraic number fields, that allow a transfer homomorphism to a commutative Krull monoid (the proof is based on theory of one-sided divisorial ideals in such orders). This result initiated much further research (e.g., [**9–11, 13, 137**]). Note that the existence of a transfer homomorphism to a commutative Krull monoid immediately implies the existence of a transfer homomorphism to a monoid of zero-sum sequences over a subset of an abelian group, which in turn opens the door for methods from additive combinatorics (see the Subsection on the Transfer Krull Property).

Acknowledgment

We would like to thank our colleagues Valentin Havlovec, Giulio Peruginelli, Andreas Reinhart, and Daniel Smertnig for their feedback and all their helpful comments.

References

[1] E. Akalan and H. Marubayashi, *Multiplicative ideal theory in noncommutative rings*, Multiplicative ideal theory and factorization theory, Springer Proc. Math. Stat., vol. 170, Springer, [Cham], 2016, pp. 1–21, DOI 10.1007/978-3-319-38855-7_1. MR3565802

[2] D. D. Anderson and D. F. Anderson, *Elasticity of factorizations in integral domains*, J. Pure Appl. Algebra **80** (1992), no. 3, 217–235, DOI 10.1016/0022-4049(92)90144-5. MR1170712

[3] D. D. Anderson, J. R. Juett, and C. P. Mooney, *Factorization of ideals*, Comm. Algebra **47** (2019), no. 4, 1742–1772, DOI 10.1080/00927872.2018.1517359. MR3975966

[4] D. D. Anderson and J. L. Mott, *Cohen-Kaplansky domains: Integral domains with a finite number of irreducible elements*, J. Algebra **148** (1992), no. 1, 17–41, DOI 10.1016/0021-8693(92)90234-D. MR1161563

[5] K. Asano, *Arithmetische Idealtheorie in nichtkommutativen Ringen* (French), Jpn. J. Math. **16** (1939), 1–36, DOI 10.4099/jjm1924.16.0_1. MR601

[6] K. Asano and K. Murata, *Arithmetical ideal theory in semigroups*, J. Inst. Polytech. Osaka City Univ. Ser. A **4** (1953), 9–33. MR59269

[7] M. Auslander and M. Bridger, *Stable module theory*, Memoirs of the American Mathematical Society, No. 94, American Mathematical Society, Providence, RI, 1969. MR269685

[8] A. Badawi and J. Coykendall (eds.), *Rings, monoids and module theory*, Springer Proceedings in Mathematics & Statistics, vol. 382, Springer, Singapore, [2021] ©2021. AUS-ICMS 2020, Sharjah, United Arab Emirates, February 6–9; Papers from the 3rd International Conference on Mathematics and Statistics held at the American University of Sharjah, DOI 10.1007/978-981-16-8422-7. MR4428449

[9] N. R. Baeth and R. Sampson, *Upper triangular matrices over information algebras*, Linear Algebra Appl. **587** (2020), 334–357, DOI 10.1016/j.laa.2019.11.014. MR4036676

[10] N. R. Baeth and D. Smertnig, *Factorization theory: from commutative to noncommutative settings*, J. Algebra **441** (2015), 475–551, DOI 10.1016/j.jalgebra.2015.06.007. MR3391936

[11] N. R. Baeth and D. Smertnig, *Arithmetical invariants of local quaternion orders*, Acta Arith. **186** (2018), no. 2, 143–177, DOI 10.4064/aa170601-13-8. MR3870710

[12] N. R. Baeth and D. Smertnig, *Lattices over Bass rings and graph agglomerations*, Algebr. Represent. Theory **25** (2022), no. 3, 669–704, DOI 10.1007/s10468-021-10040-2. MR4421814

[13] C. Barendrecht, *Finiteness of elasticities of orders in central simple algebras*, Acta Arith. **206** (2022), no. 3, 197–221, DOI 10.4064/aa211015-14-11. MR4522847

[14] A. Bashir and M. Pompili, *On transfer homomorphisms in commutative rings with zero-divisors*, Comm. Algebra **54** (2026), no. 2, 776–789, DOI 10.1080/00927872.2025.2539432. MR4996690

[15] A. Bashir and A. Reinhart, *On transfer Krull monoids*, Semigroup Forum **105** (2022), no. 1, 73–95, DOI 10.1007/s00233-022-10296-0. MR4466959

[16] J. P. Bell, K. Brown, Z. Nazemian, and D. Smertnig, *On noncommutative bounded factorization domains and prime rings*, J. Algebra **622** (2023), 404–449, DOI 10.1016/j.jalgebra.2023.01.023. MR4551907

[17] J. P. Bell, A. Heinle, and V. Levandovskyy, *On noncommutative finite factorization domains*, Trans. Amer. Math. Soc. **369** (2017), no. 4, 2675–2695, DOI 10.1090/tran/6727. MR3592524

[18] J. Brantner, A. Geroldinger, and A. Reinhart, *On monoids of ideals of orders in quadratic number fields*, Advances in rings, modules and factorizations, Springer Proc. Math. Stat., vol. 321, Springer, Cham, [2020] ©2020, pp. 11–54, DOI 10.1007/978-3-030-43416-8_2. MR4113956

[19] M. Brešar, A. Geroldinger, B. Olberding, and D. Smertnig (eds.), *Recent progress in ring and factorization theory*, Springer Proceedings in Mathematics & Statistics, vol. 477, Springer, Cham, [2025] ©2025. Graz, Austria, July 10–14, 2023, DOI 10.1007/978-3-031-75326-8. MR4928628

[20] P.-J. Cahen, M. Fontana, S. Frisch, and S. Glaz, *Open problems in commutative ring theory*, Commutative algebra, Springer, New York, 2014, pp. 353–375. MR3330228

[21] J.-L. Chabert, *Integer-valued polynomials—from combinatorics to number theory, p-adic analysis, commutative and non-commutative algebra*, American Mathematical Society Colloquium Publications, vol. 69, American Mathematical Society, Providence, RI, 2025. MR5019299

[22] J.-L. Chabert, M. Fontana, S. Frisch, S. Glaz, and K. Johnson (eds.), *Algebraic, number theoretic, and topological aspects of ring theory*, Springer, Cham, [2023] ©2023. Papers presented at "The Conference on Rings and Polynomials" held July 19–24, 2021; Including papers intended for "The Fourth International Meeting on Integer-Valued Polynomials and Related Topics" (canceled), DOI 10.1007/978-3-031-28847-0. MR4647285

[23] M. Chamarie, *Anneaux de Krull non commutatifs* (French), J. Algebra **72** (1981), no. 1, 210–222, DOI 10.1016/0021-8693(81)90318-5. MR634623

[24] G. W. Chang, V. Fadinger, and D. Windisch, *Semigroup rings as weakly Krull domains*, Pacific J. Math. **318** (2022), no. 2, 433–452, DOI 10.2140/pjm.2022.318.433. MR4474369

[25] G. W. Chang and A. Geroldinger, *On Dedekind domains whose class groups are direct sums of cyclic groups*, J. Pure Appl. Algebra **228** (2024), no. 1, Paper No. 107470, 14, DOI 10.1016/j.jpaa.2023.107470. MR4604849

[26] G. W. Chang and H. Kim, *A characterization of Krull domains in terms of their factor rings*, Comm. Algebra **51** (2023), no. 3, 1280–1292, DOI 10.1080/00927872.2022.2134404. MR4554575

[27] G. W. Chang and J. S. Oh, *The monoid of regular elements in commutative rings with zero divisors*, Comm. Algebra **50** (2022), no. 3, 1182–1198, DOI 10.1080/00927872.2021.1979028. MR4379663

[28] S. Chapman, M. Fontana, A. Geroldinger, and B. Olberding (eds.), *Multiplicative ideal theory and factorization theory*, Springer Proceedings in Mathematics & Statistics, vol. 170, Springer, [Cham], 2016. Commutative and non-commutative perspectives, DOI 10.1007/978-3-319-38855-7. MR3565801

[29] S. T. Chapman and J. Coykendall, *Half-factorial domains, a survey*, Non-Noetherian commutative ring theory, Math. Appl., vol. 520, Kluwer Acad. Publ., Dordrecht, 2000, pp. 97–115. MR1858159

[30] S. U. Chase, *Direct products of modules*, Trans. Amer. Math. Soc. **97** (1960), 457–473, DOI 10.2307/1993382. MR120260

[31] P. L. Clark, S. Gosavi, and P. Pollack, *The number of atoms in a primefree atomic domain*, Comm. Algebra **45** (2017), no. 12, 5431–5442, DOI 10.1080/00927872.2017.1310877. MR3691377

[32] L. Cossu, *Some applications of a new approach to factorization*, Recent progress in ring and factorization theory, Springer Proc. Math. Stat., vol. 477, Springer, Cham, [2025] ©2025, pp. 73–94, DOI 10.1007/978-3-031-75326-8_4. MR4928632

[33] L. Cossu and B. Olberding, *Realization of spaces of commutative rings*, J. Lond. Math. Soc. (2) **111** (2025), no. 5, Paper No. e70175, 27, DOI 10.1112/jlms.70175. MR4908204

[34] J. Coykendall, *Extensions of half-factorial domains: a survey*, Arithmetical properties of commutative rings and monoids, Lect. Notes Pure Appl. Math., vol. 241, Chapman & Hall/CRC, Boca Raton, FL, 2005, pp. 46–70, DOI 10.1201/9781420028249.ch3. MR2140686

[35] J. Coykendall and F. Gotti, *Atomicity in integral domains*, Recent progress in ring and factorization theory, Springer Proc. Math. Stat., vol. 477, Springer, Cham, [2025] ©2025, pp. 95–159, DOI 10.1007/978-3-031-75326-8_5. MR4928633

[36] J. Coykendall and C. Spicer, *Cohen-Kaplansky domains and the Goldbach conjecture*, Proc. Amer. Math. Soc. **140** (2012), no. 7, 2227–2233, DOI 10.1090/S0002-9939-2011-11086-4. MR2898686

[37] T. Dube and A. Goswami, *Ideal spaces: An extension of structure spaces of rings*, J. Algebra Appl. **22** (2023), no. 11, Paper No. 2350245, 18, DOI 10.1142/S0219498823502456. MR4663894

[38] T. Dumitrescu, *Kaplansky/Nagata-type theorems for the half factorial domains*, Comm. Algebra, to appear (2026), DOI: `https://doi.org/10.1080/00927872.2025.2610362`.

[39] T. Dumitrescu, M. Epure, and A. Gica, *Lifting multiplicative lattices to ideal systems*, Bull. Math. Soc. Sci. Math. Roumanie (N.S.) **68(116)** (2025), no. 4, 473–479. MR4990676

[40] J. Elliott, *Rings, modules, and closure operations*, Springer Monographs in Mathematics, Springer, Cham, [2019] ©2019, DOI 10.1007/978-3-030-24401-9. MR3971230

[41] E. E. Enochs and O. M. G. Jenda, *Gorenstein injective and projective modules*, Math. Z. **220** (1995), no. 4, 611–633, DOI 10.1007/BF02572634. MR1363858

[42] E. E. Enochs, O. M. G. Jenda, and B. Torrecillas, *Gorenstein flat modules* (English, with Chinese summary), Nanjing Daxue Xuebao Shuxue Bannian Kan **10** (1993), no. 1, 1–9. MR1248299

[43] A. Fabbri and O. A. Heubo-Kwegna, *Projective star operations on polynomial rings over a field*, J. Commut. Algebra **4** (2012), no. 3, 387–412, DOI 10.1216/jca-2012-4-3-387. MR3024262

[44] F. Fabsits, A. Geroldinger, A. Reinhart, and Q. Zhong, *On monoids of plus-minus weighted zero-sum sequences: the isomorphism problem and the characterization problem*, J. Commut. Algebra **16** (2024), no. 1, 1–23, DOI 10.1216/jca.2024.16.1. MR4690592

[45] A. Facchini, *Krull monoids and their application in module theory*, Algebras, rings and their representations, World Sci. Publ., Hackensack, NJ, 2006, pp. 53–71, DOI 10.1142/9789812774552_0006. MR2234300

[46] A. Facchini, *Direct-sum decompositions of modules with semilocal endomorphism rings*, Bull. Math. Sci. **2** (2012), no. 2, 225–279, DOI 10.1007/s13373-012-0024-9. MR2994204

[47] A. Facchini, M. Fontana, A. Geroldinger, and B. Olberding (eds.), *Advances in rings, modules and factorizations*, Springer Proceedings in Mathematics & Statistics, vol. 321, Springer, Cham, [2020] ©2020. Graz, Austria, February 19–23, 2018, DOI 10.1007/978-3-030-43416-8. MR4113954

[48] V. Fadinger and D. Windisch, *On the distribution of prime divisors in class groups of affine monoid algebras*, Comm. Algebra **50** (2022), no. 7, 2973–2982, DOI 10.1080/00927872.2021.2023888. MR4420327

[49] V. Fadinger-Held and D. Windisch, *Lengths of factorizations of integer-valued polynomials on Krull domains with prime elements*, Arch. Math. (Basel) **123** (2024), no. 2, 123–135, DOI 10.1007/s00013-024-02001-0. MR4778320

[50] Y. Fan, A. Geroldinger, F. Kainrath, and S. Tringali, *Arithmetic of commutative semigroups with a focus on semigroups of ideals and modules*, J. Algebra Appl. **16** (2017), no. 12, 1750234, 42, DOI 10.1142/S0219498817502346. MR3725094

[51] C. A. Finocchiaro and K. A. Loper, *Topological aspects of the ideal theory in rings of integer-valued polynomials*, Proc. Amer. Math. Soc. **152** (2024), no. 5, 1809–1819, DOI 10.1090/proc/16740. MR4728453

[52] C. A. Finocchiaro, A. Goswami, and D. Spirito, *Distinguished classes of ideal spaces and their topological properties*, Comm. Algebra **51** (2023), no. 4, 1752–1760, DOI 10.1080/00927872.2022.2141767. MR4552925

[53] M. Fontana, S. Frisch, and S. Glaz (eds.), *Commutative algebra*, Springer, New York, 2014. Recent advances in commutative rings, integer-valued polynomials, and polynomial functions. MR3307652

[54] M. Fontana, S. Frisch, S. Glaz, F. Tartarone, and P. Zanardo (eds.), *Rings, polynomials, and modules*, Springer, Cham, 2017. Including papers from the Conference "Recent Advances in Commutative Ring and Module Theory" held in Bressanone/Brixen, June 13–17, 2016, and the Conference on Rings and Polynomials held in Graz, July 3–8, 2016. MR3752128

[55] M. Fontana, E. Houston, and T. Lucas, *Factoring ideals in integral domains*, Lecture Notes of the Unione Matematica Italiana, vol. 14, Springer, Heidelberg; UMI, Bologna, 2013, DOI 10.1007/978-3-642-31712-5. MR2976403

[56] M. Fontana and K. A. Loper, *An historical overview of Kronecker function rings, Nagata rings, and related star and semistar operations*, Multiplicative ideal theory in commutative algebra, Springer, New York, 2006, pp. 169–187, DOI 10.1007/978-0-387-36717-0_11. MR2265808

[57] M. Fontana and M. Zafrullah, *On v-domains: a survey*, Commutative algebra—Noetherian and non-Noetherian perspectives, Springer, New York, 2011, pp. 145–179, DOI 10.1007/978-1-4419-6990-3_6. MR2762510

[58] L. Fuchs, *Injective modules over Mori domains*, Studia Sci. Math. Hungar. **40** (2003), no. 1-2, 33–40, DOI 10.1556/SScMath.40.2003.1-2.3. MR2002988

[59] L. Fuchs and L. Salce, *Modules over non-Noetherian domains*, Mathematical Surveys and Monographs, vol. 84, American Mathematical Society, Providence, RI, 2001, DOI 10.1090/surv/084. MR1794715

[60] S. G. Gates, J. R. Juett, L. W. Ndungu, and R. D. Roberts, *Ideal factoriality, semistar operations, and quasiprincipal ideals*, Comm. Algebra **52** (2024), no. 8, 3427–3456, DOI 10.1080/00927872.2024.2320208. MR4748976

[61] A. Geroldinger, *Non-commutative Krull monoids: A divisor theoretic approach and their arithmetic*, Osaka J. Math. **50** (2013), no. 2, 503–539. MR3080813

[62] A. Geroldinger and R. Göbel, *Half-factorial subsets in infinite abelian groups*, Houston J. Math. **29** (2003), no. 4, 841–858. MR2045657

[63] A. Geroldinger, F. Gotti, and S. Tringali, *On strongly primary monoids, with a focus on Puiseux monoids*, J. Algebra **567** (2021), 310–345, DOI 10.1016/j.jalgebra.2020.09.019. MR4159257

[64] A. Geroldinger, D. J. Grynkiewicz, G. J. Schaeffer, and W. A. Schmid, *On the arithmetic of Krull monoids with infinite cyclic class group*, J. Pure Appl. Algebra **214** (2010), no. 12, 2219–2250, DOI 10.1016/j.jpaa.2010.02.024. MR2660910

[65] A. Geroldinger, D. J. Grynkiewicz, and Q. Zhong, *Combinatorial factorization theory*, Math. Surveys Monogr., American Mathematical Society, Providence, RI, to appear.

[66] A. Geroldinger and F. Halter-Koch, *Non-unique factorizations*, Pure and Applied Mathematics (Boca Raton), vol. 278, Chapman & Hall/CRC, Boca Raton, FL, 2006. Algebraic, combinatorial and analytic theory, DOI 10.1201/9781420003208. MR2194494

[67] A. Geroldinger and W. Hassler, *Local tameness of v-Noetherian monoids*, J. Pure Appl. Algebra **212** (2008), no. 6, 1509–1524, DOI 10.1016/j.jpaa.2007.10.020. MR2391663

[68] A. Geroldinger and W. Hassler, *Arithmetic of Mori domains and monoids*, J. Algebra **319** (2008), no. 8, 3419–3463, DOI 10.1016/j.jalgebra.2007.11.025. MR2408326

[69] A. Geroldinger and F. Kainrath, *On the arithmetic of tame monoids with applications to Krull monoids and Mori domains*, J. Pure Appl. Algebra **214** (2010), no. 12, 2199–2218, DOI 10.1016/j.jpaa.2010.02.023. MR2660909

[70] A. Geroldinger, F. Kainrath, and A. Reinhart, *Arithmetic of seminormal weakly Krull monoids and domains*, J. Algebra **444** (2015), 201–245, DOI 10.1016/j.jalgebra.2015.07.026. MR3406175

[71] A. Geroldinger and M. A. Khadam, *On the arithmetic of monoids of ideals*, Ark. Mat. **60** (2022), no. 1, 67–106, DOI 10.4310/arkiv.2022.v60.n1.a4. MR4423271

[72] A. Geroldinger, S. Ramacher, and A. Reinhart, *On v-Marot Mori rings and C-rings*, J. Korean Math. Soc. **52** (2015), no. 1, 1–21, DOI 10.4134/JKMS.2015.52.1.001. MR3299367

[73] A. Geroldinger and A. Reinhart, *The monotone catenary degree of monoids of ideals*, Internat. J. Algebra Comput. **29** (2019), no. 3, 419–457, DOI 10.1142/S0218196719500097. MR3955817

[74] A. Geroldinger and M. Roitman, *On strongly primary monoids and domains*, Comm. Algebra **48** (2020), no. 9, 4085–4099, DOI 10.1080/00927872.2020.1755678. MR4124682

[75] A. Geroldinger, W. A. Schmid, and Q. Zhong, *Systems of sets of lengths: Transfer Krull monoids versus weakly Krull monoids*, Rings, polynomials, and modules, Springer, Cham, 2017, pp. 191–235. MR3751698

[76] A. Geroldinger and Q. Zhong, *A characterization of seminormal C-monoids*, Boll. Unione Mat. Ital. **12** (2019), no. 4, 583–597, DOI 10.1007/s40574-019-00194-9. MR4022295

[77] A. Geroldinger and Q. Zhong, *Factorization theory in commutative monoids*, Semigroup Forum **100** (2020), no. 1, 22–51, DOI 10.1007/s00233-019-10079-0. MR4061567

[78] A. Geroldinger and Q. Zhong, *A realization result for systems of sets of lengths*, Israel J. Math. **247** (2022), no. 1, 177–193, DOI 10.1007/s11856-021-2263-5. MR4425332

[79] R. Gilmer, *Multiplicative ideal theory*, Pure and Applied Mathematics, No. 12, Marcel Dekker, Inc., New York, 1972. MR427289

[80] R. Gilmer, *Two constructions of Prüfer domains*, J. Reine Angew. Math. **239(240)** (1969), 153–162, DOI 10.1515/crll.1969.239-240.153. MR257056

[81] S. Glaz, *Commutative coherent rings: Historical perspective and current developments*, Nieuw Arch. Wisk. (4) **10** (1992), no. 1-2, 37–56. MR1187898

[82] S. Glaz and W. V. Vasconcelos, *Flat ideals. II*, Manuscripta Math. **22** (1977), no. 4, 325–341, DOI 10.1007/BF01168220. MR472797

[83] A. Goswami, *Terminal spaces of monoids*, Commun. Korean Math. Soc. **39** (2024), no. 1, 259–266, DOI 10.4134/CKMS.c230095. MR4725948

[84] D. J. Grynkiewicz, *The characterization of finite elasticities—factorization theory in Krull monoids via convex geometry*, Lecture Notes in Mathematics, vol. 2316, Springer, Cham, 2022, DOI 10.1007/978-3-031-14869-9. MR4539835

[85] L. Guerrieri and K. A. Loper, *On the integral domains characterized by a Bezout property on intersections of principal ideals*, J. Algebra **586** (2021), 208–231, DOI 10.1016/j.jalgebra.2021.06.028. MR4285000

[86] F. Halter-Koch, *Ideal systems*, Monographs and Textbooks in Pure and Applied Mathematics, vol. 211, Marcel Dekker, Inc., New York, 1998. An introduction to multiplicative ideal theory, DOI 10.1515/9783110801194. MR1828371

[87] F. Halter-Koch, *Kronecker function rings and generalized integral closures*, Comm. Algebra **31** (2003), no. 1, 45–59, DOI 10.1081/AGB-120016775. MR1969212

[88] F. Halter-Koch, *Multiplicative ideal theory in the context of commutative monoids*, Commutative algebra—Noetherian and non-Noetherian perspectives, Springer, New York, 2011, pp. 203–231, DOI 10.1007/978-1-4419-6990-3_8. MR2762512

[89] F. Halter-Koch, *An invitation to algebraic numbers and algebraic functions*, CRC Press, Boca Raton, FL, 2020, DOI 10.1177/0962280220942558. MR4579678

[90] F. Halter-Koch, *Ideal theory of commutative rings and monoids* (A. Geroldinger and A. Reinhart, eds.), Lecture Notes in Mathematics, vol. 2368, Springer, Cham, [2025] ©2025. Edited by Alfred Geroldinger and Andreas Reinhart, DOI 10.1007/978-3-031-88878-6. MR4923090

[91] M. Harada, *Multiplicative ideal theory in hereditary orders*, J. Math. Osaka City Univ. **14** (1963), 83–106. MR168621

[92] J. R. Hedstrom and E. G. Houston, *Some remarks on star-operations*, J. Pure Appl. Algebra **18** (1980), no. 1, 37–44, DOI 10.1016/0022-4049(80)90114-0. MR578564

[93] W. Heinzer, K. A. Loper, B. Olberding, H. Schoutens, and M. Toeniskoetter, *Ideal theory of infinite directed unions of local quadratic transforms*, J. Algebra **474** (2017), 213–239, DOI 10.1016/j.jalgebra.2016.11.014. MR3595791

[94] H. Holm, *Gorenstein homological dimensions*, J. Pure Appl. Algebra **189** (2004), no. 1-3, 167–193, DOI 10.1016/j.jpaa.2003.11.007. MR2038564

[95] K. Hu and F. Wang, *Some results on Gorenstein Dedekind domains and their factor rings*, Comm. Algebra **41** (2013), no. 1, 284–293, DOI 10.1080/00927872.2011.629268. MR3010536

[96] E. Jespers and J. Okniński, *Noetherian semigroup algebras*, Algebra and Applications, vol. 7, Springer, Dordrecht, 2007. MR2301033

[97] E. Jespers and P. Wauters, *Ω-Krull rings. II*, Comm. Algebra **14** (1986), no. 5, 833–849. MR834467

[98] J. Juett, *Generalized comaximal factorization of ideals*, J. Algebra **352** (2012), 141–166, DOI 10.1016/j.jalgebra.2011.11.008. MR2862178

[99] J. R. Juett, C. P. Mooney, and L. W. Ndungu, *Unique factorization of ideals in commutative rings with zero divisors*, Comm. Algebra **49** (2021), no. 5, 2101–2125, DOI 10.1080/00927872.2020.1864390. MR4232488

[100] J. Kaczorowski, *Analytic monoids and factorization problems*, Semigroup Forum **94** (2017), no. 3, 532–555, DOI 10.1007/s00233-016-9778-9. MR3648982

[101] F. Kainrath, *Elasticity of finitely generated domains*, Houston J. Math. **31** (2005), no. 1, 43–64. MR2123000

[102] F. Kainrath, *Arithmetic of Mori domains and monoids: The global case*, Multiplicative ideal theory and factorization theory, Springer Proc. Math. Stat., vol. 170, Springer, [Cham], 2016, pp. 183–218, DOI 10.1007/978-3-319-38855-7_8. MR3565809

[103] B. G. Kang, *On the converse of a well-known fact about Krull domains*, J. Algebra **124** (1989), no. 2, 284–299, DOI 10.1016/0021-8693(89)90131-2. MR1011595

[104] H. Kim, J. W. Lim, and D. Zhou, *τ_w-Loewy modules and their applications*, Bull. Korean Math. Soc. **56** (2019), no. 6, 1617–1642, DOI 10.4134/BKMS.b190049. MR4037397

[105] M. Knebusch and D. Zhang, *Manis valuations and Prüfer extensions. I*, Lecture Notes in Mathematics, vol. 1791, Springer-Verlag, Berlin, 2002. A new chapter in commutative algebra, DOI 10.1007/b84018. MR1937245

[106] M. Knebusch and T. Kaiser, *Manis valuations and Prüfer extensions. II*, Lecture Notes in Mathematics, vol. 2103, Springer, Cham, 2014, DOI 10.1007/978-3-319-03212-2. MR3288836

[107] W. Krull, *Beiträge zur Arithmetik kommutativer Integritätsbereiche* (German), Math. Z. **41** (1936), no. 1, 665–679, DOI 10.1007/BF01180447. MR1545646

[108] W. Krull, *Beiträge zur Arithmetik kommutativer Integritätsbereiche* (German), Math. Z. **42** (1937), no. 1, 745–766, DOI 10.1007/BF01160110. MR1545707

[109] Max. D. Larsen and P. J. McCarthy, *Multiplicative theory of ideals*, Pure and Applied Mathematics, Vol. 43, Academic Press, New York-London, 1971. MR414528

[110] E. Lewis, *Factorization in the monoid of integrally closed ideals*, Comm. Algebra **53** (2025), no. 1, 259–271, DOI 10.1080/00927872.2024.2374437. MR4852403

[111] K. A. Loper and F. Tartarone, *A classification of the integrally closed rings of polynomials containing $\mathbb{Z}[X]$*, J. Commut. Algebra **1** (2009), no. 1, 91–157, DOI 10.1216/JCA-2009-1-1-91. MR2462383

[112] W. Fanggui and R. L. McCasland, *On strong Mori domains*, J. Pure Appl. Algebra **135** (1999), no. 2, 155–165, DOI 10.1016/S0022-4049(97)00150-3. MR1667555

[113] J. C. McConnell and J. C. Robson, *Noncommutative Noetherian rings*, Revised edition, Graduate Studies in Mathematics, vol. 30, American Mathematical Society, Providence, RI, 2001. With the cooperation of L. W. Small, DOI 10.1090/gsm/030. MR1811901

[114] H. Marubayashi, *Non-commutative Krull rings*, Osaka Math. J. **12** (1975), no. 3, 703–714. MR396642

[115] M. Nagata, *Local rings*, Interscience Tracts in Pure and Applied Mathematics, No. 13, Interscience Publishers (a division of John Wiley & Sons, Inc.), New York-London, 1962. MR155856

[116] B. Olberding and M. Roitman, *The minimal number of generators of an invertible ideal*, Multiplicative ideal theory in commutative algebra, Springer, New York, 2006, pp. 349–367, DOI 10.1007/978-0-387-36717-0_21. MR2265819

[117] G. Peruginelli, *Transcendental extensions of a valuation domain of rank one*, Proc. Amer. Math. Soc. **145** (2017), no. 10, 4211–4226, DOI 10.1090/proc/13574. MR3690607

[118] G. Peruginelli, *Polynomial Dedekind domains with finite residue fields of prime characteristic*, Pacific J. Math. **324** (2023), no. 2, 333–351, DOI 10.2140/pjm.2023.324.333. MR4619855

[119] G. Peruginelli, *Stacked pseudo-convergent sequences and polynomial Dedekind domains*, Algebra Number Theory **19** (2025), no. 10, 1947–1972, DOI 10.2140/ant.2025.19.1947. MR4956431

[120] M. Pompili, *Every finitely generated abelian group is the class group of a generalized cluster algebra*, J. Algebra **686** (2026), 566–594, DOI 10.1016/j.jalgebra.2025.08.017. MR4956910

[121] Y. Pu, G. Tang, and F. Wang, *w-Matlis cotorsion modules and w-Matlis domains*, Bull. Korean Math. Soc. **56** (2019), no. 5, 1187–1198, DOI 10.4134/BKMS.b180931. MR4015966

[122] Y. Pu, W. Zhao, G. Tang, F. Wang, and X. Xiao, *On w_∞-Warfield cotorsion modules and Krull domains*, Algebra Colloq. **30** (2023), no. 4, 701–712, DOI 10.1142/S1005386723000524. MR4671219

[123] Y. Pu, W. Zhao, G. Tang, F. Wang, and X. Xiao, *On 1-w-cotorsion modules and Prüfer v-multiplication domains*, J. Algebra Appl. **25** (2026), Paper No. 2650128, 14 pp.

[124] L. Qiao and F. Wang, *A half-centered star-operation on an integral domain*, J. Korean Math. Soc. **54** (2017), no. 1, 35–57, DOI 10.4134/JKMS.j150582. MR3598041

[125] B. Rago, *A characterization of half-factorial orders in algebraic number fields*, Acta Arith. **217** (2025), no. 1, 1–17, DOI 10.4064/aa230417-23-8. MR4849350

[126] B. Rago, *A characterization of transfer Krull orders in Dedekind domains with torsion class group*, Canad. Math. Bull., To appear (2026), arXiv:2411.00271.

[127] L. J. Ratliff Jr., *Characterizations of catenary rings*, Amer. J. Math. **93** (1971), 1070–1108, DOI 10.2307/2373746. MR297752

[128] A. Reinhart, *On integral domains that are C-monoids*, Houston J. Math. **39** (2013), no. 4, 1095–1116. MR3164706

[129] A. Reinhart, *A note on conductor ideals*, Comm. Algebra **44** (2016), no. 10, 4243–4251, DOI 10.1080/00927872.2015.1087539. MR3508296

[130] W. Rump, *Invertible ideals and noncommutative arithmetics*, Comm. Algebra **29** (2001), no. 12, 5673–5686, DOI 10.1081/AGB-100107952. MR1872818

[131] W. Rump, *The role of divisors in noncommutative ideal theory*, Recent progress in ring and factorization theory, Springer Proc. Math. Stat., vol. 477, Springer, Cham, [2025] ©2025, pp. 391–416, DOI 10.1007/978-3-031-75326-8_18. MR4928646

[132] W. Rump and Y. Yang, *Hereditary arithmetics*, J. Algebra **468** (2016), 214–252, DOI 10.1016/j.jalgebra.2016.08.015. MR3550864

[133] W. A. Schmid, *Some recent results and open problems on sets of lengths of Krull monoids with finite class group*, Multiplicative ideal theory and factorization theory, Springer Proc. Math. Stat., vol. 170, Springer, [Cham], 2016, pp. 323–352, DOI 10.1007/978-3-319-38855-7_14. MR3565815

[134] J.-P. Serre, *Sur la dimension homologique des anneaux et des modules noethériens* (French), Proceedings of the international symposium on algebraic number theory, Tokyo & Nikko, 1955, Science Council of Japan, Tokyo, 1956, pp. 175–189. MR86071

[135] D. Smertnig, *Sets of lengths in maximal orders in central simple algebras*, J. Algebra **390** (2013), 1–43, DOI 10.1016/j.jalgebra.2013.05.016. MR3072109

[136] D. Smertnig, *Factorizations of elements in noncommutative rings: A survey*, Multiplicative ideal theory and factorization theory, Springer Proc. Math. Stat., vol. 170, Springer, [Cham], 2016, pp. 353–402, DOI 10.1007/978-3-319-38855-7_15. MR3565816

[137] D. Smertnig, *Factorizations in bounded hereditary Noetherian prime rings*, Proc. Edinb. Math. Soc. (2) **62** (2019), no. 2, 395–442, DOI 10.1017/s0013091518000305. MR3935805

[138] J.-L. Steffan, *Longueurs des décompositions en produits d'éléments irréductibles dans un anneau de Dedekind* (French), J. Algebra **102** (1986), no. 1, 229–236, DOI 10.1016/0021-8693(86)90138-9. MR853241

[139] S. Tringali, *Structural properties of subadditive families with applications to factorization theory*, Israel J. Math. **234** (2019), no. 1, 1–35, DOI 10.1007/s11856-019-1922-2. MR4040819

[140] F. Wang and H. Kim, *Foundations of commutative rings and their modules*, Algebra and Applications, vol. 31, Springer, Singapore, [2024] ©2024. Second edition [of 3587977], DOI 10.1007/978-981-97-5284-3. MR4807670

[141] F. Wang and R. L. McCasland, *On w-modules over strong Mori domains*, Comm. Algebra **25** (1997), no. 4, 1285–1306, DOI 10.1080/00927879708825920. MR1437672

[142] F. Wang and L. Qiao, *A homological characterization of Krull domains II*, Comm. Algebra **47** (2019), no. 5, 1917–1929, DOI 10.1080/00927872.2018.1524007. MR3977710

[143] P. Wauters, *On some subsemigroups of noncommutative Krull rings*, Comm. Algebra **12** (1984), no. 13-14, 1751–1765, DOI 10.1080/00927878408823078. MR743314

[144] D. Windisch, *On local divisor class groups of complete intersections*, J. Pure Appl. Algebra **229** (2025), no. 1, Paper No. 107804, 7, DOI 10.1016/j.jpaa.2024.107804. MR4796737

[145] S. Xing, *Some examples in Gorenstein multiplicative ideal theory*, Comm. Algebra **50** (2022), no. 8, 3188–3200, DOI 10.1080/00927872.2022.2028163. MR4429455

[146] S. Xing, *Copure projective modules over FGV-domains and Gorenstein Prüfer domains*, Bull. Korean Math. Soc. **60** (2023), no. 4, 971–983, DOI 10.4134/BKMS.b220431. MR4621467

[147] S. Xing, *Gorenstein Krull domains and their factor rings*, Comm. Algebra **52** (2024), no. 8, 3419–3426, DOI 10.1080/00927872.2024.2319109. MR4748975

[148] S. Xing, L. Qiao, H. Kim, and K. Hu, *Almost Gorenstein Dedekind domains*, J. Algebra Appl. **23** (2024), no. 11, Paper No. 2450190, 16, DOI 10.1142/S0219498824501901. MR4785001

[149] S. Xing and X. Zhang, *A Gorenstein homological characterization of Krull domains*, Bull. Korean Math. Soc. **61** (2024), no. 3, 735–744, DOI 10.4134/BKMS.b230323. MR4752270

Department of Mathematics and Scientific Computing, University of Graz, 8010 Graz, Austria
Email address: alfred.geroldinger@uni-graz.at

Division of Computer Engineering, Hoseo University, Asan 31499, Republic of Korea
Email address: hkkim@hoseo.edu

Department of Mathematics, the Ohio State University, Columbus, Ohio 43210
Email address: lopera@math.ohio-state.edu

Contemporary Mathematics
Volume **836**, 2026
https://doi.org/10.1090/conm/836/16702

Extremal elasticity of quadratic orders

Kai Steve Fan and Paul Pollack

ABSTRACT. We study how large and small elasticity can be for orders belonging to a fixed quadratic field, in terms of the corresponding conductors. For example, we show that if K is an imaginary quadratic field, then the order of conductor f in K has elasticity exceeding $(\log f)^{c_1 \log\log\log f}$ for all f that are sufficiently large. On the other hand, this elasticity is smaller than $(\log f)^{c_2 \log\log\log f}$ for infinitely many f. Here c_1, c_2 are universal positive constants. The proofs borrow methods from analytic number theory previously employed to study statistics of the multiplicative groups $(\mathbb{Z}/m\mathbb{Z})^\times$.

1. Introduction

Let R be an **atomic domain**, i.e., an integral domain in which every nonzero nonunit factors as a product of irreducibles. The **elasticity** of R, denoted $\rho(R)$, is the supremum of all ratios r/s, where r and s are positive integers for which there exists an equation

$$\pi_1 \cdots \pi_r = \rho_1 \cdots \rho_s$$

with each π_i and ρ_j irreducible in R. So for instance, every unique factorization domain (not a field) has elasticity 1. Conversely, if R is an atomic domain with elasticity 1, then R might be thought of as "halfway" to unique factorization: any two factorizations of the same nonzero nonunit involve the same number of irreducibles, although the irreducibles themselves need not be pairwise associate. A domain of elasticity 1 is called a **half-factorial domain** (or **HFD**).

The study of half-factorial domains was initiated in a 1960 paper of Carlitz [**Car60**], where it is shown that the ring of integers $\mathcal{O}_K$ of the number field K is a half-factorial domain precisely when the corresponding class number $h_K = 1$ or 2. The general notion of elasticity was introduced by Valenza in a 1990 paper[1] [**Val90**] and further studied by Narkiewicz [**Nar95**] and Steffan [**Ste86**]. Collecting the results of these three papers, one has a complete determination of the elasticity of the ring of integers of an arbitrary number field.

For a finite abelian group G, we write $\operatorname{Dav} G$ for the **Davenport constant** of G, meaning the least positive integer D with the following property: Any sequence

2020 *Mathematics Subject Classification.* Primary 11R27; Secondary 11N37, 11R11, 11R65, 13A05.

The second author gratefully acknowledges the support of the National Science Foundation under award DMS-2001581.

[1] received by the journal in 1980!

©2026 American Mathematical Society

$g_1, \ldots, g_D$ of elements of G possesses a nonempty subsequence multiplying to the identity.

THEOREM A (Narkiewicz–Steffan–Valenza). *Let R be a Dedekind domain with finite class group. Suppose that every ideal class of R is represented by at least one maximal ideal of R. Then*

$$\rho(R) = \max\left\{1, \frac{1}{2}\operatorname{Dav}\operatorname{Cl}(R)\right\},$$

where $\operatorname{Cl}(R)$ *is the class group of* R.

Results of class field theory guarantee that the hypotheses of Theorem A are satisfied for any ring of S-integers in any global field. Thus, Theorem A gives a complete description of elasticities for the rings most prominent in number theory.

Of course, there are many other important rings in algebra and number theory, and there is a rich literature investigating elasticity and half-factoriality more generally. See [**And97**] and [**CC00**] for surveys. In this paper, which is a companion piece to our recent article [**FP**], we will be examining elasticity in orders of quadratic fields.

Let K be a quadratic field (a field extension of $\mathbb{Q}$ with $[K:\mathbb{Q}] = 2$). An order in K is a subring of $\mathcal{O}_K$ strictly larger than $\mathbb{Z}$. The orders in K are in one-to-one correspondence with the natural numbers (positive integers) f. Each order $\mathcal{O}$ in K has the form

$$\mathcal{O}_f := \{\alpha \in \mathcal{O}_K : \alpha \equiv a \bmod f\mathcal{O}_K \text{ for some rational integer } a\}$$

for a unique $f \in \mathbb{N}$; conversely, $\mathcal{O}_f$ as defined above is always an order in K. It is sometimes helpful to have a more explicit description of $\mathcal{O}_f$: It is well-known that for each quadratic field K, there is a unique squarefree integer D with $K = \mathbb{Q}(\sqrt{D})$. If we set

$$(1.1) \qquad \tau_D = \begin{cases} \sqrt{D} & \text{if } D \equiv 2,3 \pmod 4, \\ \frac{1}{2}(1+\sqrt{D}) & \text{if } D \equiv 1 \pmod 4, \end{cases}$$

then $\mathcal{O}_f = \mathbb{Z} + f\tau_D\mathbb{Z}$. We refer to $\mathcal{O}_f$ as the order of conductor f in K. Note that if $f \mid f'$, then $\mathcal{O}_{f'} \subseteq \mathcal{O}_f$, and that when $f = 1$, we have $\mathcal{O}_1 = \mathcal{O}_K$; we will refer to $\mathcal{O}_K$ as the maximal order in K. For more on the basic theory of quadratic orders, see the lovely book of Cox [**Cox22**] (p. 105 and following).

It was noted by Zaks [**Zak76, Zak80**] that $\mathbb{Z}[\sqrt{-3}]$, the order of conductor 2 in $K = \mathbb{Q}(\sqrt{-3})$, is a half-factorial domain. However, half-factorial quadratic orders were not systematically studied until the later work of Coykendall [**Coy01**] and Halter-Koch [**HK83**]. (This latter paper of Halter-Koch is not easy to come by; a more readily-available source for this material is [**GHK06**]; see p. 226 and following.) These papers contain a complete algebraic characterization of half-factorial quadratic orders.

The characterization in the imaginary case is surprisingly simple: $\mathbb{Z}[\sqrt{-3}]$ is the unique nonmaximal half-factorial quadratic order [**Coy01**, Theorem 2.3]. (Recall that maximal orders fall under the purview of Carlitz's 1960 theorem, so we may always restrict to the nonmaximal case.) The real case is substantially more complicated. Here there are various ways to state the characterization, some more explicit than others; the version we quote below is based on Coykendall's paper [**Coy01**] and a subsequent manuscript of Coykendall–Malcolmson–Okoh [**CMO17**].

THEOREM B. *Let $\mathcal{O}$ be the order of conductor f in the real quadratic field K. In order for $\mathcal{O}$ to be half-factorial, it is necessary that*

(i) *$\mathcal{O}_K$ be half-factorial (equivalently, $h_K = 1$ or $h_K = 2$),* and
(ii) *$f = p$ or $f = 2p$, where p is a prime inert in K; if $f = 2p$, we require $p > 2$ and both 2 and p to be inert in K.*

Conversely, suppose $\mathcal{O}_K$ is half-factorial and $f = p$ is a prime inert in K. Let ϵ be the fundamental unit of K.[2] *Then*

$$\mathcal{O} \text{ is half-factorial} \iff \epsilon \text{ generates } (\mathcal{O}_K/f\mathcal{O}_K)^\times/\langle \text{images of integers prime to } f\rangle.$$

Finally, if $f = 2p$, where 2 and p are distinct primes inert in K, then

$$\mathcal{O} \text{ is half-factorial} \iff 3 \nmid p+1, \text{ and both } \mathcal{O}_2,\ \mathcal{O}_p \text{ are half-factorial.}$$

From the algebraic standpoint, there is no room for improvement in Theorem B; the stated conditions are both necessary and sufficient. But the statistical question of how often half-factorial orders appear "in the wild" is far from settled. For this, one must study how often the conditions of Theorem B are satisfied; this is an analytic problem rather than a purely algebraic one. In [**Coy01**], Coykendall posed two conjectures in this direction:

(A) Varying both the quadratic field K and the natural number f, one encounters infinitely many half-factorial quadratic orders $\mathcal{O}_f$.
(B) Fix the field $K = \mathbb{Q}(\sqrt{2})$. Then $\mathcal{O}_f$ is a half-factorial domain for infinitely many $f \in \mathbb{N}$.

Coykendall's conjectures were recently studied by the second author (P.P.) in [**Pol24**]. The weaker Conjecture (A) is proved in full, while the stronger Conjecture (B) is demonstrated under the assumption of the **Generalized Riemann Hypothesis** (GRH).[3] The methods of [**Pol24**] were extended in [**Pol25**] to obtain various related results. Here is a sample.

THEOREM C. *Let $\mathcal{E} = \{1, \frac{3}{2}, 2, \frac{5}{2}, 3, \frac{7}{2}, \dots\} \cup \{\infty\}$.*

(i) *If $\mathcal{O}$ is any quadratic order, then $\rho(\mathcal{O}) \in \mathcal{E}$.*
(ii) *Under GRH, each element of $\mathcal{E}$ is the elasticity of infinitely many distinct orders in $\mathbb{Q}(\sqrt{2})$.*

In both [**Pol24, Pol25**], the goal is to realize a prescribed elasticity, and the conductors f are constructed accordingly. It is natural to ask what elasticities one sees if instead of constructing f towards a predetermined end, one samples f "at random" and records the results. This question was recently considered by the present authors in [**FP**].

It requires some care to settle on the right notion of "at random." Let K be a quadratic field, and let $\mathcal{O}$ be the order of conductor f in K. It follows from a general theorem of Halter-Koch (see [**HK95**, Corollary 4]) that $\rho(\mathcal{O}) = \infty$ if and only if f is divisible by a prime p that splits (completely) in K. For a given K, the Chebotarev density theorem guarantees that asymptotically half of all primes p split in K. It follows that asymptotically 100% of natural numbers f have at least

[2]While not important here, we mention for later: We always view quadratic fields as embedded in $\mathbb{C}$. If $K = \mathbb{Q}(\sqrt{D})$ is real-quadratic, we assume $\sqrt{D} > 0$, and that the fundamental unit ϵ is normalized to satisfy $\epsilon > 1$.

[3]Throughout, by GRH we mean the assertion that all nontrivial zeros of all number field zeta functions lie on the line $\Re(s) = \frac{1}{2}$.

one split prime factor, and so $\rho(\mathcal{O}_f) = \infty$ asymptotically 100% of the time. Thus, if we fix a quadratic field K, the natural problem is to study the distribution of $\rho(\mathcal{O}_f)$ with f sampled uniformly from (only) **split-free** integers, meaning integers possessing no split prime factors. This is precisely what we do in [**FP**].

THEOREM D. *Let K be a fixed imaginary quadratic field. Let Δ_K denote the discriminant of K. For almost all split-free numbers f,*

$$\rho(\mathcal{O}_f) = f/(\log f)^{\frac{1}{2}\log_3 f + \frac{1}{2}C_K + O((\log_4 f)^3/\log_3 f)},$$

where $\log_k$ denotes the kth iterate of the natural logarithm, and

$$C_K := \sum_{p>2} \frac{\log p}{(p-1)^2} - 1 - \frac{\operatorname{sgn}(\Delta_K)1_{\mathcal{D}}(\Delta_K)|\Delta_K|\log \operatorname{Rad}(|\Delta_K|)}{\varphi(|\Delta_K|)^2}.$$

*Here $\operatorname{Rad}(|\Delta_K|)$ is the product of the distinct primes dividing Δ_K, and $1_{\mathcal{D}}$ is the characteristic function of the set of **prime discriminants***

$$\mathcal{D} := \{-4, \pm 8\} \cup \left\{(-1)^{\frac{p-1}{2}}p \colon p > 2 \text{ is prime}\right\}.$$

THEOREM E (conditional on GRH). *Let K be a fixed real quadratic field. For almost all split-free numbers f,*

$$\rho(\mathcal{O}_f) = (\log f)^{\frac{1}{2} + O(1/\log_4 f)}.$$

In these theorems, **almost all** means that the proportion of split-free numbers up to x for which the estimate fails tends to 0, as $x \to \infty$. We remind the reader that $A = O(B)$ means $|A| \le C|B|$ for some **implied constant** C. In these results, as well as all the theorems appearing below, implied constants may depend on the field K.

In this paper, we continue our investigations into the distribution of elasticities of orders belonging to a fixed quadratic field. However, instead of looking for the typical size of $\rho(\mathcal{O}_f)$, we inquire into the extremal behavior. How large and how small can $\rho(\mathcal{O}_f)$ get, in terms of f?

The following are our principal results. Below, $A \ll B$ is synonymous with $A = O(B)$. That is, $|A| \le C|B|$, for a constant C that may depend on K. The notation "$A \gg B$" indicates that $B \ll A$.

THEOREM 1.1 (Maximal order, imaginary case). *Let K be a fixed imaginary quadratic field. Then $\rho(\mathcal{O}_f) \ll f$ for all split-free numbers f. Conversely, if p is a prime inert or ramified in K, then $\rho(\mathcal{O}_p) \gg p$.*

THEOREM 1.2 (Minimal order, imaginary case). *There are universal positive constants c_1, c_2, and f_0 for which the following holds: Let K be a fixed imaginary quadratic field. For all split-free numbers $f > f_0$, we have*

$$\rho(\mathcal{O}_f) > (\log f)^{c_1 \log\log\log f}. \tag{1.2}$$

On the other hand, there is a sequence of split-free numbers f tending to infinity along which

$$\rho(\mathcal{O}_f) < (\log f)^{c_2 \log\log\log f}. \tag{1.3}$$

THEOREM 1.3 (Maximal order, real case). *Let K be a real quadratic field. Then*

$$\rho(\mathcal{O}_f) \ll f/\log f$$

for all split-free $f > 1$. In the opposite direction, GRH implies that for every $\epsilon > 0$, there are infinitely many primes p, inert in K, with

$$\rho(\mathcal{O}_p) > p^{\frac{1}{4}-\epsilon}. \tag{1.4}$$

THEOREM 1.4 (Minimal order, real case). *Assume GRH. For every real quadratic field K, there is a constant C_K with the property that $\rho(\mathcal{O}_f) < C_K$ for infinitely many split-free numbers f.*

The reader will have noticed the large gap between the (rigorous) upper bound in Theorem 1.3 and the (GRH-conditional) lower bound. As we indicate in a remark following the proof, we believe the upper bound is sharp; in fact, we conjecture that there are infinitely many primes p, inert in K, with $\rho(\mathcal{O}_p) \gg p/\log p$.

2. The class group, its principal part, and the pre-class group

In this section we set up machinery from [**FP**] that will play an important role in our proofs.

Let K be a quadratic field. For each natural number f, we let $I_K(f)$ denote the group of fractional ideals of K generated by the nonzero ideals of $\mathcal{O}_K$ comaximal with $f\mathcal{O}_K$. We write $P_{K,\mathbb{Z}}(f)$ for the subgroup of $I_K(f)$ generated by principal ideals $\alpha\mathbb{Z}_K$, where $\alpha \in \mathcal{O}_K$ satisfies $\alpha \equiv a \pmod{f\mathcal{O}_K}$ for an integer a coprime to f. The class group $\mathrm{Cl}(\mathcal{O}_f)$ is defined as the quotient $I_K(f)/P_{K,\mathbb{Z}}(f)$. Note that when $f = 1$, our definition recovers the usual definition of the class group of $\mathcal{O}_K = \mathcal{O}_1$.

The next result, which appears as [**FP**, Lemma 2.1], is a variant of Theorem A for quadratic orders.

PROPOSITION 2.1. *Let K be a quadratic field. For each split-free $f \in \mathbb{N}$,*

$$\frac{1}{2}\operatorname{Dav}\mathrm{Cl}(\mathcal{O}_f) \le \rho(\mathcal{O}_f) \le \max\left\{1, \frac{1}{2}\operatorname{Dav}\mathrm{Cl}(\mathcal{O}_f) + \frac{3}{2}\Omega(f)\right\}.$$

(Here $\Omega(\cdot)$ counts the total number of prime factors, with multiplicity; e.g., $\Omega(-12) = \Omega(30) = 3$.) Proposition 2.1 is not as precise as Theorem A; except in the case $f = 1$, it does not determine the exact value of $\rho(\mathcal{O}_f)$. Nevertheless, it will suffice to obtain our statistical results.

In order to access $\operatorname{Dav}\mathrm{Cl}(\mathcal{O}_f)$, we find it helpful to "pull apart" the class group. We define the **principal part of the class group**, denoted $\mathrm{PrinCl}(\mathcal{O}_f)$, by

$$\mathrm{PrinCl}(\mathcal{O}_f) := (\mathcal{O}_K/f\mathcal{O}_K)^\times/\langle\text{images of integers prime to } f, \text{ units of } \mathcal{O}_K\rangle.$$

Write P_K for $P_{K,\mathbb{Z}}(1)$ (the group of principal fractional ideals). The exact sequence

$$(\mathbb{Z}/f\mathbb{Z})^\times \times \mathcal{O}_K^\times \xrightarrow{\mu} (\mathcal{O}_K/f\mathcal{O}_K)^\times \xrightarrow{\iota} (I_K(f) \cap P_K)/P_{K,\mathbb{Z}}(f) \longrightarrow 1$$

allows us to identify $\mathrm{PrinCl}(\mathcal{O}_f)$ with the subgroup $(I_K(f)\cap P_K)/P_{K,\mathbb{Z}}(f)$ of $\mathrm{Cl}(\mathcal{O}_f)$; here μ and ι are the maps defined by

$$\mu((a \bmod f, \eta)) := a\eta \bmod f\mathcal{O}_K \quad \text{and} \quad \iota(\alpha \bmod f\mathcal{O}_K) = [\alpha\mathcal{O}_K].$$

This identification explains the term "principal part of the class group."

With the obvious maps, there is a short exact sequence

$$\begin{array}{ccccccccc} 1 & \longrightarrow & (I_K(f) \cap P_K)/P_{K,\mathbb{Z}}(f) & \longrightarrow & I_K(f)/P_{K,\mathbb{Z}}(f) & \longrightarrow & I_K/P_K & \longrightarrow & 1. \\ & & \wr\| & & \| & & \| & & \\ & & \mathrm{PrinCl}(\mathcal{O}_f) & & \mathrm{Cl}(\mathcal{O}_f) & & \mathrm{Cl}(\mathcal{O}_K) & & \end{array}$$

(Exactness at the last position is not obvious; for this one uses that each ideal class in $\mathcal{O}_K$ has a representative comaximal with $f\mathcal{O}_K$.) Thus, we may view $\mathrm{PrinCl}(\mathcal{O}_f)$ as a subgroup of $\mathrm{Cl}(\mathcal{O}_f)$ for which

$$[\mathrm{Cl}(\mathcal{O}_f) : \mathrm{PrinCl}(\mathcal{O}_f)] = \#\mathrm{Cl}(\mathcal{O}_K) = h_K. \tag{2.1}$$

To apply Proposition 2.1 requires knowledge of $\mathrm{Dav}\,\mathrm{Cl}(\mathcal{O}_f)$. Equation (2.1) is helpful here; it implies that

$$\mathrm{Dav}\,\mathrm{PrinCl}(\mathcal{O}_f) \le \mathrm{Dav}\,\mathrm{Cl}(\mathcal{O}_f) \le h_K\,\mathrm{Dav}\,\mathrm{PrinCl}(\mathcal{O}_f). \tag{2.2}$$

(See [**FP**, Lemma 2.4].) In our results, an ambiguity up to a factor depending on K is acceptable, and (2.2) allows us to always work with $\mathrm{PrinCl}(\mathcal{O}_f)$ rather than $\mathrm{Cl}(\mathcal{O}_f)$.

We will understand $\mathrm{PrinCl}(\mathcal{O}_f)$ by viewing it as arising from a two-stage construction. First, we quotient $(\mathcal{O}_K/f\mathcal{O}_K)^\times$ (only) by the image of the integers prime to f; we call this the **pre-class group**. That is,

$$\mathrm{PreCl}(\mathcal{O}_f) := (\mathcal{O}_K/f\mathcal{O}_K)^\times/\langle \text{images of integers prime to } f\rangle.$$

(The observant reader will have noticed that this group appeared already in Theorem B.) Second, we take the quotient of $\mathrm{PreCl}(\mathcal{O}_f)$ by the image $\mathcal{U}_f$ of $\mathcal{O}_K^\times$ in $\mathrm{PreCl}(\mathcal{O}_f)$; then $\mathrm{PreCl}(\mathcal{O}_f)/\mathcal{U}_f \cong \mathrm{PrinCl}(\mathcal{O}_f)$.

It is productive to view these pre-class groups $\mathrm{PreCl}(\mathcal{O}_f)$ as close cousins of the unit groups $(\mathbb{Z}/m\mathbb{Z})^\times$; this principle is exploited heavily in [**FP**] (and implicitly in [**Pol24, Pol25**]). Let $\psi(f) := \#\mathrm{PreCl}(\mathcal{O}_f)$. By the Chinese remainder theorem,

$$\mathrm{PreCl}(\mathcal{O}_f) \cong \prod_{p^k \| f} \mathrm{PreCl}(\mathcal{O}_{p^k}). \tag{2.3}$$

Thus, ψ is a multiplicative function of f. Furthermore, since the $\varphi(p^k)$ integers in $[1, p^k]$ that are prime to p^k have distinct images in $\mathcal{O}_K/p^k\mathcal{O}_K$,

$$\psi(p^k) = \frac{1}{\varphi(p^k)}(\#\mathcal{O}_K/p^k\mathcal{O}_K)^\times = \frac{1}{p^{k-1}(p-1)}(\#\mathcal{O}_K/p^k\mathcal{O}_K)^\times.$$

Now

$$\#(\mathcal{O}_K/p^k\mathcal{O}_K)^\times = N(p^k\mathcal{O}_K)\prod_{P|p}\left(1-\frac{1}{N(P)}\right) = p^{2k}\prod_{P|p}\left(1-\frac{1}{N(P)}\right).$$

Plugging the result of this formula into the last display, considering separately the cases when p is ramified, split, or inert, we find after a short calculation that

$$\psi(p^k) = p^k - \left(\frac{\Delta_K}{p}\right)p^{k-1} = p^k\left(1-\left(\frac{\Delta_K}{p}\right)\frac{1}{p}\right),$$

where $\left(\frac{\cdot}{\cdot}\right)$ is the Kronecker symbol. Hence,

$$\psi(f) = f\prod_{p|f}\left(1-\left(\frac{\Delta_K}{p}\right)\frac{1}{p}\right),$$

in close analogy with the familiar formula $\varphi(f) = f\prod_{p|f}(1-\frac{1}{p})$ for Euler's φ-function.

The groups $\mathrm{PreCl}(\mathcal{O}_f)$ mirror $(\mathbb{Z}/m\mathbb{Z})^\times$ not only in their size but also in their structure. For instance, we have the following result.

LEMMA 2.2. *Let K be a quadratic field, and let p be a rational prime with $p > 3$. Then $\mathrm{Cl}(\mathcal{O}_{p^k})$ is cyclic for each natural number k.*

Lemma 2.2 is an analogue of Gauss's classical theorem that $(\mathbb{Z}/p^k\mathbb{Z})^\times$ is cyclic for each prime power p^k with $p > 2$.

PROOF OF LEMMA 2.2. In the case where the prime $p > 3$ is ramified or inert in K, this result appears as [**FP**, Lemma 2.5], where it is deduced from related structure theorems of Halter-Koch [**HK72**]. Suppose now that p is split, say $p\mathcal{O}_K = P_1P_2$ with the P_i distinct maximal ideals of $\mathcal{O}_K$. Then $\mathcal{O}_K/p^k\mathcal{O}_K \cong \mathcal{O}_K/P_1^k \times \mathcal{O}_K/P_2^k$, and each $\mathcal{O}_K/P_i^k \cong \mathbb{Z}/p^k\mathbb{Z}$. Furthermore,

$$\begin{aligned}\mathrm{PreCl}(\mathcal{O}_{p^k}) &= (\mathcal{O}_K/p^k\mathcal{O}_K)^\times/\langle\text{images of integers prime to } p^k\rangle\\ &\cong \frac{(\mathbb{Z}/p^k\mathbb{Z})^\times \times (\mathbb{Z}/p^k\mathbb{Z})^\times}{(\mathbb{Z}/p^k\mathbb{Z})^\times}\\ &\cong (\mathbb{Z}/p^k\mathbb{Z})^\times;\end{aligned}$$

here the $(\mathbb{Z}/p^k\mathbb{Z})^\times$ in the "denominator" of the second line is viewed as a subgroup of the "numerator" via the diagonal embedding. By the theorem of Gauss quoted above, $(\mathbb{Z}/p^k\mathbb{Z})^\times$ is cyclic, and hence $\mathrm{PreCl}(\mathcal{O}_{p^k})$ is as well. □

It is clear from the isomorphism $(\mathbb{Z}/m\mathbb{Z})^\times \cong \prod_{p^k\|m}(\mathbb{Z}/p^k\mathbb{Z})^\times$ that the exponent $\mathrm{Exp}\,(\mathbb{Z}/m\mathbb{Z})^\times$ is a divisor of $\mathrm{lcm}\{\varphi(p^k) : p^k \| m\}$, and Gauss showed in his *Disquisitiones* that the corresponding quotient is always 1 or 2. Our proof of Theorem 1.2 requires the following variant for pre-class groups. Put

$$L(f) := \mathrm{lcm}\{\psi(p^k) : p^k \| f\}. \tag{2.4}$$

PROPOSITION 2.3. *For all natural numbers f,*

$$\mathrm{Exp}\,\mathrm{PreCl}(\mathcal{O}_f) \mid L(f) \mid 12\,\mathrm{Exp}\,\mathrm{PreCl}(\mathcal{O}_f).$$

It follows from Proposition 2.3 that $L(f)/\mathrm{Exp}\,\mathrm{PreCl}(\mathcal{O}_f) \in \{1, 2, 3, 4, 6, 12\}$.

PROOF. The first divisibility is clear from the isomorphism (2.3), keeping in mind that $\#\mathrm{PreCl}(\mathcal{O}_{p^k}) = \psi(p^k)$. To prove the second, it suffices to show that

$$\psi(p^k) \mid 12\,\mathrm{Exp}\,\mathrm{PreCl}(\mathcal{O}_{p^k}) \quad \text{for each prime power } p^k,$$

for then

$$\begin{aligned}L(f) \mid \mathrm{lcm}\{12\,\mathrm{Exp}\,\mathrm{PreCl}(\mathcal{O}_{p^k}) : p^k \| f\} &= 12\,\mathrm{lcm}\{\mathrm{Exp}\,\mathrm{PreCl}(\mathcal{O}_{p^k}) : p^k \| f\}\\ &= 12\,\mathrm{Exp}\,\mathrm{PreCl}(\mathcal{O}_f),\end{aligned}$$

as desired.

When $p > 3$, we have $\psi(p^k) = \mathrm{Exp}\,\mathrm{PreCl}(\mathcal{O}_{p^k})$ (Lemma 2.2). So we may assume $p = 2$ or $p = 3$.

Suppose first that $p = 2$. Since $\psi(2^k) = (2 - (\frac{\Delta_K}{2}))2^{k-1} \mid 12$ when $k \in \{1, 2\}$, we may assume that $k \geq 3$. Let $\alpha = 1 + 2\sqrt{D}$, where D is the squarefree integer with $K = \mathbb{Q}(\sqrt{D})$. A straightforward induction shows that, for each integer $j \geq 0$,

$$\alpha^{2^j} = u_j + v_j\sqrt{D}$$

for integers u_j, v_j with u_j odd and $2^{j+1} \parallel v_j$. In particular, $\alpha^{2^{k-1}} \in \mathbb{Z}[2^k\sqrt{D}] \subseteq \mathcal{O}_{2^k}$. Since $\alpha\mathcal{O}_K$ and $2\mathcal{O}_K$ are comaximal, we conclude that α represents an element of $\mathrm{PreCl}(\mathcal{O}_{2^k})$ of order e (say) with

$$e \mid 2^{k-1}.$$

On the other hand,

$$2\alpha^{2^{k-3}} = 2u_{k-3} + 2v_{k-3}\sqrt{D} \notin \mathbb{Z}[2^k\sqrt{D}].$$

As $2\mathcal{O}_{2^k} \subseteq \mathbb{Z}[2^k\sqrt{D}]$, it must be that $\alpha^{2^{k-3}} \notin \mathcal{O}_{2^k}$. Thus,

$$e \nmid 2^{k-3}.$$

Combining the last two displays, we find that $e = 2^{k-2}$ or $e = 2^{k-1}$. Hence,

$$\psi(2^k) = (2 - (\tfrac{\Delta_K}{2}))2^{k-1} \mid 12 \cdot 2^{k-2} \mid 12e \mid 12\,\mathrm{Exp}\,\mathrm{PreCl}(\mathcal{O}_{2^k}).$$

The proof when $p = 3$ is similar, and we condense the details. Since $\psi(3^k) = (3 - (\frac{\Delta_K}{3}))3^{k-1} \mid 12$ when $k = 1$, we can assume that $k \geq 2$. In this case, one takes $\alpha' = 1 + 3\sqrt{D}$ and shows by induction that for each integer $j \geq 0$,

$$\alpha'^{3^j} = u'_j + v'_j\sqrt{D}$$

with integers u'_j, v'_j for which $3 \nmid u'_j$, $3^{j+1} \parallel v'_j$. Then $\alpha'^{3^{k-1}} \in \mathcal{O}_{3^k}$ while $\alpha'^{3^{k-2}} \notin \mathcal{O}_{3^k}$. It follows that α' represents an element in $\mathrm{PreCl}(\mathcal{O}_{3^k})$ of exact order 3^{k-1}. Hence,

$$\psi(3^k) = (3 - (\tfrac{\Delta_K}{3}))3^{k-1} \mid 12 \cdot 3^{k-1} \mid 12\,\mathrm{Exp}\,\mathrm{PreCl}(\mathcal{O}_{3^k}),$$

as desired. □

We saw above that $\mathrm{PrinCl}(\mathcal{O}_f) \cong \mathrm{PreCl}(\mathcal{O}_f)/\mathcal{U}_f$, where $\mathcal{U}_f$ is the image of $\mathcal{O}_K^\times$ inside $\mathrm{PreCl}(\mathcal{O}_f)$. Hence, setting

$$\ell(f) := \#\mathcal{U}_f,$$

we have that

$$\#\mathrm{PrinCl}(\mathcal{O}_f) = \frac{\psi(f)}{\ell(f)}.$$

We finish this section by recording two observations about $\mathcal{U}_f$ that will be relevant in the sequel:

(i) When K is imaginary quadratic, $\ell(f) \leq \#\mathcal{O}_K^\times \leq 6$. Moreover, $\mathcal{U}_f$ is cyclic, generated by (the image of) a primitive $\#\mathcal{O}_K^\times$-th root of unity.
(ii) When K is real, $\mathcal{O}_K^\times$ is generated by ± 1 and the fundamental ϵ. Since -1 projects to the identity in $\mathrm{PreCl}(\mathcal{O}_f)$, it follows that $\mathcal{U}_f$ is cyclic in this case as well, generated by the image of ϵ in $\mathrm{PreCl}(\mathcal{O}_f)$. Furthermore,

$$\ell(f) \text{ is the least integer } \ell \text{ with } \epsilon^\ell \in \mathcal{O}_f.$$

3. Maximal order in the imaginary case: Proof of Theorem 1.1

3.1. Upper bound. We start by showing that $\rho(\mathcal{O}_f) \ll f$ for each split-free f. We may (and will) assume that $f > 1$. By Proposition 2.1, for each split-free number $f > 1$,

$$\rho(\mathcal{O}_f) \leq \frac{1}{2} \operatorname{Dav} \operatorname{Cl}(\mathcal{O}_f) + \frac{3}{2}\Omega(f).$$

Clearly,

$$2^{\Omega(f)} = \prod_{p^k \,\|\, f} 2^k \leq \prod_{p^k \,\|\, f} p^k = f, \quad \text{so that} \quad \Omega(f) \ll \log f,$$

and $\log f \ll f$. Furthermore, we have from (2.2) that

$$\operatorname{Dav} \operatorname{Cl}(\mathcal{O}_f) \leq h_K \operatorname{Dav} \operatorname{PrinCl}(\mathcal{O}_f).$$

It therefore suffices to show that $\operatorname{Dav}\operatorname{PrinCl}(\mathcal{O}_f) \ll f$. This is an immediate consequence of the following result, which will also be needed later for real K.

PROPOSITION 3.1. *Let K be any quadratic field. For every natural number f,*

$$\operatorname{Dav}\operatorname{PrinCl}(\mathcal{O}_f) \ll \frac{f}{\ell(f)}. \tag{3.1}$$

The main ingredient in the proof of Proposition 3.1 is an elegant result of van Emde Boas and Kruyswijk [**EBK69**] which bounds from above the Davenport constant of a finite abelian group G in terms of $\operatorname{Exp} G$.

PROPOSITION 3.2. *Let G be a finite abelian group, and put*

$$r(G) = \frac{\#G}{\operatorname{Exp} G}.$$

Then

$$\operatorname{Dav} G \leq \#G \left(\frac{1 + \log r(G)}{r(G)}\right).$$

The function $t \mapsto \frac{1+\log t}{t}$ is decreasing on the domain $t \geq 1$, with maximum value 1. Thus, $D(G) \leq \#G$ always, with strict inequality whenever $r(G) > 1$ (that is, whenever G is not cyclic).

In order for Proposition 3.2 to be of use in the proof of (3.1), we need a handle on $r(\operatorname{PrinCl}(\mathcal{O}_f))$. This will be given to us by the next lemma. Below, we write $\operatorname{rk}_2 G$ for the 2-rank of G.

LEMMA 3.3. *For every finite abelian group G,*

$$\exp G \mid 2^{1-\operatorname{rk}_2 G} \#G.$$

Hence,

$$r(G) \geq 2^{\operatorname{rk}_2 G - 1}.$$

PROOF. The second claim is immediate from the first, so we focus on that.

If $\operatorname{rk}_2 G \leq 1$, the asserted divisibility is trivial, so we assume $\operatorname{rk}_2 G > 1$. We may also assume $G = A \oplus \mathbb{Z}/2^{r_1}\mathbb{Z} \oplus \mathbb{Z}/2^{r_2}\mathbb{Z} \oplus \cdots \oplus \mathbb{Z}/2^{r_l}\mathbb{Z}$, where $l = \operatorname{rk}_2 G$, the r_i satisfy $1 \leq r_1 \leq r_2 \leq \cdots \leq r_l$, and A has odd order. Then

$$\operatorname{Exp} G \mid 2^{r_l}\#A = 2^{1-l} \cdot 2^{(l-1)+r_l}\#A \mid 2^{1-l} \cdot 2^{r_1+r_2+\cdots+r_{l-1}+r_\ell}\#A = 2^{1-l}\#G. \quad \square$$

PROOF OF PROPOSITION 3.1. Let w denote the number of (distinct) odd primes p dividing f which are inert in K. We will bound $\mathrm{rk}_2\,\mathrm{PrinCl}(\mathcal{O}_f)$ below in terms of w and then apply Lemma 3.3.

If $p^k \parallel f$ with p odd and inert in K, then $\#\mathrm{PreCl}(\mathcal{O}_{p^k}) = \psi(p^k) = p^k + p^{k-1}$ is even. Hence, $\mathrm{rk}_2\,\mathrm{PreCl}(\mathcal{O}_{p^k}) \geq 1$, and

$$\mathrm{rk}_2\,\mathrm{PreCl}(\mathcal{O}_f) = \sum_{p^k \parallel f} \mathrm{rk}_2\,\mathrm{PreCl}(\mathcal{O}_{p^k}) \geq w.$$

As $\mathcal{U}_f$ is cyclic, $\mathrm{rk}_2\,\mathcal{U}_f \leq 1$, and

$$\mathrm{rk}_2\,\mathrm{PrinCl}(\mathcal{O}_f) = \mathrm{rk}_2\,\frac{\mathrm{PreCl}(\mathcal{O}_f)}{\mathcal{U}_f} \geq \mathrm{rk}_2\,\mathrm{PreCl}(\mathcal{O}_f) - \mathrm{rk}_2\,\mathcal{U}_f \geq w - 1.$$

Thus, by Lemma 3.3,

$$r := r(\mathrm{PrinCl}(\mathcal{O}_f)) \geq 2^{w-2}.$$

Suppose that $w \geq 2$. Then $r \geq 2^{w-2} \geq 1$, and Proposition 3.2 yields

$$\mathrm{Dav}\,\mathrm{PrinCl}(\mathcal{O}_f) \leq \frac{\psi(f)}{\ell(f)}\frac{1+\log r}{r} \leq \frac{\psi(f)}{\ell(f)}\frac{1+\log 2^{w-2}}{2^{w-2}} < \frac{4w}{2^w}\cdot\frac{\psi(f)}{\ell(f)}.$$

As f is split-free, we have $\omega(f) \leq w + 1 + \omega(\Delta_K)$. (Per the usual convention in analytic number theory, we write $\omega(\cdot)$ for the count of distinct prime factors.) Therefore,

$$\psi(f) \leq f\prod_{p \mid f}\left(1+\frac{1}{p}\right) \leq (3/2)^{\omega(f)} f \leq (3/2)^{w+1+\omega(\Delta_K)} f.$$

and

$$\mathrm{Dav}\,\mathrm{PrinCl}(\mathcal{O}_f) < \frac{4w}{2^w}\frac{\psi(f)}{\ell(f)} \leq (3/2)^{1+\omega(\Delta_K)}(4w(3/4)^w)\frac{f}{\ell(f)}.$$

In this last expression, the coefficient of $f/\ell(f)$ is bounded by a constant depending only K, establishing (3.1) when $w \geq 2$.

If $w \leq 1$, the argument is easier. In this case, $\omega(f) \leq 2 + \omega(\Delta_K)$, and

$$\begin{aligned}\mathrm{Dav}\,\mathrm{PrinCl}(\mathcal{O}_f) \leq \#\mathrm{PrinCl}(\mathcal{O}_f) &= \frac{\psi(f)}{\ell(f)} \\ &= \frac{f}{\ell(f)}\prod_{p \mid f}\left(1-\frac{1}{p}\left(\frac{\Delta_K}{p}\right)\right) \leq (3/2)^{\omega(f)}\frac{f}{\ell(f)} \ll \frac{f}{\ell(f)},\end{aligned}$$

as desired. □

3.2. Lower bound. We turn now to the (much simpler) proof that $\rho(\mathcal{O}_p) \gg p$ for all p that do not split completely in K. If $p = 2$ or $p = 3$, then $\rho(\mathcal{O}_p) \geq 1 \geq \frac{1}{3}p$, so we suppose that $p > 3$. By Proposition 2.1 and (2.2),

$$\rho(\mathcal{O}_p) \geq \frac{1}{2}\,\mathrm{Dav}\,\mathrm{PrinCl}(\mathcal{O}_p).$$

The group $\mathrm{PrinCl}(\mathcal{O}_p)$ is cyclic, as a quotient of the cyclic group $\mathrm{PreCl}(\mathcal{O}_p)$ (see Lemma 2.2). Whenever G is cyclic, $\mathrm{Dav}\,G = \#G$ (see, e.g., [**GHK06**, Lemma 1.4.9 on p. 27].) Thus, $\mathrm{Dav}\,\mathrm{PrinCl}(\mathcal{O}_p) = \#\mathrm{PrinCl}(\mathcal{O}_p) = \psi(p)/\ell(p)$, and we conclude that

$$\rho(\mathcal{O}_p) \geq \frac{\psi(p)}{2\ell(p)}. \tag{3.2}$$

Up to this point all of our reasoning has been valid whether K is real or imaginary. But if we assume K is imaginary, then $\ell(p) \leq \#\mathcal{O}_K^\times \leq 6$, leading to the conclusion that

$$\rho(\mathcal{O}_p) \geq \frac{\psi(p)}{12} = \frac{p - \left(\frac{\Delta_K}{p}\right)}{12} \geq \frac{p}{12}$$

for all non-split $p > 3$. This establishes the lower bound claimed in Theorem 1.1. In fact, we have shown that the implied constant there can be taken as $\frac{1}{12}$, independently of K.

4. Minimal order in the imaginary case: Proof of Theorem 1.2

Our proof of Theorem 1.2 is based on arguments of Erdős, Pomerance, and Schmutz [**EPS91**] used to estimate the minimal order of **Carmichael's lambda function** $\lambda(m) := \#\operatorname{Exp}(\mathbb{Z}/m\mathbb{Z})^\times$. According to the first half of Theorem 1 of [**EPS91**], we have

$$\lambda(m) > (\log m)^{\left(\frac{1}{\log 2} + o(1)\right)\log\log\log m}$$

whenever $m \to \infty$. (As usual in analytic number theory, $o(1)$ denotes a quantity tending to 0.) The second half of that theorem asserts the existence of a strictly increasing sequence of natural numbers $\{m_i\}_{i\geq 1}$ such that

$$\lambda(m_i) < (\log m_i)^{c_0 \log\log\log m_i}$$

for all $i \geq 1$.

We need an $L(f)$-analogue of the quoted results, where $L(f)$ is the function defined in (2.4).

PROPOSITION 4.1. *Let K be a quadratic field. For any sequence of natural numbers $f \to \infty$,*

$$L(f) > (\log f)^{\left(\frac{1}{\log 2} + o(1)\right)\log\log\log f},$$

uniformly in K. On the other hand, for some universal constant $c_0 > 0$, and each fixed choice of K, there is a strictly increasing sequence $\{f_i\}_{i\geq 1}$ of positive integers, each of which is a product of distinct primes inert in K, such that

$$L(f_i) < (\log f_i)^{c_0 \log\log\log f_i}$$

for all $i \geq 1$.

Taking Proposition 4.1 as shown, we can quickly conclude the proof of Theorem 1.2. We start with the lower bound (1.2). From Proposition 2.1 and (2.2), $\rho(\mathcal{O}_f) \geq \frac{1}{2}\operatorname{Dav}\operatorname{PrinCl}(\mathcal{O}_f)$. As $\operatorname{PrinCl}(\mathcal{O}_f)$ is the quotient of $\operatorname{PreCl}(\mathcal{O}_f)$ by the subgroup $\mathcal{U}_f$ of order $\#\mathcal{U}_f \leq \#\mathcal{O}_K^\times \leq 6$, we have $\operatorname{Dav}\operatorname{PrinCl}(\mathcal{O}_f) \geq \frac{1}{6}\operatorname{Dav}\operatorname{PreCl}(\mathcal{O}_f)$ (see [**FP**, Lemma 2.4]). Therefore,

$$\rho(\mathcal{O}_f) \geq \frac{1}{12}\operatorname{Dav}\operatorname{PreCl}(\mathcal{O}_f) \geq \frac{1}{12}\operatorname{Exp}\operatorname{PreCl}(\mathcal{O}_f) \geq \frac{1}{12^2}L(f),$$

invoking Proposition 2.3 at the last step. The lower bound of Proposition 4.1 thus implies that (1.2) holds for any constant $c_1 < 1/\log 2$.

The upper bound (1.3) is slightly more intricate. From Proposition 2.1, (2.2), and the bound $\Omega(f) \ll \log f$, we have

$$\rho(\mathcal{O}_f) \ll \operatorname{Dav}\operatorname{PrinCl}(\mathcal{O}_f) + \log f \tag{4.1}$$

for all split-free f. Applying the Van Emde Boas–Kruyswijk exponent bound (Proposition 3.2),

$$\begin{aligned}\operatorname{Dav}\operatorname{PrinCl}(\mathcal{O}_f) &\leq \operatorname{Exp}\operatorname{PrinCl}(\mathcal{O}_f)\left(1+\log\frac{\#\operatorname{PrinCl}(\mathcal{O}_f)}{\operatorname{Exp}\operatorname{PrinCl}(\mathcal{O}_f)}\right)\\ &\leq \operatorname{Exp}\operatorname{PrinCl}(\mathcal{O}_f)(1+\log\psi(f)).\end{aligned}$$

Now $\psi(f) \leq (3/2)^{\omega(f)} f$, so that

$$\log\psi(f) \leq \omega(f)\log\frac{3}{2} + \log f \leq \Omega(f)\log\frac{3}{2} + \log f \ll \log f.$$

We conclude that for all split-free f,

$$\begin{aligned}\operatorname{Dav}\operatorname{PrinCl}(\mathcal{O}_f) &\ll \operatorname{Exp}\operatorname{PrinCl}(\mathcal{O}_f)(1+\log f)\\ &\leq L(f)(1+\log f).\end{aligned} \tag{4.2}$$

The upper bound half of Theorem 1.2 now follows from (4.1), (4.2), and the upper bound result of Proposition 4.1. Indeed, fixing any $c_2 > c_0$, we find that (1.3) holds for all large enough f (large enough in terms of K) from the sequence $\{f_i\}$.

The remainder of this section is devoted to the proof of Proposition 4.1. We make little claim to originality here; our arguments are straightforward adaptations of those in [**EPS91**].

4.1. The lower bound in Proposition 4.1. We start by observing that $L(f) \to \infty$ whenever $f \to \infty$. Indeed, if $L(f) \leq B$ for some constant $B > 0$, then $p^k \leq 2B$ for all $p^k \parallel f$, since $L(p^k) \geq \varphi(p^k) \geq p^k/2$ for all prime powers p^k. Consequently, there are only finitely many f for which $L(f) \leq B$.

Next, a case-by-case analysis shows that L is at most 4-to-1 on prime powers. To see this, we show that every $m \in \mathbb{N}$ is the L-image of at most 4 prime powers. Since the restriction of L to powers of ramified primes is the identity function, the L-preimage of m contains at most one power of a ramified prime. Let p be the least prime for which there exists some $k \in \mathbb{N}$ with $L(p^k) = m$.

Case I (p is ramified in K). Let $L(q^l) = m$ for some $q^l \neq p^k$, where $q > p$ is unramified in K. Suppose first that q splits in K. For this case, we have $p^k = q^{l-1}(q-1)$, which implies that $l = 1$ and thus $q = p^k + 1$. Suppose now that q is inert in K. Then $p^k = q^{l-1}(q+1)$. So $l = 1$ and $q = p^k - 1$. We conclude that m is the L-image of at most 3 prime powers.

Case II (p splits in K). Again, let $L(q^l) = m$ for some $q^l \neq p^k$ with $q \geq p$. Suppose first that q is ramified in K. Then $q > p$, which leads to $p^{k-1}(p-1) \neq q^l$, contradicting $L(p^k) = L(q^l)$. If q splits in K, then $p^{k-1}(p-1) = q^{l-1}(q-1)$. Since $q^l \neq p^k$, we have $q > p$. This implies that $l = 1$ and $q = p^{k-1}(p-1) + 1$. If q is inert in K, then $q > p$ and $p^{k-1}(p-1) = q^{l-1}(q+1)$. Once again, we must have $l = 1$, from which we deduce that $q = p^{k-1}(p-1) - 1$. So as in Case I, m is the L-image of at most 3 prime powers.

Case III (p is inert in K). Suppose that $L(q^l) = m$ for some $q^l \neq p^k$ with $q \geq p$. If q is ramified in K, then $q > p$ and $p^{k-1}(p+1) = q^l$. In particular, we have $q \geq p+1$ and $q^l \mid (p+1)$. So we must have $l = 1$ and $q = p+1$. Next, if q splits in K, then $q > p$ and $p^{k-1}(p+1) = q^{l-1}(q-1)$. If $l > 1$, then the same argument shows that $l = 2$ and $q = p+1$. Of course, we get $q = p^{k-1}(p+1) + 1$ when $l = 1$. Finally, if q is inert in K, then $p^{k-1}(p+1) = q^{l-1}(q+1)$. Since $q^l \neq p^k$, we have

$q > p$. When $l > 1$, we get $l = 2$ and $q = p+1$ once again. When $l = 1$, it is obvious that $q = p^{k-1}(p+1) - 1$. In view of the fact that $q = p+1$ can occur for at most one of the three possible cases for q, we conclude that m is the L-image of at most 4 prime powers.

This concludes the proof that L is at most 4-to-1 on prime powers.

We can now complete the proof of the lower bound on $L(f)$. Let f be sufficiently large, and suppose that $L(f) = m$. Since $p^k \le 2\varphi(p^k) \le 2L(p^k)$ for all prime powers p^k, we have

$$f \le \prod_{\substack{p^k \mid f \\ L(p^k) \mid m}} p^k \le \prod_{d \mid m} \prod_{p^k \colon L(p^k) = d} p^k \le \prod_{d \mid m} (2m)^4 = (2m)^{4\tau(m)},$$

where $\tau(\cdot)$ is the divisor counting function. Using the inequality

$$\tau(m) \le 2^{(1+o(1)) \log m / \log\log m}$$

(see [**HW08**, Theorem 317, p. 345]), we deduce that

$$f \le \exp\left((4 \log(2m)) 2^{(1+o(1)) \log m / \log\log m} \right),$$

from which it follows that

$$L(f) = m \ge (\log f)^{(1/\log 2 + o(1)) \log\log\log f}.$$

As the inequality "$f \le (2m)^{4\tau(m)}$" holds independently of K, the "$o(1)$" term appearing in these last two displays decays to 0 uniformly in K.

4.2. The upper bound in Proposition 4.1. We need two lemmas. The first of these appears as Proposition 8 in [**EPS83**]. For each real number $x > 0$ and each pair of integers a, k with $k \in \mathbb{N}$ and $\gcd(a, k) = 1$, put

$$\theta(x; k, a) = \sum_{\substack{p \le x \\ p \equiv a \ (\mathrm{mod}\ k)}} \log p.$$

LEMMA 4.2. *Let $\epsilon > 0$. There are computable positive constants δ and x_0 such that*

$$\left| \theta(x; k, a) - \frac{x}{\varphi(k)} \right| < \epsilon \frac{x}{\varphi(k)}$$

for all $x \ge x_0$, all $k \in \mathbb{N} \cap [1, x^\delta]$ and all $a \in \mathbb{Z}$ with $\gcd(k, a) = 1$, except possibly for those k which are divisible by a certain integer $k_0 = k_0(x) > (\log x)^{3/2}$.

The next result is a variant of [**EPS83**, Proposition 10].

LEMMA 4.3. *Let K be a quadratic field. For all x that are sufficiently large in terms of K, there is a natural number $M_x \le x^2$ for which*

$$\#\{p \text{ inert in } K \colon (p+1) \mid M_x\} > \exp\left(C \frac{\log x}{\log\log x} \right). \tag{4.3}$$

Here C is a positive, universal constant.

PROOF. We let $\delta, x_0 > 0$ be the parameters of Lemma 4.2 corresponding to the choice $\epsilon = \frac{1}{2}$.

Without loss of generality, we may assume that x_0 is sufficiently large (in terms of K). For $x \ge x_0$, let $k_1 = k_1(x)$ denote the product of all the unramified primes

$p \le \frac{1}{2}\delta \log x$. Then $k_1 < x^{\delta}/|\Delta_K|$. Let P be the largest prime factor of Δ_K, and let $p_0 \ge P+1$ be an arbitrary common prime factor of k_0 and k_1, with the convention that $p_0 = 1$ when $\gcd(k_0, k_1)$ has no prime factors at least $P+1$. Put $k = k_1/p_0$. If $p_0 > 1$, then it is obvious that $k_0 \nmid k\Delta_K$. On the other hand, if $p_0 = 1$ and $k_0 \mid k\Delta_K$, then all of the prime factors of k_0 would be at most P. Since $k_0 > (\log x)^{3/2}$, there would exist some $p \le P$ such that $p^4 \mid k_0$. But $\gcd(k, \Delta_K) = 1$, k is squarefree, and $\Delta_K = 2^r s$, where $r \in \{0, 2, 3\}$ and $s \in \mathbb{Z}$ is odd and squarefree, so that $p^4 \nmid k\Delta_K$, a contradiction. Therefore, we have $k_0 \nmid k\Delta_K$ in both cases.

For each $d \mid k$, we have from Lemma 4.2 that

$$\begin{aligned}
\sum_{\substack{p \le x \\ p \equiv -1 \,(\mathrm{mod}\, d) \\ p \text{ inert in } K}} 1 &= \sum_{\substack{j \in (\mathbb{Z}/|\Delta_K|\mathbb{Z})^\times \\ (\Delta_K/j) = -1}} \sum_{\substack{p \le x \\ p \equiv a_j \,(\mathrm{mod}\, |\Delta_K| d)}} 1 \\
&\ge \frac{1}{\log x} \sum_{\substack{j \in (\mathbb{Z}/|\Delta_K|\mathbb{Z})^\times \\ (\Delta_K/j) = -1}} \theta(x; |\Delta_K| d, a_j) \\
&> (1-\epsilon) \frac{x}{\varphi(|\Delta_K| d) \log x} \sum_{\substack{j \in (\mathbb{Z}/|\Delta_K|\mathbb{Z})^\times \\ (\Delta_K/j) = -1}} 1 \\
&= (1-\epsilon) \frac{x}{\varphi(|\Delta_K|)\varphi(d) \log x} \cdot \frac{\varphi(|\Delta_K|)}{2} \\
&= \frac{x}{4\varphi(d) \log x},
\end{aligned}$$

where $a_j \in (\mathbb{Z}/|\Delta_K| d\mathbb{Z})^\times$ satisfies $a_j \equiv -1 \,(\mathrm{mod}\, d)$ and $a_j \equiv j \,(\mathrm{mod}\, |\Delta_K|)$.

We now use this estimate to lower bound the cardinality of the set A of pairs $(m, p) \in (\mathbb{N} \cap [1, x])^2$ satisfying the congruence $m(p+1) \equiv 0 \,(\mathrm{mod}\, k)$, where p is inert in K. To this end, we define

$$A_d := \{(m, p) \in A \colon d \mid (p+1) \text{ and } \gcd(m, k) = k/d\}$$

for each $d \mid k$. Then A is the disjoint union of A_d's over $d \mid (p+1)$. Note that the number of $m \le x$ with $\gcd(m, k) = k/d$ is at least $\varphi(d)\lfloor x/k \rfloor$. Hence, we obtain

$$\#A_d > \frac{x}{4\varphi(d) \log x} \cdot \varphi(d) \lfloor x/k \rfloor > \frac{x^2}{5k \log x}$$

and

$$\#A = \sum_{d \mid k} \#A_d > \frac{x^2}{5k \log x} \cdot \tau(k) = \frac{x^2}{5k \log x} \cdot 2^{\omega(k)} > \frac{x^2}{5k \log x} \cdot 2^{\frac{(\delta/4) \log x}{\log \log x}},$$

where we have used the prime number theorem to get

$$\omega(k) \ge \pi\left(\frac{\delta}{2} \log x\right) - \omega(\Delta_K) - 1 > \frac{(\delta/4) \log x}{\log \log x}$$

for sufficiently large x depending on Δ_K. Since $m(p+1) \in \{n \le x^2 \colon k \mid n\}$ for every pair $(m, p) \in A$, some $n \le x^2$ with $k \mid n$ must admit at least

$$\frac{\#A}{x^2/k} > \frac{1}{5 \log x} \cdot 2^{\frac{(\delta/4) \log x}{\log \log x}} > \exp\left(\frac{\delta}{6} \frac{\log x}{\log \log x}\right)$$

representations of the form $n = m(p+1)$ with $(m,p) \in A$. (We use here that $\frac{1}{4}\log 2 > \frac{1}{6}$.) The proof of (4.3) is completed by taking M_x to be this n. $\square$

We are now in a position to establish the upper bound in Proposition 4.1. Let C be as in Lemma 4.3. Put $x_i = (\log i)^{(2/C)\log\log\log i}$ and

$$g_i = \prod_{\substack{p \text{ inert in } K \\ (p+1) \mid M_{x_i}}} p.$$

By (4.3), we have

$$g_i \geq \prod_{\substack{p \text{ inert in } K \\ (p+1) \mid M_{x_i}}} 2 > \exp\left((\log 2)\exp\left(C\frac{\log x_i}{\log\log x_i}\right)\right) > i$$

for sufficiently large i. Moreover, we have $L(g_i) \mid M_{x_i}$, which implies that

$$L(g_i) \leq M_{x_i} \leq x_i^2 = (\log i)^{(4/C)\log\log\log i} < (\log g_i)^{c_0 \log\log\log g_i}$$

for sufficiently large i, where $c_0 := 4/C$. The asserted upper bound follows by extracting a strictly monotonic subsequence $\{f_i\}_{i\geq 1}$ from $\{g_i\}_{i\geq 1}$.

5. Minimal order in the real case: Proof of Theorem 1.4

Let K be a real quadratic field. We remind the reader that ϵ denotes the fundamental unit of K, and we set

$$\delta = \begin{cases} 1 & \text{if } N_{K/\mathbb{Q}}(\epsilon) = 1, \\ 2 & \text{if } N_{K/\mathbb{Q}}(\epsilon) = -1. \end{cases}$$

If p is a prime inert in K, then its associated Frobenius element is conjugation on K (the nontrivial element of $\mathrm{Gal}(K/\mathbb{Q})$). Hence,

$$\epsilon^{p+1} \equiv \epsilon^p \epsilon \equiv N_{K/\mathbb{Q}}(\epsilon) \pmod{p\mathcal{O}_K},$$

and

$$\epsilon^{\delta(p+1)} \equiv N_{K/\mathbb{Q}}(\epsilon)^\delta \equiv 1 \pmod{p\mathcal{O}_K}.$$

Thus, the order of ϵ in $(\mathcal{O}_K/p\mathcal{O}_K)^\times$ is a divisor of $\delta(p+1)$. We will base our proof of Theorem 1.4 on the following result of Roskam [**Ros00**].

PROPOSITION 5.1 (conditional on GRH). *There are infinitely many primes p, inert in K, for which the order of ϵ in $(\mathcal{O}_K/p\mathcal{O}_K)^\times$ is precisely $\delta(p+1)$.*

(In fact, Roskam shows that the order is $\delta(p+1)$ not only for infinitely many inert primes, but for a positive proportion of all inert primes. The weaker version here is sufficient for our purposes.) Theorem 1.4 is an immediate consequence of Proposition 5.1 in conjunction with the next assertion.

PROPOSITION 5.2. *If p is a prime inert in K for which ϵ has order $\delta(p+1)$ in $(\mathcal{O}_K/p\mathcal{O}_K)^\times$, then*

$$\rho(\mathcal{O}_p) \leq h_K + \frac{3}{2}.$$

PROOF. Let p be as in the proposition. Then $\epsilon^{\ell(p)} \equiv n \pmod{p\mathcal{O}_K}$ for some rational integer n prime to p. By Fermat's little theorem,

$$\epsilon^{(p-1)\ell(p)} \equiv 1 \pmod{p\mathcal{O}_K}.$$

We are assuming that ϵ has order $\delta(p+1)$ in $(\mathcal{O}_K/p\mathcal{O}_K)^\times$. Hence, the displayed congruence forces

$$p+1 \mid \delta(p+1) \mid (p-1)\ell(p).$$

Writing $(p-1)\ell(p) = (p+1)\ell(p) - 2\ell(p)$, we deduce that $p+1 \mid 2\ell(p)$. In particular,

$$\ell(p) \geq \frac{p+1}{2}.$$

By Proposition 2.1 and (2.2),

$$\rho(\mathcal{O}_p) \leq \frac{1}{2}\operatorname{Dav}\operatorname{Cl}(\mathcal{O}_p) + \frac{3}{2} \leq \frac{h_K}{2}\operatorname{Dav}\operatorname{PrinCl}(\mathcal{O}_p) + \frac{3}{2} \leq \frac{h_K}{2}\#\operatorname{PrinCl}(\mathcal{O}_p) + \frac{3}{2}.$$

The proof is completed by observing that $\#\operatorname{PrinCl}(\mathcal{O}_p) = \frac{\psi(p)}{\ell(p)} = \frac{p+1}{\ell(p)} \leq 2.$ □

6. Maximal order in the real case: Proof of Theorem 1.3

6.1. Upper bound. We start by showing that $\rho(\mathcal{O}_f) \ll f/\log f$ for each $f > 1$. By Proposition 2.1 and (2.2),

$$\rho(\mathcal{O}_f) \leq \frac{h_K}{2}\operatorname{Dav}\operatorname{PrinCl}(\mathcal{O}_f) + \frac{3}{2}\Omega(f).$$

As $\Omega(f) \ll \log f \ll \frac{f}{\log f}$, it suffices to demonstrate that $\operatorname{Dav}\operatorname{PrinCl}(\mathcal{O}_f) \ll \frac{f}{\log f}$. According to (3.1), we have $\operatorname{Dav}\operatorname{PrinCl}(\mathcal{O}_f) \ll f/\ell(f)$, so we will be done if we prove that

$$\ell(f) \gg \log f.$$

For this we may assume f is sufficiently large (bounded f can be dealt with by adjusting the implied constant). Say $K = \mathbb{Q}(\sqrt{D})$ with D squarefree, and let τ_D be defined as in (1.1), so that $1, f\tau_D$ are a $\mathbb{Z}$-basis for $\mathcal{O}_f$. If we express $\epsilon^{\ell(f)} = U + V\tau_D$, then V is a positive integer multiple of f. Writing $\epsilon^{\ell(f)} = u + v\sqrt{D}$ (with $u, v \in \mathbb{Q}$), we find that $v \geq \frac{1}{2}V \geq \frac{1}{2}f$. Therefore, using a tilde for conjugation in K,

$$f \leq 2v = \frac{\epsilon^{\ell(f)} - \tilde{\epsilon}^{\ell(f)}}{\sqrt{D}} < \frac{\epsilon^{\ell(f)} + 1}{\sqrt{D}}. \tag{6.1}$$

Hence (continuing to assume f is large),

$$\epsilon^{\ell(f)} > f\sqrt{D} - 1 > f,$$

and $\ell(f) > \frac{\log f}{\log \epsilon} \gg \log f$.

6.2. Lower bound. Now we turn to the lower bound (1.4). Let p be a prime inert in K with $p > 3$. According to (3.2),

$$\rho(\mathcal{O}_p) \geq \frac{1}{2}\frac{p+1}{\ell(p)},$$

and so it suffices to produce infinitely many inert p with

$$\frac{p+1}{\ell(p)} > 2p^{\frac{1}{4}-\epsilon}. \tag{6.2}$$

For this we borrow Lemma 6.4 from [**FP**].

LEMMA 6.1 (conditional on GRH). *Let $K = \mathbb{Q}(\sqrt{D})$ be a real quadratic field, where $D > 1$ is squarefree. Let $\eta = \epsilon$ or ϵ^2, according to whether $N_{K/\mathbb{Q}}(\epsilon) = 1$ or -1, respectively. Let q be an odd prime not dividing D, and let $y \geq 2$ be a real number. The count of primes $p \leq y$ for which*

$$(6.3) \qquad p \text{ is inert in } K, \quad p \equiv -1 \pmod q, \quad \text{and} \quad \eta^{(p+1)/q} \equiv 1 \pmod p,$$

is

$$\frac{1}{2q(q-1)} \int_2^y \frac{dt}{\log t} + O(y^{1/2} \log(qy)),$$

where the implied constant depends at most on K.

PROOF THAT (6.2) HOLDS FOR INFINITELY MANY INERT p. Fix $\epsilon \in (0, 1/2)$. For large y, we select an arbitrary prime q satisfying

$$y^{\frac{1}{4}-\frac{1}{2}\epsilon} < q \leq y^{\frac{1}{4}-\frac{1}{4}\epsilon}.$$

By the prime number theorem, there will be many choices for q once y is large enough, and none of these will divide D. Using C for a constant depending at most on K, the count of $p \leq y$ satisfying (6.3) is bounded below by

$$\begin{aligned}\frac{1}{2q(q-1)} \int_2^y \frac{dt}{\log t} - Cy^{1/2} \log y &> \frac{1}{2q(q-1)\log y} \int_2^y dt - Cy^{1/2} \log y \\ &> \frac{1}{3} \frac{y^{\frac{1}{2}+\frac{1}{2}\epsilon}}{\log y} - Cy^{1/2} \log y \\ &> y^{\frac{1}{2}+\frac{1}{4}\epsilon}.\end{aligned}$$

In particular, for all large y there is at least one such p. As $\eta^{(p+1)/q} \equiv 1 \pmod{p\mathcal{O}_K}$, and $\eta = \epsilon$ or ϵ^2, we must have $\epsilon^{(p+1)/q} \equiv \pm 1 \pmod{p\mathcal{O}_K}$. Hence, $\epsilon^{(p+1)/q} \in \mathcal{O}_p$ and $\ell(p) \mid \frac{p+1}{q}$. Therefore, $q \mid \frac{p+1}{\ell(p)}$, and

$$\frac{p+1}{\ell(p)} \geq q > y^{\frac{1}{4}-\frac{1}{2}\epsilon} > 2y^{\frac{1}{4}-\epsilon} \geq 2p^{\frac{1}{4}-\epsilon},$$

verifying (6.2).

To see that this argument produces infinitely many p, note that $p > q > y^{\frac{1}{4}-\frac{1}{2}\epsilon}$, while y can be taken arbitrarily large. □

REMARK. Let K be a real quadratic field. For each $m = 0, 1, 2, \ldots$, write $2\epsilon^m = u_m + v_m\sqrt{D}$, where $u_m, v_m \in \mathbb{Q}$. Then $u_m, v_m \in \mathbb{N}$, and v_m grows exponentially with m; in fact, $v_m \sim \epsilon^m/\sqrt{D}$, as $m \to \infty$ (compare with (6.1)).

We conjecture the existence of a constant $\delta = \delta_K > 0$ for which the following holds:

$$(6.4) \qquad v_m \text{ has an inert prime factor } p > v_m^\delta \text{ for infinitely many } m \in \mathbb{N}.$$

Suppose p, m are related as in (6.4), and that m is large. Then $p > 3$, and $p \mid v_m$ implies that $\epsilon^m \in \mathcal{O}_p$. It follows that

$$\ell(p) \leq m \ll \log v_m \ll \log p,$$

and

$$\rho(\mathcal{O}_p) \geq \frac{1}{2} \frac{p+1}{\ell(p)} \gg \frac{p}{\log p}.$$

So if our conjecture (6.4) holds, then the upper bound of Theorem 1.3(a) is sharp.

To illustrate, let $K = \mathbb{Q}(\sqrt{2})$. In this case, $\epsilon = 1 + \sqrt{2}$, each v_m is even, and $\frac{1}{2}v_m$ is commonly termed the mth Pell number. It seems plausible to conjecture that there are infinitely many Pell numbers that are themselves prime; e.g., looking at indices $m < 1000$, one finds that $\frac{1}{2}v_m$ is prime for $m = 2$, 3, 5, 11, 13, 29, 41, 53, 59, 89, 97, 101, 167, 181, 191, 523, 929 (this is OEIS sequence A096650). Among these m, the prime $\frac{1}{2}v_m$ is inert in K roughly half the time, for $m = 3$, 5, 11, 13, 29, 53, 59, 101, 181, 523. We view this as compelling evidence that when $K = \mathbb{Q}(\sqrt{2})$, our hypothesis (6.4) holds for any $\delta \in (0, 1)$.

References

[And97] D. F. Anderson, *Elasticity of factorizations in integral domains: a survey*, Factorization in integral domains (Iowa City, IA, 1996), Lecture Notes in Pure and Appl. Math., vol. 189, Dekker, New York, 1997, pp. 1–29. MR1460767

[Car60] L. Carlitz, *A characterization of algebraic number fields with class number two*, Proc. Amer. Math. Soc. **11** (1960), 391–392, DOI 10.2307/2034782. MR111741

[CC00] S. T. Chapman and J. Coykendall, *Half-factorial domains, a survey*, Non-Noetherian commutative ring theory, Math. Appl., vol. 520, Kluwer Acad. Publ., Dordrecht, 2000, pp. 97–115. MR1858159

[CMO17] J. Coykendall, P. Malcolmson, and F. Okoh, *Inert primes and factorization in extensions of quadratic orders*, Houston J. Math. **43** (2017), no. 1, 61–77. MR3647931

[Cox22] D. A. Cox, *Primes of the form $x^2 + ny^2$—Fermat, class field theory, and complex multiplication*, AMS Chelsea Publishing, Providence, RI, [2022] ©2022. Third edition [of 1028322] with solutions; With contributions by Roger Lipsett. MR4502401

[Coy01] J. Coykendall, *Half-factorial domains in quadratic fields*, J. Algebra **235** (2001), no. 2, 417–430, DOI 10.1006/jabr.2000.8505. MR1805465

[EBK69] P. van Emde Boas and D. Kruyswijk, *A combinatorial problem on finite Abelian groups III*, Report ZW 1969-008, Amsterdam (1969), 32 pp.

[EPS83] L. M. Adleman, C. Pomerance, and R. S. Rumely, *On distinguishing prime numbers from composite numbers*, Ann. of Math. (2) **117** (1983), no. 1, 173–206, DOI 10.2307/2006975. MR683806

[EPS91] P. Erdős, C. Pomerance, and E. Schmutz, *Carmichael's lambda function*, Acta Arith. **58** (1991), no. 4, 363–385, DOI 10.4064/aa-58-4-363-385. MR1121092

[FP] K. S. Fan and P. Pollack, *The typical elasticity of a quadratic order*, Preprint, arXiv:2411.09063, (2024).

[GHK06] A. Geroldinger and F. Halter-Koch, *Non-unique factorizations*, Pure and Applied Mathematics (Boca Raton), vol. 278, Chapman & Hall/CRC, Boca Raton, FL, 2006. Algebraic, combinatorial and analytic theory, DOI 10.1201/9781420003208. MR2194494

[HK72] F. Halter-Koch, *Einseinheitengruppen und prime Restklassengruppen in quadratischen Zahlkörpern* (German), J. Number Theory **4** (1972), 70–77, DOI 10.1016/0022-314X(72)90011-X. MR296045

[HK83] F. Halter-Koch, *Factorization of algebraic integers*, Grazer Math. Berichte **191** (1983).

[HK95] F. Halter-Koch, *Elasticity of factorizations in atomic monoids and integral domains* (English, with English and French summaries), J. Théor. Nombres Bordeaux **7** (1995), no. 2, 367–385, DOI 10.5802/jtnb.147. MR1378586

[HW08] G. H. Hardy and E. M. Wright, *An introduction to the theory of numbers*, 6th ed., Oxford University Press, Oxford, 2008. Revised by D. R. Heath-Brown and J. H. Silverman; With a foreword by Andrew Wiles. MR2445243

[Nar95] W. Narkiewicz, *A note on elasticity of factorizations*, J. Number Theory **51** (1995), no. 1, 46–47, DOI 10.1006/jnth.1995.1034. MR1321723

[Pol24] P. Pollack, *Half-factorial real quadratic orders*, Arch. Math. (Basel) **122** (2024), no. 5, 491–500, DOI 10.1007/s00013-024-01969-z. MR4734562

[Pol25] P. Pollack, *Maximally elastic quadratic fields*, J. Number Theory **267** (2025), 80–100, DOI 10.1016/j.jnt.2024.08.003. MR4801520

[Ros00] H. Roskam, *A quadratic analogue of Artin's conjecture on primitive roots*, J. Number Theory **81** (2000), no. 1, 93–109, DOI 10.1006/jnth.1999.2470. MR1743503

[Ste86] J.-L. Steffan, *Longueurs des décompositions en produits d'éléments irréductibles dans un anneau de Dedekind* (French), J. Algebra **102** (1986), no. 1, 229–236, DOI 10.1016/0021-8693(86)90138-9. MR853241

[Val90] R. J. Valenza, *Elasticity of factorization in number fields*, J. Number Theory **36** (1990), no. 2, 212–218, DOI 10.1016/0022-314X(90)90074-2. MR1072466

[Zak76] A. Zaks, *Half factorial domains*, Bull. Amer. Math. Soc. **82** (1976), no. 5, 721–723, DOI 10.1090/S0002-9904-1976-14130-4. MR407001

[Zak80] A. Zaks, *Half-factorial-domains*, Israel J. Math. **37** (1980), no. 4, 281–302, DOI 10.1007/BF02788927. MR599463

Department of Mathematics, University of Georgia, Athens, Georgia 30602
Email address: `Steve.Fan@uga.edu`

Department of Mathematics, University of Georgia, Athens, Georgia 30602
Email address: `pollack@uga.edu`

Contemporary Mathematics
Volume **836**, 2026
https://doi.org/10.1090/conm/836/16703

On the smallest partition associated to a numerical semigroup

Nathan Kaplan, Kaylee Kim, Cole McGeorge, Fabian Ramirez, and Deepesh Singhal

ABSTRACT. The set of hook lengths of an integer partition λ is the complement of some numerical semigroup S. There has been recent interest in studying the number of partitions with a given set of hook lengths. Very little is known about the distribution of sizes of this finite set of partitions. We focus on the problem of determining the size of the smallest partition with its set of hook lengths equal to $\mathbb{N} \setminus S$.

1. Introduction

An *integer partition* λ is a list of positive integers $(\lambda_1, \lambda_2, \ldots, \lambda_\ell)$ satisfying $\lambda_1 \geq \lambda_2 \geq \cdots \geq \lambda_\ell$. The λ_i are called the *parts* of λ. We write $|\lambda| = \lambda_1 + \cdots + \lambda_\ell$ for the *size* of λ. If $|\lambda| = n$ we say that λ is a partition of n. We can represent a partition λ using its *Young diagram*, a left-justified array of boxes with λ_i boxes in the ith row. Every box has an associated *hook*, which is the set of boxes below and to the right of the box (including the original box). The *hook length* of a box is the size of its hook, that is, the number of boxes below it, plus the number of boxes to the right of it, plus 1 for the box itself. The *hook set* of λ, denoted $\mathrm{H}(\lambda)$, is the set of hook lengths of the Young diagram of λ. Ignoring the left edge and the top edge, the boundary of the Young diagram of λ is the *profile* of λ. We give an example in Figure 1.

Our next goal is to explain the connection between hook sets of partitions and numerical semigroups. We first need to discuss numerical sets. Let $\mathbb{N} = \{0, 1, 2, \ldots\}$ denote the set of nonnegative integers. A *numerical set* T is a subset of $\mathbb{N}$ that contains 0 and has finite complement in $\mathbb{N}$. The elements of $\mathbb{N} \setminus T$ are called the *gaps* of T, denoted $\mathrm{G}(T)$. The largest of these gaps is the *Frobenius number* of T, denoted $\mathrm{F}(T)$, and the number of these gaps is the *genus* of T, denoted $\mathrm{g}(T)$. The smallest nonzero element of T is the *multiplicity* of T, denoted $\mathrm{m}(T)$. The Frobenius number of T lies in the interval between two consecutive multiples of $\mathrm{m}(T)$. That is, there is a unique nonnegative integer $\mathrm{q}(T)$, called the *depth* of T, such that $(\mathrm{q}(T) - 1)\,\mathrm{m}(T) \leq \mathrm{F}(T) < \mathrm{q}(T)\,\mathrm{m}(T)$.

2020 *Mathematics Subject Classification.* Primary 20M14, 05A17.

The authors received support from NSF Grant DMS 2154223. The second and third author received support from the UCI UROP Summer Undergraduate Research Program.

©2026 American Mathematical Society

There is a bijection between the set of integer partitions and the set of numerical sets given by the enumeration of a numerical set T. We construct this map by describing a walk starting at the origin in $\mathbb{Z}^2$ that is determined by the elements of T. Starting with $n = 0$:

- if $n \in T$, draw a line of unit length to the right,
- if $n \notin T$, draw a line of unit length up,
- repeat for $n + 1$.

For any integer $n > \mathrm{F}(T)$ we draw a line to the right. We see that this process ends with an infinite series of steps to the right. We disregard this section of the walk and see that we have constructed the profile of the Young diagram of a partition. This is the partition $\lambda(T)$, called the *enumeration of* T. This idea of studying numerical sets and numerical semigroups by associating them with walks on a grid was introduced by Bras-Amorós and de Mier [**5**]. For a more extensive discussion of this bijection see [**10**, Section 2].

Let T be a numerical set. Note that each box of $\lambda(T)$ is uniquely determined by a pair (u, x) satisfying $u \in T$, $x \in \mathrm{G}(T)$ and $u < x$. We denote this box by $\square(u, x)$ and denote its hook by hook(u, x). The size of a hook is the number of boxes in it. Note that hook(u, x) has size $|\,\mathrm{hook}(u, x)| = x - u$. We also see that

$$|\lambda(T)| = \#\{(u, x)\colon u \in T, x \in \mathrm{G}(T), u < x\} = \sum_{x \in \mathrm{G}(T)} \#\{u \in T\colon u < x\}. \tag{1.1}$$

When we list the elements of a numerical set T we always write them in increasing order. Since the complement of T is finite, we can write $T = \{0, a_1, \ldots, a_k, \rightarrow\}$, where the symbol $\rightarrow$ indicates that T contains all integers greater than a_k. We see that the partition given in Figure 1 is the enumeration of the numerical set $T = \{0, 5, 7, 9, \rightarrow\}$.

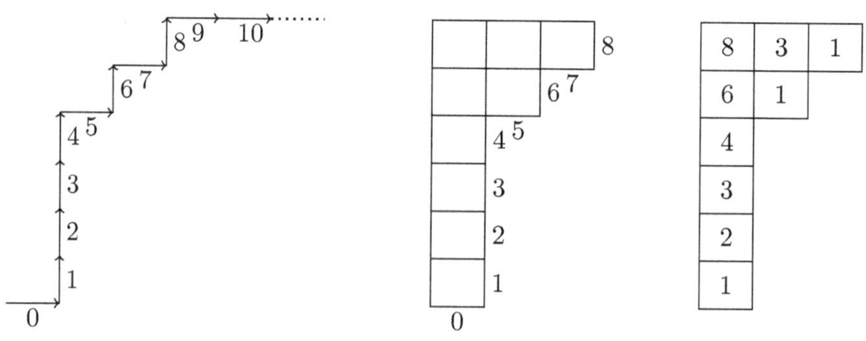

FIGURE 1. From left to right we have, the walk defined by $T = \{0, 5, 7, 9, \rightarrow\}$, the Young diagram of $\lambda(T)$, and $\lambda(T)$ where each box is labeled with its hook length.

A numerical set S that is closed under addition is a *numerical semigroup*. It is not difficult to show that the hook set of an integer partition λ is the complement of some numerical semigroup S. See for example [**15**] or [**10**, Proposition 4]. Several authors have recently studied families of integer partitions that arise from special classes of numerical semigroups. See for example [**6, 21**]. Let $\mathcal{P}(S) = \{\lambda\colon \mathrm{H}(\lambda) = \mathbb{N} \setminus S\}$ and $\mathrm{P}(S) = |\mathcal{P}(S)|$. There has been recent interest in studying $\mathrm{P}(S)$; see

[**8–10**, **19**]. However, very little is known about the distribution of the sizes of the partitions in $\mathcal{P}(S)$. We focus on a particular question about these sizes.

QUESTION 1. Let S be a numerical semigroup. What is the size of the smallest partition λ with $\mathrm{H}(\lambda) = \mathbb{N} \setminus S$?

We first observe that since $\mathrm{H}(\lambda(S)) = \mathbb{N} \setminus S$, the set $\mathcal{P}(S)$ is nonempty. See [**10**, Proposition 4] or [**16**, Proposition 1] for a proof. If S is a numerical semigroup for which $\lambda(S)$ has minimal size among all partitions in $\mathcal{P}(S)$, then we say that S is *enumeration-minimal* or *λ-minimal*. Analyzing small examples suggests that it is very common for a semigroup S to have this property. We implemented [**8**, Algorithm 5.1] using the **numericalsgps** package [**11**] in the computer algebra system GAP [**13**] to verify the following result. Computations for this paper were also done in the computer algebra system Sage [**20**]. Our code is available at [**14**].

PROPOSITION 1.1. *Let S be a numerical semigroup. If $\mathrm{F}(S) \leq 16$ or $\mathrm{g}(S) \leq 11$, then S is λ-minimal.*

One can use the **numericalsgps** package to check that there are 784 numerical semigroups S with $\mathrm{F}(S) \leq 16$ and there are 820 numerical semigroups with $\mathrm{g}(S) \leq 11$. However, not every numerical semigroup is λ-minimal.

EXAMPLE 1.2. Let $S = \{0, 9, 10, 11, 12, 13, 18, \rightarrow\}$. Then $|\lambda(S)| = 32$. The partition $\lambda = (9, 8, 2, 2, 2, 2, 2, 2, 2)$ also has $\mathrm{H}(\lambda) = \mathbb{N} \setminus S$, and $|\lambda| = 31$.

This paper came out of an attempt to understand this example. Further computation shows that the semigroup S from Example 1.2 is the only one of the 592 numerical semigroups of genus 12 that is not λ-minimal.

We now describe several of our main results. We first introduce some additional notation. A set of positive integers $n_1, \ldots, n_t$ is a *generating set* for a numerical semigroup S if S is equal to the set of linear combinations of $n_1, \ldots, n_t$ with nonnegative integer coefficients. That is,

$$S = \langle n_1, \ldots, n_t \rangle = \{x \in \mathbb{Z} \colon x = a_1 n_1 + \cdots + a_t n_t \text{ with } a_i \in \mathbb{N}\}.$$

It is not difficult to show that every numerical semigroup S has a unique generating set of minimal size. We direct the reader to [**18**] for a general reference on numerical semigroups.

We show that the semigroup S from Example 1.2 is a specialization of two different infinite families of numerical semigroups that are not λ-minimal.

THEOREM 1.3. *Let $m \geq 9$ and $S_m = \langle m, m+1, \ldots, 2m-5 \rangle$. Then S_m is not λ-minimal.*

We prove this result in Section 3 by determining the sizes of all of the partitions in $\mathcal{P}(S_m)$.

THEOREM 1.4. *Let $k \geq 4$ be a positive integer that is not prime. Then $S_k = \langle 2k+1, 2k+2, \ldots, 3k+1 \rangle$ is not λ-minimal.*

In fact, we show that if k has many proper divisors, then there are many partitions in $\mathcal{P}(S_k)$ with size less than $|\lambda(S_k)|$.

In Sections 4, 5, and 6 we prove that several families of numerical semigroups S are λ-minimal.

THEOREM 1.5. *Let m and k be positive integers with $m \geq 2$ and let s be an integer satisfying $1 \leq s \leq m-1$. The numerical semigroup*

$$S_{m,k,s} = \{0, m, 2m, \ldots, km, km+s+1, \rightarrow\}$$

is λ-minimal.

This theorem is the focus of Section 4. These numerical semigroups have previously appeared in the literature. See, for example, [**7**, Corollary 4.5], where the authors compute the size of their ideal class monoids. In [**9**], the authors show that for a fixed m and s, the cardinality of $\mathcal{P}(S_{m,k,s})$ is given by a polynomial in k.

In Section 5 we prove that several types of numerical semigroups with depth 2 are λ-minimal. A partition λ has a *Durfee square* of size k if k is the largest positive integer i such that λ has at least i parts of size greater than or equal to i. Equivalently, λ has a Durfee square of size k if k is the largest side length of a square contained in the Young diagram of λ. For example, the partition in Figure 1 has a Durfee square of size 2.

THEOREM 1.6. *Let S be a numerical semigroup of depth 2. If $\lambda(S)$ has a Durfee square of size at most 3, then S is λ-minimal.*

We note that if S is the semigroup from Example 1.2, then S has depth 2, $\lambda(S)$ has a Durfee square of size 4, and S is not λ-minimal.

We use the term *small elements* of a numerical semigroup S for the positive elements of S that are less than $\mathrm{F}(S)$. Some authors include 0 and also the smallest element of S larger than $\mathrm{F}(S)$ in the set of small elements of S. The set we call the small elements of S together with 0 is sometimes called the set of *sporadic elements* of S or the set of *left elements* of S.

The number of small elements of S is one less than the number of columns of the Young diagram of $\lambda(S)$. If S is a numerical semigroup with k small elements, then $\lambda(S)$ has a Durfee square of size at most $k+1$. Therefore, Theorem 1.6 implies that a numerical semigroup of depth 2 with at most 2 small elements is λ-minimal. We prove something stronger.

THEOREM 1.7. *Let S be a numerical semigroup of depth 2. If S has at most 3 small elements, then S is λ-minimal.*

The numerical semigroup in Example 1.2 has 5 small elements and is not λ-minimal. We suspect that all numerical semigroups of depth 2 with 4 small elements are λ-minimal, although we cannot prove this. Moreover, we suspect that for any n, there are only finitely many numerical semigroups of depth 2 that have n small elements and are not λ-minimal.

In Section 6, we focus on numerical semigroups of small type. The idea of considering numerical semigroups in terms of this invariant goes back to an influential paper of Fröberg, Gottlieb, and Häggkvist [**12**]. The set of *pseudo-Frobenius numbers* of S is

$$\mathrm{PF}(S) = \{P \in \mathbb{N} \setminus S \colon P + s \in S \text{ for all } s \in S \setminus \{0\}\}.$$

The number of pseudo-Frobenius numbers of S is the *type* of S, denoted $\mathrm{t}(S)$. At the beginning of Section 6, we explain how earlier results imply that numerical semigroups of type 1 or 2 are λ-minimal. We extend this to numerical semigroups of type 3.

THEOREM 1.8. *Let S be a numerical semigroup with* $\mathrm{t}(S) = 3$. *Then S is λ-minimal.*

It is not possible to extend this kind of result to semigroups of larger type. We will show in Section 3 that the semigroups S_m from Theorem 1.3 all have $\mathrm{t}(S_m) = 4$, so there are infinitely many semigroups of type 4 that are not λ-minimal. We also note that the semigroups $S_{5,k,s}$ in Theorem 1.5 have type 4, so there are infinitely many semigroups of type 4 that are λ-minimal.

We describe the connection between partitions with a given set of hook lengths and numerical sets with a given atom monoid. Antokoletz and Miller in [**2**] define the *atom monoid* of a numerical set T as

$$\mathrm{A}(T) = \{x \in \mathbb{N} \colon x + T \subseteq T\}.$$

The atom monoid $\mathrm{A}(T)$ is sometimes denoted by $T - T$ or $T : T$. It is not difficult to see that $\mathrm{A}(T) \subseteq T$ and that $\mathrm{A}(T)$ is closed under addition. Therefore, $\mathrm{A}(T)$ is a numerical semigroup contained in T. Moreover, if S is a numerical semigroup, then $\mathrm{A}(S) = S$. If T is a numerical set with $\mathrm{A}(T) = S$, we say that T is a *numerical set associated to* S. Numerical sets associated to S are in bijection with partitions λ with $\mathrm{H}(\lambda) = \mathbb{N} \setminus S$ [**15**, Corollary 1]. In fact, $\mathcal{P}(S) = \{\lambda(T) : \mathrm{A}(T) = S\}$. This correspondence is also discussed in [**10**, Section 2]. Marzuola and Miller introduced the *Anti-Atom Problem*, which asks for the number of numerical sets associated to a given numerical set S [**16**]. This is equivalent to computing $\mathrm{P}(S)$.

The set of *normalized ideals* of S is

$$\mathfrak{I}_0(S) = \{T \subseteq \mathbb{N} \colon 0 \in T,\ T + S \subseteq T\}.$$

For a discussion of normalized ideals of numerical semigroups, see [**4**, Section 3]. Further results in this direction are found in [**3, 7**]. Note that all numerical sets associated to S are normalized ideals of S. In fact a numerical set T is a normalized ideal of S if and only if $S \subseteq A(T)$.

The main focus of [**8**] is on computing $\mathrm{P}(S)$. The results about $\mathrm{P}(S)$ proven in that paper are all phrased in terms of numerical sets associated to a numerical semigroup. The authors study a poset defined in terms of a subset of the gaps of S and show that numerical sets associated to S correspond to certain order ideals of this poset. In Section 2, we prove that a particular class of order ideals corresponding to numerical sets associated to S cannot produce partitions of size smaller than $|\lambda(S)|$.

2. The void poset and order ideals

A main result of [**8**] is that numerical sets associated to a numerical semigroup S correspond to certain order ideals of a poset constructed from a subset of the gaps of S.

DEFINITION 2.1 (Void and void poset). Let S be a numerical semigroup. The *void* of S is the set

$$\mathrm{M}(S) = \{x \in \mathrm{G}(S) \colon \mathrm{F}(S) - x \in \mathrm{G}(S)\}.$$

The partial order on $\mathrm{M}(S)$ is defined as follows. Given $x, y \in \mathrm{M}(S)$, we say $x \preccurlyeq y$ if and only if $y - x \in S$. The partially ordered set $(\mathrm{M}(S), \preccurlyeq)$ is the *void poset* of S.

The void of a numerical semigroup S is called the set of *missing pairs* of S in [**10**]. The elements of the void are referred to as *holes of second type* in [**17**], as *h-gaps* in [**1**], and simply as *holes* in the `numericalsgps` pacakage [**11**].

We recall a basic property of $(\mathrm{M}(S), \preccurlyeq)$.

PROPOSITION 2.2 ([**8**, Lemma 2.1 and Proposition 2.2]). *The maximal elements of* $(\mathrm{M}(S), \preccurlyeq)$ *are the elements of* $\mathrm{PF}(S) \setminus \{\mathrm{F}(S)\}$ *and the minimal elements of* $(\mathrm{M}(S), \preccurlyeq)$ *are the elements* $\mathrm{F}(S) - P$ *where* $P \in \mathrm{PF}(S) \setminus \{\mathrm{F}(S)\}$.

Therefore, we can determine $\mathrm{t}(S)$ by counting the maximal elements of $(\mathrm{M}(S), \preccurlyeq)$ and adding one for $\mathrm{F}(S)$.

DEFINITION 2.3 (The dual of a numerical set). Let T be a numerical set. The *dual* of T is

$$T^* = \{x \in \mathbb{Z} \colon \mathrm{F}(T) - x \notin T\}.$$

Marzuola and Miller in [**16**, Proposition 1] show that if T is a numerical set associated to S, then $S \subseteq T \subseteq S^*$. It is shown in [**10**, Lemma 3] that $S^* = S \cup \mathrm{M}(S)$. The numerical set S^* is also called the *canonical ideal* of S. These results imply that numerical sets associated to S are of the form $S \cup I$ where $I \subseteq \mathrm{M}(S)$.

A numerical semigroup S is called symmetric if $s \in S$ if and only if $\mathrm{F}(S) - s \notin S$. It is easy to see that a numerical is symmetric if and only if it coincides with its canonical ideal. This is equivalent to $\mathrm{M}(S) = \varnothing$. In fact, symmetric numerical semigroups are the only ones for which $\mathrm{t}(S) = 1$ [**12**]. They are also the only numerical semigroups for which $\mathrm{P}(S) = 1$.

A main result of [**8**] is to characterize the subsets $I \subseteq \mathrm{M}(S)$ that give numerical sets associated to S.

DEFINITION 2.4 (Order ideal). Let S be a numerical semigroup and $I \subseteq \mathrm{M}(S)$. If $x \in I$ and $x \preccurlyeq y$ implies $y \in I$, then I is an *order ideal* of $(\mathrm{M}(S), \preccurlyeq)$.

The following result gives an important property of $\mathrm{A}(S \cup I)$ for $I \subseteq \mathrm{M}(S)$.

PROPOSITION 2.5 ([**8**, Proposition 2.4]). *Let S be a numerical semigroup and $I \subseteq \mathrm{M}(S)$. Then $S \subseteq \mathrm{A}(S \cup I)$ if and only if I is an order ideal of* $(\mathrm{M}(S), \preccurlyeq)$.

This proposition says that a subset $I \subseteq \mathrm{M}(S)$ is an order ideal of $(\mathrm{M}(S), \preccurlyeq)$ if and only if $S \cup I$ is a normalized ideal of S. In order to characterize numerical sets associated to S, one must determine the conditions on an order ideal $I \subseteq \mathrm{M}(S)$ that imply $\mathrm{A}(S \cup I) \subseteq S$.

DEFINITION 2.6 (Ideal triangle). Let $P, x, y \in \mathrm{M}(S)$ with $P + x + y = \mathrm{F}(S)$. We call (P, x, y) an *ideal triangle* of S.

Given an order ideal $I \subseteq \mathrm{M}(S)$ and an ideal triangle (P, x, y) we say that I *satisfies* (P, x, y) if $P, x \in I$ and $\mathrm{F}(S) - y \notin I$.

The notion of a *Frobenius triangle* introduced in [**8**, Definition 3.1] is a special case of an ideal triangle.

DEFINITION 2.7 (Frobenius triangle). An ideal triangle (P, x, y) in which $P \in \mathrm{PF}(S)$ is called a *Frobenius triangle* of S.

We recall the characterization of numerical sets associated to a numerical semigroup S.

THEOREM 2.8 ([**8**, Theorem 3.9]). *Let S be a numerical semigroup, $I \subseteq \mathrm{M}(S)$, and $T = S \cup I$. Then T is a numerical set associated to S if and only if*

(1) *I is an order ideal of $(\mathrm{M}(S), \preccurlyeq)$, and*
(2) *for each $P \in I \cap \mathrm{PF}(S)$, one of the following conditions is satisfied: $2P \notin S$, $\mathrm{F}(S) - P \in I$, or there is a Frobenius triangle (P, x, y) that is satisfied by I.*

Moreover, if $S \cup I$ is a numerical set associated to S, then for all $P \in I \cap \mathrm{PF}(S)$ either $\mathrm{F}(S) - P \in I$ or there is a Frobenius triangle (P, x, y) that is satisfied by I.

The *special gaps* of S are the gaps $h \in \mathrm{G}(S)$ for which $\{h\} \cup S$ is a numerical semigroup. It is easy to see that the special gaps of S are precisely the $P \in \mathrm{PF}(S)$ for which $2P \in S$. For more information, see [**18**, Chapter 3]. Thus, condition (2) of Theorem 2.8 says that all special gaps P of S that are in I satisfy: either $\mathrm{F}(S) - P \in I$, or I satisfies some Frobenius triangle (P, x, y).

In [**8**] it is shown that there is a class of order ideals of $(\mathrm{M}(S), \preccurlyeq)$ that always satisfy the hypotheses of Theorem 2.8.

DEFINITION 2.9 (Self-dual order ideal). Let S be a numerical semigroup and $I \subseteq \mathrm{M}(S)$ be an order ideal of $(\mathrm{M}(S), \preccurlyeq)$. If $x \in I$ implies $\mathrm{F}(S) - x \in I$, then I is called *self-dual*.

PROPOSITION 2.10 ([**8**, Proposition 3.2]). *If S is a numerical semigroup and I is a self-dual order ideal of $(\mathrm{M}(S), \preccurlyeq)$, then $S \cup I$ is a numerical set associated to S.*

Self-dual order ideals of $(\mathrm{M}(S), \preccurlyeq)$ give numerical sets associated to S, but we now show that the partition associated to such a numerical set cannot have size smaller than $|\lambda(S)|$.

PROPOSITION 2.11. *Let S be a numerical semigroup, and let $T = S \cup I$ be a numerical set associated to S where I is a self-dual order ideal of $(\mathrm{M}(S), \preccurlyeq)$. Then $|\lambda(S)| \le |\lambda(T)|$.*

We apply the following lemma in the proof of this proposition.

LEMMA 2.12. *Let S be a numerical semigroup and let I be an order ideal of $(\mathrm{M}(S), \preccurlyeq)$. Suppose $T = S \cup I$ is a numerical set associated to S. Then*

$$|\lambda(T)| = |\lambda(S)| + |A| - |B|,$$

where

$$A = \{(i, h) \colon i \in I,\ h \in \mathrm{G}(T),\ i < h\} \quad \textit{and} \quad B = \{(s, i) \colon s \in S,\ i \in I,\ s < i\}.$$

PROOF. We recall the discussion in the paragraph ending with equation (1.1). Each box in the Young diagram of $\lambda(T)$ is $\square(t, h)$ for some $t < h$, with $t \in T$ and $h \in \mathrm{G}(T)$. We can determine $|\lambda(T)|$ by counting such pairs. We have

$$|\lambda(S)| = \#\{(s, g) \colon s \in S,\ g \in \mathrm{G}(S),\ s < g\}$$
$$|\lambda(T)| = \#\{(t, h) \colon t \in T,\ h \in \mathrm{G}(T),\ t < h\}.$$

Since $T = S \cup I$ for some $I \subseteq \mathrm{M}(S)$, we have

$$|\lambda(T)| = \#\{(s, h) \colon s \in S,\ h \in \mathrm{G}(T),\ s < h\} + \#\{(i, h) \colon i \in I,\ h \in \mathrm{G}(T),\ i < h\}.$$

That is,

$$|\lambda(T)| = \#\{(s, h) \colon s \in S,\ h \in \mathrm{G}(T),\ s < h\} + |A|.$$

We note that $\mathrm{G}(T) = \mathrm{G}(S) \setminus I$. This implies

$|\lambda(T)| = \#\{(s,g)\colon s \in S,\ g \in \mathrm{G}(S),\ s < g\} - \#\{(s,i)\colon s \in S,\ i \in I,\ s < i\} + |A|,$

completing the proof. □

PROOF OF PROPOSITION 2.11. Let $\mathrm{F}(S) = F$ and define the sets A and B as in the statement of Lemma 2.12. Consider the function $f\colon B \to A$ defined by $f(s,i) = (F-i, F-s)$. Because I is self-dual and $i \in I$, we know that $F - i \in I$. Since $s \in S$, it is clear that $F - s \in \mathrm{G}(S) \setminus \mathrm{M}(S)$. This implies $F - s \notin T$. Since $s < i$, we have $F - i < F - s$. So f is well-defined. It is easy to see that f is injective. Hence, $|\lambda(S)| \leq |\lambda(T)|$. □

Theorem 2.8 gives a characterization of the order ideals I of $(\mathrm{M}(S), \preccurlyeq)$ for which $T = S \cup I$ is a numerical set associated to S. In particular, such an order ideal gives a numerical set corresponding to S if and only if for each $P \in I \cap \mathrm{PF}(S)$ one of two conditions is satisfied. We show that a similar statement holds for other elements of I that are not necessarily pseudo-Frobenius numbers of S.

PROPOSITION 2.13. *Let S be a numerical semigroup and I be an order ideal of $(\mathrm{M}(S), \preccurlyeq)$ such that $T = S \cup I$ is a numerical set associated to S. For each $x \in I$ one of the following conditions holds:*

(1) $\mathrm{F}(S) - x \in I$, *or*
(2) *there is an ideal triangle (x, y, z) satisfied by I.*

PROOF. Let $\mathrm{F}(S) = F$. Since the elements of $\mathrm{PF}(S) \setminus \{F\}$ are the maximal elements of $(\mathrm{M}(S), \preccurlyeq)$, we know that there is some $P \in \mathrm{PF}(S) \setminus \{F\}$ such that $x \preccurlyeq P$. Since I is an order ideal, we know that $P \in I$. Theorem 2.8 implies that either $F - P \in I$ or I satisfies a Frobenius triangle of the form (P, y, z).

(1) Suppose $F - P \in I$. Note that $x \preccurlyeq P$ implies $F - P \preccurlyeq F - x$, so $F - x \in I$.
(2) Suppose I satisfies a Frobenius triangle (P, y, z). Let $s = P - x$. We know that $s \in S$ since $x \preccurlyeq P$. Let $y_1 = y + s$. We want to show that $y_1 \in \mathrm{M}(S)$.

First, if $y_1 \in S$, then $(F - z) - x = (P + y) - x = y + s = y_1$. This means $x \preccurlyeq F - z$. However, $x \in I$ and $F - z \notin I$. This contradicts the fact that I is an order ideal. Thus, $y_1 \notin S$.

On the other hand, if $F - y_1 \in S$, then $F - y = (F - y_1) + s \in S$. This contradicts the fact that $y \in \mathrm{M}(S)$. Thus, $F - y_1 \notin S$.

We conclude that $y_1 \in \mathrm{M}(S)$ and hence (x, y_1, z) is an ideal triangle. Since $y \in I$, we know that $y_1 = y + s \in I$. Also, $F - z \notin I$ since I satisfies the Frobenius triangle (P, y, z). Therefore, I satisfies the ideal triangle (x, y_1, z). □

3. Numerical semigroups that are not λ-minimal

The focus of this section is on numerical semigroups that are not λ-minimal. We first return to the semigroup S from Example 1.2.

EXAMPLE 3.1. Let $S = \langle 9, 10, 11, 12, 13\rangle$ and consider the order ideal $I = \{1, 14, 16\} \subseteq \mathrm{M}(S)$. It is not difficult to check that $T = S \cup I$ is a numerical set associated to S for which

$$|\lambda(S)| = 32 > 31 = |\lambda(T)|.$$

Hence, S is not λ-minimal.

We now consider an infinite family of numerical semigroups containing the semigroup $S = \langle 9, 10, 11, 12, 13\rangle$.

LEMMA 3.2. *Let $m \geq 9$ and $S = \langle m, m+1, m+2, \dots, 2m-5\rangle$. Then*

$$\mathrm{G}(S) = \{1, 2, \dots, m-1, 2m-4, 2m-3, 2m-2, 2m-1\}.$$

We see that

$$\mathrm{PF}(S) = \{2m-4, 2m-3, 2m-2, 2m-1\},$$

and in particular, $\mathrm{F}(S) = 2m-1$.

PROOF. The smallest element of S that is not one of its minimal generators is $2m$, so it is clear that $\{1, 2, \dots, m-1, 2m-4, 2m-3, 2m-2, 2m-1\} \subseteq \mathrm{G}(S)$. Since $m \geq 9$, we have $3m-1 \leq (2m-5)+(2m-5)$. Adding pairs of minimal generators shows that S contains $2m$, $2m+1$, $\dots$, $3m-5$, $3m-4$, $3m-3$, $3m-2$, $3m-1$. Thus, S contains m consecutive positive integers. Adding multiples of m shows that S contains all positive integers larger than $2m$. Therefore,

$$\mathrm{G}(S) = \{1, 2, \dots, m-1, 2m-4, 2m-3, 2m-2, 2m-1\}.$$

Once we know $\mathrm{G}(S)$ it is easy to determine $\mathrm{PF}(S)$. □

Before proving Theorem 1.3 we introduce some terminology related to integer partitions.

DEFINITION 3.3. Let $\lambda = (\lambda_1, \dots, \lambda_\ell)$ be a partition. The conjugate $\tilde{\lambda}$ of λ is $(\lambda_1', \dots, \lambda_r')$ where $\lambda_i' = \#\{j \colon \lambda_j \geq i\}$. A partition λ is *self-conjugate* if $\lambda = \tilde{\lambda}$.

It is easy to see that the Young diagram of $\tilde{\lambda}$ has columns of length λ_1, λ_2, $\dots$, λ_ℓ. It is clear that $|\lambda| = |\tilde{\lambda}|$, and moreover that the multiset of hook lengths of λ is the same as the multiset of hook lengths of $\tilde{\lambda}$.

We recall a result about the relationship between the enumeration of T and the enumeration of T^*.

PROPOSITION 3.4 ([**10**, Proposition 12]). *For any numerical set T, $\widetilde{\lambda(T)} = \lambda(T^*)$, and therefore $\mathrm{A}(T) = \mathrm{A}(T^*)$.*

Theorem 1.3 follows from the following more precise result.

THEOREM 3.5. *Let $m \geq 9$ and $S = \langle m, m+1, \dots, 2m-5\rangle$. Then we have $|\lambda(S)| = 5m-13$, $\mathrm{P}(S) = 6$, and*

$$\{|\lambda(T)| : T \in \mathcal{P}(S)\} = \{4m-5, 5m-13, 5m-7\}.$$

That is, S is not λ-minimal. In fact, we have

$$\lim_{m\to\infty} \left(|\lambda(S)| - \min\{|\lambda(T)| \colon T \in \mathcal{P}(S)\}\right) = \infty$$

and

$$\lim_{m\to\infty} \frac{\min\{|\lambda(T)| \colon T \in \mathcal{P}(S)\}}{|\lambda(S)|} = \frac{4}{5}.$$

PROOF. The description of $\mathrm{G}(S)$ from Lemma 3.2 implies that

$$\mathrm{M}(S) = \{1, 2, 3, 2m-4, 2m-3, 2m-2\}.$$

Theorem 2.8 implies that the only order ideals I of $(\mathrm{M}(S), \preccurlyeq)$ that give numerical sets associated to S are

$\varnothing$, $\mathrm{M}(S)$, $\{1, 2m-4, 2m-2\}$, $\{2, 2m-4, 2m-3\}$, $\{1, 2m-4\}$, $\{1, 2, 2m-4, 2m-3\}$.

For each of these order ideals we consider $T = S \cup I$ and construct the Young diagram of $\lambda(T)$. We will see that these order ideals come in conjugate pairs.

The first pair of order ideals is $\varnothing$ and $\mathrm{M}(S)$. The Young diagrams of the numerical sets corresponding to these order ideals are:

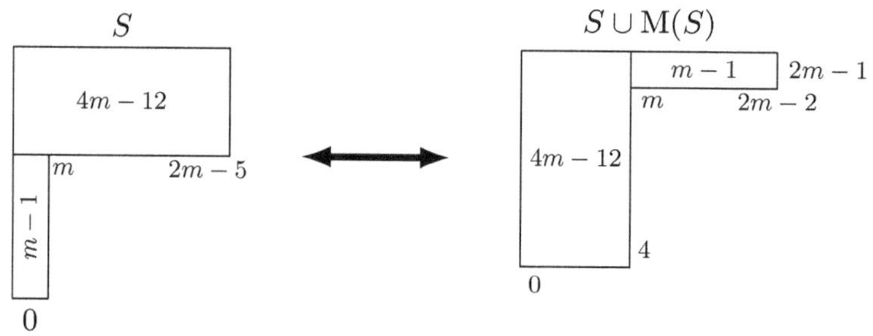

We compute $|\lambda(S)| = |\lambda(S \cup \mathrm{M}(S))| = 5m-13$.

The next pair of order ideals is $\{1, 2m-4, 2m-2\}$ and $\{2, 2m-4, 2m-3\}$. The Young diagrams of the numerical sets corresponding to these order ideals are:

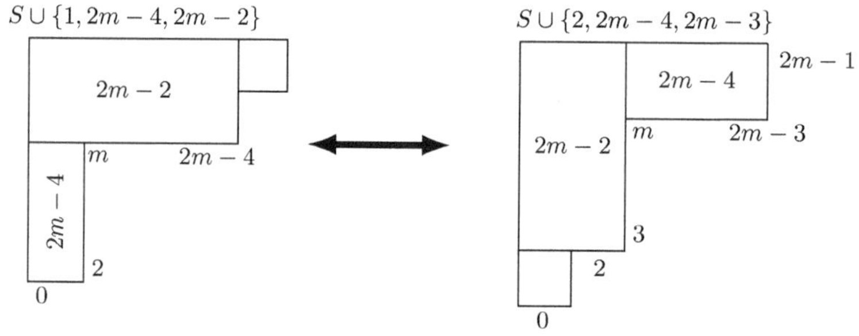

We compute $|\lambda(S \cup \{1, 2m-4, 2m-2\})| = |\lambda(S \cup \{2, 2m-4, 2m-3\})| = 4m-5$.

Lastly, we consider the pair of order ideals $\{1, 2m-4\}$ and $\{1, 2, 2m-4, 2m-3\}$. The Young diagrams of the numerical sets corresponding to these order ideals are:

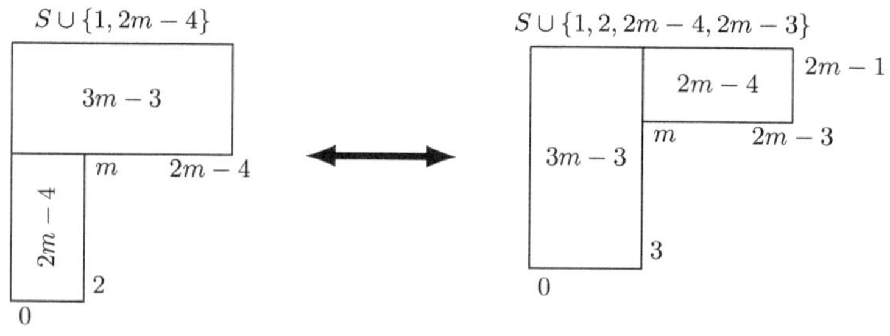

We compute $|\lambda(S \cup \{1, 2m-4\})| = |\lambda(S \cup \{1, 2, 2m-4, 2m-3\})| = 5m-7$. The final statement of the theorem follows by taking a limit as $m \to \infty$. □

The semigroup $S = \langle 9, 10, 11, 12, 13 \rangle$ from Example 1.2 also generalizes in a different direction to give a second infinite family of numerical semigroups that are not λ-minimal.

THEOREM 3.6. *Let $k \geq 4$ and $S = \langle 2k+1, 2k+2 \dots, 3k+1 \rangle$. Then $|\lambda(S)| = k^2 + 4k$.*

Let l be a divisor of k. Consider

$$T = S \cup \{1, 2, \dots, l-1\} \cup \{3k+1 < x \leq 4k+1 \colon x \not\equiv 1 \pmod{l}\}.$$

Then $\mathrm{A}(T) = S$ and

$$|\lambda(T)| = \frac{3k^2 l - k^2 + 4kl^3 + 3kl^2 + kl - 2l^4 + 2l^3}{2l^2}.$$

PROOF. Note that $\mathrm{G}(S) = \{1, 2, \dots, 2k\} \cup \{3k+2, \dots, 4k+1\}$. This implies $\mathrm{F}(S) = 4k+1$ and $\mathrm{M}(S) = \{1, 2, \dots, k-1\} \cup \{3k+2, \dots, 4k\}$.

By equation (1.1), we have

$$|\lambda(S)| = \sum_{y \in \mathrm{G}(S)} \#\{s \in S \colon s < y\} = 2k \cdot 1 + k(k+2) = k^2 + 4k.$$

Similarly,

$$\begin{aligned} \mathrm{G}(T) &= \{l, l+1, \dots, 2k\} \cup \{3k+1 < y \leq 4k+1 \colon y \equiv 1 \pmod{l}\} \\ &= \{l, l+1, \dots, 2k\} \cup \{3k+1+ly_1 \colon 1 \leq y_1 \leq k/l\}. \end{aligned}$$

So

$$\begin{aligned} |\lambda(T)| &= \sum_{y \in \mathrm{G}(T)} \#\{x \in T \colon x < y\} \\ &= (2k - (l-1))l + \sum_{y_1=1}^{k/l} (l + (k+1) + (l-1)y_1) \\ &= (2k - l + 1)l + (k+1+l)\frac{k}{l} + (l-1)\frac{\frac{k}{l}(1+\frac{k}{l})}{2}. \end{aligned}$$

Putting this expression over a common denominator gives the formula for $|\lambda(T)|$ from the statement of the theorem.

We now show that $\mathrm{A}(T) = S$. The relations in $(\mathrm{M}(S), \preccurlyeq)$ are $a \preccurlyeq 3k+1+b$, for $1 \leq b \leq a \leq k-1$. Thus,

$$I = \{1, 2, \dots, l-1\} \cup \{3k+1 < x \leq 4k+1 \colon x \not\equiv 1 \pmod{l}\}$$

is an order ideal of the void poset. The pseudo-Frobenius numbers in I are

$$I \cap \mathrm{PF}(S) = \{3k+1 < x \leq 4k+1 \colon x \not\equiv 1 \pmod{l}\}.$$

Given $x \in I \cap \mathrm{PF}(S)$, say $1 - x \equiv a \pmod{l}$ for $1 \leq a \leq l-1$ (since $x \not\equiv 1 \pmod{l}$). Note that $a \in I$, and $x + a$ is the smallest positive integer larger than x that is congruent to 1 modulo l. This implies $3k+1 \leq x+a \leq 4k+1$. We see that $x + a \notin I$.

- If $x + a = 4k+1$, then $a = \mathrm{F}(S) - x \in I$.
- If $x + a < 4k+1$, then $x + a \in \mathrm{M}(S)$, so $(x, a, \mathrm{F}(S) - x - a)$ is a Frobenius triangle. Moreover, $a \in I$ and $x + a = \mathrm{F}(S) - (\mathrm{F}(S) - x - a) \notin I$. Therefore, I satisfies the Frobenius triangle $(x, a, \mathrm{F}(S) - x - a)$.

By Theorem 2.8, it follows that $\mathrm{A}(T) = S$. □

COROLLARY 3.7. *Let $k = 12l_1^2$ and $l = 3l_1$ in Theorem 3.6. Then we have*

$$|\lambda(T)| = 2\sqrt{3}|\lambda(S)|^{3/4} + O(|\lambda(S)|^{1/2}).$$

PROOF. In this case $|\lambda(S)| = k^2+4k = 144l_1^4+O(l_1^2)$ and $|\lambda(T)| = 144l_1^3+O(l_1^2)$. The result follows. □

The following corollary is a more precise version of Theorem 1.4 from the introduction.

COROLLARY 3.8. *Let $k \geq 4$ and $S = \langle 2k+1, 2k+2, \ldots, 3k+1\rangle$.*

If $l \notin \{1,k\}$, then the construction of Theorem 3.6 gives a numerical set T with $\lambda(T) \in \mathcal{P}(S)$ and $|\lambda(T)| < |\lambda(S)|$. Therefore, if k is not prime, then S is not λ-minimal.

Let N be a positive integer. For any k with sufficiently many divisors, S has the property that there are at least N partitions in $\mathcal{P}(S)$ with size less than $|\lambda(S)|$.

PROOF. Let $\frac{k}{l} = u$, so $u > 1$. Then,

$$|\lambda(S)| - |\lambda(T)| = (l-1)(u-1)\Big(lu - l - \frac{u}{2}\Big).$$

Since $l \notin \{1,k\}$, we have $(l-1) \geq 1$ and $(u-1) \geq 1$. We see that

$$lu - l - \frac{u}{2} = \Big(l - \frac{1}{2}\Big)(u-1) - \frac{1}{2} \geq \frac{3}{2}\cdot 1 - \frac{1}{2} = 1.$$

The second part of the statement follows directly from the first. □

4. The staircase family

We now turn our attention to a family of numerical semigroups that has received recent interest in relation to the Anti-Atom Problem.

DEFINITION 4.1 (Staircase family of numerical semigroups). Let m and k be positive integers with $m \geq 2$ and let s be an integer satisfying $1 \leq s \leq m-1$. A numerical semigroup of the form

$$S_{m,k,s} = \{0, m, 2m, \ldots, km, km+s+1 \rightarrow\}$$

is said to be in a *staircase family.*

In this section, we will show that every numerical semigroup of the form $S_{m,k,s}$ is λ-minimal. We begin by describing the pseudo-Frobenius numbers, void, and void poset of a numerical semigroup in the staircase family. When the semigroup S is clear from context, we write F instead of $\mathrm{F}(S)$.

LEMMA 4.2. *Let $S = S_{m,k,s}$ be a numerical semigroup in a staircase family. Then*

1. *The pseudo-Frobenius elements of S are precisely the elements of the set $\mathrm{PF}(S) = \{(k-1)m+s+1, \ldots, (k-1)m+(m-1), km+1, \ldots, km+s\}$. In particular, $\mathrm{t}(S) = m-1$ and $\mathrm{F}(S) = km+s$.*
2. *Suppose $x, y \in \mathrm{M}(S)$ with $x \leq y$. Then $x \preccurlyeq y$ in $(\mathrm{M}(S), \preccurlyeq)$ if and only if $m \mid (y-x)$.*
3. *The void poset of S, $(\mathrm{M}(S), \preccurlyeq)$, consists of $m-2$ chains of elements $x_1 \preccurlyeq x_2 \preccurlyeq \cdots \preccurlyeq x_n$ where $n \in \{k, k+1\}$. In particular, $x_1, \ldots, x_n$ are positive elements of a nonzero congruence class modulo m that does not contain $\mathrm{F}(S)$.*

PROOF. The formula for $\mathrm{F}(S)$ is clear from the description of the elements of S. Furthermore, $x \in \mathrm{M}(S)$ if and only if $F-x \notin S$. Since $F-x<F$, this is equivalent to saying that $F-x \not\equiv 0 \pmod{m}$. That is, every element of $\mathrm{G}(S)$ not congruent to F modulo m is an element of $\mathrm{M}(S)$. Furthermore, let $x, y \in \mathrm{M}(S)$ with $x \preccurlyeq y$. We see that $y-x<F$, so $y-x \in S$ if and only if $x \equiv y \pmod{m}$. Considering the set of elements of $\mathrm{M}(S)$ in each residue class modulo m, we conclude that each set has either k or $k+1$ elements. Using that fact that pseudo-Frobenius numbers not equal to F are exactly the maximal elements of the void poset, we conclude they are indeed the elements described above. □

Suppose that S is a numerical semigroup and T is a numerical set associated to S. We describe a way to divide the Young diagram of $\lambda(T)$ into sections. Let $m=\mathrm{m}(S)$, $F=\mathrm{F}(S)$, and $q=\mathrm{q}(S)$. For any b satisfying $0 \leq b \leq q-1$, bm is an element of S with $bm<F$. In particular, $bm \in T$. We draw a vertical line in $\lambda(T)$ to the left of the step on the profile of $\lambda(T)$ corresponding to bm. For any a satisfying $0 \leq a \leq q-1$, $F-am \notin S$ and $F-am \notin \mathrm{M}(S)$. Therefore, $F-am \notin T$. We draw a horizontal line in $\lambda(T)$ above the step on the profile of $\lambda(T)$ corresponding to $F-am$. This divides the Young diagram of $\lambda(T)$ into sections indexed by pairs (a, b) satisfying $0 \leq a, b \leq q-1$ and $a+b \leq q-1$. We let

$$\lambda(T)_{a,b}=\left\{\square(u, x): u \in T, x \in \mathrm{G}(T), u<x,\left\lfloor\frac{u}{m}\right\rfloor=b,\left\lfloor\frac{F-x}{m}\right\rfloor=a\right\}.$$

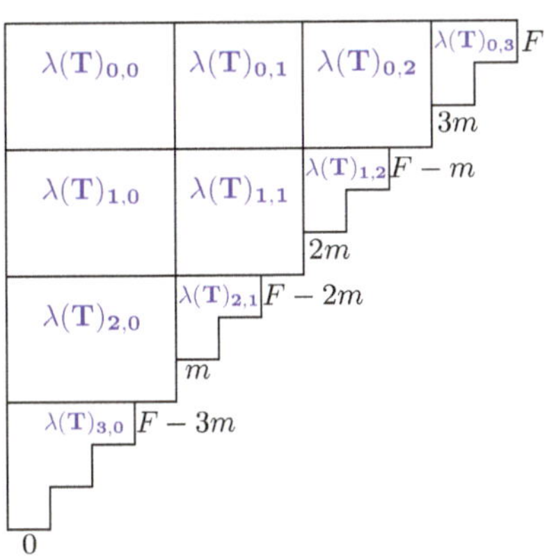

We denote the set of hook lengths that appear in the boxes of $\lambda(T)_{a,b}$ by

$$\mathcal{H}(T)_{a,b}=\{|\operatorname{hook}(u, x)|: \square(u, x) \in \lambda(T)_{a,b}\}.$$

It is clear that $|\mathcal{H}(T)_{a,b}| \leq |\lambda(T)_{a,b}|$.

We can now give the proof of Theorem 1.5. We want to show that if T is a numerical set associated to $S=S_{m,k,s}$, then $|\lambda(S)| \leq |\lambda(T)|$. We do this by showing that for each section we have $|\lambda(S)_{a,b}| \leq |\mathcal{H}(T)_{a,b}|$.

PROOF OF THEOREM 1.5. Let $S=S_{m,k,s}$, so $\mathrm{m}(S)=m$, $F=\mathrm{F}(S)=km+s$, and $\mathrm{q}(S)=k+1$. Note that each section of the Young diagram of $\lambda(S)$ is contained in a single column. We see that

$$|\lambda(S)_{a,b}|= \begin{cases}m-1 & \text{if } a+b<k, \\ s & \text{if } a+b=k.\end{cases}$$

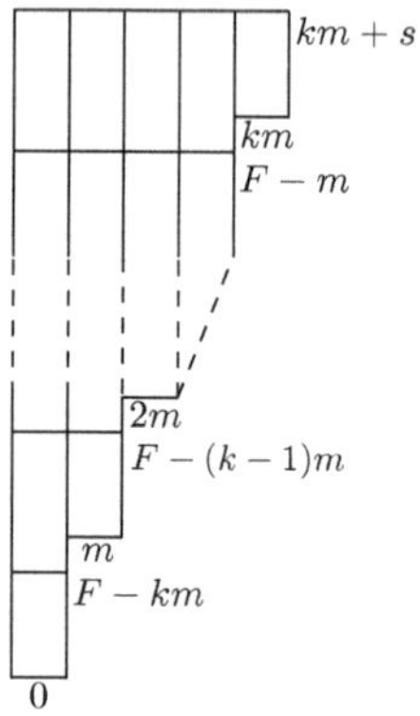

FIGURE 2. Young diagram of $S_{m,k,s}$

Let T be a numerical set associated to S. Our goal is to show that in each section $|\lambda(T)_{a,b}| \geq |\lambda(S)_{a,b}|$. Since $|\lambda(T)_{a,b}| \geq |\mathcal{H}(T)_{a,b}|$, it is enough to show that $|\mathcal{H}(T)_{a,b}| \geq |\lambda(S)_{a,b}|$.

In the case where $a+b<k$, we want to show that $|\mathcal{H}(T)_{a,b}| \geq m-1$. We will do this by showing that for each $\ell \in [1, m-1]$, $\mathcal{H}(T)_{a,b}$ contains a hook length congruent to ℓ modulo m. Note that $\lambda(T)_{a,b}$ contains $\square(u,x)$ for

$$u \in [bm, (b+1)m-1] \cap T, \qquad x \in [F-am-(m-1), F-am] \cap \mathrm{G}(T).$$

In the case where $a+b=k$, we want to show that $|\mathcal{H}(T)_{a,b}| \geq s$. We will do this by showing that for each $\ell \in [1,s]$, $\lambda(T)_{a,b}$ contains a box with hook length congruent to ℓ modulo m. Since $F-am = bm+s$, we have that $\lambda(T)_{a,b}$ contains $\square(u,x)$, where $u<x$ and

$$u \in [bm, bm+s] \cap T, \qquad x \in [bm, bm+s] \cap \mathrm{G}(T).$$

Consider $\ell \in [1, m-1]$. We will show that $\lambda(T)_{a,b}$ contains a box with hook length congruent to ℓ modulo m. If $a+b=k$, then we include the additional requirement that $\ell \leq s$. Say $F-\ell \equiv \alpha \pmod{m}$ with $0 \leq \alpha \leq m-1$ and $x_1 = F-am-\alpha$. So $x_1 \equiv \ell \pmod{m}$ and $x_1 \in [F-am-(m-1), F-am]$.

First, consider the case where $x_1 \notin T$. The box $\square(bm, x_1)$ has hook length congruent to ℓ modulo m. If $a+b<k$, then this box is clearly in $\lambda(T)_{a,b}$. If $a+b=k$, then $\ell \leq s$, so $\alpha \leq s$. Therefore, $x_1 \in [bm, bm+s]$ and we see that $\square(bm, x_1)$ is in $\lambda(T)_{a,b}$.

Next, consider the case where $x_1 \in T$. Suppose $T = S \cup I$ for an order ideal I. Since $x_1 \equiv \ell \not\equiv 0 \pmod{m}$, we know that $x_1 \notin S$. This means $x_1 \in I \subseteq \mathrm{M}(S)$, so $x_1 \not\equiv F \pmod{m}$, and hence $\alpha \neq 0$. Since I is an order ideal, the description of $(\mathrm{M}(S), \preccurlyeq)$ from Lemma 4.2 implies that I contains the pseudo-Frobenius number of S that is congruent to x_1 modulo m. We call this pseudo-Frobenius number P. Then by Theorem 2.8 we have two cases.

(1) Suppose $F-P \in T$. Proposition 2.2 implies that $F-P$ is a minimal element of $\mathrm{M}(S)$ and Lemma 4.2 implies that T contains all positive integers congruent to $F-P$ modulo m. Note that $bm+\alpha \equiv F-P \pmod{m}$ and the box $\square(bm+\alpha, F-am)$ has hook length congruent to ℓ modulo m. If $a+b<k$, then this box is clearly in $\lambda(T)_{a,b}$. In the case where $a+b=k$, we have $0<\ell \leq s$, which implies $0 \leq \alpha < s$, so $bm+\alpha \in [bm, bm+s]$. Moreover,

$F - am = bm + s$, so $F - am \in [bm, bm + s]$ and $bm + \alpha < F - am$. Thus, $\square(bm + \alpha, F - am)$ is in $\lambda(T)_{a,b}$.

(2) Suppose I satisfies a Frobenius triangle (P, y, z). We have $y \in T$, $F - z \notin T$, and $P + y + z = F$. Lemma 4.2 implies that $F - P < m$. Therefore, $y + z = F - P < m$. Since $y, z \in \mathrm{M}(S)$, Proposition 2.2 implies that both y and z are minimal elements of $(\mathrm{M}(S), \preccurlyeq)$. Since $y \in I$ and I is an order ideal, Lemma 4.2 implies that T contains all positive integers congruent to y modulo m. Since z is a minimal element of $(\mathrm{M}(S), \preccurlyeq)$, Proposition 2.2 implies that $F - z$ is a maximal element of $(\mathrm{M}(S), \preccurlyeq)$. Since $F - z \notin I$ and I is an order ideal, Lemma 4.2 implies that T does not contain any positive integer less than $\mathrm{F}(S)$ that is congruent to $F - z$ modulo m.

The box $\square(bm+y, F-am-z)$ has hook length $F-am-z-bm-y \equiv P \equiv \ell \pmod{m}$. If $a + b < k$, then this box is clearly in $\lambda(T)_{a,b}$. In the case where $a+b = k$, we know that $1 \leq \ell \leq s$, which implies $F-P \leq s$. Also, we are in the case $x_1 \in T$, so $x_1 \not\equiv F \pmod{m}$, that is, $\ell \neq s$. Recall that $F - am = bm + s$. We see that $y+z = F-P < s$, so $bm+y, F-am-z \in [bm, bm+s]$. Moreover, $bm + y < F - am - z$ as $y + z < s$. Hence, $\square(bm + y, F - am - z)$ is in $\lambda(T)_{a,b}$.

For sections with $a + b < k$, we have shown that

$$|\lambda(T)_{a,b}| \geq |\mathcal{H}(T)_{a,b}| \geq m - 1 = |\lambda(S)_{a,b}|.$$

For sections with $a + b = k$, we have shown that

$$|\lambda(T)_{a,b}| \geq |\mathcal{H}(T)_{a,b}| \geq s = |\lambda(S)_{a,b}|.$$

This completes the proof. □

5. Families of numerical semigroups with depth 2

In this section, we consider several families of numerical semigroups with depth 2 and show that they are λ-minimal.

5.1. Depth 2 and Durfee squares. Our goal is to analyze numerical semigroups of depth 2 with a Durfee square of fixed size. We provide a sufficient condition on the large gaps of such numerical semigroups to ensure that they are λ-minimal. We then show that if S has depth 2 and the Durfee square of $\lambda(S)$ has size at most 3, this condition is always satisfied, which allows us to conclude that all such numerical semigroups S are λ-minimal.

If the Durfee square of $\lambda(S)$ has size n, we will partition the Young diagram of $\lambda(S)$ into a union of n hooks. If T is any numerical set associated to S, then we find n disjoint hooks of corresponding lengths in the Young diagram of $\lambda(T)$.

Recall that each hook is uniquely determined by a pair (a, x) satisfying $a \in T$, $x \in \mathrm{G}(T)$, and $a < x$. Moreover, two hooks $\mathrm{hook}(a_1, x_1)$ and $\mathrm{hook}(a_2, x_2)$ with $a_1 \leq a_2$ are disjoint if either $a_1 < a_2 < x_2 < x_1$ or $a_1 < x_1 < a_2 < x_2$.

THEOREM 5.1. *Let S be a numerical semigroup with depth 2 for which $\lambda(S)$ has a Durfee square of size n. Let $F = \mathrm{F}(S)$ and suppose that F, $F - \alpha_1$, $F - \alpha_2$, $F - \alpha_3$, ..., $F - \alpha_{n-1}$ are the n largest gaps of S in descending order.*

Suppose $\alpha_1, \ldots, \alpha_{n-1}$ have the property that $\alpha_i = \alpha_j + \alpha_k$ implies $j = k = i-1$. Then S is λ-minimal.

PROOF. Since the Durfee square of $\lambda(S)$ has size n, the Young diagram of $\lambda(S)$ can be partitioned as a disjoint union n hooks. Suppose that the n smallest elements of S are $0 < F - \beta_1 < \cdots < F - \beta_{n-1}$. Note that $\mathrm{m}(S) = F - \beta_1$. Therefore, $\lambda(S)$ is the disjoint union of the hooks

$$\text{hook}(0, F),\ \text{hook}(F - \beta_1, F - \alpha_1), \ldots, \text{hook}(F - \beta_{n-1}, F - \alpha_{n-1}).$$

These hooks have sizes $F, \beta_1 - \alpha_1, \ldots, \beta_{n-1} - \alpha_{n-1}$, and therefore

$$|\lambda(S)| = F + \sum_{i=1}^{n-1} (\beta_i - \alpha_i).$$

Note that $0 < \alpha_1 < \cdots < \alpha_{n-1} < \beta_{n-1} < \cdots < \beta_1 < \frac{F}{2}$, where the last inequality follows from the assumption that S has depth 2. Figure 3 illustrates these hooks.

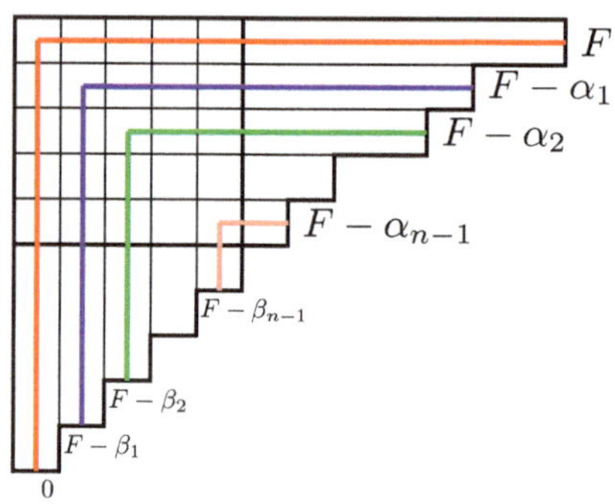

FIGURE 3. Decomposition of Young diagram into n hooks

We now consider the set of ideal triangles of S that could potentially contain $F - \alpha_i$. Suppose $(F - \alpha_i, u, v)$ is an ideal triangle of S. Then $F - v$ is a gap of S and $F > F - v = (F - \alpha_i) + u > F - \alpha_i$. This implies $F - v = F - \alpha_{j_1}$ for some j_1. Similarly $F - u = F - \alpha_{j_2}$, hence $v = \alpha_{j_1}$ and $u = \alpha_{j_2}$. If $(F - \alpha_i, \alpha_{j_1}, \alpha_{j_2})$ is an ideal triangle, then $\alpha_i = \alpha_{j_1} + \alpha_{j_2}$. The assumption in the statement of the theorem implies $j_1 = j_2 = i - 1$.

Let T be a numerical set associated to S. We will find n disjoint hooks in $\lambda(T)$, of sizes at least F, $\beta_1 - \alpha_1$, …, $\beta_{n-1} - \alpha_{n-1}$. We choose the first hook to correspond to the pair $(0, F)$. This hook has size F. None of $F - \alpha_1$, …, $F - \alpha_{n-1}$ are elements of S. Proposition 2.13 implies that for each i satisfying $1 \le i \le n-1$ one of the following holds

- $F - \alpha_i \notin T$;
- $F - \alpha_i, \alpha_i \in T$;
- $F - \alpha_i, \alpha_{i-1} \in T$, $2\alpha_{i-1} = \alpha_i$, and $F - \alpha_{i-1} \notin T$.

In the first case, we choose the $(i+1)^{\text{st}}$ hook in our list to be $\text{hook}(F - \beta_i, F - \alpha_i)$. Since $F - \beta_i \in S \subseteq T$ and $F - \beta_i < F - \alpha_i$, we see that $\text{hook}(F - \beta_i, F - \alpha_i)$ is a hook in the Young diagram of $\lambda(T)$. This hook has size $\beta_i - \alpha_i$.

In the second case, we choose the $(i+1)^{\text{st}}$ hook in our list to be $\text{hook}(\alpha_i, \beta_i)$. Since $F - \beta_i \in S$, we see that β_i is a gap of S that is not in $\mathrm{M}(S)$. Therefore, $\beta_i \notin T$. Moreover, $\alpha_i < \beta_i$, so this is indeed a valid hook. It has size $\beta_i - \alpha_i$.

If the third condition holds and the second does not, then we choose the $(i+1)^{\text{st}}$ hook in our list to be $\text{hook}(\alpha_{i-1}, \beta_i)$. As in the previous case, we see that $\beta_i \notin T$. Since $\alpha_{i-1} < \alpha_i < \beta_i$, we see that $\text{hook}(\alpha_{i-1}, \beta_i)$ is a valid hook in the Young diagram of $\lambda(T)$. This hook has size $\beta_i - \alpha_{i-1} > \beta_i - \alpha_i$.

Consider distinct $1 \leq i_1, i_2 \leq n-1$. We will show that the hooks in our list corresponding to i_1 and i_2 are disjoint.

(1) Suppose that the hooks are $\text{hook}(F-\beta_{i_1}, F-\alpha_{i_1})$, $\text{hook}(F-\beta_{i_2}, F-\alpha_{i_2})$. Assume $i_1 < i_2$. Then we have $F-\beta_{i_1} < F-\beta_{i_2} < F-\alpha_{i_2} < F-\alpha_{i_1}$.
(2) Suppose that the hooks are $\text{hook}(F-\beta_{i_1}, F-\alpha_{i_1})$, $\text{hook}(\alpha_{i_2}, \beta_{i_2})$. Then we have $\alpha_{i_2} < \beta_{i_2} < F-\beta_{i_1} < F-\alpha_{i_1}$.
(3) Suppose that the hooks are $\text{hook}(\alpha_{i_1}, \beta_{i_1})$, $\text{hook}(\alpha_{i_2}, \beta_{i_2})$. Without loss of generality, we may assume that $i_1 < i_2$. We see that $\alpha_{i_1} < \alpha_{i_2} < \beta_{i_2} < \beta_{i_1}$.
(4) Suppose that the hooks are $\text{hook}(F-\beta_{i_1}, F-\alpha_{i_1})$, $\text{hook}(\alpha_{i_2-1}, \beta_{i_2})$. Then we have $\alpha_{i_2-1} < \beta_{i_2} < F-\beta_{i_1} < F-\alpha_{i_1}$.
(5) Suppose that the hooks are $\text{hook}(\alpha_{i_1}, \beta_{i_1})$, $\text{hook}(\alpha_{i_2-1}, \beta_{i_2})$. For this case to happen, it is necessary that $F-\alpha_{i_1} \in T$ and $F-\alpha_{i_2-1} \notin T$. This implies $i_1 \neq i_2 - 1$.

   We first consider the sub-case where $i_2 < i_1$. Then we have $\alpha_{i_2-1} < \alpha_{i_1} < \beta_{i_1} < \beta_{i_2}$.

   Next, consider the sub-case where $i_2 > i_1$. Since $i_2 - i_1 \neq 1$, this implies $i_2 - 1 > i_1$. In this case, we have $\alpha_{i_1} < \alpha_{i_2-1} < \beta_{i_2} < \beta_{i_1}$.
(6) Suppose that the hooks are $\text{hook}(\alpha_{i_1-1}, \beta_{i_1})$, $\text{hook}(\alpha_{i_2-1}, \beta_{i_2})$. Without loss of generality, we may assume that $i_1 < i_2$. We see that $\alpha_{i_1-1} < \alpha_{i_2-1} < \beta_{i_2} < \beta_{i_1}$.

Therefore, in all cases, the chosen hooks are disjoint from each other. This shows that

$$|\lambda(T)| \geq F + \sum_{i=1}^{n-1} (\beta_i - \alpha_i) = |\lambda(S)|. \qquad \square$$

Proof of Theorem 1.6. If $n \leq 3$, then $\alpha_1, \ldots, \alpha_{n-1}$ always satisfy the condition that $\alpha_i = \alpha_j + \alpha_k$ implies $j = k = i-1$. Applying Theorem 5.1 completes the proof. $\square$

If we consider collections $\alpha_1, \ldots, \alpha_{n-1}$ for which $\alpha_i = \alpha_j + \alpha_k$ does not necessarily imply $j = k = i-1$, then the same arguments do not work. Let S be one of the numerical semigroups from the statement of Theorem 3.5. The depth of S is 2 and $\lambda(S)$ has a Durfee square of size 4. The four largest gaps of S are

$$2m-1, 2m-2, 2m-3, 2m-4,$$

with $\text{F}(S) = 2m-1$. So $(\alpha_1, \alpha_2, \alpha_3) = (1, 2, 3)$. We have that $\alpha_3 = \alpha_1 + \alpha_2$. Therefore, Theorem 5.1 does not apply, and as was shown in Theorem 3.5, S is not λ-minimal. This example also shows that Theorem 1.6 is sharp in the sense that there are infinitely many numerical semigroups S of depth 2 for which $\lambda(S)$ has a Durfee square of size 4 that are not λ-minimal.

5.2. Depth 2 and few small elements. Our work in this section is motivated by some results of Singhal and Lin [**19**] that build on earlier work of Marzuola and Miller [**16**]. They show that when considering all numerical sets T with a fixed Frobenius number, $\text{A}(T)$ is very likely to be of depth 2 with $0, 1, 2,$ or 3 small elements. We show that in each of these cases, $\text{A}(T)$ is λ-minimal.

We begin by noting that the number of small elements and the Durfee square size of a numerical semigroup are closely related.

LEMMA 5.2. *If a numerical semigroup S has k small elements, then the size of the Durfee square of $\lambda(S)$ is at most $k+1$.*

This follows directly from noting that the number of columns of $\lambda(S)$ is one more than the number of small elements of S. We demonstrate this with Example 5.3.

EXAMPLE 5.3. Consider the numerical semigroup $S = \{0, 5, 6, 10, \rightarrow\}$. The Young diagram of $\lambda(S)$ is shown in Figure 4.

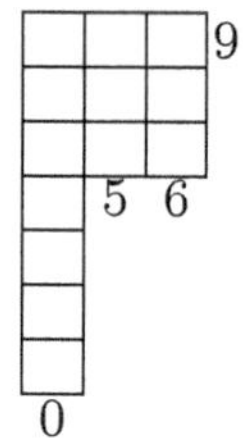

FIGURE 4. The Young diagram of $\lambda(S)$

The small elements of S are 5 and 6, and $\lambda(S)$ has a Durfee square of size 3.

COROLLARY 5.4. *Numerical semigroups with depth 2 and at most 2 small elements are λ-minimal.*

PROOF. If S has at most 2 small elements, then Lemma 5.2 implies that $\lambda(S)$ has a Durfee square of size at most 3. Since S has depth 2, Theorem 1.6 implies that S is λ-minimal. $\square$

In this section, we only consider numerical semigroups of depth 2. This significantly simplifies the notation for the sections of the Young diagram of $\lambda(T)$ that we introduced in Section 4. Let $\lambda(T)_1 = \lambda(T)_{1,0}$, $\lambda(T)_2 = \lambda(T)_{0,0}$, and $\lambda(T)_3 = \lambda(T)_{0,1}$. We also denote $\mathcal{H}(T)_1 = \mathcal{H}(T)_{1,0}$, $\mathcal{H}(T)_2 = \mathcal{H}(T)_{0,0}$, and $\mathcal{H}(T)_3 = \mathcal{H}(T)_{0,1}$.

LEMMA 5.5. *If S is a numerical semigroup of depth 2 and T is a numerical set associated to S, then*

$$\mathcal{H}(\lambda(T)) \cap [\mathrm{F}(S) - \mathrm{m}(S) + 1, \mathrm{F}(S)] \subseteq \mathcal{H}(T)_2.$$

Moreover, if $T = S$, then

$$\mathcal{H}(\lambda(S)) \cap [\mathrm{F}(S) - \mathrm{m}(S) + 1, \mathrm{F}(S)] = \mathcal{H}(S)_2$$

and $|\mathcal{H}(S)_2| = |\lambda(S)_2|$.

PROOF. Let $m = \mathrm{m}(S)$ and $F = \mathrm{F}(S)$. Recall that $\lambda(T)_1$ consists of the $\square(u, x)$ for which $0 \le u < m$ and $x \le F - m$. This means that every hook length in $\mathcal{H}(T)_1$ is at most $F - m$. Similarly, $\lambda(T)_3$ consists of the $\square(u, x)$ for which $m \le u$ and $F - m < x \le F$. This means that every hook length in $\mathcal{H}(T)_3$ is at most $F - m$. Therefore, hook lengths larger than $F - m$ can only occur in $\mathcal{H}(T)_2$. The first part of the lemma follows.

Next, suppose that $T = S$. Note that $[0, m-1] \cap S = \{0\}$. This means that $\lambda(S)_2$ consists of $\square(0, x)$ for $x \in \mathrm{G}(S)$ with $F - m < x \le F$. This shows that every hook length in $\mathcal{H}(S)_2$ is at least $F-m+1$. It follows that $\mathcal{H}(S)_2$ has precisely those hook lengths that are larger than $F-m$. Finally, from the description

$$\lambda(S)_2 = \{\square(0,x) : x \in \mathrm{G}(S), F-m < x\},$$

we see that there are no repeated hook lengths in $\lambda(S)_2$. This implies that $|\mathcal{H}(S)_2| = |\lambda(S)_2|$. □

COROLLARY 5.6. *If S is a numerical semigroup of depth 2 and T is a numerical set associated to S, then $|\lambda(S)_2| \leq |\lambda(T)_2|$.*

PROOF. Since $\mathrm{A}(T) = S$, we know that $\mathcal{H}(\lambda(T)) = \mathcal{H}(\lambda(S))$. Therefore, Lemma 5.5 implies that

$$\mathcal{H}(S)_2 = \mathcal{H}(\lambda(S)) \cap [F-m+1, F] = \mathcal{H}(\lambda(T)) \cap [F-m+1, F] \subseteq \mathcal{H}(T)_2.$$

This shows that $|\lambda(S)_2| = |\mathcal{H}(S)_2| \leq |\mathcal{H}(T)_2| \leq |\lambda(T)_2|$. □

THEOREM 5.7. *If S is a numerical semigroup of depth 2 with 3 small elements, then S is λ-minimal.*

PROOF. Let $m = \mathrm{m}(S)$. Define a and b so that the three small elements of S are $m < F-a < F-b$. So $b < a < F-m < \frac{F}{2}$. If $b \leq 3$, then the size of the Durfee square of $\lambda(S)$ is at most 3. Applying Theorem 1.6 completes the proof in this case. Therefore, assume $b \geq 4$. Let T be a numerical set associated to S.

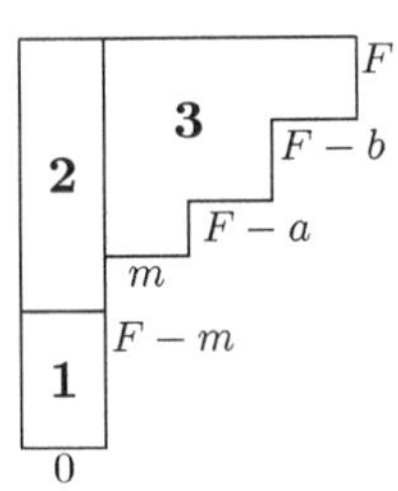

FIGURE 5. Young diagram of $S = \{0, m, F-a, F-b, F+1 \rightarrow\}$

By Corollary 5.6, we know that $|\lambda(S)_2| \leq |\lambda(T)_2|$. Therefore, it suffices to show that

$$|\lambda(S)_1| + |\lambda(S)_3| \leq |\lambda(T)_1| + |\lambda(T)_3|.$$

Note that $|\lambda(S)_1| = F-m$. Since $b \geq 3$, we know that $F-1, F-2 \in \mathrm{G}(S)$. Therefore, $|\lambda(S)_3| = (F-m)+(a-1)+(b-2)$. We will construct 4 disjoint hooks in $\lambda(T)_1 \cup \lambda(T)_3$ of sizes at least $F-m$, $F-m$, $a-1$ and $b-2$.

Notice that there are no ideal triangles of S containing $F-1$ and the only possibility for an ideal triangle of S that contains $F-2$ is $(F-2, 1, 1)$.

First, we know that $0, m \in T$ and $F-m, F \notin T$, so we have hooks $\mathrm{hook}(0, F-m) \subseteq \lambda(T)_1$ and $\mathrm{hook}(m, F) \subseteq \lambda(T)_3$. They both have size $F-m$.

Since $F-1 \notin S$, and $F-1$ is not contained in any ideal triangle of S, Theorem 2.13 implies that we have two possibilities:

- $F-1 \notin T$. In this case, we take the hook $\mathrm{hook}(F-a, F-1) \subseteq \lambda(T)_3$. Note that this is disjoint from $\mathrm{hook}(m, F)$ as $m < F-a < F-1 < F$.

- $F-1, 1 \in T$. In this case, we take the hook $\text{hook}(1,a) \subseteq \lambda(T)_1$. Note that $a \notin T$ as $a \notin S \cup \mathrm{M}(S)$. Therefore, this is a valid hook. Moreover, this hook is disjoint from $\text{hook}(0, F-m)$ as $0 < 1 < a < F-m$. In either case, we have constructed a hook of size $a-1$.

Since $F-2 \notin S$ and the only possibility for an ideal triangle of S containing $F-2$ is $(F-2,1,1)$, Theorem 2.13 implies that we have three possibilities:

- $F-2 \notin T$. In this case, we take the hook $\text{hook}(F-b, F-2) \subseteq \lambda(T)_3$. Note that this is disjoint from $\text{hook}(m, F)$ as $m < F-b < F-2 < F$. It is also disjoint from $\text{hook}(F-a, F-1)$ or $\text{hook}(1,a)$, whichever was chosen.
- $F-2, 2 \in T$. In this case, we take the hook $\text{hook}(2,b) \subseteq \lambda(T)_1$. Note that $b \notin T$ as $b \notin S \cup \mathrm{M}(S)$. Therefore, this is a valid hook. Moreover, this is disjoint from $\text{hook}(0, F-m)$ as $0 < 2 < b < F-m$. It is also disjoint from $\text{hook}(F-a, F-1)$ or $\text{hook}(1,a)$, whichever was chosen.
- $F-2, 1 \in T$ and $F-1 \notin T$. In this case, we take the hook $\text{hook}(1,b) \subseteq \lambda(T)_1$. This is disjoint from $\text{hook}(0, F-m)$ as $0 < 1 < b < F-m$. Since $F-1 \notin T$, the previous hook chosen must have been $\text{hook}(F-a, F-1)$ which is disjoint from this hook. In any of these cases, we have constructed a hook of size $b-2$.

In all cases we constructed 4 disjoint hooks in $\lambda(T)_1 \cup \lambda(T)_3$ of sizes at least $F-m$, $F-m$, $a-1$, and $b-2$. This implies that

$$|\lambda(S)_1| + |\lambda(S)_3| = (F-m) + (F-m) + (a-1) + (b-2) \leq |\lambda(T)_1| + |\lambda(T)_3|.$$

We conclude that $|\lambda(S)| \leq |\lambda(T)|$ and S is λ-minimal. □

6. Numerical semigroups of small type

We begin this section by showing how earlier results imply that numerical semigroups of type at most 2 are λ-minimal.

PROPOSITION 6.1. *Let S be a numerical semigroup with* $\mathrm{t}(S) \leq 2$. *Then S is λ-minimal.*

Proposition 1.1 in [**16**] implies that if $\mathrm{t}(S) = 1$, then $\mathrm{P}(S) = 1$. Theorem 2.5 in [**8**] says that if $\mathrm{t}(S) = 2$, then $\mathrm{P}(S) = 2$. Therefore, Proposition 6.1 follows from the next result.

PROPOSITION 6.2. *Let S be a numerical semigroup with* $\mathrm{P}(S) \leq 2$. *Then S is λ-minimal.*

PROOF. If $\mathrm{P}(S) = 1$, then $\lambda(S)$ is the only partition in $\mathcal{P}(S)$, so it is clearly the smallest one. Suppose $\mathrm{P}(S) = 2$. Then [**10**, Proposition 11] implies that $\lambda(S)$ is not self-conjugate. So, the two partitions in $\mathcal{P}(S)$ are $\lambda(S)$ and its conjugate, which clearly have the same size. □

The main goal of this section is to prove Theorem 1.8. We will also give some remarks on the λ-minimality of numerical semigroups of type 4.

We recall the discussion of the dual of a numerical set from Section 2. If T is a numerical set, then the dual of T is

$$T^* = \{x \in \mathbb{Z} \colon \mathrm{F}(T) - x \notin T\}.$$

Proposition 3.4 says that $\mathrm{A}(T) = \mathrm{A}(T^*)$. Let $S = \mathrm{A}(T)$. Therefore, $T = S \cup I$ and $T^* = S \cup J$ for some order ideals I, J of $(\mathrm{M}(S), \preccurlyeq)$. We now explain how these order ideals are related.

PROPOSITION 6.3. *Let S be a numerical semigroup with $\mathrm{F}(S) = F$. Let T be a numerical set such that $T = S \cup I$ for some order ideal I of $(\mathrm{M}(S), \preccurlyeq)$. Then $T^* = S \cup I^*$ where*

$$I^* = \{x \in \mathrm{M}(S) \colon F - x \notin I\}.$$

Moreover, I^* is an order ideal of $(\mathrm{M}(S), \preccurlyeq)$.

PROOF. Note that

$$\begin{aligned} T^* &= \{x \in \mathbb{Z} : F - x \notin S \cup I\} \\ &= \{x \in \mathrm{M}(S) : F - x \notin S \cup I\} \cup \{x \in \mathbb{Z} : x \notin \mathrm{M}(S), F - x \notin S \cup I\}. \end{aligned}$$

If $x \in \mathrm{M}(S)$, then $F - x \notin S$. So we know that

$$\{x \in \mathrm{M}(S) : F - x \notin S \cup I\} = \{x \in \mathrm{M}(S) : F - x \notin I\} = I^*.$$

Moreover, if $x \notin \mathrm{M}(S)$, then $x \in S$ or $F - x \in S$. This means that if $x \notin \mathrm{M}(S)$ and $F - x \notin S \cup I$, then $x \in S$. Also, if $x \in S$, we automatically have that $x \notin \mathrm{M}(S)$ and $F - x \notin S \cup I$. Therefore,

$$\{x \in \mathbb{Z} : x \notin \mathrm{M}(S), F - x \notin S \cup I\} = S.$$

Now, it is left to see that I^* is an order ideal of $(\mathrm{M}(S), \preccurlyeq)$. Suppose $x \in I^*$ and $x \preceq y$. Then $F - x \notin I$ and $F - y \preccurlyeq F - x$. Since I is an order ideal, this implies that $F - y \notin I$. So $y \in I^*$, and I^* is an order ideal. □

The main result of [**8**, Section 6] is to determine $\mathrm{P}(S)$ for any numerical semigroup S with $\mathrm{t}(S) = 3$. We build on these results to show that all of these semigroups are λ-minimal. For the remainder of this section, S is a numerical semigroup of type 3 where $\mathrm{PF}(S) = \{P, Q, F\}$ with $P < Q < F$.

LEMMA 6.4 ([**8**, Lemma 6.2]). *If $P + Q - F \in S$ and $Q - P \notin \mathrm{M}(S)$, then $\mathrm{P}(S) = 2$.*

Proposition 6.1 implies that numerical semigroups of this kind are λ-minimal.

LEMMA 6.5 ([**8**, Lemma 6.1]). *If $P + Q - F \notin S$, then $\mathrm{P}(S) = 4$. Moreover, $(\mathrm{M}(S), \preccurlyeq)$ has 4 order ideals, each of which is self-dual.*

Proposition 2.11 implies that numerical semigroups of this kind are λ-minimal.

LEMMA 6.6 ([**8**, Lemma 6.4]). *Suppose $P + Q - F \in S$ and $Q - P \in \mathrm{M}(S)$. If $S \cup I$ is a numerical set associated to S, then I is $\varnothing$, $\mathrm{M}(S)$, or one of the following:*

$$\begin{aligned} I_1 &= \{x \in \mathrm{M}(S) \colon Q - P \preccurlyeq x\}, \\ I_2 &= \{x \in \mathrm{M}(S) \colon F - Q \preccurlyeq x\}. \end{aligned}$$

(Note that it is possible to have $I_1 = I_2$.)

LEMMA 6.7 ([**8**, Section 6]). *For the order ideals I_1 and I_2 of Lemma 6.6, we have that $I_1^* = I_2$.*

PROPOSITION 6.8. *If $P + Q - F \in S$ and $Q - P \in \mathrm{M}(S)$, then S is λ-minimal.*

PROOF. Since the ideals $\varnothing$ and $\mathrm{M}(S)$ correspond to $\lambda(S)$ and its conjugate, we may disregard these. We claim that $|\lambda(S)| \leq |\lambda(S \cup I_1)|$. By Lemma 2.12, it suffices to show that $|B| \leq |A|$, where

$$A = \{(i, g) : i \in I_1,\ g \notin S \cup I_1,\ i < g\},$$
$$B = \{(s, i) : s \in S,\ i \in I_1,\ s < i\}.$$

Consider the map $f : B \to A$ defined by

$$(s, i) \mapsto (Q - i, Q - s).$$

To confirm that this map is well-defined, we must show that $Q - i \in I_1$ and $Q - s \notin S \cup I_1$.

In order to show $Q - i \in I_1$, we first show that $Q - i \in \mathrm{M}(S)$. Note that $Q - (Q - P) = P \notin S$, which means that $Q - P \not\preccurlyeq Q$. Now, since $i \in I_1$ we see that $Q - P \preccurlyeq i$. It follows that $i \not\preccurlyeq Q$, that is $Q - i \notin S$. Since P, Q are the maximal elements of $(\mathrm{M}(S), \preccurlyeq)$ and $i \not\preccurlyeq Q$, we see that $i \preccurlyeq P$. Therefore,

$$(Q - i) - (Q - P) = P - i \in S.$$

Now, if $F - (Q - i) \in S$, then

$$F - (Q - P) = (F - (Q - i)) + ((Q - i) - (Q - P)) \in S.$$

However, this contradicts the fact that $Q - P \in \mathrm{M}(S)$. Therefore, $F - (Q - i) \notin S$, and so $Q - i \in \mathrm{M}(S)$. We have shown that $Q - i \in \mathrm{M}(S)$ and $Q - P \preccurlyeq Q - i$. It follows that $Q - i \in I_1$.

We now show that $Q - s \notin S \cup I_1$. Note that $Q - s \notin S$ as otherwise $Q = (Q - s) + s \in S$. If $Q - s \in I_1$, then $Q - P \preccurlyeq Q - s$. This means that $(Q - s) - (Q - P) = P - s \in S$. But $(P - s) + s = P \notin S$, so this is a contradiction. Hence, $Q - s \notin S \cup I_1$.

Since f is trivially injective, we conclude that $|B| \leq |A|$. So $|\lambda(S)| \leq |\lambda(S \cup I_1)|$. By Lemma 6.7 and Proposition 3.4, $|\lambda(S \cup I_1)| = |\lambda(S \cup I_2)|$. Hence, S is λ-minimal. □

Collecting these results completes the proof of Theorem 1.8.

We conclude by discussing numerical semigroups of type 4. Theorem 1.3 gives an infinite family of semigroups that are not λ-minimal. Lemma 3.2 implies that each of these semigroups has type 4.

Consider the staircase family numerical semigroups

$$S_{5,k,4} = \{0, 5, 10, \ldots, 5k, 5(k + 1) \rightarrow\}.$$

By Theorem 1.5, these semigroups are λ-minimal. Lemma 4.2 implies that for any positive integer k, $\mathrm{t}(S_{5,k,4}) = 4$. Thus, there are also infinitely many semigroups of type 4 that are λ-minimal.

Acknowledgments

The authors thank the referee for many helpful suggestions that significantly improved the paper. We thank Ian Farish, Erik Imathiu-Jones, Matilda LaFortune, and Victoria Wiest for many helpful discussions. The first author thanks Maria Monks Gillespie for conversations many years ago that helped get this project started.

References

[1] F. Aicardi and L. G. Fel, *Gaps in nonsymmetric numerical semigroups*, Israel J. Math. **175** (2010), 85–112, DOI 10.1007/s11856-010-0003-3. MR2607538

[2] E. Antokoletz and A. Miller, *Symmetry and factorization of numerical sets and monoids*, J. Algebra **247** (2002), no. 2, 636–671, DOI 10.1006/jabr.2001.8994. MR1877867

[3] V. Barucci and F. Khouja, *On the class semigroup of a numerical semigroup*, Semigroup Forum **92** (2016), no. 2, 377–392, DOI 10.1007/s00233-014-9679-8. MR3472022

[4] S. Bonzio and P. A. García-Sánchez, *The poset of normalized ideals of numerical semigroups with multiplicity three*, Comm. Algebra **53** (2025), no. 9, 3610–3624, DOI 10.1080/00927872.2025.2464956. MR4924514

[5] M. Bras-Amorós and A. de Mier, *Representation of numerical semigroups by Dyck paths*, Semigroup Forum **75** (2007), no. 3, 677–682, DOI 10.1007/s00233-007-0717-7. MR2353289

[6] H. E. Burson, H. Nam, and S. Sisneros-Thiry, *On integer partitions corresponding to numerical semigroups*, Results Math. **78** (2023), no. 5, Paper No. 193, 30, DOI 10.1007/s00025-023-01974-8. MR4622154

[7] L. Casabella, M. D'Anna, and P. A. García-Sánchez, *Apéry sets and the ideal class monoid of a numerical semigroup*, Mediterr. J. Math. **21** (2024), no. 1, Paper No. 7, 28, DOI 10.1007/s00009-023-02550-8. MR4670095

[8] A. Chen, N. Kaplan, L. Lawson, C. O'Neill, and D. Singhal, *Enumerating numerical sets associated to a numerical semigroup*, Discrete Appl. Math. **341** (2023), 218–231, DOI 10.1016/j.dam.2023.08.010. MR4635583

[9] A. Chen, N. Kaplan, L. Lawson, C. O'Neill, and D. Singhal, *On the asymptotic growth of the number of associated numerical sets*, In preperation (2025).

[10] H. Constantin, B. Houston-Edwards, and N. Kaplan, *Numerical sets, core partitions, and integer points in polytopes*, Combinatorial and additive number theory. II, Springer Proc. Math. Stat., vol. 220, Springer, Cham, 2017, pp. 99–127, DOI 10.1007/978-3-319-68032-3_7. MR3754898

[11] M. Delgado, P. A. Garcia-Sanchez, and J. Morais, *NumericalSgps, a package for numerical semigroups, Version 1.4.0*, `https://gap-packages.github.io/numericalsgps`, 2024, GAP package.

[12] R. Fröberg, C. Gottlieb, and R. Häggkvist, *On numerical semigroups*, Semigroup Forum **35** (1987), no. 1, 63–83, DOI 10.1007/BF02573091. MR880351

[13] The GAP Group, *GAP – Groups, Algorithms, and Programming, Version 4.14.0*, 2024.

[14] N. Kaplan, K. Kim, C. McGeorge, F. Ramirez, and D. Singhal, *LambdaMinimality*, `https://github.com/DDeepuS/LambdaMinimality`, 2025, GitHub repository, accessed 2025-05-19.

[15] W. J. Keith and R. Nath, *Partitions with prescribed hooksets*, J. Comb. Number Theory **3** (2011), no. 1, 39–50. MR2908181

[16] J. Marzuola and A. Miller, *Counting numerical sets with no small atoms*, J. Combin. Theory Ser. A **117** (2010), no. 6, 650–667, DOI 10.1016/j.jcta.2010.03.002. MR2645183

[17] D. P. Patil and G. Tamone, *On the type sequences of some one dimensional rings*, Univ. Iagel. Acta Math. **45** (2007), 119–130. MR2453951

[18] J. C. Rosales and P. A. García-Sánchez, *Numerical semigroups*, Developments in Mathematics, vol. 20, Springer, New York, 2009, DOI 10.1007/978-1-4419-0160-6. MR2549780

[19] D. Singhal and Y. Lin, *Density of numerical sets associated to a numerical semigroup*, Comm. Algebra **49** (2021), no. 10, 4291–4303, DOI 10.1080/00927872.2021.1918136. MR4296838

[20] The Sage Developers, *SageMath, the Sage Mathematics Software System (version 10.6beta8)*, 2025, `https://www.sagemath.org`, DOI: 10.5281/zenodo.8042260.

[21] N. Tutaş, H. İ. Karakaş, and N. Gümüşbaş, *Young tableaux and Arf partitions*, Turkish J. Math. **43** (2019), no. 1, 448–459, DOI 10.3906/mat-1807-181. MR3909306

Department of Mathematics, University of California, Irvine, 340 Rowland Hall, Irvine, California 92697-3875
Email address: nckaplan@math.uci.edu

Department of Mathematics, University of California, Irvine, 340 Rowland Hall, Irvine, California 92697-3875
Email address: kayleesk@uci.edu

Department of Mathematics, University of California, Irvine, 340 Rowland Hall, Irvine, California 92697-3875
Email address: cmcgeorg@uci.edu

Department of Mathematics, University of California, Irvine, 340 Rowland Hall, Irvine, California 92697-3875
Email address: fabiar3@uci.edu

Department of Mathematics, University of California, Irvine, 340 Rowland Hall, Irvine, California 92697-3875
Email address: singhald@uci.edu

Contemporary Mathematics
Volume **836**, 2026
https://doi.org/10.1090/conm/836/16704

When the divisibility poset of the ideal class monoid of a numerical semigroup is a lattice

S. Bonzio and P. A. García-Sánchez

Abstract. We characterize numerical semigroups for which the divisibility poset of its ideal class monoid is a lattice, and study the irreducible elements of such a lattice with respect to meet and join in this poset, as well as with respect to union and intersection.

1. Introduction

A *numerical semigroup* is a co-finite submonoid of the monoid of non-negative integers under addition, denoted by $\mathbb{N}$ in this manuscript. The co-finite condition is equivalent to saying that the greatest common divisor of the elements of the semigroup is one.

A (relative) *ideal* of S is a non-empty subset I of integers such that $I+S=I$ and $z+I\subseteq S$ for some integer z. This last condition is equivalent to the existence of $\min(I)$. An ideal I is said to be *normalized* if $\min(I)=0$. The set of normalized ideals of S is denoted by $\mathfrak{I}_0(S)$.

If I and J are ideals of a numerical semigroup S, then $I+J=\{i+j : i\in I, j\in J\}$ is also an ideal of S (see [**1**, Chapter 3] for the basic properties of ideals of numerical semigroups). The set of ideals of S under this operation is a monoid, and its identity element is S. Notice that $(\mathfrak{I}_0(S),+)$ is a submonoid of this monoid.

Observe that if I is an ideal of S and $z\in\mathbb{Z}$, then $z+I=(z+S)+I$ is also an ideal of S. For I and J ideals of S, we write $I\sim J$ if $I=z+J$ for some $z\in\mathbb{Z}$. We denote by $\mathcal{C}\ell(S)$ the quotient of the set of ideals of S by the binary relation $\sim$, and by $[I]$ the equivalence class of I under $\sim$. We may think of $\mathcal{C}\ell(S)$ as the set of ideals modulo principal ideals.

Given I and J ideals of S, it is easy to see that $[I]+[J]=[I+J]$ is well defined, and so $(\mathcal{C}\ell(S),+)$ is a monoid, which is isomorphic to $(\mathfrak{I}_0(S),+)$, via the isomorphism $[I]\mapsto -\min(I)+I$ (see, for instance, [**4**]). If S and T are numerical semigroups such that $(\mathcal{C}\ell(S),+)$ and $(\mathcal{C}\ell(T),+)$ are isomorphic, then S and T must be equal [**7**]. The set $\mathcal{C}\ell(S)$ was introduced in [**2**], where the first bounds on its cardinality were given together with some properties related to the reduction number of its elements and the generators of $(\mathcal{C}\ell(S),+)$.

If I and J are normalized ideals of a numerical semigroup S, and there exists $K\in\mathfrak{I}_0(S)$ such that $I+K=J$, then we write $I\preceq J$. The binary relation $\preceq$ is

2020 *Mathematics Subject Classification.* Primary 20M14, 20M12.
Key words and phrases. Ideal class monoid, numerical semigroup, poset, lattice.

©2026 American Mathematical Society

an order relation on $\mathfrak{I}_0(S)$, and thus $(\mathfrak{I}_0(S), \preceq)$ is a poset. In the same way, we write $[I] \preceq [J]$ if there exists an ideal K of S such that $[I] + [K] = [J]$. Hence, $(\mathcal{C}\ell(S), \preceq)$ becomes a poset isomorphic to $(\mathfrak{I}_0(S), \preceq)$; we will refer to these posets as the divisibility posets of $\mathcal{C}\ell(S)$ and $\mathfrak{I}_0(S)$, respectively. In [**3**], the poset $(\mathfrak{I}_0(S), \preceq)$ is studied for S a numerical semigroup of embedding dimension three, and it is shown that it is always a lattice.

One of the goals of this manuscript is to characterize the numerical semigroups S for which $(\mathfrak{I}_0(S), \preceq)$ is a lattice (Section 4); the other is to study irreducible elements with respect to the different operations defined on $\mathfrak{I}_0(S)$, when $(\mathfrak{I}_0(S), \preceq)$ is a lattice (Section 5).

The reader will notice that some computations with specific numerical semigroups and ideals are omitted. These computations can be found in the file titled *lattice-normalized-ideals-ns.ipynb* on the repository `https://github.com/numerical-semigroups/ideal-class-monoid` and were performed with the help of the `GAP` [**6**] package `numericalsgps` [**5**].

2. Preliminaries

2.1. Numerical semigroups. Let S be a numerical semigroup. The set $\mathbb{N} \setminus S$ is known as the set of gaps of S, denoted here by $\mathrm{G}(S)$, and its cardinality is the genus of S. The largest integer not belonging to S is known as the *Frobenius number* of S, which we denote by $\mathrm{F}(S)$. The *conductor* of S is the least integer c such that $c + \mathbb{N} \subseteq S$. We will use $\mathrm{C}(S)$ to refer to the conductor of S; it follows easily that $\mathrm{C}(S) = \mathrm{F}(S) + 1$. The least positive integer belonging to S is known as the *multiplicity* of S, denoted by $\mathrm{m}(S)$.

Given $A \subseteq \mathbb{N}$ the smallest submonoid of $\mathbb{N}$ that contains A is

$$\langle A \rangle = \left\{ \sum\nolimits_{i=1}^{n} a_i : n \in \mathbb{N}, a_1, \dots, a_n \in A \right\}.$$

Clearly, if S is a numerical semigroup, then $\langle S \rangle = S$. If $A \subseteq S$ is such that $\langle A \rangle = S$, then we say that A is a *set of generators* of S, or simply that A generates S. We say that A is a *minimal set of generators* of S if no proper subset of A generates S. It is well known that S admits a unique minimal set of generators, which is $S^* \setminus (S^* + S^*)$, with $S^* = S \setminus \{0\}$ (see for instance [**1**, Corollary 2]), moreover this set cannot have two elements congruent modulo $\mathrm{m}(S)$, and so the cardinality of the minimal set of generators, known as the embedding dimension of S, is always smaller than the multiplicity of S. The elements of $S^* \setminus (S^* + S^*)$ are known as minimal generators of S. If follows easily that $s \in S$ is a minimal generator if and only if $S \setminus \{s\}$ is a numerical semigroup.

Associated to a numerical semigroup S one can define the order induced (on $\mathbb{Z}$) by S as $a \leq_S b$ if $b - a \in S$, for any $a, b \in \mathbb{Z}$.

If X is a finite set of integers, then $X + S$ is an ideal of S. If $X = \{x\}$ for some $x \in \mathbb{Z}$, then we write $x + S$ instead of $\{x\} + S$. Notice that every ideal I of S is of this form, since it suffices to take $X = \mathrm{Minimals}_{\leq_S}(I)$, which is know as a *minimal generating set* of I, and its elements are called *minimal generators* of I. There cannot be two different minimal generators congruent modulo the multiplicity of S. In particular, the minimal generating set of an ideal I of S has at most cardinality $\mathrm{m}(S)$.

The union of two ideals of a numerical semigroup is again an ideal, and the same holds for the intersection (see [**1**, Chapter 3] for an introduction of the basic properties of ideals of numerical semigroups).

2.2. Posets and lattices. A poset $(P, \leq)$ is a set equipped with a (partial) order relation $(\leq)$. Given two elements a, b in a poset $(P, \leq)$, b is a *cover* of a if $a < b$ (that is $a \leq b$ and $a \neq b$) and for every c such that $a \leq c \leq b$, then either $c = a$ or $c = b$.

A poset $(P, \leq)$ is a lattice if, for every pair of elements $x, y \in P$, there exist the infimum and the supremum of $\{x, y\}$: in such a case they are (binary) operations on the set P usually denoted by $x \wedge y$ and $x \vee y$, and referred to as *meet* and *join*, respectively. A lattice can be equivalently defined as a set equipped with two idempotent, associative, commutative and absorptive operations $\wedge$ and $\vee$.

A *meet semilattice* $(S, \wedge)$ is a set S with an associative, commutative and idempotent operation $\wedge$. On every meet semilattice one can define the partial order relation $x \leq y$ if $x \wedge y = x$. Similarly, $(S, \vee)$ is a *join semilattice* if $\vee$ is an associative, commutative and idempotent operation on S; in this case, we can define the partial order relation $x \leq y$ if $x \vee y = y$. In a join (respectively meet) semilattice the element $x \vee y$ (respectively $x \wedge y$) is the supremum (respectively infimum) of the set $\{x, y\}$ with respect to the induced order $\leq$.

Given a poset $(P, \leq)$ and $X \subseteq P$, we indicate by $\uparrow X$ and $\downarrow X$ the set of upper and lower bounds, respectively, of X, namely $\uparrow X = \{a \in P : x \leq a,\ \text{for every}\ x \in X\}$ and $\downarrow X = \{a \in P : a \leq x,\ \text{for every}\ x \in X\}$. For $X = \{x\}$, we will write $\uparrow x$ and $\downarrow x$ instead of $\uparrow\{x\}$ and $\downarrow\{x\}$, respectively.

Let $(P, \leq)$ be a lattice, and let $p \in P$. We say that p is $\wedge$-*irreducible* if it cannot be expressed as $p = q \wedge r$ with $q, r \in P \setminus \{p\}$. Analogously, we say that p is $\vee$-*irreducible* if there is no $q, r \in P \setminus \{p\}$ such that $p = q \vee r$.

The next result is a well known characterization of $\wedge$ and $\vee$-irreducible elements in a finite lattice in terms of covers (see for instance [**9**, Theorem 3.9] and the observation just afterwards). Nevertheless we include its proof for the reader's convenience.

PROPOSITION 1. *Let $(P, \leq)$ be a finite lattice, and let $p \in P$. Then, p is $\wedge$-irreducible if and only if p has at most one cover. Dually, p is $\vee$-irreducible if and only if p covers at most one element in P.*

PROOF. Suppose that p is $\wedge$-irreducible and that q and r are covers of p, with $q \neq r$. Then, $p \leq q \wedge r \leq q$. As q covers p, either $p = q \wedge r$ or $q \wedge r = q$. The first case is not possible, since p is irreducible, while the second leads to $q = q \wedge r \leq r$, which cannot be possible since then $p \leq q \leq r$, and then either $q = r$ or $q = p$ (r is a cover of p).

For the other implication, if p has no covers, then, since P is finite, $\uparrow p = \{p\}$, and this forces p to be $\wedge$-irreducible. Suppose that q is the unique cover of p. If $p = r \wedge s$ with $r, s \in P \setminus \{p\}$, then $p \leq r$ and $p \leq s$, which yields $q \leq r$ and $q \leq s$, since q is the unique cover of p. In particular, $q \leq r \wedge s$, leading to $r \wedge s = p < q \leq r \wedge s$, a contradiction. □

3. Normalized ideals of numerical semigroups

Recall that a relative ideal I of a numerical semigroup S is a *normalized* ideal I if $\min(I) = 0$. The set $\mathcal{I}_0(S)$ of normalized ideals of S is always finite and forms a (complete) lattice under the operations of $\cap$ and $\cup$ (see [**4**]). Moreover, it can be turned into a poset upon considering the following partial order relation:

$$I \preceq J \text{ if there exists } K \in \mathcal{I}_0(S) \text{ such that } I + K = J.$$

We will write $I \prec J$ if $I \preceq J$ and $I \neq J$ (see the paragraph preceding Remark 5.1 in [**4**]).

REMARK 2. Let S be a numerical semigroup, and let $I, J \in \mathfrak{I}_0(S)$. If $I \preceq J$, then $J = I+K$ for some $K \in \mathfrak{I}_0(S)$. Thus, $I = I+0 \subseteq I+K = J$, and consequently $I \subseteq J$.

The map $\mathcal{C\ell}(S) \to \mathfrak{I}_0(S)$, $[I] \mapsto -\min(I) + I$ is a monoid isomorphism which preserves the order $\preceq$, and consequently the posets $(\mathcal{C\ell}(S), \preceq)$ and $(\mathfrak{I}_0(S), \preceq)$ are isomorphic.

Let m be the multiplicity of S. If I is an ideal of S, and $x \in I$, then $x + ks \in I$ for every non-negative integer k. It follows that the set $\mathrm{Ap}(I) = \{x \in I : x - m \notin I\}$ generates I as an ideal, and it has precisely m elements, one per each congruence class modulo m. The set $\mathrm{Ap}(I)$ is known as the *Apéry set* of I (with respect to m). Thus, $\mathrm{Ap}(I) = \{w_0, w_1, \dots, w_{m-1}\}$, where $w_i = \min(I \cap (i + m\mathbb{N}))$.

As $I \subseteq \mathbb{N}$, we deduce that $\mathrm{Ap}(I) \subseteq \mathbb{N}$, and consequently for every $i \in \{0, \dots, m-1\}$, $w_i = mx_i + i$ for some non-negative integer x_i; $i = w_i \bmod m$, the remainder of the division of w_i by m. The tuple $(x_1, \dots, x_{m-1})$ is known as the *Kunz coordinates* of I (see [**4**, Section 4]). Notice that we are omitting x_0, which is equal to 0, as $0 = \min(I)$. In the sequel we will write $I = (x_1, \dots, x_{m-1})_{\mathcal{K}}$ to denote that $(x_1, \dots, x_{m-1})$ are the Kunz coordinates of I.

It is well known (see [**4**, Theorem 4.4]) that $(x_1, \dots, x_{m-1}) \in \mathbb{N}^{m-1}$ are the Kunz coordinates of an ideal I of a numerical semigroup $S = (k_1, \dots, k_{m-1})_{\mathcal{K}}$ if and only if $(x_1, \dots, x_{m-1})$ fulfills the following inequalities

$$\begin{cases} x_i \leq k_i, \text{for all } i \in \{1, \dots, m-1\}, \\ x_i + k_j + \lfloor \frac{i+j}{m} \rfloor \geq x_{(i+j) \bmod m}, \text{for all } i, j \in \{1, \dots, m-1\}, \end{cases} \tag{1}$$

where $\lfloor q \rfloor = \max\{z \in \mathbb{Z} : z \leq q\}$ for every $q \in \mathbb{Q}$.

Let $n \in \mathbb{N}$, and let $i = n \bmod m$. Then, $n = km + i$ for some $k \in \mathbb{N}$. We know that w_i is the minimum element in I congruent with i modulo m. Hence, $n \in I$ if and only if $n \geq w_i$, or equivalently $k \geq x_i$. In particular, $n \notin I$ if and only if $k \in \{0, \dots, x_i - 1\}$, and this holds for every congruence class modulo m. Thus, the number of non-negative integers not belonging to I is

$$|\mathbb{N} \setminus (x_1, \dots, x_{m-1})_{\mathcal{K}}| = x_1 + \cdots + x_{m-1}. \tag{2}$$

With this in mind it is easy to show that if J is an ideal with Kunz coordinates $(y_1, \dots, y_{m-1})_{\mathcal{K}}$, then

$$I \subseteq J \text{ if and only if } (y_1, \dots, y_{m-1}) \leq (x_1, \dots, x_{m-1}) \tag{3}$$

with respect to the usual partial order on $\mathbb{N}^{m-1}$, and

$$(x_1, \dots, x_{m-1})_{\mathcal{K}} \cap (y_1, \dots, y_{m-1})_{\mathcal{K}} = (\max(\{x_1, y_1\}), \dots, \max(\{x_{m-1}, y_{m-1}\}))_{\mathcal{K}},$$
$$(x_1, \dots, x_{m-1})_{\mathcal{K}} \cup (y_1, \dots, y_{m-1})_{\mathcal{K}} = (\min(\{x_1, y_1\}), \dots, \min(\{x_{m-1}, y_{m-1}\}))_{\mathcal{K}}.$$

Addition requires more effort, but can be derived by translating [**4**, Proposition 4.8] to Kunz coordinates. If $I+J=(z_1,\dots,z_{m-1})_{\mathcal{K}}$, then for every $i\in\{1,\dots,m-1\}$

$$(4)\quad z_i=\min(\{x_{i_1}+y_{i_2}+\lfloor\tfrac{i_1+i_2}{m}\rfloor : i_1,i_2\in\{0,\dots,m-1\}, i_1+i_2\equiv i \pmod{m}\}).$$

Let $I\in\mathfrak{I}_0(S)$. Then, $x\in I\setminus\{0\}$ is a minimal generator of I if and only if $I\setminus\{x\}\in\mathfrak{I}_0(S)$ (see [**7**, Lemma 9]).

The following result relates the divisibility poset of normalized ideals of a numerical semigroup S with that of $S\setminus\{a\}$ with a a minimal generator of S larger than the Frobenius number and the multiplicity of S (illustrated in Figure 1).

PROPOSITION 3. *Let S be a numerical semigroup with multiplicity m and Kunz coordinates $(k_1,\dots,k_{m-1})$. Let a be a minimal generator of S, and set $i=a \bmod m$. If $a>\max\{m,\mathrm{F}(S)\}$, then*

- $\mathfrak{I}_0(S)\subsetneq\mathfrak{I}_0(S\setminus\{a\})$;
- $\mathfrak{I}_0(S\setminus\{a\})\setminus\mathfrak{I}_0(S)$ *is the set of ideals of* $\mathfrak{I}_0(S\setminus\{a\})$ *not containing* a, *which is the set of ideals of* $\mathfrak{I}_0(S\setminus\{a\})$ *whose ith Kunz coordinate is* k_i+1.

Moreover, for every $I=(x_1,\dots,x_{i-1},k_i+1,x_{i+1},\dots,x_{m-1})_{\mathcal{K}}\in\mathfrak{I}_0(S\setminus\{a\})$, $I+S=I\cup\{a\}$, *that is,*

$$(x_1,\dots,x_{i-1},k_i+1,x_{i+1},\dots,x_{m-1})_{\mathcal{K}}+S=(x_1,\dots,x_{i-1},k_i,x_{i+1},\dots,x_{m-1})_{\mathcal{K}}.$$

In particular, $I+S$ *covers* I.

PROOF. First, observe that the multiplicity of S and $S'=S\setminus\{a\}$ is the same, and if $S'=(k'_1,\dots,k'_{m-1})_{\mathcal{K}}$, then $k_j=k'_j$ for $i\neq j$, and $k'_i=k_i+1$.

Notice that if $I\in\mathfrak{I}_0(S)$, then $I=I+0\subseteq I+S\setminus\{a\}\subseteq I+S=I$, and so $I\in\mathfrak{I}_0(S\setminus\{a\})$. Also, the ith Kunz coordinate of S' is k_i+1, which by (1) means that $S'\notin\mathfrak{I}_0(S)$. This proves that $\mathfrak{I}_0(S)\subsetneq\mathfrak{I}_0(S\setminus\{a\})$.

Next, we prove that $\mathfrak{I}_0(S\setminus\{a\})\setminus\mathfrak{I}_0(S)$ is the set of ideals of $\mathfrak{I}_0(S\setminus\{a\})$ not containing a. If $I=(x_1,\dots,x_{m-1})_{\mathcal{K}}$ is an ideal of $\mathfrak{I}_0(S\setminus\{a\})$ and $a\notin I$, then $x_i>k_i$ and by (1), I is not an ideal of $\mathfrak{I}_0(S)$. For the other inclusion, let $I\in\mathfrak{I}_0(S\setminus\{a\})\setminus\mathfrak{I}_0(S)$. If $a\in I$, then $S\subseteq I$ and $a+x\in S\subseteq I$ for all $x\in I$ (recall that $a>\mathrm{F}(S)$), which proves that $a+I\subseteq I$, and consequently $I+S=I+((S\setminus\{a\})\cup\{a\})\subseteq I$, yielding $I\in\mathfrak{I}_0(S)$, a contradiction. Thus, $a\notin I$.

Now, we show that the ideals $I=(x_1,\dots,x_{m-1})_{\mathcal{K}}$ of $\mathfrak{I}_0(S\setminus\{a\})$ not containing a are precisely those with $x_i=k_i+1$. If I does not contain a, then $x_i>k_i$ (because $a=k_im+i\notin I$), and as I is an ideal of $S\setminus\{a\}$, by (1), we have $x_i\leq k'_i=k_i+1$. This forces $x_i=k_i+1$. The converse also holds: if $x_i=k_i+1$, then I does not contain a.

The semigroup S is an ideal of $S\setminus\{a\}$ (it is an oversemigroup of $S\setminus\{a\}$, and thus it is an idempotent ideal of $S\setminus\{a\}$ by [**4**, Proposition 5.14]). Let $I=(x_1,\dots,x_{m-1})_{\mathcal{K}}\in\mathfrak{I}_0(S\setminus\{a\})$ with its ith coordinate equal to k_i+1. By (4), for $j\neq i$, the jth coordinate of $I+S$ is the minimum of $\{x_j,k'_j\}\cup\{x_l+k'_n : l+n\equiv j \pmod{m}, l+n<m\}\cup\{x_j+k'_n+1 : l+n\equiv j \pmod{m}, l+n>m\}$. Notice that by (1), this minimum is precisely x_j. The ith coordinate of $I+S$ is the minimum of $\{k_i+1,k_i\}\cup\{x_l+k'_n : l+n\equiv i \pmod{m}, l+n<m\}\cup\{x_l+k'_n+1 : l+n\equiv i \pmod{m}, l+n>m\}$. This minimum is k_i, and so $I+S=(x_1,\dots,x_{i-1},k_i,x_{i+1},\dots,x_{m-1})_{\mathcal{K}}$. It follows that $(I+S)\setminus I=\{k_im+i\}=\{a\}$. Thus, $I=(I+S)\setminus\{a\}$ and $I\cup\{a\}=I+S$.

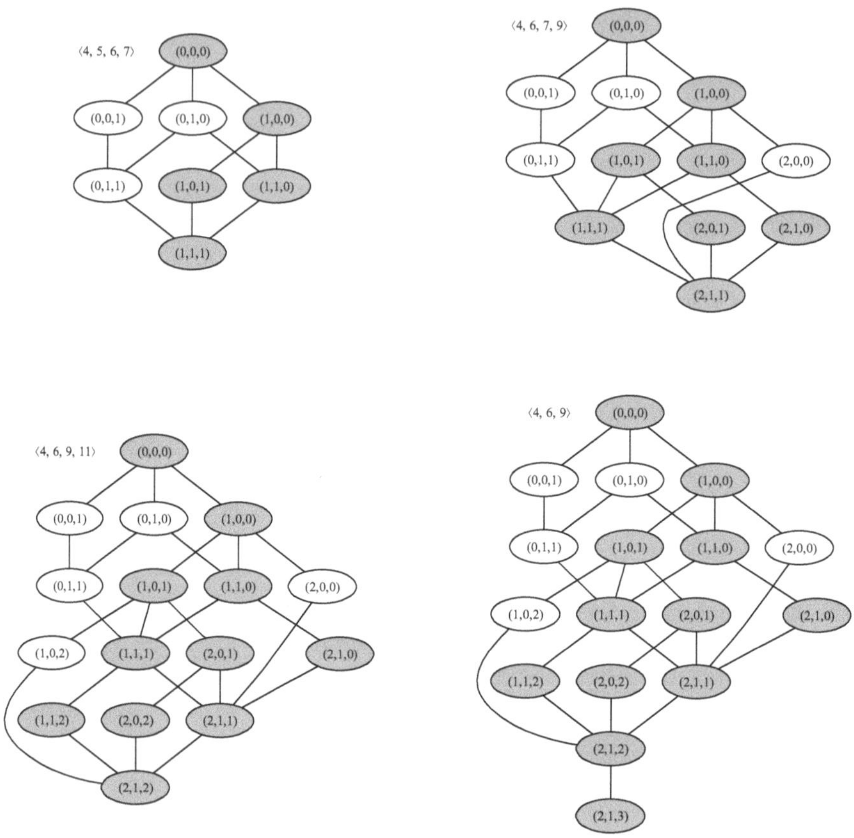

FIGURE 1. Evolution of $\mathfrak{I}_0(S)$ after removing a minimal generator.

As $I \cup \{a\} = (I + S)$, there cannot be another ideal J with $I + S \prec J \prec I$, because that would lead to $I + S \subsetneq J \subsetneq I$, and $|(I + S) \setminus I|$ would be greater than two, a contradiction. This proves that $I + S$ covers I. □

The following example shows that the condition $a > f$ in the previous result is necessary.

EXAMPLE 4. Let $S = \langle 3, 19, 23 \rangle$, and let $a = 19$. Then, $i = 1$. The set of Kunz coordinates of the ideals in $\mathfrak{I}_0(S) \subset \mathfrak{I}_0(S \setminus \{a\})$ is

$$\{(0,6), (1,7), (6,0), (6,1), (6,2), (6,3), (6,4), (6,5), (6,6), (6,7)\}.$$

The ideals not covered by Proposition 3 are $(0,6)_{\mathcal{K}}$ and $(1,7)_{\mathcal{K}}$.

REMARK 5. Let S and a be as in Proposition 3. Then, for every $I \in \mathfrak{I}_0(S \setminus \{a\})$, we have $I + S \in \mathfrak{I}_0(S)$. If $I \in \mathfrak{I}_0(S \setminus \{a\}) \setminus \mathfrak{I}_0(S)$, then by Proposition 3, $I + S \in \mathfrak{I}_0(S)$, and if $I \in \mathfrak{I}_0(S)$, then $I + S = I \in \mathfrak{I}_0(S)$.

REMARK 6. Observe that in Proposition 3, if I is an ideal of $S \setminus \{a\}$ not containing a, then $I = (I + S) \setminus \{a\}$ and $I + S \in \mathfrak{I}_0(S)$. Also, if J is another ideal of $S \setminus \{a\}$ with $I \preceq J$, then there exists $K \in \mathfrak{I}_0(S \setminus \{a\})$ such that $I + K = J$. Consequently, $I + S + K = I + S + K + S = J + S$. By Remark 5, $K + S \in \mathfrak{I}_0(S)$, and so $I + S \preceq J + S$ in $\mathfrak{I}_0(S)$. If, in addition, $a \in J$, then J is an ideal of S and $I + S \preceq J$.

Recall that a numerical semigroup S is called *ordinary* when its multiplicity m coincides with the conductor of S, that is, $S = \mathbb{N} \setminus \{1, 2, \ldots, m-1\}$. In the remaining part of the manuscript, we will denote by H_m the ordinary numerical semigroup with multiplicity m.

REMARK 7. By [**4**, Proposition 3.4], the set of ideals $\mathfrak{I}_0(S)$ of a numerical semigroup is in bijective correspondence with the set of antichains in the poset $(\mathrm{G}(S), \leq_S)$. For H_m, we have $\mathrm{G}(H_m) = \{1, 2, \ldots, m-1\}$ and therefore $|\mathfrak{I}_0(H_m)| = 2^{m-1}$. Moreover, if $I \in \mathfrak{I}_0(H_m)$ has Kunz coordinates $(x_1, \ldots, x_{m-1})_{\mathcal{K}}$, then $x_i \in \{0, 1\}$, for every $i \in \{1, \ldots, m-1\}$, and $x_i = 0$ if and only if $i \in I$. In other words, the poset $(\mathfrak{I}_0(H_m), \subseteq)$ is a Boolean algebra for every $m \in \mathbb{N}$ (see (3) and the formulas of intersection and union following it).

REMARK 8. The equation (4) expressing the Kunz coordinates of the sum of two ideals $I = (x_1, \ldots, x_{m-1})_{\mathcal{K}}$ and $J = (y_1, \ldots, y_{m-1})_{\mathcal{K}}$ can be significantly simplified when S is ordinary. Indeed, if $I + J = (z_1, \ldots, z_{m-1})_{\mathcal{K}}$, then for every $i \in \{1, \ldots, m-1\}$,

$$z_i = \min(\{x_{i_1} + y_{i_2} : i_1, i_2 \in \{0, \ldots, i\}, i_1 + i_2 = i\}). \tag{5}$$

Notice that if $i_1, i_2 \in \{0, \ldots, m-1\}$ are such that $i_1 + i_2 = i + m$, then $\lfloor \frac{i_1+i_2}{m} \rfloor = 1$, and consequently $x_i + y_0 + \lfloor \frac{i+0}{m} \rfloor = x_i \leq 1 = \lfloor \frac{i_1+i_2}{m} \rfloor \leq x_{i_1} + y_{i_2} + \lfloor \frac{i_1+i_2}{m} \rfloor$.

The following result extends, in some sense, the content of Proposition 3 to the divisibility posets $\mathfrak{I}_0(H_m)$ and $\mathfrak{I}_0(H_{m+1})$.

PROPOSITION 9. *For every $m \in \mathbb{N}$, the following facts hold.*

(1) $\mathfrak{I}_0(H_m) \subsetneq \mathfrak{I}_0(H_{m+1})$.
(2) *For every $I = (x_1, \ldots, x_m)_{\mathcal{K}} \in \mathfrak{I}_0(H_{m+1})$, $I \in \mathfrak{I}_0(H_m)$ if and only if $x_m = 0$. In particular, $I = (x_1, \ldots, x_m)_{\mathcal{K}} \in \mathfrak{I}_0(H_{m+1}) \setminus \mathfrak{I}_0(H_m)$ if and only if $x_m = 1$.*
(3) *For every $I \in \mathfrak{I}_0(H_{m+1}) \setminus \mathfrak{I}_0(H_m)$, $I + H_m \in \mathfrak{I}_0(H_m)$.*

PROOF. Let $I \in \mathfrak{I}_0(H_m)$. Then, $I = I + 0 \subseteq I + H_{m+1} \subseteq I + H_m = I$, which yields $I = I + H_{m+1}$ and consequently $I \in H_{m+1}$. Notice that $H_{m+1} + H_m = H_m \neq H_{m+1}$, and so $H_{m+1} \in \mathfrak{I}_0(H_{m+1}) \setminus \mathfrak{I}_0(H_m)$. This proves (1).

Now, take $I = (x_1, \ldots, x_m)_{\mathcal{K}} \in \mathfrak{I}_0(H_{m+1})$. We know by the paragraph after (1) that $m = 0 \times (m+1) + m \in I$ if and only if $x_m \leq 0$, or equivalently, $x_m = 0$.

If $I \in \mathfrak{I}_0(H_m)$, then $m \in I$, which forces $x_m = 0$. For the converse, suppose that $x_m = 0$. Then, $m \in I$, and so $H_m \subseteq I$, and this yields $I + H_m = I + (H_{m+1} \cup \{m\}) = I \cup (m + I) = I$. Hence, $I \in \mathfrak{I}_0(H_{m+1})$. This proves (2), since the second part of this assertion follows from the fact that $x_m \in \{0, 1\}$.

Let $I \in \mathfrak{I}_0(H_{m+1}) \setminus \mathfrak{I}_0(H_m)$ with $I = (x_1, \ldots, x_{m-1}, 1)_{\mathcal{K}}$ (in virtue of (2)). Clearly, H_m has Kunz coordinates $(1, \ldots, 1, 0)_{\mathcal{K}}$ as an element of $\mathfrak{I}_0(H_{m+1})$. In light of Remark 8, $I + H_m = (x_1, \ldots, x_{m-1}, 1)_{\mathcal{K}} + (1, \ldots, 1, 0)_{\mathcal{K}} = (x_1, \ldots, x_{m-1}, 0)_{\mathcal{K}}$. Hence, $I + H_m \in \mathfrak{I}_0(H_m)$ by (2). □

REMARK 10. Let m be a positive integer, and let $I = (x_1, \ldots, x_{m-1})_{\mathcal{K}} \in \mathfrak{I}_0(H_m)$. Then, using (3) and Remark 7, $I^c = (x'_1, \ldots, x'_{m-1})_{\mathcal{K}}$ is a cover of I with respect to inclusion if and only if there exists a unique $i \in \{1,\dots,m-1\}$ such that $x_i = 1$ and $x'_i = 0$ and $x'_j = x_j$, for all $j \in \{1,\dots m-1\} \setminus \{i\}$. This can be seen as a rephrasing of [**7**, Lemma 10] for the particular case of ordinary semigroups with the use of (2).

4. When the divisibility poset of normalized ideals is a lattice

In [**3**, Theorem 20], we proved that for S a numerical semigroup with multiplicity three, the poset $(\mathfrak{I}_0(S), \preceq)$ is a lattice, providing also an explicit description of its lattice operations.

In this section, we characterize numerical semigroups S fulfilling that $(\mathfrak{I}_0(S), \preceq)$ is a lattice. We prove that $(\mathfrak{I}_0(S), \preceq)$ is a lattice if and only if the multiplicity of S does not exceed four.

From the content of [**3**, Theorem 20], one might reasonably think that $I \cup J$ or $I + J$ could act as $I \vee J$, but this is not the case in general. There are ideals I and J for which $I \vee J$ exists, however it is neither $I \cup J$ nor $I + J$, as shown by the following.

EXAMPLE 11. Let $S = \langle 4, 9\rangle$. Then, $(\mathfrak{I}_0(S), \preceq)$ is a join semilattice. We also have that

- $(\{0,1,2\}+S) \vee (\{0,1,6,7\}+S) = (\{0,1,2\}+S) \cup (\{0,1,6,7\}+S) \neq (\{0,1,2\}+S) + (\{0,1,6,7\}+S)$,
- $(\{0,1,2\}+S) \vee (\{0,2,5\}+S) = (\{0,1,2\}+S) + (\{0,2,5\}+S) = \mathbb{N} \neq (\{0,1,2\}+S) \cup (\{0,2,5\}+S)$,
- $(\{0,1,2\}+S) \vee (\{0,1,6\}+S) = \{0,1,2,7\}+S$, which is not equal to $(\{0,1,2\}+S) + (\{0,1,6\}+S)$ nor to $(\{0,1,2\}+S) \cup (\{0,1,6\}+S)$.

The following example will be central to prove that for any numerical semigroup S with multiplicity greater than four $(\mathfrak{I}_0(S), \preceq)$ is not a lattice.

EXAMPLE 12. The set of minimal upper bounds for $\{\{0,1\}+H_5, \{0,1,3\}+H_5\}$ is $\{\{0,1,2,3\}+H_5, \{0,1,3,4\}+H_5\}$, and so there is no supremum for the pair of elements $\{0,1\}+H_5$ and $\{0,1,3\}+H_5$ in $\mathfrak{I}_0(H_5)$.

Let S be a numerical semigroup, and let T be an oversemigroup of S, that is, an idempotent element in $\mathfrak{I}_0(S)$ [**4**, Proposition 5.14]. Define

$$C_T = \{I \in \mathfrak{I}_0(S) : I + T = I\} \text{ and } \uparrow\! T = \{I \in \mathfrak{I}_0(S) : T \preceq I\}.$$

Recall that $\uparrow\! T = C_T = \mathfrak{I}_0(T)$ [**3**, Lemma 7].

PROPOSITION 13. *Let S be a numerical semigroup with multiplicity greater than or equal to five. Then, $(\mathfrak{I}_0(S), \preceq)$ is not a join semilattice nor a meet semilattice.*

PROOF. Since $(\mathfrak{I}_0(S), \preceq)$ is a finite poset, by [**9**, Theorem 2.4], it suffices to show that $(\mathfrak{I}_0(S), \preceq)$ is not a meet semilattice.

As $\mathrm{m}(S) \geq 5$, we have that H_5 is an oversemigroup of S. The poset $(\mathfrak{I}_0(H_5), \preceq)$ is not a join semilattice (see Example 12) and, in particular, there is no supremum for the elements $\{0,1\}+H_5$ and $\{0,1,3\}+H_5$; since these ideals are in $\uparrow\! H_5 = \mathfrak{I}_0(H_5)$, its supremum in $\mathfrak{I}_0(S)$ should also be in $\uparrow\! H_5$. □

A sort of converse to Remark 6 is covered in the following technical lemma.

LEMMA 14. *Let S be a numerical semigroup with multiplicity m and let a be a minimal generator of S with $m \neq a > \mathrm{F}(S)$. Let I and J be two ideals of S such that a is a minimal generator of I and of J as ideals of $S \setminus \{a\}$. If $I+K=J$ for some ideal K of S, then a is a minimal generator of K (as an ideal of $S \setminus \{a\}$) and $(I \setminus \{a\})+(K \setminus \{a\}) = J \setminus \{a\}$.*

PROOF. Let $i = a \bmod m$. If $(k_1, \ldots, k_{m-1})$ are the Kunz coordinates of S, then $a = k_i m + i$.

If a is not a minimal generator of K as an ideal of $S \setminus \{a\}$, then $a = x+s$ for some $x \in K$ and $s \in S \setminus \{a\}$. But then $x \in J$ (as $K \preceq J$, we have $K \subseteq J$), which contradicts the fact that a is a minimal generator of J.

As a is a minimal generator of I, J and K, by [**7**, Lemma 9], we deduce that $I \setminus \{a\}, J \setminus \{a\}, K \setminus \{a\} \in \mathfrak{I}_0(S \setminus \{a\})$. Notice that $I \setminus \{a\}, J \setminus \{a\}, K \setminus \{a\} \notin \mathfrak{I}_0(S)$. For instance, $(I \setminus \{a\})+S \neq I \setminus \{a\}$ because $a = 0+a \in (I \setminus \{a\})+S$ and $a \notin I \setminus \{a\}$.

By Proposition 3, if $I \setminus \{a\} = (x_1, \ldots, x_{m-1})_{\mathcal{K}}$, $K \setminus \{a\} = (y_1, \ldots, y_{m-1})_{\mathcal{K}}$, and $J \setminus \{a\} = (z_1, \ldots, z_{m-1})_{\mathcal{K}}$, then $x_i = y_i = z_i = k_i + 1$. If $I = (x_1', \ldots, x_{m-1}')_{\mathcal{K}}$, $K = (y_1', \ldots, y_{m-1}')_{\mathcal{K}}$ and $J = (z_1', \ldots, z_{m-1}')_{\mathcal{K}}$ (seen as elements of $\mathfrak{I}_0(S)$), then $(x_1', \ldots, x_{m-1}') = (x_1, \ldots, x_{m-1}) - \mathbf{e}_i$, $(y_1', \ldots, y_{m-1}') = (y_1, \ldots, y_{m-1}) - \mathbf{e}_i$, and $(z_1', \ldots, z_{m-1}') = (z_1, \ldots, z_{m-1}) - \mathbf{e}_i$, where $\mathbf{e}_i$ is the ith row of the $(m-1) \times (m-1)$ identity matrix. Also, $(I \setminus \{a\})+S = I$, $(J \setminus \{a\})+S = J$, and $(K \setminus \{a\})+S = K$.

By hypothesis,

$$I+K = (x_1', \ldots, x_{m-1}')_{\mathcal{K}} + (y_1', \ldots, y_{m-1}')_{\mathcal{K}} = (z_1', \ldots, z_{m-1}')_{\mathcal{K}} = J.$$

By (4), $z_l' = \min(\{x_{l_1}' + y_{l_2}' + \lfloor \frac{l_1+l_2}{m} \rfloor : l_1, l_2 \in \{0, \ldots, m-1\}, l_1+l_2 \equiv l \pmod{m}\})$. If we want to prove that

$$\begin{aligned}(I \setminus \{a\})+(K \setminus \{a\}) &= (x_1, \ldots, x_{m-1})_{\mathcal{K}} + (y_1, \ldots, y_{m-1})_{\mathcal{K}} \\ &= (z_1, \ldots, z_{m-1})_{\mathcal{K}} \\ &= J \setminus \{a\},\end{aligned}$$

then we have to show that for every $l \in \{1, \ldots, m-1\}$,

$$z_l = \min\left(\left\{x_{l_1} + y_{l_2} + \left\lfloor \frac{l_1+l_2}{m} \right\rfloor : l_1, l_2 \in \{0, \ldots, m-1\}, l_1+l_2 \equiv l \pmod{m}\right\}\right).$$

First, observe that since $a \notin J \setminus \{a\}$, $a \neq u+v$ for any $u \in I \setminus \{a\}$ and $v \in K \setminus \{a\}$, and consequently $a \notin \mathrm{Ap}(I \setminus \{a\}) + \mathrm{Ap}(K \setminus \{a\})$. Hence, $x_{i_1} + y_{i_2} + \lfloor \frac{i_1+i_2}{m} \rfloor > k_i$ for all $i_1, i_2 \in \{1, \ldots, m-1\}$ such that $i_1+i_2 \equiv i \pmod{m}$. This, in particular, implies that $x_{i_1} + y_{i_2} + \lfloor \frac{i_1+i_2}{m} \rfloor \geq k_i + 1$ for all $i_1, i_2 \in \{1, \ldots, m-1\}$ such that $i_1+i_2 \equiv i \pmod{m}$. Therefore, $z_i = k_i + 1 = x_i + y_0 = x_0 + y_i = x_i = y_i = \min(\{x_{i_1} + y_{i_2} + \lfloor \frac{i_1+i_2}{m} \rfloor : i_1, i_2 \in \{0, \ldots, m-1\}, i_1+i_2 \equiv i \pmod{m}\})$.

Now, take $l \in \{1, \ldots, m-1\} \setminus \{i\}$. Then,

$$z_l' = \min(\{x_{l_1}' + y_{l_2}' + \lfloor \frac{l_1+l_2}{m} \rfloor : l_1, l_2 \in \{0, \ldots, m-1\}, l_1+l_2 \equiv l \pmod{m}\}).$$

It follows that

$$\begin{aligned} z_l' &= \min(\{x_{l_1}' + y_{l_2}' + \lfloor \tfrac{l_1+l_2}{m} \rfloor : l_1, l_2 \in \{0, \ldots, m-1\} \setminus \{i\}, l_1+l_2 \equiv l \pmod{m}\}) \\ &= \min(\{x_{l_1} + y_{l_2} + \lfloor \tfrac{l_1+l_2}{m} \rfloor : l_1, l_2 \in \{0, \ldots, m-1\} \setminus \{i\}, l_1+l_2 \equiv l \pmod{m}\}) \\ &= \min(\{x_{l_1} + y_{l_2} + \lfloor \tfrac{l_1+l_2}{m} \rfloor : l_1, l_2 \in \{0, \ldots, m-1\}, l_1+l_2 \equiv l \pmod{m}\}) \\ &= z_l. \end{aligned}$$

The first equality follows by applying (1) to I and K:

- if $l_1 = i$, then $x'_{l_1} + y'_{l_2} = k_i + y'_{l_2} + \lfloor \frac{l_1+l_2}{m} \rfloor \geq y'_{(i+l_2) \bmod m} = x'_0 + y'_l$;
- if $l_2 = i$, then $x'_{l_1} + y'_{l_2} = x'_{l_1} + k_i + \lfloor \frac{l_1+l_2}{m} \rfloor \geq x'_{(i+l_2) \bmod m} = x'_l + y'_0$.

While the third equality holds by applying (1) to $I \setminus \{a\}$ and $K \setminus \{a\}$:

- if $l_1 = i$, then $x_{l_1} + y_{l_2} = (k_i + 1) + y_{l_2} + \lfloor \frac{l_1+l_2}{m} \rfloor \geq y_{(i+l_2) \bmod m} = x_0 + y_l$;
- if $l_2 = i$, then $x_{l_1} + y_{l_2} = x_{l_1} + (k_i + 1) + \lfloor \frac{l_1+l_2}{m} \rfloor \geq x_{(i+l_2) \bmod m} = x_l + y_0$.

This proves that $I \setminus \{a\} + K \setminus \{a\} = J \setminus \{a\}$. □

Notice that the previous result is false if a is chosen to be the multiplicity of S, as shown by the following example, inspired by the Hasse diagram of H_4 drawn in Figure 1.

EXAMPLE 15. Let $S = \langle 3, 4, 5 \rangle$ and let $a = 3 = \mathrm{m}(S)$. Set $T = S \setminus \{3\} = \langle 4, 5, 6, 7 \rangle$.

Let $I = \{0, 2\} + S = \{0, 2, 3\} + T$ and $J = \{0, 1, 2\} + S = \{0, 1, 2, 3\} + T = \mathbb{N}$. Then, 3 is a minimal generator of I and J as ideals of T. Take $K = \{0, 1\} + S = \{0, 1, 3\} + T$. Then, $I + K = J$, and so $I \preceq J$. However, $I \setminus \{3\} = \{0, 2\} + T$, $J \setminus \{3\} = \{0, 1, 2\} + T$, $K \setminus \{3\} = \{0, 1\} + T$, and $(I \setminus \{3\}) + (K \setminus \{3\}) \neq (J \setminus \{3\})$. As a matter of fact, $I \setminus \{3\} \npreceq J \setminus \{3\}$.

Notice that $I = (1, 0, 0)_{\mathcal{K}} = (1, 0)_{\mathcal{K}}$, $I \setminus \{3\} = (1, 0, 1)_{\mathcal{K}}$, $J = (0, 0, 0)_{\mathcal{K}} = (0, 0)_{\mathcal{K}}$, and $J \setminus \{3\} = (0, 0, 1)_{\mathcal{K}}$.

We now have all the ingredients to prove that the lattice property on $\mathfrak{I}_0(S)$ is preserved once we remove a minimal generator of S different from its multiplicity and larger than its Frobenius number.

THEOREM 16. *Let S be a numerical semigroup such that $(\mathfrak{I}_0(S), \preceq)$ is a lattice, and let a be a minimal generator of S, with $m \neq a > \mathrm{F}(S)$. Then, $(\mathfrak{I}_0(S \setminus \{a\}), \preceq)$ is a lattice as well.*

PROOF. Let $I, J \in \mathfrak{I}_0(S \setminus \{a\})$. We distinguish several cases, depending on whether a belongs or not to I and J.

If $a \in I \cap J$, then by Proposition 3, I and J are in $\mathfrak{I}_0(S)$, and so $I \vee J$ exists, because all upper bounds of I and J are in $\uparrow S = \mathfrak{I}_0(S)$, and $(\mathfrak{I}_0(S), \preceq)$ is a lattice by hypothesis.

Suppose that $a \in I$ and $a \notin J$. Let $K \in \mathfrak{I}_0(S \setminus \{a\})$ be such that $I \preceq K$ and $J \preceq K$. Hence, $I \subseteq K$ and consequently $a \in K$. By Proposition 3, $I, K \in \mathfrak{I}_0(S)$ and $J + S \preceq K + S = K$. As $J + S \in \mathfrak{I}_0(S)$ (by Proposition 3) and $(\mathfrak{I}_0(S), \preceq)$ is a lattice, $I \vee (J + S) \preceq K$. This proves that $I \vee J = I \vee (J + S)$.

The case $a \notin I$ and $a \in J$ is similar: $I \vee J = (I + S) \vee J$.

Now, suppose that $a \notin I \cup J$. If every $K \in \mathfrak{I}_0(S \setminus \{a\})$ with $I \preceq K$ and $J \preceq K$ is in $\mathfrak{I}_0(S)$, then $I + S \preceq K + S = K$ and $J + S \preceq K + S = K$. By using that $(\mathfrak{I}_0(S), \preceq)$ is a lattice, we deduce that $(I + S) \vee (J + S) \preceq K$, and so $I \vee J = (I + S) \vee (J + S)$.

Thus, the remaining case is when there is $K \in \mathfrak{I}_0(S \setminus \{a\}) \setminus \mathfrak{I}_0(S)$ ($a \notin K$ in virtue of Proposition 3), such that $I \preceq K$ and $J \preceq K$. Notice that in this setting a is a minimal generator of $K + S = K \cup \{a\}$ as an ideal of $S \setminus \{a\}$.

As $(\mathfrak{I}_0(S) \preceq)$ is a lattice and both $I + S$ and $J + S$ are ideals of S, there exists $M = (I + S) \vee (J + S)$, which is an ideal of S. Since $I \preceq K$ and $J \preceq K$, we deduce that $I + S \preceq K + S$ and $J + S \preceq K + S$; therefore $M \preceq K + S = K \cup \{a\}$.

Notice that $a \in M$ (M contains $I + S = I \cup \{a\}$). Suppose that a is not a minimal generator of M (as an ideal of $S \setminus \{a\}$). Then, there exists $g \in M$ and $s \in S \setminus \{a\}$ with $s \neq 0$ and $a = g + s$. But then $g \in K$, and so $a = g + s \in K$, a contradiction. Therefore, a is a minimal generator of M, and consequently the set $M \setminus \{a\}$ is an ideal of $S \setminus \{a\}$. Now, we can apply Lemma 14 to $I + S \preceq M$ and $J + S \preceq M$, to derive that $I \preceq M \setminus \{a\}$ and $J \preceq M \setminus \{a\}$; also, from $M \preceq K + S$ to deduce that $M \setminus \{a\} \preceq K$. This proves that $M \setminus \{a\} \preceq K$, for every upper bound of I and J (with respect to $\preceq$) not containing a. If K is an upper bound of I and J containing a, then $I + S \preceq K + S = K$ and $J + S \preceq K + S = K$, and so $M \preceq K$. Since $(M \setminus \{a\}) + S = M$, we deduce that $M \setminus \{a\} \preceq M \preceq K$. This proves that $M \setminus \{a\} = I \vee J$. □

COROLLARY 17. *Let S be a numerical semigroup with multiplicity four. Then, $(\mathfrak{I}_0(S), \preceq)$ is a lattice.*

PROOF. Let H_4 be the ordinary numerical semigroup of multiplicity 4 (we are using the same notation introduced in Section 3). The Hasse diagram of $(\mathfrak{I}_0(H_4), \preceq)$ is shown in Figure 1 (first diagram on the left; Kunz coordinates are used to display the vertices of the diagram). Clearly, $(\mathfrak{I}_0(H_4), \preceq)$ is a lattice. If $S = H_4$, then we are done. Otherwise, S is strictly contained in H_4. Let $S = S_1 \subset \cdots \subset S_k = H_4$ be the sequence of numerical semigroups obtained by applying [**1**, Lemma 6]. Then, for every $i \in \{1, \ldots, k-1\}$, $S_{i+1} = S_i \cup \{g_i\}$, with $g_i = \max(H_4 \setminus S_i) = \mathrm{F}(S_i)$, and $S_i = S_{i+1} \setminus \{g_i\}$. Observe that g_i is a minimal generator of S_{i+1} greater than four; also $g_i = \max(H_4 \setminus S_i) > \max(H_4 \setminus S_{i+1}) = \mathrm{F}(S_{i+1}) = g_{i+1}$. Thus, we can apply Theorem 16 to obtain that $(\mathfrak{I}_0(S_i), \preceq)$ is a lattice for all i. □

If S has multiplicity two, then [**4**, Example 5.9] ensures that $(\mathfrak{I}_0(S), \preceq)$ is a totally ordered set, and thus a lattice. Also, recall that [**3**, Theorem 20] states that if S is a numerical semigroup with multiplicity three, then $(\mathfrak{I}_0(S), \preceq)$ is a lattice. Putting all this together, we obtain the following consequence.

COROLLARY 18. *Let S be a numerical semigroup. Then, $(\mathfrak{I}_0(S), \preceq)$ is a lattice if and only if $\mathrm{m}(S) \leq 4$.*

The following shows that, in general, the lattice $(\mathfrak{I}_0(S), \preceq)$ is not distributive (for both multiplicity three and four).

REMARK 19. It can be shown that the set

$$\{H_4, \{0,3\} + H_4, \{0,2,3\} + H_4, \{0,1,2\} + H_4, \mathbb{N}\}$$

is a sublattice of $\mathfrak{I}_0(H_4)$ isomorphic to a a pentagon, and thus $\mathfrak{I}_0(H_4)$ cannot be a distributive lattice. Every numerical semigroup T with multiplicity larger than three will have H_4 as an oversemigroup. Thus, if $\mathfrak{I}_0(T)$ is a lattice, it will contain the sublattice $\mathfrak{I}_0(T)$, and thus $\mathfrak{I}_0(T)$ will not be a distributive lattice.

REMARK 20. Let S be a numerical semigroup with multiplicity three in which $\preceq$ does not equal $\subseteq$. Then, by [**3**, Proposition 12], $\{4,5\} \cap S = \emptyset$. Thus, $T = \langle 3,7,8 \rangle = \{0,3\} \cup (6+\mathbb{N})$ is an oversemigroup of S. Let $D = \{T, \{0,4\}+T, \{0,4,5\}+T, \{0,1,5\}+T, \{0,1\}+T\}$ (see Figure 2). Then, D is a sublattice of $\mathfrak{I}_0(T)$ isomorphic to a pentagon, and consequently neither $\mathfrak{I}_0(T)$ nor $\mathfrak{I}_0(S)$ are distributive.

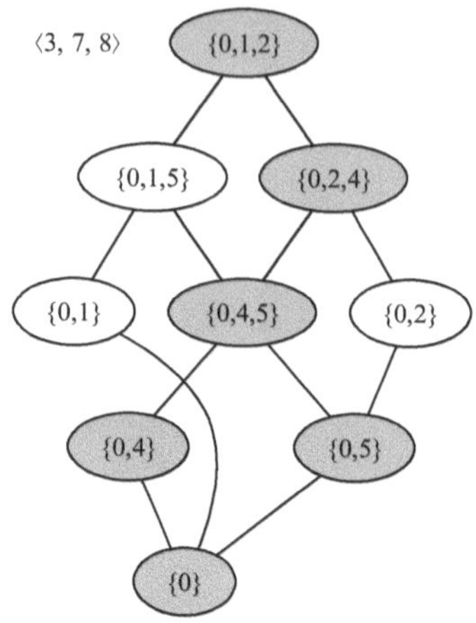

FIGURE 2. The non-distributive lattice $(\mathcal{I}_0(T), \preceq)$, for $T = \langle 3, 7, 8\rangle$.

5. Irreducibility

Let S be a numerical semigroup. As $\mathfrak{I}_0(S)$ has finitely many elements, if $(\mathfrak{I}_0(S), \preceq)$ is lattice (by Corollary 18 this is the case if and only if $m(S) \leq 4$), then any of its elements is a join of $\vee$-irreducible elements, and also a meet of $\wedge$-irreducible elements.

Observe that the containment poset of $\mathfrak{I}_0(S)$, $(\mathfrak{I}_0(S), \subseteq)$, is a lattice where meet equals intersection of ideals and join is the union of ideals. Thus, it also makes sense to think about $\cap$- and $\cup$-irreducible ideals.

An ideal $I \in \mathfrak{I}_0(S) \setminus \{S\}$ is *+-irreducible* if $I = J + K$ implies $I = J$ or $I = K$. In [**4**], these ideals are called irreducible. We use the + prefix to distinguish them from $\vee$-irreducibles, $\cap$-irreducibles, $\cup$-irreducibles and $\wedge$-irreducibles.

Define $\mathfrak{P}_0(S) = \{\{0, g\} + S : g \text{ is a gap of } S\}$. In light of [**2**, Proposition 3.1], we know that if I is in $\mathfrak{P}_0(S)$, then I is +-irreducible. The converse does not hold (see for instance [**2**, Example 3.7] or [**4**, Example 5.3]).

REMARK 21. Let S be a numerical semigroup. By Proposition 1, I is $\cup$-irreducible if and only if I is the cover of at most one ideal in $\mathfrak{I}_0(S)$. By [**7**, Lemma 11], the number of ideals covered by I equals the number of non-zero minimal generators of I. Thus, I is $\cup$-irreducible if and only if $I \in \mathfrak{P}_0(S) \cup \{S\}$.

Next, we see that, in our context, +-irreducible implies $\vee$-irreducible.

LEMMA 22. *Let S be a numerical semigroup with $m(S) \leq 4$, and let $I \in \mathfrak{I}_0(S)$. If I is +-irreducible, then it is $\vee$-irreducible.*

PROOF. Suppose that $I \in \mathfrak{I}_0(S)$ is +-irreducible and, to the contrary, that $I = J \vee K$ with $J, K \in \mathfrak{I}_0(S) \setminus \{I\}$ (observe that $J \vee K$ exists as $(\mathfrak{I}_0(S), \preceq)$ is a lattice by Corollary 18). As $J \preceq J \vee K = I$ and $K \preceq J \vee K = I$, there exists $N, M \in \mathfrak{I}_0(S)$ such that $I = J + N = K + M$. By taking into account that I is +-irreducible and $J \neq I \neq K$, we deduce that $I = N = M$, and consequently

$I = J + I = K + I$. Thus, $J + K + I = J + I = I$, and so $J + K \preceq I$. Also, $I = J \vee K \preceq J + K$, since $J + K$ is an upper bound for $\{J, K\}$ with respect to $\preceq$, and thus $J + K \preceq I \preceq J + K$, yielding $I = J + K$, contradicting that I is $+$-irreducible. □

Hence, for every $I \in \mathfrak{I}_0(S)$, $I \in \mathfrak{P}_0(S)$ implies I is $+$-irreducible, which in turn forces I to be $\vee$-irreducible (for $m(S) \leq 4$). As we mentioned above, the converse of the first implication does not hold. The following example illustrates that $\vee$-irreducibility does not imply $+$-irreducibility, and thus none of these implications can be reversed in general.

EXAMPLE 23. There are ideals that are $\vee$-irreducible but are not $+$-irreducible. Consider, for instance, $S = \langle 4, 7, 9 \rangle$, and let $I = \{0, 1, 2\} + S$. Then, $I = (\{0, 1\} + S) + (\{0, 1\} + S)$ is $\vee$-irreducible.

For multiplicity three, the above implications can be reversed.

PROPOSITION 24. *Let S be a numerical semigroup with multiplicity three and $I \in \mathfrak{I}_0(S) \setminus S$. Then, the following are equivalent:*

(1) *I is $+$-irreducible.*
(2) *I is $\vee$-irreducible;*
(3) *I is $\cup$-irreducible;*
(4) *$I \in \mathfrak{P}_0(S)$.*

PROOF. We already know that the first assertion implies the second (Lemma 22), the fourth implies the first (by [**2**, Proposition 3.1]), and the third is equivalent to the fourth (see Remark 21). Thus, it suffices to prove that the second statement implies the fourth.

Suppose that $I \notin \mathfrak{P}_0(S)$. Since S has multiplicity three, $I = \{0, g_1, g_2\} + S$, for some g_1, g_2 gaps of S, such that neither $g_2 - g_1 \in S$ nor $g_1 - g_2 \in S$ (otherwise, $I = \{0, g_1\} + S$ or $I = \{0, g_2\} + S$, leading to $I \in \mathfrak{P}_0(S)$). Consider $J = \{0, g_1\} + S$ and $K = \{0, g_2\} + S$. By [**7**, Lemma 8] and using the fact that $\{g_2 - g_1, g_2 - g_1\} \cap S = \emptyset$, we deduce that $J \not\subseteq K$ and $K \not\subseteq J$. Therefore, J and K are incomparable elements in the lattice $(\mathfrak{I}_0(S), \preceq)$ and so, by [**3**, Theorem 20], $J \vee K = J + K = \{0, g_1, g_2, g_1 + g_2\} + S$. Notice that $\{0 \bmod 3, g_1 \bmod 3, g_2 \bmod 3\} = \{0, 1, 2\}$, and so $g_1 + g_2$ is either congruent with 0, g_1 or g_2 modulo three. This implies that $\{0, g_1, g_2, g_1 + g_2\} + S = \{0, g_1, g_2\} + S = I$, showing that $I = J + K$ and thus I is not $\vee$-irreducible. □

From the proof of this last result we derive the following consequence.

COROLLARY 25. *Let S be a numerical semigroup with multiplicity three and let $I \in \mathfrak{I}_0(S)$. Then, I is either $+$-irreducible or there exist two $+$-irreducible ideals J and K such that $I = J \vee K$.*

Let S be a numerical semigroup and let $I \in \mathfrak{I}_0(S)$. The *Frobenius number* of I can be defined, exactly as for numerical semigroups, as $\mathrm{F}(I) = \max(\mathbb{Z} \setminus I)$.

LEMMA 26. *Let S be a numerical semigroup and let $I \in \mathfrak{I}_0(S) \setminus \mathbb{N}$. Then, $\bar{I} = I \cup \{\mathrm{F}(I)\} \in \mathfrak{I}_0(S)$.*

PROOF. We start by proving that $\bar{I} \in \mathfrak{I}_0(S)$, that is, $\bar{I} + S = \bar{I}$ and $0 \in \bar{I}$. The second condition follows easily since $0 \in I \subset \bar{I}$. As $I + S = I$, in order to prove that $\bar{I}+S=\bar{I}$, it suffices to show that $\mathrm{F}(I)+S\subseteq\bar{I}$. Clearly, $\mathrm{F}(I)+0=\mathrm{F}(I)\in\bar{I}$, and for every $s\in S\setminus\{0\}$, $\mathrm{F}(I)+s>\mathrm{F}(I)$, and thus $\mathrm{F}(I)+s\in I\subseteq\bar{I}$. □

LEMMA 27. *Let S be a numerical semigroup and let $I\in\mathfrak{I}_0(S)$ with a unique cover with respect to inclusion. For every $J\in\mathfrak{I}_0(S)\setminus\{I\}$, if $I\preceq J$, then $\bar{I}\preceq J$, where $\bar{I}=I\cup\{\mathrm{F}(I)\}$.*

PROOF. Notice that as I has a unique cover (with respect to inclusion) and $\bar{I}=I\cup\{\mathrm{F}(I)\}$ covers I, we deduce that $\bar{I}$ is precisely the unique cover of I.

Suppose that $I\prec J$ for some $J\in\mathfrak{I}_0(S)$. As $I\subseteq J$, and $I\neq J$, $\bar{I}\subseteq J$, and in particular $\mathrm{F}(I)\in J$. From $I\preceq J$, we deduce that there exists $K\in\mathfrak{I}_0(S)$ such that $I+K=J$. Let us show that $\bar{I}+K=J$. Clearly, $J=I+K\subseteq\bar{I}+K$. For the other inclusion, take $k\in K$ and $i\in\bar{I}$. If $i\in I$, then $i+k\in I+K=J$. If $i=\mathrm{F}(I)$ and $k=0$, then $i+k=\mathrm{F}(I)$, which we already know belongs to J. Finally, for $i=\mathrm{F}(I)$ and $k\in K\setminus\{0\}$, we get that $i+k>\mathrm{F}(I)$, and so $i+k\in I\subseteq J$. □

PROPOSITION 28. *Let S be a numerical semigroup such that $(\mathfrak{I}_0(S),\preceq)$ is a lattice, and let $I\in\mathfrak{I}_0(S)$. If I is $\cap$-irreducible, then I is $\wedge$-irreducible.*

PROOF. As I is $\cap$-irreducible, we know by Proposition 1 that I has a unique cover, which by Lemma 26 must be $\bar{I}=I\cup\{\mathrm{F}(I)\}$. Suppose that $I=J\wedge K$ with $J,K\in\mathfrak{I}_0(S)$. Then, $I\preceq J$ and $I\preceq K$, and by Lemma 27 we deduce that $\bar{I}\preceq J$ and $\bar{I}\preceq K$. But then $\bar{I}\preceq J\wedge K=I$, yielding $\bar{I}\subseteq I$, a contradiction. □

The other implication is not true as the following example shows.

EXAMPLE 29. Let $S=\langle 3,7\rangle$. Then, $I=\{0,4,8\}+S=(\{0,1\}+S)\cap(\{0,2,4\}+S)$ is $\wedge$-irreducible and it is not $\cap$-irreducible.

Acknowledgments

The research was carried out thanks to the hospitality offered by the Institute of Mathematics (IMAG) of the University of Granada and the staff exchange program Erasmus+ funded by the European Community.

The authors would like to thank the anonymous referee for their suggestions and comments.

S. Bonzio acknowledges the support by the Italian Ministry of Education, University and Research through the PRIN 2022 project DeKLA ("Developing Kleene Logics and their Applications", project code: 2022SM4XC8) and the PRIN Pnrr project "Quantum Models for Logic, Computation and Natural Processes (Qm4Np)" (cod. P2022A52CR). He also acknowledges the Fondazione di Sardegna for the support received by the project MAPS (grant number F73C23001550007), the University of Cagliari for the support by the StartUp project "GraphNet". Finally, he gratefully acknowledges also the support of the INDAM GNSAGA (Gruppo Nazionale per le Strutture Algebriche, Geometriche e loro Applicazioni).

P. A. García-Sánchez is partially supported by the grant number ProyExcel_00868 (Proyecto de Excelencia de la Junta de Andalucía) and by the Junta de Andalucía Grant Number FQM–343. He also acknowledges financial support from the grant PID2022-138906NB-C21 funded by MICIU/AEI/10.13039/501100011033 and by ERDF "A way of making Europe", and from the Spanish Ministry of Science and Innovation (MICINN), through the "Severo Ochoa and María de Maeztu Programme for Centres and Unities of Excellence" (CEX2020-001105-M).

References

[1] A. Assi, M. D'Anna, and P. A. García-Sánchez, *Numerical semigroups and applications*, RSME Springer Series, vol. 3, Springer, Cham, [2020] ©2020. Second edition [of 3558713], DOI 10.1007/978-3-030-54943-5. MR4230109

[2] V. Barucci and F. Khouja, *On the class semigroup of a numerical semigroup*, Semigroup Forum **92** (2016), no. 2, 377–392, DOI 10.1007/s00233-014-9679-8. MR3472022

[3] S. Bonzio and P. A. García-Sánchez, *The poset of normalized ideals of numerical semigroups with multiplicity three*, Comm. Algebra **53** (2025), no. 9, 3610–3624, DOI 10.1080/00927872.2025.2464956. MR4924514

[4] L. Casabella, M. D'Anna, and P. A. García-Sánchez, *Apéry sets and the ideal class monoid of a numerical semigroup*, Mediterr. J. Math. **21** (2024), no. 1, Paper No. 7, 28, DOI 10.1007/s00009-023-02550-8. MR4670095

[5] M. Delgado, P. A. García-Sánchez, and J. Morais, NumericalSgps, A package for numerical semigroups, Version 1.3.1 dev (2023), Refereed GAP package, `https://gap-packages.github.io/numericalsgps`.

[6] The GAP-Group, GAP – Groups, Algorithms, and Programming, Version 4.12.2, (2022), `https://www.gap-system.org`.

[7] P. A. García-Sánchez, *The isomorphism problem for ideal class monoids of numerical semigroups*, Semigroup Forum **108** (2024), no. 2, 365–376, DOI 10.1007/s00233-024-10429-7. MR4740715

[8] S. Li, *Counting numerical semigroups by Frobenius number, multiplicity, and depth*, Comb. Theory **3** (2023), no. 3, Paper No. 6, 29. MR4683613

[9] J. N. Nation, *Notes on Lattice Theory*, Lecture Notes, 2017 (avaliable at `https://math.hawaii.edu/~jb/math618/Nation-LatticeTheory.pdf`).

[10] J. C. Rosales and P. A. García-Sánchez, *Numerical semigroups*, Developments in Mathematics, vol. 20, Springer, New York, 2009, DOI 10.1007/978-1-4419-0160-6. MR2549780

Departament of Mathematics and Computer Science, University of Cagliari, via Ospedale 72, 09124 Cagliari, Italy

Email address: `stefano.bonzio@unica.it`

Departamento de Álgebra and IMAG, Universidad de Granada, E-18071 Granada, España

Email address: `pedro@ugr.es`

Contemporary Mathematics
Volume **836**, 2026
https://doi.org/10.1090/conm/836/16705

On numerical semigroup elements and the ℓ_0 and ℓ_∞ norms of their factorizations

Sogol Cyrusian, Alex Domat, Christopher O'Neill, Vadim Ponomarenko, Eric Ren, and Mayla Ward

Abstract. A numerical semigroup S is a cofinite, additively-closed subset of $\mathbb{Z}_{\geq 0}$ that contains 0, and a factorization of $x \in S$ is a k-tuple $z = (z_1, \ldots, z_k)$ where $x = z_1a_1 + \cdots + z_ka_k$ expresses x as a sum of generators of the semigroup $S = \langle a_1, \ldots, a_k \rangle$. Much of the study of non-unique factorization centers on the factorization length $z_1 + \cdots + z_k$, which coincides with the ℓ_1-norm of z as the k-tuple. In this paper, we study the ℓ_∞-norm and ℓ_0-norm of factorizations, viewed as alternative notions of length, with particular focus on the generalizations $\Delta_\infty(x)$ and $\Delta_0(x)$ of the delta set $\Delta(x)$ from classical factorization length. We prove that the ∞-delta set $\Delta_\infty(x)$ is eventually periodic as a function of $x \in S$, classify $\Delta_\infty(S)$ and the 0-delta set $\Delta_0(S)$ for several well-studied families of numerical semigroups, and identify families of numerical semigroups demonstrating $\Delta_\infty(S)$ and $\Delta_0(S)$ can be arbitrarily long intervals and can avoid arbitrarily long subintervals.

1. Introduction

A *numerical semigroup* is a cofinite, additively closed set $S \subseteq \mathbb{Z}_{\geq 0}$ containing 0. We often specify a numerical semigroup via a list of generators, i.e.,

$$S = \langle a_1, \ldots, a_k \rangle = \{z_1a_1 + \cdots + z_ka_k : z_i \in \mathbb{Z}_{\geq 0}\}.$$

As ubiquitous mathematical objects, numerical semigroups arise in countless settings across the mathematics spectrum; see [**3, 20**] for a thorough introduction. Most notably for this manuscript, numerical semigroups arise frequently in factorization theory [**13**] and discrete optimization [**22**].

A *factorization* of an element $x \in S$ is an expression

$$x = z_1a_1 + \cdots + z_ka_k$$

of x with each $z_i \in \mathbb{Z}_{\geq 0}$. The *support* and *length* of a factorization z are

$$\mathrm{supp}(z) = \{i : z_i > 0\}, \qquad \text{and} \qquad \ell_1(z) = z_1 + \cdots + z_k,$$

respectively. We denote by

$$\mathsf{Z}(x) = \{z \in \mathbb{Z}^k_{\geq 0} : x = z_1a_1 + \cdots + z_ka_k\} \qquad \text{and} \qquad \mathcal{L}(x) = \{\ell_1(z) : z \in \mathsf{Z}(x)\}$$

the *set of factorizations* and *length set* of x, respectively. Factorization lengths are a cornerstone of factorization theory, and numerous combinatorial invariants derived

2020 *Mathematics Subject Classification.* Primary 20M14, 05E40.

©2026 American Mathematical Society

from length sets used to quantify and compare the non-uniqueness of factorizations across rings and semigroups [**12**]. One of the more popular such invariants is the *delta set*, which is defined on semigroup elements as

$$\Delta(x) = \{c_i - c_{i-1} : i = 2, \ldots, r\} \qquad \text{where} \qquad \mathcal{L}(x) = \{c_1 < \cdots < c_r\},$$

and defined on semigroups as $\Delta(S) = \bigcup_{x \in S} \Delta(x)$. For numerical semigroups, $\Delta(x)$ is known to be eventually periodic as a function of x [**8**], and $\Delta(S)$ is more varied than for some other well-studied families of semigroups [**4**], such as Krull monoids [**7**].

In this paper, we study the *0-length* and *∞-length* of factorizations z, which are

$$\ell_0(z) = |\operatorname{supp}(z)| \qquad \text{and} \qquad \ell_\infty(z) = \max(z_1, \ldots, z_k),$$

respectively. For each $p \in \{0, 1, \infty\}$, we define the *p-length set* of x as

$$\mathcal{L}_p(x) = \{\ell_p(z) : z \in \mathsf{Z}(x)\}$$

(when $p = 1$, we recover the classical definitions). In discrete optimization, factorizations achieving minimal 0-length are known as *sparse solutions* and have been studied in the context of numerical semigroups [**1**] as well as for more general semigroups [**5, 15**]. Additionally, the asymptotic behavior of ∞-length was recently studied in [**9**], along with the extremal ℓ_p-norms of factorizations for $p \in [1, \infty) \cap \mathbb{Z}$.

In this paper, we study the *p-delta set* of x, defined as

$$\Delta_p(x) = \{c_i - c_{i-1} : i = 2, \ldots, r\} \qquad \text{where} \qquad \mathcal{L}_p(x) = \{c_1 < \cdots < c_r\},$$

and the *p-delta set* of S, defined as $\Delta_p(S) = \bigcup_{x \in S} \Delta_p(x)$.

The contributions of this manuscript are two-fold. First, we prove several structural results about the set $\mathcal{L}_\infty(x)$ for large elements $x \in S$. Our results are reminiscent of the *structure theorem for sets of length*, which drives much of the study of factorization theory [**11, 12**] and a specialized version of which was recently proven for numerical semigroups [**16**]. We derive as a consequence that $\Delta_\infty(x)$ is an eventually periodic function of $x \in S$ (Theorem 2.6), a result that is also known for the classical delta set [**8**] and joins a vast literature of eventual periodicity results for large numerical semigroup elements [**18**].

Second, we characterize $\Delta_\infty(S)$ and $\Delta_0(S)$ for several well-studied families of numerical semigroups, and demonstrate via explicit families of numerical semigroups that $\Delta_\infty(S)$ and $\Delta_0(S)$ can each be arbitrarily long intervals and in general can contain arbitrarily long "gaps". Our results lead us to make the following conjecture.

CONJECTURE 1.1. *For every finite set $D \subset \mathbb{Z}_{\geq 1}$ with $1 \in D$, there exist numerical semigroups S and S' with $\Delta_0(S) = D$ and $\Delta_\infty(S') = D$.*

This part of our work is motivated by the *delta set realization problem* [**10**], which makes an analogous claim for the classical delta set $\Delta(S)$. The delta set realization problem is known to be difficult, in part because proving a given integer lies outside of $\Delta(S)$ necessitates a large amount of control over the factorization structure of S; see [**4**] for examples. Given this, and the technical nature of our arguments in Sections 3 and 4, we suspect Conjecture 1.1 to be difficult in general.

2. A structure theorem for sets of ∞-length

NOTATION 2.1. Throughout this paper, $S = \langle a_1, \ldots, a_k \rangle$ denotes a numerical semigroup with minimal generators $a_1 < a_2 < \cdots < a_k$. Additionally, throughout this section,

$$A = a_1 + \cdots + a_k, \qquad g_i = \gcd(\{a_j : i \neq j\}), \qquad \text{and} \qquad S_i = \langle \tfrac{1}{g_i} a_j : j \neq i \rangle$$

for each i. Additionally, for each i, fix $a_i' \in \mathbb{Z}$ with $a_i' a_i \equiv 1 \bmod g_i$, let

$$\mathsf{Z}(x, i) = \{z \in \mathsf{Z}(x) : z_i = \ell_\infty(z)\} \qquad \text{and} \qquad \mathcal{L}_\infty(x, i) = \{\ell_\infty(z) : z \in \mathsf{Z}(x, i)\},$$

and let

$$L_\infty(x, i) = \max \mathcal{L}_\infty(x, i) \qquad \text{and} \qquad l_\infty(x, i) = \min \mathcal{L}_\infty(x, i).$$

This section contains several structural results concerning the sets $\mathcal{L}_\infty(x)$, $\mathcal{L}_\infty(x, i)$, and $\Delta_\infty(x)$ for large $x \in S$. We briefly outline these results here.

- We prove in Theorem 2.3 that each $\mathcal{L}_\infty(x, i)$ forms what is known as an *almost arithmetic sequence (AAP)* (i.e., an arithmetic sequence with some missing values near either end), a central ingredient to the classical *structure theorem for sets of length* [**11**].
- We prove that in the AAP description of $\mathcal{L}_\infty(x, i)$, the "missing values" near either end depend only on the equivalence class of x modulo certain products of the a_i's and g_i's (Theorem 2.4). This result is reminiscent of [**16**, Theorem 4.2], a more detailed version of the structure theorem for sets of length recently proven for numerical semigroups.
- Proposition 2.5 and Theorem 2.6 are the culmination of these results, collecting the conclusions drawn about $\Delta_\infty(x)$ for large x and $\Delta_\infty(S)$.

The depiction in Figure 1 illustrates how the structure of each $\mathcal{L}_\infty(x, i)$ for large x contributes to that of $\mathcal{L}_\infty(x)$ and $\Delta_\infty(x)$.

Recall that the *Frobenius number* of S is $\mathsf{F}(S) = \max(\mathbb{Z}_{\geq 0} \setminus S)$, and the *Apéry set* of S with respect to a nonzero element $m \in S$ is

$$\mathrm{Ap}(S; m) = \{n \in S : n - m \notin S\}.$$

It is known $\mathrm{Ap}(S; m) = \{0, w_1, \ldots, w_{m-1}\}$, where each $w_i \equiv i \bmod m$ is the smallest element of S in its equivalence class modulo m.

LEMMA 2.2. *For ever $x \in S$, the following inequalities hold:*

(a) $\frac{1}{A} x \leq l_\infty(x) \leq \frac{1}{A} x + a_k$; *and*
(b) *for each* i, $\frac{1}{a_i} x - k a_k \leq L_\infty(x, i) \leq \frac{1}{a_i} x$.

PROOF. Letting $z \in \mathsf{Z}(x)$ with $\ell_\infty(z) = l_\infty(x)$, we see

$$x = z_1 a_1 + \cdots + z_k a_k \leq \ell_\infty(z) a_1 + \cdots + \ell_\infty(z) a_k = l_\infty(x) A.$$

Next, write $x = a + qA$ for $a \in \mathrm{Ap}(S; A)$. We claim $l_\infty(a) \leq a_k$. Indeed, by way of contradiction, fix a factorization $z \in \mathsf{Z}(a)$ with $\ell_\infty(z) = l_\infty(a)$, and assume some $z_j > a_k$. Some $z_i' = 0$ since $a \in \mathrm{Ap}(S; A)$, so trading a_i copies of a_j for a_j copies of a_i yields a factorization $z' \in \mathsf{Z}(a)$ with strictly fewer copies of a_j and no new coordinates larger than a_k. After applying such a trade to each maximal entry in z, we obtain a factorization z' with $\ell_\infty(z') < \ell_\infty(z) = l_\infty(a)$, which is a contradiction.

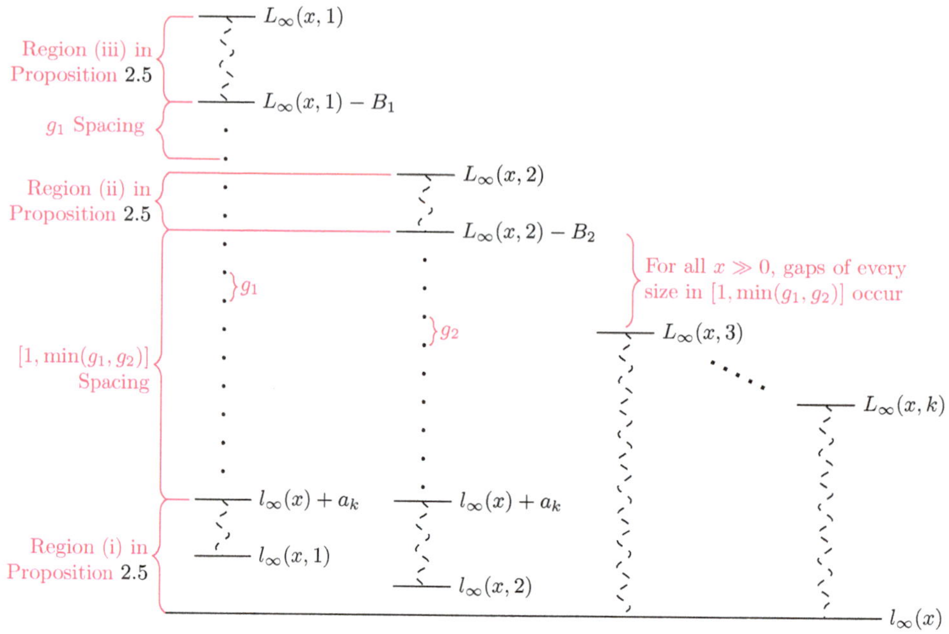

FIGURE 1. Diagram of ∞-length set elements for large $x \in S$, where the B_i are defined in Theorem 2.3

Letting $z \in \mathsf{Z}(a)$ with $\ell_\infty(z) = l_\infty(a)$, [**9**, Theorem 2.6] implies the factorization $z'' = (z_1+q, \ldots, z_k+q)$ of $x = a+qA$ has $l_\infty(x) = \ell_\infty(z'') \le q + a_k$. Thus,

$$l_\infty(x) \le q + a_k = \tfrac{1}{A}(x-a) + a_k \le \tfrac{1}{A}x + a_k,$$

Proceeding to part (b), suppose $z \in \mathsf{Z}(x)$ satisfies $\ell_\infty(z) = z_i = L_\infty(x,i)$. Then

$$L_\infty(x,i)a_i \le L_\infty(x,i) + \sum_{j\ne i} z_j a_j = x.$$

Additionally, we must have $z_j < a_i$ for $j \ne i$, as otherwise one could trade a_i copies of a_j for a_j copies of a_i and contradict the maximality of z_i. As such,

$$x \le L_\infty(x,i)a_i + \sum_{j\ne i} z_j a_j < L_\infty(x,i)a_i + \sum_{j\ne i} a_i a_k \le L_\infty(x,i)a_i + k a_i a_k$$

from which the last remaining inequality is immediately obtained. □

THEOREM 2.3. *For each $i = 1, \ldots, k$, there exists $B_i \in \mathbb{Z}$ such that*

$$[\tfrac{1}{A}x + a_k, \tfrac{1}{a_i}x - B_i] \cap (g_i\mathbb{Z} + a_i'x) \subseteq \mathcal{L}_\infty(x,i) \subseteq g_i\mathbb{Z} + a_i'x.$$

PROOF. The second containment holds since $\ell \in \mathcal{L}_\infty(x,i)$ implies $x - \ell a_i \in g_i S_i$. Define

$$B_i = \tfrac{1}{a_i} g_i(F(S_i)+1)$$

and fix $\ell \in g_i\mathbb{Z} + a_i'x$ with $\frac{1}{A}x + a_k \le \ell \le \frac{1}{a_i}x - B_i$. Since

$$x - \ell a_i = a_i(\tfrac{1}{a_i}x - \ell) \ge g_i(F(S_i)+1),$$

we have $x - \ell a_i \in g_i S_i$. Moreover, we claim $x - \ell a_i$ has a factorization in $g_i S_i$ of ∞-length at most ℓ. Indeed, notice that

$$x - \ell a_i + a_k(A - a_i) \le x + a_k A - \ell a_i = (\tfrac{1}{A}x + a_k)A - \ell a_i \le \ell(A - a_i),$$

from which we obtain

$$\tfrac{1}{A-a_i}(x - \ell a_i) + \max(\{a_j : j \ne i\}) \le \tfrac{1}{A-a_i}(x - \ell a_i) + a_k \le \ell.$$

Now applying Lemma 2.2 to $g_i S_i$ implies $x - \ell a_i$ has a factorization in $g_i S_i$ of ∞-length at most ℓ, which completes the proof. □

THEOREM 2.4. *Fix $B, B' > 0$. For all $x \gg 0$, we have*

$$\mathcal{L}_\infty(x + a_i, i) \cap [\tfrac{1}{a_i}(x + a_i) - B, \infty) = 1 + \big(\mathcal{L}_\infty(x, i) \cap [\tfrac{1}{a_i}x - B, \infty)\big)$$

for each i, as well as

$$\mathcal{L}_\infty(x + A) \cap [0, \tfrac{1}{A}(x + A) + B'] = 1 + \big(\mathcal{L}_\infty(x) \cap [0, \tfrac{1}{A}x + B']\big)$$

In particular, these hold whenever $x > a_i^2 C + a_i B$ and $x > \frac{1}{a_i}A(A - a_i)B'$, respectively, where $C = \lceil \frac{1}{a_j}(B + 1) \rceil$.

PROOF. One can readily check $z \in \mathsf{Z}(x, i)$ implies $z + e_i \in \mathsf{Z}(x + a_i, i)$, which shows one containment in the first equality. For the converse direction, we first claim any factorization $z \in \mathsf{Z}(x, i)$ with $\ell_\infty(z) = z_i \ge \frac{1}{a_i}(x + a_i) - B$ has $z_i > z_j$ for all $j \ne i$. Indeed, if $z_j = z_i$ for some j, then

$$z_j \ge \tfrac{1}{a_i}x - B > a_i C + B - B \ge a_i C,$$

so trading $a_i C$ copies of a_j in z for $a_j C$ copies of a_i yields a factorization in $\mathsf{Z}(x, i)$ with i-th coordinate

$$z_i + a_j C = z_i + a_j \lceil \tfrac{1}{a_j}(B + 1) \rceil > \tfrac{1}{a_i}x - B + B = \tfrac{1}{a_i}x$$

which is impossible by Lemma 2.2(b). Having now proven the claim, we conclude any $z' \in \mathsf{Z}(x + a_i, i)$ has $z' - e_i \in \mathsf{Z}(x, i)$, and the first equality is proven.

For the second equality, any $z \in \mathsf{Z}(x)$ has $z' = z + e_1 + \cdots + e_k \in \mathsf{Z}(x + A)$. For the reverse containment, we claim any $z \in \mathsf{Z}(x)$ with $\ell_\infty(z) \le \frac{1}{A}x + B'$ has no zero entries. Indeed, if some $z_i = 0$, then

$$\begin{aligned} x &= \sum_{j \ne i} z_j a_j \le \sum_{j \ne i} (\tfrac{1}{A}x + B')a_j = (\tfrac{1}{A}x + B')(A - a_i) = x - \tfrac{1}{A}a_i x + (A - a_i)B' \\ &< x - (A - a_i)B' + (A - a_i)B' = x \end{aligned}$$

a contradiction. As such, any factorization $z' \in \mathsf{Z}(x + A)$ with $\ell_\infty(z') \le \frac{1}{A}x + B' + 1$ has $z' - e_1 - \cdots - e_k \in \mathsf{Z}(x)$, thereby completing the proof. □

PROPOSITION 2.5. *For all $x \gg 0$, we have $[1, \min(g_1, g_2)] \cup \{g_1\} \subseteq \Delta_\infty(x)$. Moreover, if $\ell < \ell'$ are successive elements of $\mathcal{L}_\infty(x)$ with*

$$\ell' - \ell \notin [1, \min(g_1, g_2)] \cup \{g_1\},$$

then at least one of ℓ and ℓ' lies in one of the following intervals:

(i) $[\frac{1}{A}x, \frac{1}{A}x + a_k]$; (ii) $[\frac{1}{a_2}x - B_2, \frac{1}{a_2}x]$; *or* (iii) $[\frac{1}{a_1}x - B_1, \frac{1}{a_1}x]$.

PROOF. First, if x is large enough that $(\frac{1}{a_1}x - B_1) - \frac{1}{a_2}x > 3g_1$, then by Lemma 2.2(b),

$$\mathcal{L}_\infty(x) \cap (\tfrac{1}{a_2}x, \tfrac{1}{a_1}x - B_1) = \mathcal{L}_\infty(x, 1) \cap (\tfrac{1}{a_2}x, \tfrac{1}{a_1}x - B_1)$$

contains at least 2 lengths, and any two consecutive lengths therein must have difference $g_1 \in \Delta_\infty(x)$ by Theorem 2.3.

Analogously, if x is large enough that $(\frac{1}{a_2}x - B_2) - \frac{1}{a_3}x > 2g_1g_2$, then by Lemma 2.2(b) and Theorem 2.3,

$$\begin{aligned}\mathcal{L}_\infty(x) \cap (\tfrac{1}{a_3}x, \tfrac{1}{a_2}x - B_2) &= \big(\mathcal{L}_\infty(x, 1) \cup \mathcal{L}_\infty(x, 2)\big) \cap (\tfrac{1}{a_3}x, \tfrac{1}{a_2}x - B_2)\\ &= \big((g_1\mathbb{Z} + a_1'x) \cup (g_2\mathbb{Z} + a_2'x)\big) \cap (\tfrac{1}{a_3}x, \tfrac{1}{a_2}x - B_2),\end{aligned}$$

within which successive elements achieve each difference in $[1, \min(g_1, g_2)]$ by the Chinese Remainder Theorem since $\gcd(g_1, g_2) = \gcd(a_1, \ldots, a_k) = 1$.

For the final claim, by Theorem 2.3, aside from the three claimed intervals, the only subinterval of $[\frac{1}{A}x, \frac{1}{a_1}x]$ not containing an arithmetic sequences of step size $\min(g_1, g_2)$ is $[\frac{1}{a_2}x, \frac{1}{a_1}x - B_1]$, whose lengths form an arithmetic sequence of step size g_1. □

THEOREM 2.6. *If $x \gg 0$, then $\Delta_\infty(x+p) = \Delta_\infty(x)$, where $p = \operatorname{lcm}(a_1, g_1a_2, A)$.*

PROOF. We begin by considering the intervals (i), (ii), and (iii) in Proposition 2.5. Let

$$\begin{aligned}R_1(x) &= [\tfrac{1}{A}x, \tfrac{1}{A}x + (a_k + g_1)] \cap \mathcal{L}_\infty(x),\\ R_2(x) &= [\tfrac{1}{a_2}x - (B_2 + g_1), \tfrac{1}{a_2}x + g_1] \cap \mathcal{L}_\infty(x),\\ R_3(x) &= [\tfrac{1}{a_1}x - (B_1 + g_1), \tfrac{1}{a_1}x] \cap \mathcal{L}_\infty(x).\end{aligned}$$

Applying Theorem 2.4 with $B' = a_k + g_1$ gives $R_3(x+p) = R_3(x) + \frac{1}{A}p$, and letting $B = B_1 + g_1$, we have

$$R_1(x) = [\tfrac{1}{a_1}x - B, \tfrac{1}{a_1}x] \cap \mathcal{L}_\infty(x, 1)$$

by Lemma 2.2(b), so $R_1(x+p) = R_1(x) + \frac{1}{a_1}p$. Additionally, by Theorem 2.3,

$$R_2(x) = \big([\tfrac{1}{a_2}x - B, \tfrac{1}{a_2}x] \cap \mathcal{L}_\infty(x, 2)\big) \cup \big([\tfrac{1}{a_2}x - B, \tfrac{1}{a_2}x + g_1] \cap (g_1\mathbb{Z} + a_1'x)\big)$$

for $B = B_2 + g_1$. Clearly $g_1\mathbb{Z} + a_1'(x+p) = g_1\mathbb{Z} + a_1'x$, and since $g_1 \mid \frac{1}{a_2}p$,

$$\big(g_1\mathbb{Z} + a_1'(x+p)\big) + \tfrac{1}{a_2}p = g_1\mathbb{Z} + a_1'x$$

as well. As such, $R_2(x+p) = R_2(x) + \frac{1}{a_2}p$ once again by Theorem 2.4.

Lastly, by Proposition 2.5 any successive lengths in $\mathcal{L}_\infty(x)$ or $\mathcal{L}_\infty(x+p)$ not residing in one of the above intervals must have difference in $[1, \min(g_1, g_2)] \cup \{g_1\}$, which is a subset of both $\mathcal{L}_\infty(x)$ and $\mathcal{L}_\infty(x+p)$ by Proposition 2.5. □

COROLLARY 2.7. *Fix $B > 0$. For all $x \gg 0$, we have*

$$\ell \in \mathcal{L}_\infty(x) \cap [\tfrac{1}{a_1}x - B, \tfrac{1}{a_1}x] \qquad \textit{if and only if} \qquad x - \ell a_1 \in g_1S_1 \cap [0, a_1B].$$

In particular, $\Delta(g_1S_1 \cap (a_1\mathbb{Z} + j)) \subseteq \Delta_\infty(S)$ for each j.

PROOF. As in the proof of Theorem 2.3, we have $\ell \in \mathcal{L}_\infty(x,1)$ if and only if (i) $x-\ell a_1 \in g_1 S_1$ and (ii) there exists a factorization of $x-\ell a_1$ in $g_1 S_1$ with ∞-length at most ℓ. Tracing through the proof of Theorem 2.3, so long as condition (i) holds, condition (ii) holds whenever $\ell > \frac{1}{A}x + a_k$, which is certainly the case if $\ell > \frac{1}{a_1}x - B$ for $x \gg 0$. Moreover,

$$x - \ell a_1 < x - (\tfrac{1}{a_1}x - B)a_1 = a_1 B.$$

As such, Lemma 2.2(b) implies the frst claim, and the second claim then follows upon unraveling definitions. □

3. Some families of ∞-delta sets

In this section, we examine the set $\Delta_\infty(S)$ for several families of numerical semigroups. We characterize the ∞-delta set for supersymmetric numerical semigroups [**6**], and numerical semigroups whose generators form an arithmetic sequence [**17**] or a geometric sequence [**19, 23**]; in each case, the ∞-delta set turns out to be an interval. We also demonstrate that $\Delta_\infty(S)$ can be an arbitrarily long interval (Theorem 3.2) and have arbitrarily long gaps (Theorem 3.3).

Many of our arguments in this section and the next utilize trades, presentations, and Betti elements. We briefly review the relevant concepts here, though the reader is encouraged to see [**3**, Chapter 5] and [**21**] for a thorough introduction.

Define an equivalence relation $\sim$ on $\mathbb{Z}^k_{\geq 0}$ that sets $z \sim z'$ whenever $z, z' \in \mathsf{Z}(x)$ are factorizations of the same element $x \in S$. We call each relation $z \sim z'$ between facatorizations of disjoint support a *trade* of S, and sometimes identify the difference $z - z' \in \mathbb{Z}^k$ with the trade $z \sim z'$. A *presentation* of S is a collection ρ of trades with the property that for any $x \in S$ and any $z, z' \in S$, there exists a chain of factorizations

$$z = y_1 \sim y_2 \sim \cdots \sim y_r = z'$$

wherein $y_i - y_{i-1} \in \rho$ or $y_{i-1} - y_i \in \rho$ for each pair of sequential factorizations y_{i-1} and y_i. A presentation is *minimal* if it is minimal with respect to containment among all presentations for S. It is known that any two minimal presentations ρ and ρ' for S have the same number of trades, and in fact the set

$$\mathrm{Betti}(S) = \{z_1 a_1 + \cdots + z_k a_k : z - z' \in \rho\}$$

of *Betti elements* is independent of the choice of minimal presentation ρ.

THEOREM 3.1. *Let $S = \langle a_1, \ldots, a_k\rangle$ with $a_1 < \cdots < a_k$.*

(a) *If $a < b$ are coprime and each $a_i = a^{k-1-i}b^{i-1}$, then $\Delta_\infty(S) = \{1, 2, \ldots, b\}$.*

(b) *If $p_1, \ldots, p_k \in \mathbb{Z}_{\geq 1}$ are pairwise coprime with $p_1 > \cdots > p_k$, $T = p_1 \cdots p_k$, and each $a_i = \frac{1}{p_i}T$, then $\Delta_\infty(S) = \{1, 2, \ldots, p_1\}$.*

PROOF. For part (a), the trades $be_i \sim ae_{i+1}$ for $i = 1, \ldots, k-1$ form a minimal presentation for S by [**14**, Theorem 8], so $\max \Delta_\infty(S) \leq b$. Now, if $1 \leq c \leq a$, then

$$\mathsf{Z}((b + a - c)a_1) = \{(b + a - c)e_1, (a - c)e_1 + ae_2\}$$

so $c \in \Delta_\infty(S)$. Moreover, if $a < c \leq b$, then

$$\mathsf{Z}(ca_2) = \{(be_1 + (c - a)e_2, ce_2\},$$

so again $c \in \Delta_\infty(S)$. Thus, $\Delta_\infty(S) = \{1, 2, \ldots, b\}$.

For part (b), the trades $p_{i+1}e_i \sim p_i e_{i+1}$ for $i = 1, \ldots, k-1$ form a minimal presentation for S by [**6**], so $\max \Delta_\infty(S) \leq p_1$. Using a similar argument to part (a), we have $\Delta_\infty((p_1+p_2-c)a_1) = \{c\}$ whenever $1 \le c \le a$ and $\Delta_\infty(ca_2) = \{c\}$ whenever $p_2 < c \le p_1$. Thus, $\Delta_\infty(S) = \{1, 2, \ldots, p_1\}$. □

THEOREM 3.2. *Let $S = \langle a, a+d, \ldots, a+kd\rangle$ with $2 \le k < a$ and $\gcd(a,d) = 1$, and write $a - 1 = qk + r$ for $q, r \in \mathbb{Z}_{\ge 0}$ with $0 \le r < k$. Then*

$$\Delta_\infty(S) = \{1, 2, \ldots, q+d+1\}.$$

PROOF. In what follows, write $a_i = a + ik$ for $i \in [0, k]$, and for $x \in S$, write factorizations $z \in \mathsf{Z}(x)$ as $z = (z_0, \ldots, z_k) = z_0e_0 + \cdots + z_ke_k$. Before beginning the proof, we recall some facts about arithmetical numerical semigroups; see [**2, 17**]. Each $x \in S$ has $\ell \in \mathcal{L}_1(x)$ if and only if

$$x = \ell a + bd \qquad \text{with} \qquad 0 \le b \le k\ell,$$

as for any factorization $z \in \mathsf{Z}(x)$ with $\ell_1(z) = \ell$, we can write

$$x = (z_0 + \cdots + z_k)a + (z_1 + 2z_2 + \cdots + kz_k)d$$

with $b = z_1 + 2z_2 + \cdots + kz_k$. Moreover,

$$\ell - \lceil \tfrac{1}{k} b\rceil \ge z_0 = \ell - (z_1 + \cdots + z_k) \ge \ell - b,$$

with equality on the right if $z_2 = \cdots = z_k = 0$.

We now proceed with the proof. First, suppose $1 \le G \le d$. We see

$$x = (a+d)a + (d-G)(a+d) = (a+d-G)(a+d)$$

are factorizations $z, z' \in \mathsf{Z}(x)$, respectively, with ∞-lengths $\ell_\infty(z) = a+d$ and $\ell_\infty(z') = a+d-G$. Now, since $\ell_\infty(z') = \ell_1(z') = z'_1$, any factorization $z'' \in \mathsf{Z}(x)$ with $\ell_\infty(z'') > \ell_\infty(z')$ must have $\ell_1(z'') > \ell_1(z')$. This means $\ell_1(z'') \ge \ell_1(z)$, and letting $b = z''_1 + 2z''_2 + \cdots + kz''_k$, any such factorization must have

$$z''_0 \ge \ell_1(z'') - b \ge (a+2d-G) - (d-G) = a+d$$

by the first paragraph above. As such, $G \in \Delta_\infty(x)$.

Next, suppose $d \le G \le d+q+1$. We see

$$x = (a+G)a = (G-d)a + a(a+d)$$

are factorizations $z, z' \in \mathsf{Z}(x)$, respectively, with $\ell_\infty(z) = a+G$ and $\ell_\infty(z') = a$. Fix a factorization $z'' \in \mathsf{Z}(x)$ with $\ell_\infty(z'') < \ell_\infty(z)$, and let $b = z''_1 + 2z''_2 + \cdots + kz''_k$. Since $z = (a+G)e_1$, we must have $\ell_1(z'') < \ell_1(z) = a+G$. As such, we have $\ell_1(z'') \le a+G-d = \ell_1(z')$ and $b \ge a$, meaning

$$z''_0 \le \ell_1(z'') - \lceil \tfrac{1}{k} b\rceil \le (a+G-d) - \lceil \tfrac{1}{k} a\rceil \le a+G-d-q-1 \le a.$$

Additionally, if $z''_j \ge a$ for some $j \ge 1$, then

$$\sum_{i \ne j} z''_i a_i = x - z''_j a_j < x - aa_j \le x - a(a+d) = (G-d)a \le (q+1)a,$$

and all factorizations of such an element have equal 1-length. As such, since

$$x = (G-jd)a + a(a+jd),$$

we must have $\ell_1(z'') = \ell_1(z') - jd$, meaning z'' coincides with the above factorization. Thus $\ell_\infty(z'') \le a$, thereby ensuring $G \in \Delta_\infty(x)$. □

THEOREM 3.3. *Fix $m \ge 3$. If $S = \langle 3, 3m+1, 3m+2\rangle$, then*

$$\Delta_\infty(S) = \{1, 2, \ldots, m+1\} \cup \{2m, 2m+1\}.$$

PROOF. Since S has max embedding dimension (see [**20**, Chapter 3]):

(i) the trades

$$(2m+1)e_1 \sim e_2 + e_3, \qquad me_1 + e_3 \sim 2e_2, \qquad \text{and} \qquad (m+1)e_1 + e_2 \sim 2e_3$$

comprise a minimal presentation for S;

(ii) for each $x \in S$, the unique factorization $z = (a, b, c) \in \mathsf{Z}(x)$ with $\ell_\infty(z)$ maximal in $\mathcal{L}_\infty(x)$ is also the unique factorization with $b + c \le 1$; and

(iii) for each $a \ge 0$, the factorizations $(a, 1, 1)$ and $(a + 2m - 1, 0, 0)$ of the element $x = 3(a + 2m - 1)$ have the two highest ∞-lengths in $\mathcal{L}_\infty(x)$.

Fix $x \in S$ and $z = (a, b, c) \in \mathsf{Z}(x)$. Suppose z does not have maximal ∞-length, and let G be minimal with $\ell_\infty(z) + G \in \mathcal{L}_\infty(x)$. By (iii), if $b = c = 1$, then $G \ge 2m$. Otherwise, by (ii) either $b \ge 2$ or $c \ge 2$. If $b \ge 2$, then fixing $q \in \mathbb{Z}$ with $b - 2q \in \{0, 1\}$ and performing the second trade in (i) q times yields a chain of factorizations

$$(a, b, c) \sim (a + m, b - 2, c + 1) \sim \cdots \sim (a + qm, b - 2q, c + q),$$

wherein each factorization differs in ∞-length from the previous factorization by at most m, and the final factorization in which has strictly larger ∞-length than z. As such, we have $G \le m$. By an analogous argument, if $c \ge 2$, then $G \le m + 1$. This proves $m + 2, \ldots, 2m - 1 \notin \Delta_\infty(S)$.

Now, by (i), we have $\max \Delta_\infty(S) \le 2m + 1$. We can see by inspection that

$$\mathsf{Z}(6m+3) = \{(2m+1, 0, 0), (0, 1, 1)\} \quad \text{and} \quad \mathsf{Z}(6m+6) = \{(2m+2, 0, 0), (1, 1, 1)\},$$

so $2m, 2m + 1 \in \Delta_\infty(S)$. Also by inspection,

$$\mathsf{Z}(6m+8) = \{(m+2, 0, 1), (2, 2, 0)\} \quad \text{and} \quad \mathsf{Z}(6m+10) = \{(m+3, 1, 0), (2, 0, 2)\},$$

so $m, m + 1 \in \Delta_\infty(S)$. Lastly, for each $G \in \{1, \ldots, m - 1\}$, we have

$$(0, 0, G+1), (m+1, 1, G-1) \in \mathsf{Z}(x)$$

for $x = (G + 1)(3m + 2)$. Fix a factorization $z = (a, b, c) \in \mathsf{Z}(x)$. Since

$$3a = x - (3m+1)b + (3m+2)c = (G + 1 - b - c)(3m+2) + b,$$

we must have $b + c \le G + 1$. If $b + c = G + 1$, then $a = \frac{1}{3}b < G + 1$, so $\ell_\infty(z) \le G + 1$. If $b + c \le G$, then

$$3a \ge (G + 1 - b - c)(3m + 2) \ge 3m + 2$$

so $\ell_\infty(z) \ge a \ge m + 1$. This proves $m - G \in \Delta_\infty(S)$. □

4. Some families of 0-delta sets

In a similar vein to the prior section, in Theorems 4.2 and 4.3 we characterize $\Delta_0(S)$ for numerical semigroups S residing in several well-studied families, including maximal embedding dimension numerical semigroups [**20**, Chapter 3], supersymmetric numerical semigroups [**6**], 3-generated numerical semigroups [**20**, Chapter 10], and numerical semigroups generated by generalized arithmetic sequences [**17**]; in each case, the 0-delta set turns out to be either $\{1\}$ or $\{1, 2\}$. We also identify two families of numerical semigroups achieving notable extremal behavior (Theorems 4.4 and 4.5). First, we demonstrate that the structure of $\mathcal{L}_0(x)$ for large $x \in S$ differs substantially from that of $\mathcal{L}_\infty(x)$ detailed in Section 2.

THEOREM 4.1. *For all $x \gg 0$, we have $\Delta_0(x) = \{1\}$. In particular, $1 \in \Delta_0(S)$.*

PROOF. For fixed $I \subseteq \{1, \ldots k\}$ nonempty, and letting $d = \gcd(a_i : i \in I)$, $x \in S$ has a factorization $z \in \mathsf{Z}(x)$ with $\operatorname{supp}(z) = I$ if and only if $d \mid x$ and

$$\tfrac{1}{d}(x - \sum_{i \in I} a_i) > \mathsf{F}(\langle \tfrac{1}{d} a_i : i \in I \rangle).$$

As such, for $x \gg 0$, if x has a factorization with support I, then x also has a factorization with support J for every $J \supseteq I$. Thus, $\mathcal{L}_0(x)$ is an interval and $\Delta_0(x) = \{1\}$. □

THEOREM 4.2. *The following hold.*

(a) *If S has a minimal presentation ρ in which every trade if between factorizations with singleton support, then $\Delta_0(S) = 1$. In particular, this occurs whenever S is supersymmetric or generated by a geometric sequence.*
(b) *If $S = \langle m, a_1, \ldots, a_{m-1} \rangle$ with $m \geq 3$ and each $a_i \equiv i \bmod m$ (i.e., S is maximal embedding dimension), then $\Delta_0(S) = \{1, 2\}$.*
(c) *If $S = \langle a, ah + d, ah + 2d, \ldots, ah + kd \rangle$ with $h \geq 1$, $2 \leq k < a$, and $\gcd(a, d) = 1$ (i.e., S is generated by a generalized arithmetic sequence), then $\Delta_0(S) = \{1, 2\}$.*

PROOF. Part (a) follows from the fact that any two factorizations of an element $x \in S$ are connected by a chain of factorizations in which successive factorizations z, z' differ by a trade in ρ, and thus satisfy $|\ell_0(z) - \ell_0(z')| \leq 1$. As such, $\Delta_0(x) = \{1\}$. The claims about supersymmetric numerical semigroups and semigroups generated by geometric sequences immediately follow [**6, 14**].

For part (b), by [**20**, Theorem 8.30] S has a minimal presentation in which each trade has the form

$$e_i + e_j \sim e_k + ce_0 \qquad \text{with} \qquad i + j \equiv k \bmod m \qquad \text{and} \qquad c \in \mathbb{Z}_{\geq 1},$$

so by similar reasoning to part (a), $\Delta_0(S) \subseteq \{1, 2\}$. Moreover, since $m \geq 3$, applying the trade with $i = 1$ and $j = 2$ to the factorization $z = e_0 + \cdots + e_{m-1}$ yields a factorization z' with $\ell_0(z') = m - 2$. Moreover, no other factorization z'' can have $\ell_0(z'') = m - 1$, as then the trade $z \sim z''$ would be between distinct factorizations for a minimal generator of S.

For part (c), in the minimal presentation for S presented in [**17**, Theorem 2.16], each trade is between factorizations with 0-length at most 2, so $\Delta_0(S) \subseteq \{1, 2\}$. Moreover, writing $a - 1 = qk + r$ with $0 \leq r < k$, the minimal presentation in [**17**] also implies

$$x = a + (ah + (r + 1)d) + q(ah + kd) = a(d + h(q + 1))$$

are the only two factorizations of x, so $\Delta_0(x) = \{2\}$. □

We next characterize $\Delta_0(S)$ when S is 3-generated. Recall that an expression

$$S = t'S' + t''S'' \qquad \text{with} \qquad S' = \langle b_1, \ldots, b_r \rangle \qquad \text{and} \qquad S'' = \langle c_1, \ldots, c_{k-r} \rangle$$

is called a *gluing* if $t' \in S'' \setminus \{c_1, \ldots, c_{k-r}\}$, $t'' \in S' \setminus \{b_1, \ldots, b_r\}$, and $\gcd(t', t'') = 1$; see [**20**, Chapter 9] for more on gluings. Note that such an expression for S need not be unique. In particular, if $S = \langle a_1, a_2, a_3 \rangle$, then there can be up to 3 such expressions for S as a gluing, each of the form $S = \langle a_i \rangle + t'S'$ for some $i \in \{1, 2, 3\}$.

THEOREM 4.3. *Suppose $S = \langle a_1, a_2, a_3 \rangle$. If S has at most 1 expression as a gluing, then $\Delta_0(S) = \{1, 2\}$. Otherwise, $\Delta_0(S) = \{1\}$.*

PROOF. If S has at least 2 distinct expressions $S = \langle a_i \rangle + t'S' = \langle a_j \rangle + t''S''$ as a gluing, then we can write

$$S = \langle t'b_1, t't''b_2, t''b_3 \rangle.$$

Since $t'b_1 \in S' = \langle t'b_2, b_3 \rangle$, there exist $z_2, z_3 \in \mathbb{Z}_{\geq 0}$ with $t'b_1 = z_2 t'b_2 + z_3 b_3$, and since $\gcd(t', b_3) = 1$, we must have $t' \mid z_2$. As such, $b_1 = z_1 b_2 + z_2 b_3$ and thus $b_1 \in \langle b_2, b_3 \rangle$. By similar reasoning, we know $b_3 \in \langle b_1, b_2 \rangle$. Assuming $b_1 \leq b_3$ without loss of generality, this is only possible if $b_1 = b_3$ or $b_2 \mid b_1$. In particular, $t'b_1$ has a factorization in S' with singleton support. As such, by [**20**, Theorem 9.2], S has a minimal presentation within which every factorization has singleton support, so $\Delta_0(S) = \{1\}$ by Theorem 4.2(a).

Conversely, suppose $S = \langle a_1 \rangle + tS'$ with $S' = \langle b_1, b_2 \rangle$ is the only expression of S as a gluing. Then writing $a_1 = z_1 b_1 + z_2 b_2$, we cannot have $z_2 = 0$, as otherwise

$$S = \langle z_1 b_1, t'b_1, t'b_2 \rangle = \langle t'b_2 \rangle + b_1 \langle z_1, t' \rangle$$

is a second expression of S as a gluing. Analogously, $z_1 > 0$. As such,

$$x = (t+1)a_1 = a_1 + z_2 t'b_1 + z_3 t'b_2$$

has $\mathcal{L}_0(x) = \{1, 3\}$, so $\Delta_0(x) = \{2\}$.

This leaves the case where S cannot be expressed as a gluing. By [**20**, Section 10.3], S has a unique minimal presentation comprised of trades

$$c_1 e_1 \sim r_{12} e_2 + r_{13} e_3, \qquad c_2 e_2 \sim r_{21} e_1 + r_{23} e_3, \qquad \text{and} \qquad c_3 e_3 \sim r_{31} e_1 + r_{32} e_2$$

where each $r_{ij} > 0$ and each $c_k = r_{ik} + r_{jk}$ for $\{i, j, k\} = \{1, 2, 3\}$. We consider cases.

- If $r_{ij} = r_{ik} = 1$ for some i, then $x = a_1 + a_2 + a_3$ has at least one factorization without full support, and any such factorization must have singleton support, so $\Delta_0(x) = \{2\}$.
- If $r_{ji} \geq 2$ and $r_{ki} \geq 2$ for some i, then

  $$x = (c_i + 1)a_i = a_i + r_{ij} a_j + r_{ik} a_k$$

  are the only factorizations of x, so $\Delta_0(x) = \{2\}$.
- If $r_{ji} = r_{ki} = 1$ for some i, then either $r_{jk} \geq 2$ and $r_{kj} \geq 2$, in which case

  $$x = (c_j + 1)a_j = a_j + r_{ji} a_i + r_{jk} a_k$$

  are the only factorizations of x and $\Delta_0(x) = \{2\}$, or $r_{jk} = 1$ or $r_{kj} = 1$, meaning we are in the first case above.
- In all other cases, up to reordering i, j, and k, we have $r_{ij} = r_{jk} = r_{ki} = 1$ while $r_{ji}, r_{kj}, r_{ik} \geq 2$. In this case,

  $$x = (c_j + 1)a_j = a_j + r_{ji} a_i + a_k$$

  are the only factorizations of x, so $\Delta_0(x) = \{2\}$.

In all cases above, we conclude $\Delta_0(S) = \{1, 2\}$. □

Thus far, all semigroups S presented have $\max \Delta_0(S) \leq 2$. We close by presenting two families of numerical semigroups exhibiting more interesting behavior: one demonstrating $\Delta_0(S)$ can be an arbitrarily large interval (Theorem 4.4), and another demonstrating $\Delta_0(S) \setminus [1, \max \Delta_0(S)]$ can be arbitrarily large (Theorem 4.5).

THEOREM 4.4. *For each $k \geq 2$, there exists a numerical semigroup S such that $\Delta_0(S) = \{1, 2, \ldots, k-1\}$.*

PROOF. Fix distinct primes p_1, p_2 with $p_1, p_2 > k$. Let $S_2 = \langle p_1, p_2 \rangle$, so $\Delta_0(S_2) = \{1\}$. Proceeding inductively, assume $S_{i-1} = \langle a_1, \ldots, a_{i-1} \rangle$ has Betti elements $b_1, \ldots, b_{i-2}$ with $\mathsf{Z}(b_1) = \{p_2 e_1, p_1 e_2\}$ and for each $j \geq 2$,

$$\mathsf{Z}(b_j) = \{p_{j+1} e_{j+1}, (k + 1 - j)e_1 + e_2 + \cdots + e_j\}$$

for some prime p_{j+1}. Since each $j \leq k$, we have $\Delta_0(b_j + a_{j+1}) = \{j\}$ for each j. Letting

$$a_i = (k + 1 - i)a_1 + a_2 + \cdots + a_{i-1},$$

we see (i) the above factoriation of a_i is not preceded (under the component-wise partial order) by any factorizations of $b_1, \ldots, b_{i-2}$ (meaning a_i is uniquely factorable in S_{i-1}), and (ii) the above factorization of a_i does not precede a factorization of any b_j. As such, choosing a prime $p_i > a_i$, the semigroup

$$S_i = p_i S_{i-1} + \langle a_i \rangle$$

is a gluing, so we have $\mathrm{Betti}(S_i) = \{p_i b_1, \ldots, p_i b_{i-2}, p_i a_i\}$ and

$$\mathsf{Z}(p_i a_i) = \{p_i e_i, (k + 1 - i)e_1 + e_2 + \cdots + e_{i-1}\}.$$

This ensures $\Delta_0((p_i + 1)a_i) = \{i - 1\}$ and $\Delta_0(S_i) = \{1, 2, \ldots, i - 1\}$. Thus, at the conclusion of this process, the semigroup S_k has $\Delta_0(S_k)$ as claimed. □

THEOREM 4.5. *For $k \geq 16$, there is a numerical semigroup $S = \langle a_1, \ldots, a_{k+1} \rangle$ with $\Delta_0(S) \cap [\frac{7}{8}k, k] = \{k - 1, k\}$.*

PROOF. Let $S_2 = \langle 2, 3 \rangle$. Next, for each $i = 3, \ldots, k$, let

$$S_i = 2S_{i-1} + \langle 2a_{i-2} + a_{i-1} \rangle \qquad \text{where} \qquad S_{i-1} = \langle a_1, \ldots, a_{i-1} \rangle.$$

Lastly, let

$$S = S_{k+1} = 2S_k + \langle a_1 + \cdots + a_k \rangle \qquad \text{where} \qquad S_k = \langle a_1, \ldots, a_k \rangle.$$

As each S_i is easily shown to be a gluing, the trades

$$2e_2 \sim 3e_1, \qquad 2e_{k+1} \sim e_1 + \cdots + e_k, \qquad \text{and} \qquad 2e_i \sim 2e_{i-2} + e_{i-1} \text{ for } 3 \leq i \leq k$$

form a minimal presentation ρ of S.

In what follows, write $S = \langle a_1, \ldots, a_{k+1} \rangle$. We see by inspection that

$$\mathsf{Z}(3a_{k+1}) = \{3e_{k+1}, e_1 + \cdots + e_{k+1}\},$$

since no other trades in ρ can be performed, so in particular $\Delta_0(3a_{k+1}) = \{k\}$ and $\Delta_0(2a_{k+1}) = \{k - 1\}$. We claim every other $x \in S$ with $\Delta_0(x)$ nonempty has $\max \Delta_0(x) \leq \frac{7}{8}k$. Indeed, any two factorizations of x can be connected by a sequence of trades in ρ, and of such trades, the only one that can yield a change in 0-length of more than 2 is the trade $2e_{k+1} \sim e_1 + \cdots + e_k$. As such, consider factorizations $z, z' \in \mathsf{Z}(x)$ of the form

$$z = u + e_1 + \cdots + e_k \qquad \text{and} \qquad z' = u + 2e_{k+1}$$

for some $u \in \mathbb{Z}^k_{\geq 0}$. By way of contradiction, suppose $\ell_0(u) \leq \frac{1}{8}k$, so that

$$\ell_0(z) - \ell_0(z') \geq k - \tfrac{1}{8}k = \tfrac{7}{8}k.$$

First, suppose $u_i \geq 1$ for some $i \leq \frac{1}{2}k$, and fix j maximal with $i + 2j \leq k$. Performing the trade

$$2e_i + e_{i+1} + e_{i+3} + \cdots + e_{i+2j-1} \sim 2e_{i+2j}$$

to z yields a factorization z'' in which $j \geq \frac{1}{4}k$ entries are strictly smaller than in z. However, since $\ell_0(u) \leq \frac{1}{8}k$, at least $\frac{1}{8}k$ entries of z'' must be zero. As such,

$$\ell_0(z) - \ell_0(z'') \leq \tfrac{1}{2}k \qquad \text{and} \qquad \ell_0(z'') - \ell_0(z') \leq \tfrac{7}{8}k.$$

Next, suppose $u_i \geq 1$ for $\frac{1}{2}k < i \leq k$. Performing the trade

$$\begin{aligned} 2e_i + e_{i-1} + e_{i-3} + e_{i-5} + \cdots &\sim 2e_{i-1} + 2e_{i-3} + 2e_{i-5} + \cdots \\ &\sim e_{i-2} + 2e_{i-4} + 3e_{i-6} + \cdots \end{aligned}$$

to z yields a factorization z'' in which at least $\frac{1}{4}k$ entries are strictly smaller than in z. As in the previous case, we obtain

$$\ell_0(z) - \ell_0(z'') \leq \tfrac{1}{2}k \qquad \text{and} \qquad \ell_0(z'') - \ell_0(z') \leq \tfrac{7}{8}k.$$

Lastly, since $x \neq 2a_{k+1}, 3a_{k+1}$, the only remaining case is when $u = ca_{k+1}$ with $c \geq 2$. In this case, one may perform the trade

$$\begin{aligned} 4e_{k+1} &\sim 2e_1 + 2e_2 + \cdots + 2e_k \\ &\sim 3e_{k-1} + 4e_{k-3} + 5e_{k-5} + \cdots \end{aligned}$$

to obtain a factorization with at least $\frac{1}{2}k$ zero entries, thus completing the proof. □

References

[1] I. Aliev, J. A. De Loera, T. Oertel, and C. O'Neill, *Sparse solutions of linear Diophantine equations*, SIAM J. Appl. Algebra Geom. **1** (2017), no. 1, 239–253, DOI 10.1137/16M1083876. MR3633776

[2] J. Amos, S. T. Chapman, N. Hine, and J. Paixão, *Sets of lengths do not characterize numerical monoids*, Integers **7** (2007), A50, 8. MR2373112

[3] A. Assi, M. D'Anna, and P. A. García-Sánchez, *Numerical semigroups and applications*, RSME Springer Series, vol. 3, Springer, Cham, [2020] ©2020. Second edition [of 3558713], DOI 10.1007/978-3-030-54943-5. MR4230109

[4] C. Bowles, S. T. Chapman, N. Kaplan, and D. Reiser, *On delta sets of numerical monoids*, J. Algebra Appl. **5** (2006), no. 5, 695–718, DOI 10.1142/S0219498806001958. MR2269412

[5] A. M. Bruckstein, D. L. Donoho, and M. Elad, *From sparse solutions of systems of equations to sparse modeling of signals and images*, SIAM Rev. **51** (2009), no. 1, 34–81, DOI 10.1137/060657704. MR2481111

[6] L. Bryant and J. Hamblin, *The maximal denumerant of a numerical semigroup*, Semigroup Forum **86** (2013), no. 3, 571–582, DOI 10.1007/s00233-012-9448-5. MR3053780

[7] S. T. Chapman, F. Gotti, and R. Pelayo, *On delta sets and their realizable subsets in Krull monoids with cyclic class groups*, Colloq. Math. **137** (2014), no. 1, 137–146, DOI 10.4064/cm137-1-10. MR3271229

[8] S. T. Chapman, R. Hoyer, and N. Kaplan, *Delta sets of numerical monoids are eventually periodic*, Aequationes Math. **77** (2009), no. 3, 273–279, DOI 10.1007/s00010-008-2948-4. MR2520501

[9] S. Chapman, E. B. Dugan, S. Gaskari, E. Lycan, S. Mendoza de la Cruz, C. O'Neill, and V. Ponomarenko, *Some asymptotic results on p-lengths of factorizations for numerical semigroups and arithmetical congruence monoids*, Houston J. Math. **50** (2024), no. 3, 509–523. MR4946534

[10] S. Colton and N. Kaplan, *The realization problem for delta sets of numerical semigroups*, J. Commut. Algebra **9** (2017), no. 3, 313–339, DOI 10.1216/JCA-2017-9-3-313. MR3685046

[11] A. Geroldinger, *A structure theorem for sets of lengths*, Colloq. Math. **78** (1998), no. 2, 225–259, DOI 10.4064/cm-78-2-225-259. MR1659136

[12] A. Geroldinger, *Sets of lengths*, Amer. Math. Monthly **123** (2016), no. 10, 960–988, DOI 10.4169/amer.math.monthly.123.10.960. MR3593640

[13] A. Geroldinger and F. Halter-Koch, *Non-unique factorizations*, Pure and Applied Mathematics (Boca Raton), vol. 278, Chapman & Hall/CRC, Boca Raton, FL, 2006. Algebraic, combinatorial and analytic theory, DOI 10.1201/9781420003208. MR2194494

[14] C. Kiers, C. O'Neill, and V. Ponomarenko, *Numerical semigroups on compound sequences*, Comm. Algebra **44** (2016), no. 9, 3842–3852, DOI 10.1080/00927872.2015.1087013. MR3503387

[15] S. Li and G. Ge, *Deterministic construction of sparse sensing matrices via finite geometry*, IEEE Trans. Signal Process. **62** (2014), no. 11, 2850–2859, DOI 10.1109/TSP.2014.2318139. MR3225149

[16] G. Moskowitz and C. O'Neill, *The structure theorem for sets of lengths for numerical semigroups*, J. Commut. Algebra **17** (2025), no. 1, 31–43, DOI 10.1216/jca.2025.17.31. MR4906150

[17] M. Omidali and F. Rahmati, *On the type and the minimal presentation of certain numerical semigroups*, Comm. Algebra **37** (2009), no. 4, 1275–1283, DOI 10.1080/00927870802278701. MR2510984

[18] C. O'Neill and R. Pelayo, *Factorization invariants in numerical monoids*, Algebraic and geometric methods in discrete mathematics, Contemp. Math., vol. 685, Amer. Math. Soc., Providence, RI, 2017, pp. 231–249, DOI 10.1090/conm/685. MR3625578

[19] D. C. Ong and V. Ponomarenko, *The Frobenius number of geometric sequences*, Integers **8** (2008), A33, 3. MR2425631

[20] J. C. Rosales and P. A. García-Sánchez, *Numerical semigroups*, Developments in Mathematics, vol. 20, Springer, New York, 2009, DOI 10.1007/978-1-4419-0160-6. MR2549780

[21] J. C. Rosales and P. A. García-Sánchez, *Finitely generated commutative monoids*, Nova Science Publishers, Inc., Commack, NY, 1999. MR1694173

[22] H. M. Salkin and C. A. de Kluyver, *The knapsack problem: a survey*, Naval Res. Logist. Quart. **22** (1975), 127–144, DOI 10.1002/nav.3800220110. MR403658

[23] A. Tripathi, *On the Frobenius problem for geometric sequences*, Integers **8** (2008), A43, 5. MR2472061

Mathematics Department, University of California Santa Barbra, Santa Barbara, California 93106
Email address: sogol@ucsb.edu

Mathematics Department, Trinity College, Hartford, Connecticut 06106
Email address: alexander.domat@trincoll.edu

Mathematics Department, San Diego State University, San Diego, California 92182
Email address: cdoneill@sdsu.edu

Mathematics Department, San Diego State University, San Diego, California 92182
Email address: vponomarenko@sdsu.edu

Mathematics Department, Arizona State University, Tempe, Arizona 85287
Email address: reneric.2002@gmail.com

Mathematics Department, Western Washington University, Bellingham, Washington 98225
Email address: maylacward@gmail.com

Contemporary Mathematics
Volume **836**, 2026
https://doi.org/10.1090/conm/836/16706

On three families of dense Puiseux monoids

Scott T. Chapman, Felix Gotti, Marly Gotti, and Harold Polo

Abstract. A positive monoid is a submonoid of the nonnegative cone of a linearly ordered abelian group. The positive monoids of rank 1 are called Puiseux monoids, and their atomic structure, arithmetic of lengths, and factorization behavior have been systematically investigated for about ten years. Each Puiseux monoid can be realized as an additive submonoid of the nonnegative cone of $\mathbb{Q}$. We say that a Puiseux monoid is dense if it is isomorphic to an additive submonoid of $\mathbb{Q}_{\geq 0}$ that is dense in $\mathbb{R}_{\geq 0}$ with respect to the Euclidean topology. Every non-dense Puiseux monoid is known to be a bounded factorization monoid. However, the atomic structure as well as the arithmetic and factorization properties of dense Puiseux monoids turn out to be quite interesting. In this paper, we study the atomic structure and some arithmetic and factorization aspects of three families of dense Puiseux monoids.

1. Introduction

A submonoid of the nonnegative cone of a linearly ordered abelian group is often called a positive monoid. In particular, rank-one positive monoids are known as Puiseux monoids, a term introduced when their atomicity was first studied by the second and third authors in [**31**, **34**]. Since then, not only the atomic structure but also the factorization behavior and the arithmetic of lengths of Puiseux monoids have been actively studied by several authors over the last ten years (see the recent papers [**13**] and references therein). As the Grothendieck group of any Puiseux monoid is a rank-one torsion-free abelian group, every Puiseux monoid is isomorphic to a monoid of consisting of rationals under the standard addition [**20**, Section 24] and, as a result, every nontrivial Puiseux monoid is isomorphism to a monoid consisting of nonnegative rationals under the standard addition [**26**, Theorem 2.9]. The interested reader can find a survey on the atomicity of Puiseux monoids in [**11**] by the first three authors.

2020 *Mathematics Subject Classification.* Primary 11Y05, 20M13; Secondary 06F05, 20M14.
Key words and phrases. Length-finite factorization, integral domains, finite factorization, bounded factorization, length-factoriality, atomic domain.

While working on this paper, the second author was supported first by the UC Berkeley Chancellor Fellowship and then by the NSF (under the awards DMS-1903069 and DMS-2213323). The fourth author was first funded by the UF Mathematics Department Fellowship and then by the UC Presidential Fellowship.

©2026 American Mathematical Society

Albeit a natural generalization of numerical monoids, Puiseux monoids exhibit a complex and interesting atomic/arithmetic structure. For instance, for any prescribed nonnegative integer n, there exists a non-atomic Puiseux monoid with exactly n atoms [**31**, Proposition 5.4], and there are also non-atomic Puiseux monoids with infinitely many atoms [**31**, Example 3.5]. In addition, there are atomic Puiseux monoids having full systems of sets of lengths, as proved in [**33**] by the second author (the elasticity of Puiseux monoids was studied in [**36**] by O'Neill and the second author). Moreover, as Theorem 3.6 states, there are atomic Puiseux monoids whose sets of atoms are dense in the nonnegative part of the real line under the Euclidean topology.

The remarkable variety of unexpected atomic and factorization behaviors exhibited by Puiseux monoids have provided a rich source of (counter)examples, which have proven instrumental over recent decades in testing the sharpness of theorems and disproving conjectures within commutative algebra and factorization theory. For instance, a Puiseux monoid is the most relevant ingredient in Grams' construction of the first known atomic domain not satisfying the ACCP (see [**38**]), while Puiseux monoids have been essential in fully addressing Gilmer's question [**26**, page 189] about the ascent of atomicity to monoid algebras (see [**19**, Section 5] and [**37**]). Also, in the recent paper [**27**], Gonzalez, Panpaliya, and the second author generalized the class of Puiseux monoids in [**19**, Section 5] to argue that neither quasi-atomicity nor almost atomicity (two weaker notions of atomicity introduced by Boynton and Coykendall back in 2015) ascend to monoid algebras.

Let M be a Puiseux monoid, and assume that M is a subset of $\mathbb{Q}_{\geq 0}$, the set consisting of nonnegative rationals. Since addition is continuous in the real line under the Euclidean topology, the subspace topology inherited by M is intrinsically linked to its algebraic structure. In particular, if one aims to study the atomic decomposition of elements of M, examining neighborhoods of 0 might provide substantial insight. Indeed, if 0 is not a limit point of M,[1] then one can readily show that M is atomic [**28**, Proposition 4.5]. However, when 0 is a limit point of M, the atomic structure of M might become considerably intricate and significantly different for distinct classes of Puiseux monoids. When 0 is a limit point of M, the additive closedness of M implies that M is dense in the nonnegative part of the real line under the Euclidean topology. Thus, M is dense in the nonnegative part of the real line if and only if 0 is a limit point of M. This observation motivates the following definition.

DEFINITION 1.1. A Puiseux monoid is *dense* if it is dense in the nonnegative part of the real line under the Euclidean topology.

Dense Puiseux monoids serve as valuable tools to differentiate between classes in the nested diagram of atomicity introduced by Anderson, Anderson, and Zafrullah in their landmark paper [**4**], where they coined and studied the bounded and finite factorization properties as two weaker variants of the unique factorization property. Additionally, dense Puiseux monoids help distinguish the nested classes of AP-ness investigated by Anderson and Quintero in [**5**]. We dedicate this paper to examining some atomic and factorization aspects of three families of dense Puiseux monoids, which have been selected based on significant examples from recent literature (see

[1]We say that p is a limit point of the Puiseux monoid M if for all $\epsilon > 0$ the open interval $(p - \epsilon, p + \epsilon)$ contains a point of M different from p.

[**10, 22, 37, 43**]). The class of dense Puiseux monoids has also been studied by Bras-Amorós and the third author in [**8**].

In Section 2, which is the background section, we introduce most of the standard notation and terminology as well as the non-standard results we shall be using later.

In Section 3, we provide some basic results related to dense Puiseux monoids and their monoid homomorphisms. We conclude the section producing atomic Puiseux monoids whose sets of atoms are dense in $\mathbb{R}_{\geq 0}$, the set consisting nonnegative real numbers.

In Section 4, we briefly discuss some examples of dense Puiseux monoids generated by unit fractions whose denominators have a a fixed length ℓ in the multiplicative monoid $\mathbb{N}$ of positive integers (the length of $n \in \mathbb{N}$ is the number of prime factors of n counting repetitions).

In Section 5, we first study the class of p-adic Puiseux monoids (for any p in $\mathbb{P}$, the set of primes): for each $p \in \mathbb{P}$, a p-adic Puiseux monoid is a submonoid of the valuation Puiseux monoid

$$\mathbb{N}_0\left[\frac{1}{p}\right] := \left\{f\left(\frac{1}{p}\right) : f(x) \in \mathbb{N}_0[x]\right\},$$

where $\mathbb{N}_0[x]$ is the semiring of polynomials with nonnegative integer coefficients. We establish necessary and sufficient conditions for p-adic Puiseux monoids to be atomic. Then we consider, for each function $f\colon \mathbb{P} \to \mathbb{N}_0$ such that $p \nmid f(p)$ for any $p \in \mathbb{P}$ with $f(p) \neq 0$, the Puiseux monoid

$$M_f := \sum_{p \in \mathbb{P}} \frac{f(p)}{p}\mathbb{N}_0, \tag{1.1}$$

which is an internal sum of p-adic monoids, each of them isomorphic to $\mathbb{N}_0$. Clearly, M_f is dense if and only if the infimum of $\{\frac{f(p)}{p} : p \in \mathbb{P}\}$ is 0. We prove that elements of M_f can be decomposed as a canonical sum, and then we use such a canonical decomposition to show that M_f satisfies the ascending chain condition on principal ideals (ACCP). Finally, for each finite nonempty subset P consisting of primes, we consider the following dense p-adic Puiseux monoid:

$$M_P := \sum_{p \in P} \mathbb{N}_0\left[\frac{1}{p}\right], \tag{1.2}$$

which is the internal sum of valuation monoids. We prove that elements in M_P can also be decomposed as a canonical sum, and then we use this to settle some divisibility properties taking place in M_P.

In Section 6, which is the final section of the paper, we continue the investigation of the Puiseux monoids $M_q := \langle q^n : n \in \mathbb{N}_0 \rangle$ carried out in [**10**] by the first three authors.[2] For each $q \in \mathbb{Q}_{>0}$, the Puiseux monoid M_q is closed under multiplication: indeed, M_q is the additive monoid of the semiring $\mathbb{N}_0[q]$, and so

$$M_q = \{f(q) : f \in \mathbb{N}_0[x]\}.$$

The additive monoids M_q, as well as the more general monoid M_α, where $\alpha \in \mathbb{C}$ is an algebraic parameter, have received significant attention in recent literature (see [**7, 15**] and references therein). As we did with the monoids in (1.1) and (1.2), we establish a canonical sum decomposition for elements in M_q, and then we use it

[2]The atomic structure of $\mathbb{N}_0[q]$ was first explored in [**34**] by the second and third authors.

to prove that for any non-unit fraction $q \in \mathbb{Q}_{>0}$, the monoid M_q satisfies the length-finite factorization property, that is, M_q is atomic and, for any pair $(r, \ell) \in M_q \times \mathbb{N}_0$, the element r has only finitely many factorizations of length ℓ. In the second part of Section 6, we investigate the algebraic and atomic structure of Puiseux monoids of the form

$$M_Q := \langle q^n : (q, n) \in Q \times \mathbb{N}_0 \rangle,$$

where Q is a finite nonempty set consisting of positive rationals. Observe that M_Q specializes to M_q when the set Q is the singleton $\{q\}$. It is clear that M_Q is dense if and only if $\min Q < 1$. After having established a canonical sum decomposition for the monoids M_Q, we conclude the paper considering their atomicity, factorizations, and multiplicative closedness.

Although the first version of this paper was written back in 2017, one of the primary motivations of writing this new version was to include the existence of a canonical sum decomposition of any of the studied families of Puiseux monoids. It turns out that the Puiseux monoids used in applications to commutative algebra and factorization theory have certain canonical decomposition. For instance, the Puiseux monoid

$$G := \Big\langle \frac{1}{2^n p_n} : n \in \mathbb{N}_0 \Big\rangle,$$

where p_n is the n-th odd prime, was used by Grams [**38**] to produced the first known atomic domain not satisfying the ACCP, a localization R of the monoid algebra $F[G]$. Every element $q \in G$ admit the following unique sum decomposition

$$r = c_0(r) + \sum_{n \in \mathbb{N}} c_n(r) \frac{1}{2^n p_n}, \tag{1.3}$$

where $c_0(r) \in \mathbb{N}_0\left[\frac{1}{2}\right]$ and all but finitely many of the nonnegative integer coefficients $c_n(r)$ equal zero. The sum decomposition in (1.3), referred here as 'canonical sum decomposition', played a crucial role in the proof that the Grams' domain R is atomic. Recent application of Puiseux monoids throughout their canonical sum decompositions can be found in [**13**, Section 4], [**35**, Section 3], and [**37**, Sections 3 and 4]. The canonical sum decompositions found here for the Puiseux monoids in the families we are interested in play an important role in understanding their corresponding atomic and factorization structure: see the proofs of Theorems 5.9 and 6.2.

2. Preliminary

2.1. General notation. The symbols $\mathbb{P}$, $\mathbb{N}$, and $\mathbb{N}_0$ denote the sets of standard primes, positive integers, and nonnegative integers, respectively. For any $r \in \mathbb{R}$ and $S \subseteq \mathbb{R}$, we denote the set $\{s \in S : s \geq r\}$ by $S_{\geq r}$. For any $m, n \in \mathbb{Z}$ with $m \leq n$, we let $[\![m, n]\!]$ denote the discrete interval from m to n. We say that a sequence consisting of real numbers is *finitely supported* if only finitely many terms of the sequence are different from zero, whence whether a real sequence is finitely supported is invariant under modification of finitely many terms.

For any nonzero rational $q \in \mathbb{Q}^\times$, we call the unique relatively primes $\mathsf{n}(q), \mathsf{d}(q) \in \mathbb{Z}$ such that $q = \mathsf{n}(q)/\mathsf{d}(q)$ and $\mathsf{d}(q) > 0$ the *numerator* and the *denominator* of q, respectively. We say that a positive rational is a *unit fraction* if its numerator is 1. For each set Q consisting of nonzero rationals, we call

$$\mathsf{n}(Q) := \{\mathsf{n}(q) : q \in Q\} \quad \text{and} \quad \mathsf{d}(Q) := \{\mathsf{d}(q) : q \in Q\}$$

the *numerator set* and *denominator set* of Q, respectively. For a prime p, the *p-adic valuation* on $\mathbb{Q}$ is the map $v_p\colon \mathbb{Q} \to \mathbb{Z} \cup \{\infty\}$ defined as follows: $v_p(0) = \infty$ and $v_p(q) = v_p(\mathsf{n}(q)) - v_p(\mathsf{d}(q))$ for any $q \neq 0$, where for $n \in \mathbb{N}$ the value $v_p(n)$ is the exponent of the maximal power of p dividing n. Thus,

$$v_p(q_1 + \cdots + q_n) \geq \min\{v_p(q_1), \ldots, v_p(q_n)\}$$

for all $q_1, \ldots, q_n \in \mathbb{Q}_{>0}$.

2.2. Commutative monoids. Throughout this paper, we tacitly assume that every monoid mentioned is commutative and cancellative. Also, unless we specify otherwise, each monoid M in this paper is written additively ($+$ and 0 denote the binary operation and the identity element of M, respectively). Let M be a monoid. A *submonoid* of M is a subset containing 0 and closed under addition, while a map $\phi\colon M \to M'$ is a *monoid homomorphism* if M' is a monoid and $\phi(b+c) = \phi(b)+\phi(c)$ for all $b, c \in M$. We let $M^\bullet$ denote the set of nonzero elements of M. Also, we let $\mathscr{U}(M)$ denote the group of invertible elements of M, and we say that M is *reduced* if the group $\mathscr{U}(M)$ is trivial. For any $b, c \in M$, we say that c *divides* b in M and write $c \mid_M b$ if there exists $d \in M$ such that $b = c + d$, in which case we write $c \mid_M b$. The monoid M is called a *valuation monoid* if for all $b, c \in M$ either $b \mid_M c$ or $c \mid_M b$. For any subset S of M, we let $\langle S \rangle$ denote the submonoid of M generated by S, and we say that M is *finitely generated* if $M = \langle S \rangle$ for some finite subset S of M.

Since M is cancellative, it can be minimally embedded into an abelian group $\mathscr{G}(M)$ and such an abelian group, which is unique up to isomorphism, is called the *Grothendieck group* of M. The monoid M is said to be *torsion-free* if its Grothendieck group is torsion-free. The *rank* of M is defined to be the rank of its Grothendieck group as a $\mathbb{Z}$-module or, equivalently, the dimension of the $\mathbb{Q}$-vector space $\mathbb{Q} \otimes_{\mathbb{Z}} \mathscr{G}(M)$. A monoid is called a *positive* if it is isomorphic to a submonoid of the nonnegative cone of a linearly ordered abelian group (i.e., an abelian group with a total order relation compatible with its operation). Clearly, every positive monoid is reduced. Following [**31**], we refer to a rank-1 positive monoid as a *Puiseux monoid.*

An element $a \in M \setminus \mathscr{U}(M)$ is an *atom* if whenever $a = b + c$ for some $b, c \in M$, either $b \in \mathscr{U}(M)$ or $c \in \mathscr{U}(M)$. The set of atoms of M is denoted by $\mathscr{A}(M)$. Observe that $\mathscr{A}(M)$ is a subset of any generating set of M when M is reduced. If $\mathscr{A}(M)$ is the empty set, then M is said to be *antimatter* (the term antimatter was coined by Coykendall, Dobbs, and Mullins [**18**] in the setting of integral domains). An element $b \in M$ is said to be *atomic* if either b is invertible or b can be written as a sum of finitely many atoms (allowing repetitions). Following Cohn [**16**], we say that M is atomic provided that every element of M is atomic. A subset I of M is called an *ideal* of M provided that $I + M \subseteq I$, and an ideal of the form $b + M$ for some $b \in M$ is called a *principal ideal.* The monoid M is said to satisfy the *ascending chain condition on principal ideals* (ACCP) provided that for any sequence $(I_n)_{n \geq 1}$ of principal ideals of M that is increasing under set-inclusion, there exists $m \in \mathbb{N}$ such that $I_n = I_m$ for all $n \geq m$. It is well known and not hard to show that every monoid that satisfies the ACCP is atomic.

Assume throughout the rest of this section that M is an atomic monoid. The free commutative monoid on $\mathscr{A}(M/\mathscr{U}(M))$ is denoted by $\mathsf{Z}(M)$, and the elements of $\mathsf{Z}(M)$ are called *factorizations.* Let $\phi\colon \mathsf{Z}(M) \to M/\mathscr{U}(M)$ be the unique monoid homomorphism fixing $a + \mathscr{U}(M)$ for all $a \in \mathscr{A}(M)$. We say that $z = a_1 \cdots a_\ell \in \mathsf{Z}(M)$ is a *factorization* of $b \in M$ if $\phi(z) = b + \mathscr{U}(M)$, in which case ℓ is referred to as the *length* of z and is denoted by $|z|$. For each $b \in M$, set

$$\mathsf{Z}(b) := \phi^{-1}(b + \mathscr{U}(M)) \subseteq \mathsf{Z}(M).$$

If $\mathsf{Z}(b)$ is a singleton for all $b \in M$, then M is called a *unique factorization monoid* (UFM). More generally, if $\mathsf{Z}(b)$ is finite for all $b \in M$, then M is called a *finite factorization monoid* (FFM). Every finitely generated monoid is an FFM [**23**, Proposition 2.7.8]. For each $b \in M$, set

$$\mathsf{L}(b) := \{|z| : z \in \mathsf{Z}(b)\}.$$

If $\mathsf{L}(b)$ is finite for all $b \in M$, then M is called a *bounded factorization monoid* (BFM). It follows from the definitions that every FFM is a BFM, and it is well known that every BFM satisfies the ACCP [**40**, Corollary 1]. The system $\{\mathsf{L}(b) : b \in M\}$ has been fairly investigated during the past few years for Puiseux monoids M (see [**3, 9, 24**] for numerical monoids and [**22, 28, 36**] for Puiseux monoids). For each pair $(b, \ell) \in M \times \mathbb{N}_0$, we set

$$\mathsf{Z}_\ell(b) := \{z \in \mathsf{Z}(b) : |z| = \ell\}.$$

If $\mathsf{Z}_\ell(b)$ is finite for all pairs $(b, \ell) \in M \times \mathbb{N}_0$, then M is called a *length-finite factorization monoid* (LFFM). It follows from the definitions that a monoid is an FFM if and only if it is both a BFM and an LFFM, whence we can think of the property of being an LFFM and that of being a BFM properties to be complementary with respect to the property of being an FFM. The notion of an LFFM was introduced by Geroldinger and Zhong [**25**], and it was further investigated by Jiang, Kanungo, and Kim [**41**] recently.

One of the most elementary families of atomic monoids is the class of numerical monoids. A *numerical monoid* is a co-finite submonoid of the additive monoid $\mathbb{N}_0$. Each numerical monoid has a unique minimal generating set, which is finite. Thus, every numerical monoid is an FFM. If $\{a_1, \dots, a_n\}$ is the minimal generating set of a numerical monoid N, then $\mathscr{A}(N) = \{a_1, \dots, a_n\}$ and $\gcd(a_1, \dots, a_n) = 1$. Let N be a numerical monoid and, for $n \in \mathbb{N}$, let $\mathscr{A}(N) = \{a_1, \dots, a_n\}$ be the minimal generating set of N. The *Frobenius number* of N, denoted in this paper by $\mathsf{f}(N)$, is the largest integer that does not belong to N. Although a general explicit formula for $\mathsf{f}(N)$ in terms of $\{a_1, \dots, a_n\}$ is unknown, it is not difficult to show that $\mathsf{f}(N) = a_1a_2 - a_1 - a_2$ when $n = 2$, and a formula when $n = 3$ has been established by Tripathi in [**45**].

3. Density of Puiseux monoids

In the first part of this section, we briefly discuss homomorphisms and isomorphisms between Puiseux monoids. As previously noted, the density of a Puiseux monoid is determined by its behavior in any neighborhood of 0, or equivalently, around any generating set. These and other related matters are also discussed in this section.

3.1. Homomorphisms and isomorphisms. Since the standard operation of addition is continuous with respect to the Euclidean topology, the additive operations of Puiseux monoids and their homomorphisms are also continuous. This significantly restricts the set of homomorphisms between Puiseux monoids. We proceed to describe the homomorphisms between Puiseux monoids. Although the following lemma is well known and was proved in [44] by Sasaki and Tamura, we include a brief proof here for the sake of completeness. After that, we propose an isomorphism criterion, which will be helpful later.

LEMMA 3.1. *The homomorphisms between two Puiseux monoids are precisely those given by nonnegative rational multiplication.*

PROOF. Let M and M' be two Puiseux monoids, and let $\phi\colon M \to M'$ be a monoid homomorphism. If ϕ is the zero homomorphism, then it is multiplication by 0. Therefore let us assume that ϕ is not the zero homomorphism, which implies that $M \neq \{0\}$. As $N = M \cap \mathbb{N}_0$ is a nontrivial submonoid of the additive monoid of nonnegative integers, it has a nonempty minimal set of generators, namely, $\{a_1, \dots, a_k\}$. Because ϕ is not the zero homomorphism, there exists $j \in [\![1,k]\!]$ such that $\phi(a_j) \neq 0$. Set $b := \frac{\phi(a_j)}{a_j}$. For $q \in M^\bullet$ and $n_1, \dots, n_k \in \mathbb{N}_0$ such that $\mathsf{n}(q) = n_1a_1 + \dots + n_ka_k$, the fact that $a_i\phi(a_j) = \phi(a_ia_j) = a_j\phi(a_i)$ for every $i \in [\![1,k]\!]$ guarantees that

$$\phi(q) = \frac{1}{\mathsf{d}(q)}\phi(\mathsf{n}(q)) = \frac{1}{\mathsf{d}(q)}\sum_{i=1}^{k} n_i\phi(a_i) = \frac{1}{\mathsf{d}(q)}\sum_{i=1}^{k} n_ia_i\frac{\phi(a_j)}{a_j} = qb.$$

Hence the homomorphism ϕ consists in multiplying by the positive rational b. On the other hand, it follows immediately that, for all $r \in \mathbb{Q}_{>0}$, the map $M \to M'$ defined via the assignments $q \mapsto rq$ for all $q \in M$ is a monoid homomorphism provided that $rM \subseteq M'$. □

The following corollary follows immediately.

COROLLARY 3.2. *Two Puiseux monoids M and M' are isomorphic if and only if there exists $q \in \mathbb{Q}_{>0}$ such that $M' = qM$.*

The rest of this section is dedicated to establishing the existence of infinitely many non-isomorphic Puiseux monoids whose sets of atoms are dense in $\mathbb{R}_{\geq 0}$. If two Puiseux monoids M and M' are isomorphic, then we write $M \cong M'$, and otherwise we write $M \ncong M'$.

LEMMA 3.3. *Let P and Q be infinite sets of primes such that $P \cap Q$ is empty. Let $M_P := \langle a_p : p \in P\rangle$ and $M_Q := \langle b_q : q \in Q\rangle$ be Puiseux monoids such that for all $(p,q) \in P \times Q$ and $\mathsf{d}(a_p)$ and $\mathsf{d}(b_q)$ are powers of p and q, respectively. Then $M_P \ncong M_Q$.*

PROOF. Suppose, by way of contradiction, that $M_P \cong M_Q$. In light of Corollary 3.2, there exists $r \in \mathbb{Q}_{>0}$ such that $M_Q = rM_P$. If p is a prime in P such that $p \nmid \mathsf{n}(r)$, then rb_p is an element of M_P such that $\mathsf{d}(rb_p)$ is a power of p and, therefore, $p \in Q$. However, this contradicts the fact that P and Q are disjoint sets. □

3.2. General facts on the density of Puiseux monoids. As mentioned in the introduction, a Puiseux monoid is dense if and only if it is topologically dense in $\mathbb{R}_{\geq 0}$ with respect to the inherited standard topology. We proceed to show that a Puiseux monoid M is dense if and only if 0 is a limit point of $M^\bullet$.

PROPOSITION 3.4. *A Puiseux monoid is dense if and only if 0 is a limit point of its subset of nonzero elements.*

PROOF. The direct implication follows immediately. To argue the reverse implication, assume that M is a Puiseux monoid such that 0 is a limit point of $M^\bullet$. Let $(r_n)_{n\geq 1}$ be a sequence in $M^\bullet$ converging to 0. Fix any $\ell \in \mathbb{R}_{>0}$, and let us show that ℓ is a limit point of M. Fix $\epsilon > 0$, and note that because $\lim_{n\to\infty} r_n = 0$, we can take $n \in \mathbb{N}$ such that $r_n < \min\{\ell, \epsilon\}$. Let m be the maximum positive integer such that $\ell - mr_n > 0$ and set $r := mr_n$. By the maximality of m,

$$0 < \ell - r = \ell - (m+1)r_n + r_n \leq r_n < \epsilon.$$

Thus, for any arbitrary $\epsilon \in \mathbb{R}_{>0}$, we have found $r \in M \setminus \{\ell\}$ such that $|\ell - r| < \epsilon$ and, therefore, we can conclude that ℓ is a limit point of $M^\bullet$. Hence M is a dense Puiseux monoid. □

The following corollary follows directly from the fact that a Puiseux monoid M is a BFM and so an atomic monoid provided that 0 is not a limit point of $M^\bullet$ (see [**28**, Proposition 4.5]).

COROLLARY 3.5. *Every non-atomic Puiseux monoid is topologically dense in* $\mathbb{R}_{\geq 0}$.

To generate a dense Puiseux monoid, it suffices to take a sequence of positive rationals having 0 as a limit point. Let $\mathscr{P} = \{P_n : n \in \mathbb{N}\}$ be a collection of infinite subsets of primes in $\mathbb{P}$ such that $P_i \cap P_j$ is empty for $i \neq j$. Now for each $j \in \mathbb{N}$, consider the Puiseux monoid $M_j = \langle \frac{1}{p} : p \in P_j \rangle$. Because every P_j is infinite, each M_j is dense. Moreover, $M_i \not\cong M_j$ for $i \neq j$; this is an immediate consequence of Lemma 3.3. Hence we can conclude that there are countably many non-isomorphic dense Puiseux monoids. We proceed to produce a class of atomic Puiseux monoids whose sets of atoms are dense in $\mathbb{R}_{\geq 0}$.

THEOREM 3.6. *There are infinitely many non-isomorphic Puiseux monoids whose sets of atoms are dense in* $\mathbb{R}_{\geq 0}$.

PROOF. Let S be the set consisting of all the nonzero elements of the valuation Puiseux monoid $\mathbb{N}_0[\frac{1}{p}]$, the Puiseux monoid consisting of all nonnegative p-adic integers. It is clear that S is dense in $\mathbb{R}$ is dense in $\mathbb{R}_{\geq 0}$ for every $p \in \mathbb{P}$.

Fix $\ell \in \mathbb{R}_{>0}$ and let us verify that ℓ is a limit point of S in $\mathbb{R}_{\geq 0}$. For this, fix $\epsilon > 0$ and then take $m, n \in \mathbb{N}$ such that $\frac{1}{p^n} < \epsilon$ and $\frac{m}{p^n} < \ell \leq \frac{m+1}{p^n}$. It follows immediately that the element $s := \frac{m}{p^n}$ of S satisfies the inequality $|\ell - \frac{m}{p^n}| < \epsilon$. Since ϵ was arbitrarily taken, ℓ is a limit point of S. It is obvious that 0 is also a limit point of S. Hence S is dense in $\mathbb{R}_{\geq 0}$.

Now take $(r_n)_{n\geq 1}$ to be a sequence of positive rationals whose underlying set R dense in $\mathbb{R}_{\geq 0}$. Also, consider the collection $\mathscr{P} := \{P_n : n \in \mathbb{N}\}$ of infinite sets of primes such that $P_i \cap P_j$ is empty for any $i, j \in \mathbb{N}$ with $i \neq j$. For each $j \in \mathbb{N}$, let $(p_{j,k})_{k\geq 1}$ be a sequence of pairwise distinct primes whose underlying set is P_j.

For each index $j \in \mathbb{N}$, we proceed to construct a Puiseux monoid M_j such that $\mathscr{A}(M_j)$ is dense in $\mathbb{R}_{\geq 0}$. Fix an index $j \in \mathbb{N}$. Observe that, for each $k \in \mathbb{N}$, the set

$$\left\{\frac{m}{p_{j,k}^n} : m, n \in \mathbb{N} \text{ and } p_{j,k} \nmid m\right\}$$

is dense in $\mathbb{R}_{\geq 0}$ and, as a consequence, there exist $m_k, n_k \in \mathbb{N}$ such that the inequality $\left|r_k - \frac{m_k}{p_{j,k}^{n_k}}\right| < \frac{1}{k}$ holds. Now consider the following Puiseux monoid:

$$M_j = \left\langle \frac{m_k}{p_{j,k}^{n_k}} : k \in \mathbb{N} \right\rangle. \tag{3.1}$$

Because distinct generators in (3.1) have powers of distinct primes in their denominators, it follows that M_j is an atomic monoid such that

$$\mathscr{A}(M_j) = \left\{ \frac{m_k}{p_{j,k}^{n_k}} : k \in \mathbb{N} \right\}.$$

Finally, we are led to verify that the set $\mathscr{A}(M_j)$ is dense in $\mathbb{R}_{\geq 0}$. To complete this, take $x \in \mathbb{R}_{\geq 0}$ and then fix $\epsilon > 0$. Since R is dense in $\mathbb{R}_{\geq 0}$, there exists $k \in \mathbb{N}$ large enough such that $\frac{1}{k} < \frac{\epsilon}{2}$ and $|x - r_k| < \frac{\epsilon}{2}$. Consequently, $\left|r_k - \frac{m_k}{p_{j,k}^{n_k}}\right| < \frac{1}{k} < \frac{\epsilon}{2}$, which implies that

$$\left|x - \frac{m_k}{p_{j,k}^{n_k}}\right| < |x - r_k| + \left|r_k - \frac{m_k}{p_{j,k}^{n_k}}\right| < \epsilon.$$

This means that $\mathscr{A}(M_j)$ is dense in $\mathbb{R}_{\geq 0}$, as desired. Since the collection $\mathscr{P}$ consists of pairwise disjoint sets of primes, it follows from Lemma 3.3 that the Puiseux monoids in $\{M_j : j \in \mathbb{N}\}$ are pairwise non-isomorphic. □

4. k-prime reciprocal monoids

For a nonempty set P consisting of primes, it seems that the Puiseux monoid $M_P := \langle \frac{1}{p} : p \in P \rangle$ was first considered in the setting of commutative ring theory [**4**, Example 2.1] (for $P = \mathbb{P}$). In the mentioned example, M_P is the monoid of exponents of the monoid algebra $F[M]$ (where F was taken to be a field), which was proved to be an integral domain that satisfies the ACCP but is not a BFD (see [**30**, Proposition 4.2]). Our main purpose in this section is to discuss a natural generalization of the monoids M_P and then provide insight into the atomicity of this generalized family. We note that another generalization of the prime reciprocal Puiseux monoid was recently considered in [**22**, Proposition 3.1] by Geroldinger and the second author in connection with the length-sets of monoid algebras not having the bounded factorization property.

DEFINITION 4.1. For an infinite set P consisting of primes in $\mathbb{P}$ and $k \in \mathbb{N}$, a *k-prime reciprocal (Puiseux) monoid* over P is a Puiseux monoid generated by a set S such that $\mathsf{d}(s)$ is the product of k distinct primes in P for all $s \in S$.

For a set S and a nonnegative integer n, we let $\binom{S}{n}$ denote the collection of subsets of S of cardinality n. The *elementary* k-prime reciprocal monoid over an infinite set of primes P is the monoid

$$M_{P,k} = \langle S_k \rangle, \text{ where } S_k = \left\{ \frac{1}{p_1 \cdots p_k} : \{p_1, \dots, p_k\} \in \binom{P}{k} \right\}.$$

We call the elements of S_k the *elementary generators* of $M_{P,k}$. We now show that while the elementary 1-prime reciprocal monoid is atomic, for each $k \geq 2$, every elementary k-prime reciprocal monoid is antimatter.

THEOREM 4.2. *Let P be an infinite set consisting of primes and, for $k \in \mathbb{N}$, consider the elementary k-prime reciprocal monoid $M_{P,k}$. Then the following statements hold.*

(1) *If* $k = 1$, *then* $M_{P,k}$ *satisfies the ACCP and so is atomic.*
(2) *If* $k \geq 2$, *then* $M_{\mathbb{P},k}$ *is antimatter.*

PROOF.

(1) This is well known (for instance, see (see [**6**, Example 3.3] or [**22**, Proposition 3.1]).

(2) First, we show that for every $N \in \mathbb{N}$ and $p, q \in \mathbb{P}$ with $p \neq q$ there exist $m, n \in \mathbb{N}$ and $p', q' \in \mathbb{P}$ with $q' > p' > N$ such that

$$p'q' = mqq' + npp' + pq. \tag{4.1}$$

To argue this, fix $p, q \in \mathbb{P}$ with $p \neq q$. Therefore $\gcd(p, q) = 1$, and so it follows from Dirichlet's theorem on arithmetic progressions of primes that there exists $m \in \mathbb{N}$ such that $p' := mq + p$ is a prime greater than N. Dirichlet's theorem comes into play again to yield $n \in \mathbb{N}$ such that $q' := np' + q$ is a also a prime. Therefore one finds that

$$p'q' = mqq' + pq' = mqq' + p(np' + q) = mqq' + npp' + pq.$$

Now consider the elementary 2-prime reciprocal monoid $M_2 := M_{\mathbb{P},2}$. If p and q are distinct primes, then by the argument given above, there exist $m, n \in \mathbb{N}$ and $p', q' \in \mathbb{P}$ with $q' > p' > \max\{p, q\}$ such that the identity (4.1) holds. After dividing both sides of (4.1) by $pqp'q'$, we obtain that

$$\frac{1}{pq} = m\frac{1}{pp'} + n\frac{1}{qq'} + \frac{1}{p'q'}.$$

As a consequence, no element of M_2 can be an atom, which means that M_2 is an antimatter Puiseux monoid.

At this point we are in a position to check the more general fact that, for each $k \geq 2$, the elementary k-prime reciprocal monoid $M_k := M_{\mathbb{P},k}$ is antimatter. To do this, fix an arbitrary elementary generator $(p_1 \cdots p_k)^{-1}$ of M_k. As before, there exist $m, n \in \mathbb{N}$ and $p', q' \in \mathbb{P}$ satisfying that $q' > p' > \max\{p_1, \dots, p_k\}$ and

$$\frac{1}{p_1 p_2} = m\frac{1}{p_1 p'} + n\frac{1}{p_2 q'} + \frac{1}{p'q'}. \tag{4.2}$$

Multiplying both sides of (4.2) by $(p_3 \cdots p_k)^{-1}$, one obtains the following identity:

$$\frac{1}{p_1 p_2 \cdots p_k} = m\frac{1}{p_1 p' p_3 \cdots p_k} + n\frac{1}{p_2 q' p_3 \cdots p_k} + \frac{1}{p'q'p_3 \cdots p_k}.$$

Hence no element of M_k can be an atom and, as a result, the Puiseux monoid M_k is antimatter. □

The next example sheds some light upon the fact a monoid generated by infinitely many elementary generators of the elementary k-prime reciprocal monoid may not be antimatter.

EXAMPLE 4.3. Let $(p_n)_{n \geq 1}$ be a sequence whose underlying set is $\mathbb{P}$. For $k \in \mathbb{N}$, consider the k-prime reciprocal monoid

$$M = \left\langle a_n := \prod_{i=1}^{k} \frac{1}{p_{nk+i}} : n \in \mathbb{N}_0 \right\rangle.$$

Because, for each index $n \in \mathbb{N}$, the prime $p_{nk+1} \mid \mathsf{d}(a_m)$ if and only if $m = n$, we can use the p_{nk+1}-adic valuation map to argue that a_n is an atom of M. As a consequence, M is an atomic monoid whose sets of atoms is

$$\mathscr{A}(M) = \{a_n : n \in \mathbb{N}_0\}.$$

Moreover, since the terms of the sequence $(\mathsf{d}(a_n))_{n \ge 0}$ are pairwise relatively primes, the Puiseux monoid M must satisfy the ACCP [**22**, Proposition 3.1]. Finally, we can see that M does not have the bounded factorization property: for each $n \in \mathbb{N}$, we can write $1 = \mathsf{d}(a_n)a_n$ (because $\mathsf{n}(a_n) = 1$) and so $\{\mathsf{d}(a_n) : n \in \mathbb{N}\} \subseteq \mathsf{L}(1)$ (indeed, it is not difficult to verify that this inclusion is an equality).

As Example 4.4 illustrates, a k-prime reciprocal monoid generated by multiples of the elementary generators of $M_{\mathbb{P},k}$ may be atomic.

EXAMPLE 4.4. Let $(p_n)_{n \ge 1}$ be a sequence whose underlying set is $\mathbb{P}$. For $k \in \mathbb{N}$, consider the following k-prime reciprocal monoid:

$$M = \Big\langle a_S := \sum_{s \in S} \frac{1}{p_s} : S \in \binom{\mathbb{N}}{k} \Big\rangle.$$

Observe that M is a submonoid of $M_{\mathbb{P},k}$. For each subset $S \in \binom{\mathbb{N}}{k}$, the element $\frac{1}{p_s}$ belongs to the prime reciprocal monoid $M_{\mathbb{P}} = \langle \frac{1}{p} : p \in \mathbb{P} \rangle$ for every $s \in S$, whence $a_S = \sum_{s \in S} \frac{1}{p_s} \in M_{\mathbb{P}}$. Therefore M is a submonoid of $M_{\mathbb{P}}$. It follows from Theorem 4.2 that $M_{\mathbb{P}}$ satisfies the ACCP, and so the fact that $M_{\mathbb{P}}$ is a reduced monoid ensures that every submonoid of $M_{\mathbb{P}}$ also satisfies the ACCP. In particular, M satisfies the ACCP and must be atomic.

5. p-adic Puiseux monoids and their internal sums

For a fixed $p \in \mathbb{P}$, the submonoids of the antimatter valuation Puiseux monoid $\mathbb{N}_0\left[\frac{1}{p}\right]$ will play a fundamental role in this section. Let us introduce some convenient terminology.

DEFINITION 5.1. Let p be a prime. We say that a Puiseux monoid M is *p-adic* if M is a submonoid of $\mathbb{N}_0\left[\frac{1}{p}\right]$ or, equivalently, $v_q(M) \subseteq \mathbb{N}_0$ for every $q \in \mathbb{P} \setminus \{p\}$.

We use the term *p-adic monoid* as a shorthand for p-adic Puiseux monoid. Throughout this section, each time we define a p-adic monoid by specifying a sequence of generators $(r_n)_{n \ge 1}$, we shall implicitly assume that $(\mathsf{d}(r_n))_{n \ge 1}$ diverges to infinity; this assumption comes without loss of generality because in order to generate a Puiseux monoid, we only need to repeat each denominator finitely many times. On the other hand, $\lim \mathsf{d}(r_n) = \infty$ does not affect the generality of the results we prove in this section. To see this, if $(\mathsf{d}(r_n))_{n \ge 1}$ is a bounded sequence, then the p-adic monoid generated by $(r_n)_{n \ge 1}$ is finitely generated and, therefore, isomorphic to a numerical monoid.

5.1. Atomicity of p-adic monoids.

Although $\mathbb{N}_0\left[\frac{1}{p}\right]$ is antimatter, it contains plenty of submonoids with diverse and complex atomic behavior. In this section, we delve into the atomicity of p-adic monoids.

Strongly bounded Puiseux monoids were considered in [**34**]. A set Q consisting of rationals is called *strongly bounded* if $\mathsf{n}(Q)$ is bounded, and a Puiseux monoid is called *strongly bounded* if it can be generated by a strongly bounded set of rationals.

Strongly bounded p-adic monoids happen to have only finitely many atoms, as the next proposition indicates.

PROPOSITION 5.2. *A strongly bounded p-adic monoid has only finitely many atoms.*

PROOF. For $p \in \mathbb{P}$, let M be a strongly bounded p-adic monoid. Let $(r_n)_{n\ge 1}$ be a generating sequence for M with underlying set R satisfying that $\mathsf{n}(R) = \{n_1, \dots, n_k\}$ for some k, n_1, $\dots$, $n_k \in \mathbb{N}$. For each $i \in [\![1,k]\!]$, set $R_i := \{r_n : \mathsf{n}(r_n) = n_i\}$ and $M_i = \langle R_i \rangle$. The fact that $R \subseteq M_1 \cup \dots \cup M_k$, along with $\mathscr{A}(M) \cap M_i \subseteq \mathscr{A}(M_i)$, implies that

$$\mathscr{A}(M) \subseteq \bigcup_{i=1}^{k} \mathscr{A}(M_i).$$

Thus, showing that $\mathscr{A}(M)$ is finite amounts to verifying that $|\mathscr{A}(M_i)| < \infty$ for each $i \in [\![1,k]\!]$. Fix $i \in [\![1,k]\!]$. If M_i is finitely generated, then $|\mathscr{A}(M_i)| < \infty$. Let us assume, therefore, that M_i is not finitely generated. This means that there exists a strictly increasing sequence $(\alpha_n)_{n\ge 1}$ such that $M_i = \big\langle \frac{n_i}{p^{\alpha_n}} : n \in \mathbb{N} \big\rangle$. Because $\frac{n_i}{p_n^{\alpha}} = p^{\alpha_{n+1}-\alpha_n}\big(\frac{n_i}{p^{\alpha_{n+1}}}\big)$, the monoid M_i satisfies that $|\mathscr{A}(M_i)| = 0$. Hence we conclude that $\mathscr{A}(M)$ is finite. □

We are now in a position to give a necessary condition for the atomicity of p-adic monoids.

THEOREM 5.3. *For $p \in \mathbb{P}$, let M be a p-adic monoid with $\mathscr{A}(M) = \{r_n : n \in \mathbb{N}\}$ and $\lim r_n = 0$. If M is atomic, then $\lim \mathsf{n}(r_n) = \infty$ (i.e., the set $\mathsf{n}(\mathscr{A}(M))$ is not bounded).*

PROOF. Set $a_n := \mathsf{n}(r_n)$ and $p^{\alpha_n} := \mathsf{d}(r_n)$ for every natural n. Suppose, by way of contradiction, that $\lim a_n \neq \infty$. Then there exists $m \in \mathbb{N}$ such that $a_n = m$ for infinitely many $n \in \mathbb{N}$. For each positive divisor d of m we define the Puiseux monoid

$$M_d := \langle S_d \rangle, \quad \text{where} \quad S_d = \left\{ \frac{a_{k_n}}{p^{\alpha_{k_n}}} : a_{k_n} = m \text{ or } \gcd(m, a_{k_n}) = d \right\}.$$

Observe that $\mathscr{A}(M)$ is included in the union of the M_d. In addition, for each positive divisor d of m, the inclusion $\mathscr{A}(M) \cap M_d \subseteq \mathscr{A}(M_d)$ holds and so

$$\mathscr{A}(M) \subseteq \bigcup_{d \mid m} \mathscr{A}(M_d). \tag{5.1}$$

Because $\mathscr{A}(M)$ contains infinitely many atoms, the inclusion (5.1) implies the existence of a positive divisor d of m such that $|\mathscr{A}(M_d)| = \infty$. Set $N_d = \frac{1}{d}M_d$. Since d divides $\mathsf{n}(q)$ for all $q \in M_d$, it follows that N_d is also a p-adic monoid. In addition, the fact that N_d is isomorphic to M_d implies that $|\mathscr{A}(N_d)| = |\mathscr{A}(M_d)| = \infty$. After setting $b_n := \frac{a_{k_n}}{d}$ and $\beta_n = \alpha_{k_n}$ for every $n \in \mathbb{N}$ such that either $a_{k_n} = m$ or $\gcd(m, a_{k_n}) = d$, we see that

$$N_d = \left\langle \frac{b_n}{p^{\beta_n}} : n \in \mathbb{N} \right\rangle.$$

As $a_n = m$ for infinitely many $n \in \mathbb{N}$, the sequence $(\beta_n)_{n\ge 1}$ is an infinite subsequence of $(\alpha_n)_{n\ge 1}$ and, therefore, it diverges to infinity. In addition, as $\lim \frac{a_n}{p^{\alpha_n}} = 0$, it follows that $\lim \frac{b_n}{p^{\beta_n}} = 0$.

Now we argue that $\mathscr{A}(N_d)$ is finite, yielding the desired contradiction. Take $m' = \frac{m}{d}$. Since there are infinitely many $n \in \mathbb{N}$ such that $b_n = m'$, it is guaranteed that $\frac{m'}{p^n} \in N_d$ for every $n \in \mathbb{N}$. In addition, $\gcd(m', b_n) = 1$ for each $b_n \neq m'$. If $b_n \neq m'$ for only finitely many n, then N_d is strongly bounded and Proposition 5.2 ensures that $\mathscr{A}(N_d)$ is finite. Suppose otherwise that $\gcd(b_n, m') = 1$ (i.e., $b_n \neq m'$) for infinitely many $n \in \mathbb{N}$. For a fixed index $i \in \mathbb{N}$ with $b_i \neq m'$ take an index $j \in \mathbb{N}$ such that $\gcd(b_j, m') = 1$ and large enough so that $b_i p^{\beta_j - \beta_i} > b_j m'$ (the existence of such an index j is guaranteed by the fact that $\lim \frac{b_n}{p^{\beta_n}} = 0$). Since $b_i p^{\beta_j-\beta_i} > b_j m' > \mathsf{f}(\langle b_j, m'\rangle)$, we can take $c, c' \in \mathbb{N}$ such that $b_i p^{\beta_j - \beta_i} = c b_j + c' m'$, which means that

$$\frac{b_i}{p^{\beta_i}} = c\frac{b_j}{p^{\beta_j}} + c'\frac{m'}{p^{\beta_j}}.$$

As $\frac{b_j}{p^{\beta_j}}, \frac{m'}{p^{\beta_j}} \in N_d^\bullet$, it follows that $\frac{b_i}{p^{\beta_i}} \notin \mathscr{A}(N_d)$. Because the index i was arbitrarily taken, N_d is antimatter. In particular, $\mathscr{A}(N_d)$ is finite, which leads to a contradiction. ◻

With the notation as in the statement of Theorem 5.3, the p-adic monoid M may fail to be atomic even when the conditions $\lim r_n = 0$ and $\lim \mathsf{n}(r_n) = \infty$ hold simultaneously. The next example sheds some light upon this observation.

EXAMPLE 5.4. For each $n \in \mathbb{N}_0$, we let F_n denote the n-th Fermat number: $F_n := 2^{2^n} + 1$. Then we consider the following Puiseux monoid:

$$M := \Big\langle \frac{F_n - 2}{2^{2^{n+1}}}, \frac{F_n}{2^{2^{n+1}}} : n \in \mathbb{N} \Big\rangle. \tag{5.2}$$

Observe that, for each $n \in \mathbb{N}$,

$$\frac{1}{2^{2^n - 1}} = \frac{(2^{2^n} - 1) + (2^{2^n} + 1)}{2^{2^{n+1}}} = \frac{F_n - 2}{2^{2^{n+1}}} + \frac{F_n}{2^{2^{n+1}}} \in M.$$

This clearly implies that $\frac{1}{2^n} \in M$ for every $n \in \mathbb{N}$. Therefore the monoid M contains the valuation Puiseux monoid $\mathbb{N}_0\big[\frac{1}{2}\big]$. We proceed to argue that

$$\mathscr{A}(M) = \Big\{ \frac{F_n - 2}{2^{2^{n+1}}} : n \in \mathbb{N} \Big\}. \tag{5.3}$$

Since M is reduced, $\mathscr{A}(M)$ must be contained in the defining generating set

$$\Big\{ \frac{F_n - 2}{2^{2^{n+1}}}, \frac{F_n}{2^{2^{n+1}}} : n \in \mathbb{N} \Big\}$$

of M. On the other hand, for each $n \in \mathbb{N}$, we see that $\frac{1}{2^{2^{n+1}}} \in \mathbb{N}_0\big[\frac{1}{2}\big] \subseteq M$ and so the element $\frac{F_n}{2^{2^{n+1}}}$ is not an atom of M because

$$\frac{F_n}{2^{2^{n+1}}} = 2\frac{1}{2^{2^{n+1}}} + \frac{F_n - 2}{2^{2^{n+1}}}.$$

Thus, $\mathscr{A}(M) \subseteq \big\{ \frac{F_n - 2}{2^{2^{n+1}}} : n \in \mathbb{N} \big\}$. Now fix an arbitrary $k \in \mathbb{N}$, and let us verify that $\frac{F_k - 2}{2^{2^{k+1}}} \in \mathscr{A}(M)$. Since the sequence $\big(\frac{F_n-2}{2^{2^{n+1}}}\big)_{n \ge 1}$ is strictly decreasing, $\frac{F_j - 2}{2^{2^{j+1}}} \nmid_M \frac{F_k - 2}{2^{2^{k+1}}}$

for any $j \in [\![1, k-1]\!]$, whence proving that $\frac{F_k - 2}{2^{2^{k+1}}}$ is an atom amounts to writing

$$(5.4) \qquad \frac{F_k - 2}{2^{2^{k+1}}} = \sum_{j=k}^{n} b_j \frac{F_j - 2}{2^{2^{j+1}}}$$

for some index $n \in \mathbb{N}$ with $n \geq k$ and coefficients $b_k, \dots, b_n \in \mathbb{N}_0$ with $b_k \in \{0, 1\}$ and showing that $b_k = 1$. Observe that we can rewrite (5.4) as follows

$$(5.5) \qquad -b_k \frac{F_k - 2}{2^{2^{k+1}}} + \frac{1}{2^{2^{k+1}}} \prod_{i=0}^{k-1} F_i = \sum_{j=k+1}^{n} b_j \frac{1}{2^{2^{j+1}}} \prod_{i=0}^{j-1} F_i.$$

Let p be an (odd) prime divisor of F_k, and observe that the p-adic valuation of the right-hand side of (5.5) is positive as each summand has a factor F_k. This implies that $b_k \neq 0$ as otherwise the p-adic valuation of the left-hand side of (5.5) would be zero. Hence the set of atoms of M is that described in (5.3), also it is clear that $\lim \frac{F_n - 2}{2^{2^{n+1}}} = 0$ while $\lim(F_n - 2) = \infty$. Finally, let us verify that M is not atomic by showing that we cannot write $1 \in \mathbb{N}_0\left[\frac{1}{2}\right] \subset M$ as a sum of finitely many atoms. Indeed, for each $n \in \mathbb{N}$ the presence of the factor $F_0 = 3$ in $\frac{F_n - 2}{2^{2^{n+1}}} = 2^{-2^{n+1}} \prod_{i=0}^{n-1} F_i$ ensures that the 3-adic valuation of every atom of M is positive and so the same holds for every atomic element of M.

Next we establish both a necessary and a sufficient condition for the atomicity of p-adic monoids having generating sets whose numerators are powers of the same prime.

THEOREM 5.5. *Let p and q be two different primes, and let $M = \langle r_n : n \in \mathbb{N} \rangle$ be a p-adic monoid such that $\mathsf{n}(r_n)$ is a power of q for every $n \in \mathbb{N}$. Then the following statements hold.*

(1) *If M is atomic, then $\lim \mathsf{n}(r_n) = \infty$.*
(2) *If $\lim \mathsf{n}(r_n) = \infty$ and $(r_n)_{n \geq 1}$ is decreasing, then M is atomic.*

PROOF.

(1) Define the sequences $(\alpha_n)_{n \geq 1}$ and $(\beta_n)_{n \geq 1}$ such that $p^{\alpha_n} = \mathsf{d}(r_n)$ and $q^{\beta_n} = \mathsf{n}(r_n)$. Suppose, by way of contradiction, that $\lim \mathsf{n}(r_n) \neq \infty$. Therefore there is a natural j such that $\mathsf{n}(\mathsf{r_n}) = q^j$ for infinitely many $n \in \mathbb{N}$. This implies that $\frac{q^j}{p^n} \in M$ for every $n \in \mathbb{N}$. Thus, for every $x \in M^\bullet$ such that $\mathsf{n}(x) = q^m \geq q^j$, one can write

$$x = \frac{q^m}{\mathsf{d}(x)} = pq^{m-j} \frac{q^j}{p\mathsf{d}(x)} \notin \mathscr{A}(M).$$

As a result, every $a \in \mathscr{A}(M)$ satisfies that $\mathsf{n}(a) < q^j$. This immediately implies that $\mathscr{A}(M)$ is finite. Because M is atomic with $|\mathscr{A}(M)| < \infty$, it must be finitely generated, which is a contradiction.

(2) Consider the subsequence $(k_n)_{n \geq 1}$ of positive integers satisfying that $\mathsf{n}(r_{k_n}) < \mathsf{n}(r_i)$ for every $i > k_n$. It follows immediately that the sequence $(\mathsf{n}(r_{k_n}))_{n \geq 1}$ is increasing. We claim that the Puiseux monoid $M := \langle r_{k_n} : n \in \mathbb{N} \rangle$. Suppose that $j \notin \{k_n : n \in \mathbb{N}\}$. Because $\lim \mathsf{n}(r_n) = \infty$ there are only finitely many indices $i \in \mathbb{N}$ such that $\mathsf{n}(r_i) \leq \mathsf{n}(r_j)$, and it is easy to see that the maximum of such indices, say m, belongs to $\{k_n : n \in \mathbb{N}\}$. As $r_i = p^{\alpha_m - \alpha_i} q^{\beta_i - \beta_m} r_m$, it follows that

$r_i \in \langle r_{k_n} : n \in \mathbb{N}\rangle$. Hence $M = \langle r_{k_n} : n \in \mathbb{N}\rangle$. Therefore it suffices to show that $r_{k_n} \in \mathscr{A}(M)$ for every $n \in \mathbb{N}$. If

$$\frac{q^{\beta_{k_n}}}{p^{\alpha_{k_n}}} = \sum_{i=1}^{t} c_i \frac{q^{\beta_{k_i}}}{p^{\alpha_{k_i}}}, \tag{5.6}$$

for some t, $c_1, \dots, c_t \in \mathbb{N}_0$, then $t \geq n$, $c_1 = \cdots = c_{n-1} = 0$, and $c_n \in \{0, 1\}$. If $c_n = 0$, then by applying the q-adic valuation map to both sides of (5.6) we immediately obtain a contradiction. Thus, $c_n = 1$, which implies that r_{k_n} is an atom. Hence M is atomic. □

5.2. Sum of p-adic: Back to the elementary 1-prime reciprocal. Let M be a Puiseux monoid. It follows that $\{\mathsf{n}(q) : q \in M\}$ is a submonoid of M that is contained in $\mathbb{N}_0$: we call this submonoid the *numerator submonoid* of M. Notice that the numerator submonoid of a Puiseux monoid M is $\mathbb{N}_0 \cap M$.

Let $f \colon \mathbb{P} \to \mathbb{N}_0$ be a function such that $p \nmid f(p)$ if $f(p) \neq 0$. Then we can consider the internal sum $\sum_{p \in \mathbb{P}} \frac{f(p)}{p}\mathbb{N}_0$ inside $\mathbb{Q}_{\geq 0}$ of the free commutative (Puiseux) monoids $\frac{f(p)}{p}\mathbb{N}_0$ over the set of primes $\mathbb{P}$, which is the Puiseux monoid M_f defined via f as follows:

$$M_f := \Big\langle \frac{f(p)}{p} : p \in \mathbb{P} \Big\rangle. \tag{5.7}$$

We call M_f the Puiseux monoid *induced* by the function f. The *support* of f is defined to be $\operatorname{supp} f := \{p \in \mathbb{P} : f(p) \neq 0\}$. One can readily check that M_f is an atomic monoid with set of atoms

$$\mathscr{A}(M_f) = \Big\{ \frac{f(p)}{p} : p \in \operatorname{supp} f \Big\}. \tag{5.8}$$

One important behavior of the monoid M_f is that every element of M_f has an elegant and helpful sum decomposition, as we show in the following proposition.

PROPOSITION 5.6. *Let $f \colon \mathbb{P} \to \mathbb{N}_0$ be a function with support P such that $p \nmid f(p)$ for any $p \in P$, and let M_f be the Puiseux monoid induced by f. Each $q \in M_f$ can be uniquely written as follows:*

$$q = c_0(q) + \sum_{p \in P} c_p(q) \frac{f(p)}{p}, \tag{5.9}$$

where $c_0(q) \in \mathsf{n}(M)$ and all but finitely many terms of the coefficient sequence $(c_p(q))_{p \in P}$ equal zero and $c_p(q) \in [\![0, p-1]\!]$ for all $p \in P$.

PROOF. To simplify notation, set $M := M_f$ and $N := \mathsf{n}(M_f)$. To argue the existence of the sum decomposition in (5.9), first observe that in light of the atomicity of M, consider the set $\mathscr{S}_q$ of pairs $(c_0, (c_p)_{p \in P})$, where $c_0 \in N$ and $(c_p)_{p \in P}$ is a sequence whose terms are nonnegative integers such that

$$q = c_0 + \sum_{p \in P} c_p \frac{f(p)}{p}. \tag{5.10}$$

Observe that $\mathscr{S}_q$ is nonempty because M is atomic with $\mathscr{A}(M) = \{\frac{f(p)}{p} : p \in P\}$ and $0 \in N$. It is clear that the first entry of any pair in $\mathscr{S}_q$ is at most q. Among all such pairs, choose $(c_0(q), (c_p(q))_{p \in P})$ maximizing the value of the first entry. Let us check that with this choice $c_p(q) < p$ for all $p \in P$: indeed, if $c_p(q) \geq p$ for some $p \in P$, after replacing $c_0(q)$ and $c_p(q))$ respectively by $c_0(q) + f(p) \in N$ and $c_p(q) - p \in \mathbb{N}_0$, we obtain another sequence in $\mathscr{S}_q$ whose first term is strictly larger than $c_0(q)$, which is not possible. Hence, the chosen pair $(c_0(q), (c_p(q))_{p \in P})$ yields a sum decomposition as that in (5.9).

For the uniqueness of the sum decomposition, we can assume that the pair $(c_0'(q), (c_p'(q))_{p \in P})$ yields a sum decomposition of q as that on the right-hand side of (5.9). Therefore, for each $p \in P$, after clearing denominators in

$$(c_0(q) - c_0'(q)) + \sum_{p \in P} (c_p(q) - c_p'(q)) \frac{f(p)}{p} = 0,$$

we obtain that $p \mid c_p(q) - c_p'(q)$ and so $c_p(q) = c_p'(q)$. Thus, $c_0(q) = c_0'(q)$. Hence we find that every $q \in M$ can be uniquely written as in (5.9), which concludes our proof. □

The sum decomposition in (5.9) turns out to be quite helpful, so we introduce the following convenient terminology.

DEFINITION 5.7. Let $f\colon \mathbb{P} \to \mathbb{N}_0$ be a function with support P such that $p \nmid f(p)$ for any $p \in P$, and let M_f be the Puiseux monoid induced by f. For each element $q \in M_f$, we call the right-hand side of (5.9) the *canonical sum decomposition* of q in M_f.

- We let $c_0 : M_f \to \mathsf{n}(M_f)$ be the function defined by the assignments $c_0\colon q \mapsto c_0(q)$ for all $q \in M_f$, where $c_0(q)$ is as in (5.9).
- For each $p \in P$, we let $c_p\colon M_f \to \mathbb{N}_0$ be the function defined by the assignments $c_p\colon q \mapsto c_p(q)$ for all $q \in M_f$, where $c_p(q)$ is as in (5.9).

The coefficients in the canonical sum decomposition of M_f have the following desirable behavior with respect to divisibility.

PROPOSITION 5.8. *Let $f\colon \mathbb{P} \to \mathbb{N}_0$ be a function with support P such that $p \nmid f(p)$ for any $p \in P$, and let M_f be the Puiseux monoid induced by f. For $r, s \in M_f$ such that $r \mid_{M_f} s$, the following statements hold.*

(1) $c_0(r) \mid_{\mathsf{n}(M_f)} c_0(s)$.
(2) *If $c_p(r) > c_p(s)$ for some $p \in P$, then $c_0(r) + f(p) \le c_0(s)$.*

PROOF. For simplicity, set $M := M_f$ and $N := \mathsf{n}(M_f)$. Write $s = q + r$ for some $q \in M$. Observe that for each $p \in P$, the inequalities $0 \le c_p(q), c_p(r) < p$ allow us to pick $b_p \in \{0, 1\}$ such that $0 \le c_p(q) + c_p(r) - b_p p < p$. Now write

$$\begin{aligned} s = q + r &= c_0(q) + c_0(r) + \sum_{p \in P} (c_p(q) + c_p(r)) \frac{f(p)}{p} \\ &= \Big(c_0(q) + c_0(r) + \sum_{p \in P} b_p f(p) \Big) + \sum_{p \in P} \big(c_p(q) + c_p(r) - b_p \big) \frac{f(p)}{p}. \end{aligned}$$

(1) From the fact that $f(p) \in N$ for all $p \in P$, we deduce $c_0(q) + c_0(r) + \sum_{p \in P} b_p f(p) \in N$. In addition, $c_p(q) + c_p(r) - b_p \in [\![0, p-1]\!]$ for all $p \in P$, whence the uniqueness of the canonical sum decomposition (5.9) ensures that

$$c_0(s) = c_0(q) + c_0(r) + \sum_{p \in P} b_p f(p) \quad \text{and} \quad c_p(s) = c_p(q) + c_p(r) - b_p$$

for all $p \in P$. Thus, since $f(p) \in N$ for all $p \in P$, the equality $c_0(s) = c_0(q) + c_0(r) + \sum_{p \in P} b_p f(p)$ implies that $c_0(r) \mid_N c_0(s)$.

(2) Now assume that $c_p(r) > c_p(s)$ for some $p \in P$. Then $c_p(s) + b_p = c_p(q) + c_p(r) \geq c_p(r) > c_p(s)$, which implies that $b_p = 1$. Therefore

$$c_0(r) + f(p) = c_0(r) + b_p f(p) \leq c_0(q) + c_0(r) + \sum_{p \in P} b_p f(p) = c_0(s). \qquad \square$$

We proceed to show an application of the canonical sum decompositions of Puiseux monoids M_f induced by functions $f \colon \mathbb{P} \to \mathbb{N}_0$. We argue that such monoids satisfy the ACCP.

THEOREM 5.9. *Let $f \colon \mathbb{P} \to \mathbb{N}_0$ be a function with support P such that $p \nmid f(p)$ for any $p \in P$, and let M_f be the Puiseux monoid induced by f. Then M_f satisfies the ACCP.*

PROOF. For simplicity, set $M := M_f$ and $N := \mathsf{n}(M_f)$. Since N is a numerical monoid, it must satisfy the ACCP.

To argue that M satisfies the ACCP, fix an ascending chain $(q_n + M)_{n \geq 1}$ of principal ideals of M. For each $n \in \mathbb{N}$, the fact that $q_{n+1} \mid_M q_n$, along with part (1) of Proposition 5.8 ensures that $c_0(q_{n+1}) \mid_N c_0(q_n)$. Thus, $(q_n + N)_{n \geq 1}$ is an ascending chain of principal ideals in N, and so it must stabilize. Hence, after dropping finitely many terms from $(q_n + M)_{n \geq 1}$, we can assume that $c_0(q_n) = c_0(q_1)$ for every $n \in \mathbb{N}$.

Now consider the function $\sigma \colon M \to \mathbb{N}_0$ defined as follows: $\sigma(q) := \sum_{p \in P} c_p(q)$ for all $q \in M$. It follows now from part (2) of Proposition 5.8 that $c_p(q_{n+1}) \leq c_p(q_n)$ for every $n \in \mathbb{N}$. Therefore, if the strict inclusion $q_n + M \subsetneq q_{n+1} + M$ holds for some $n \in \mathbb{N}$, then the equality $c_0(q_{n+1}) = c_0(q_n)$ implies that $\sigma(q_n) > \sigma(q_{n+1})$. This, along with the fact that $\sigma(q) \geq 0$ for all $q \in M$, ensures that the set $\{n \in \mathbb{N} : q_n + M \subsetneq q_{n+1} + M\}$ is finite, which is equivalent to the fact that the chain $(q_n + M)_{n \geq 1}$ stabilizes. Thus, M satisfies the ACCP. $\square$

It turns out that we can make choice of the function f inducing Puiseux monoids M_f with various factorization behavior.

EXAMPLE 5.10. Consider the function $f \colon \mathbb{P} \to \mathbb{N}_0$ defined as $f(p) = 1$ for every $p \in \mathbb{P}$. The Puiseux monoid induced by f is the 1-prime reciprocal $M_{\mathbb{P}}$:

$$M_f = M_{\mathbb{P}} = \Big\langle \frac{1}{p} : p \in \mathbb{P} \Big\rangle.$$

We have seen in (5.8) that $\mathscr{A}(M) = \big\{\frac{1}{p} : p \in \mathbb{P}\big\}$. Also, it follows from Theorem 5.9 or part (1) of Theorem 4.2 that the Puiseux monoid M_f satisfies the ACCP. However, M_f is not a BFM: for instance, for each $p \in \mathbb{P}$, the element $1 \in M_f$ can be written as the sum of p copies of the atom $\frac{1}{p}$, and so the inclusion $\mathbb{P} \subseteq \mathsf{L}(1)$ holds (indeed, we can readily check that $\mathsf{L}(1) = \mathbb{P}$ [**22**, Proposition 3.1]).

Let us make a choice of f such that M_f is an FFM that is neither a UFM nor a finitely generated monoid.

EXAMPLE 5.11. Now consider the function $f \colon \mathbb{P} \to \mathbb{N}_0$ defined as $f(p) = p - 1$ for every $p \in \mathbb{P}$. Then the function f induces the following Puiseux monoid:

$$M_f = \Big\langle 1 - \frac{1}{p} : p \in \mathbb{P} \Big\rangle,$$

whose set of atoms is $\{\frac{1}{p} : p \in \mathbb{P}\}$ as pointed out in (5.8), and so M_f cannot be finitely generated. As the set of atoms can be listed increasingly, it follows from [**28**, Theorem 5.6] that M_f is an FFM. Finally, M_f is not a UFM as for any distinct $p, q \in \mathbb{P}$ the inclusion $\{q(p-1), p(q-1)\} \subseteq \mathsf{L}((p-1)(q-1))$ holds: in fact, the element $(p-1)(q-1)$ can be written in M_f as the sum of $q(p-1)$ copies of the atom $1 - \frac{1}{q}$ or as the sum of $p(q-1)$ copies of the atom $1 - \frac{1}{p}$.

We conclude this section with two examples of dense Puiseux monoids that are the internal sum of p-adic monoids. The first of such examples is a monoid that satisfies the ACCP but it not a BFM.

EXAMPLE 5.12. Let $f \colon \mathbb{P} \to \mathbb{N}_0$ be a bounded function whose support P is infinite such that $p \nmid f(p)$ for any $p \in P$. We claim that M_f is a monoid that satisfies the ACCP but is not a BFM. The fact that M_f satisfies the ACCP follows from Theorem 5.9. To argue that M_f is not a BFM, take $n \in \mathbb{N}$ such that the set $L := \{p \in P : f(p) = n\}$ is infinite (which can be done because f is bounded), and observe that n is an element of M_f such that $L \subseteq \mathsf{L}(n)$: indeed, $n = p\frac{f(p)}{p}$ for all $p \in L$. Since L is an infinite, so is $\mathsf{L}(n)$, which ensures that M_f is not a BFM.

This last example is a Puiseux monoid that has the bounded factorization property but not the finite factorization property. Although this monoid is also the internal sum of p-adic monoids, we were not able to find a canonical sum decomposition.

EXAMPLE 5.13. Let P be the set of odd primes, and consider the Puiseux monoid

$$M := \langle A \rangle, \quad \text{where} \quad A = \left\{ \frac{\lfloor p/2 \rfloor}{p}, \frac{p - \lfloor p/2 \rfloor}{p} : p \in P \right\}.$$

Clearly, M is an atomic Puiseux monoid. In addition, it is not hard to verify that $\mathscr{A}(M) = A$. Observe that the element $1 \in M$ has infinitely many factorizations in M: indeed,

$$1 = \frac{\lfloor p/2 \rfloor}{p} + \frac{p - \lfloor p/2 \rfloor}{p}$$

for every $p \in P$. Hence M is not an FFM. Also, notice that $a \geq \frac{1}{3}$ for every $a \in \mathscr{A}(M)$. This implies that no element $q \in M$ can be the sum of more than $\lfloor 3q \rfloor$ atoms: $\mathsf{L}(q) \subseteq [\![1, \lfloor 3q \rfloor]\!]$. Because $|\mathsf{L}(q)| < \lfloor 3q \rfloor$ for all $q \in M$, the Puiseux monoid M is a BFM that is not an FFM.

5.3. Sum of p-adic monoids: Valuation monoids. Next we discuss the internal sum of p-adic valuation monoids. Throughout this section, for any $p \in \mathbb{P}$, we let M_p denote the underlying additive monoid of the rational cyclic semiring $\mathbb{N}_0\left[\frac{1}{p}\right]$. Therefore, for each $p \in \mathbb{P}$, the monoid M_p is the Puiseux monoid consisting of all nonnegative p-adic rationals, and so M_p is a rank-1 valuation monoid. In addition, for a finite nonempty set P consisting of primes, we let M_P denote the internal sum over P of the Puiseux monoids M_p:

$$M_P := \sum_{p \in P} M_p. \tag{5.11}$$

The Puiseux monoids M_P are the central objects we study in this section. It turns out that, inside these Puiseux monoids, each element has a convenient decomposition as a canonical sum, which we will describe in detail, as we proceed.

PROPOSITION 5.14. *Let P be a finite nonempty set consisting of primes, and let M_P be as in* (5.11). *Then each $q \in M_P$ can be written uniquely as follows:*

$$q = c_0 + \sum_{(p,n) \in P \times \mathbb{N}} c_{p,n} \frac{1}{p^n}, \tag{5.12}$$

where $c_0 \in \mathbb{N}_0$ and, for each $p \in P$, the sequence $(c_{p,n})_{n \ge 1}$ consists of nonnegative integer coefficients almost all being zero such that $0 \le c_{p,n} < p$ for every $n \in \mathbb{N}$.

PROOF. Fix $q \in M_P$. The existence and uniqueness of the sum decomposition (5.12) when $q = 0$ is clear. Thus, we assume that $q > 0$.

Because M_P is the internal sum of the monoids M_p, we can write $q = \sum_{p \in P} m_p$, where $m_p \in M_p$ for every $p \in P$. For each $p \in P$ such that $m_p > 0$, we can write $m_p = \frac{n(p)}{p^{e(p)}}$ for some $n(p) \in \mathbb{N}$ and $e(p) \in \mathbb{N}_0$ such that $p \nmid n(p)$. Thus, after assuming that $\sum_{n \in \mathbb{N}_0} c'_{p,n} p^n$ is the representation of $n(p)$ in base p, we obtain that

$$\begin{aligned} m_p &= \frac{n(p)}{p^{e(p)}} \\ &= \frac{1}{p^{e(p)}} \sum_{n \in \mathbb{N}_0} c'_{p,n} p^n \\ &= \sum_{n \ge e(p)} c'_{p,n} p^{n-e(p)} + \sum_{n < e(p)} c'_{p,n} \frac{1}{p^{e(p)-n}} \\ &= c_{p,0} + \sum_{n=1}^{e(p)} c_{p,n} \frac{1}{p^n}, \end{aligned}$$

where $c_{p,0} := \sum_{n \ge e(p)} c'_{p,n} p^{n-e(p)} \in \mathbb{N}_0$ and $c_{p,k} := c'_{p,e(p)-k}$ for every $k \in [\![1, e(p)]\!]$. Hence the existence of (5.12) follows after taking $c_0 := \sum_{p \in P} c_{p,0}$ and $c_{p,n} = 0$ for all $(p,n) \in P \times \mathbb{N}$ with $n > e(p)$.

We proceed to argue the uniqueness of (5.12). For this, suppose that we can write some $q \in M_P$ with $q > 0$ as in (5.12) and also as follows:

$$q = b_0 + \sum_{(p,n) \in P \times \mathbb{N}} b_{p,n} \frac{1}{p^n}, \tag{5.13}$$

where, for each $p \in P$, the sequence $(b_{p,n})_{n \ge 1}$ consists of nonnegative integers almost all being zero such that $b_{p,n} \in [\![0, p-1]\!]$ for every $n \in \mathbb{N}$. Suppose, towards a contradiction, that the sum decompositions of q in (5.12) and (5.13) are not equal. In this case, we can pick $p \in P$ such that the sequences $(b_{p,n})_{n \ge 1}$ and $(c_{p,n})_{n \ge 1}$ are distinct even though

$$\sum_{n \in \mathbb{N}} b_{p,n} \frac{1}{p^n} = \sum_{n \in \mathbb{N}} c_{p,n} \frac{1}{p^n}.$$

Therefore, after taking $m := \max\{n \in \mathbb{N} : b_{p,n} \ne c_{p,n}\}$, we see that

$$c_{p,m} - b_{p,m} = \sum_{n=1}^{m-1} (b_{p,n} - c_{p,n}) p^{m-n},$$

which implies that $p \mid c_{p,m} - b_{p,m}$. However, this contradicts that $0 \le b_{p,m}, c_{p,m} < p$. Thus, the sum decomposition on the right-hand side of (5.12) is unique. $\square$

With notation as in Proposition 5.14, we will see that the sum decomposition in (5.12) is quite helpful to understand divisibility aspects in the monoids M_P. Based on this, we introduce the following terminology.

DEFINITION 5.15. Let P be a finite nonempty set consisting of primes, and let M_P be the Puiseux monoid in (5.11). For each $q \in M_P$, we call the right-hand side of (5.12) the *canonical sum decomposition* of q in M_P.

- We let $c_0 \colon M_P \to \mathbb{N}_0$ be the function defined as follows: $c_0(q) = c_0$ for all $q \in M_P$, where c_0 is as in (5.12).
- For each pair $(p, n) \in P \times \mathbb{N}$, we let $c_{p,n} \colon M_P \to \mathbb{N}_0$ be the function defined as follows: $c_{p,n}(q) = c_{p,n}$ for all $q \in M_P$, where $c_{p,n}$ is as in (5.12).

Let us take a look at some basic properties of the canonical sum decomposition we have just introduced for the Puiseux monoids M_P introduced in (5.11).

PROPOSITION 5.16. *Let P be a finite nonempty set of primes, and let M_P be the monoid defined in* (5.11). *Then the following statements hold.*

(1) *For each $p \in P$ and $q \in M_P$,*

$$v_p(q) \geq 0 \quad \text{if and only if} \quad c_{p,n}(q) = 0 \quad \text{for every} \quad n \in \mathbb{N}.$$

(2) *For each $(p, n) \in P \times \mathbb{N}$ and for all $q, r \in M_P$ with* $\gcd(\mathsf{d}(q), \mathsf{d}(r)) = 1$,

$$c_{p,n}(q + r) = c_{p,n}(q) + c_{p,n}(r).$$

(3) *For all $q, r \in M_P$ with* $\gcd(\mathsf{d}(q), \mathsf{d}(r)) = 1$,

$$c_0(q + r) = c_0(q) + c_0(r).$$

PROOF.

(1) Observe that there exists $n \in \mathbb{N}$ such that $c_{p,n}(q) > 0$ if and only if $v_p(q) = -m$, where $m := \max\{n \in \mathbb{N} : c_{p,n}(q) > 0\}$.

(2) Fix a pair $(p, n) \in P \times \mathbb{N}$, and take $q, r \in M_P$ with $\gcd(\mathsf{d}(q), \mathsf{d}(r)) = 1$. In light of part (1), the fact that p is a prime factor of at most one of the positive integers $\mathsf{d}(q)$ and $\mathsf{d}(r)$ implies that either $c_{p,n}(q) = 0$ or $c_{p,n}(r) = 0$. Hence $c_{p,n}(q + r) = c_{p,n}(q) + c_{p,n}(r)$.

(3) As before, take $q, r \in M_P$ with $\gcd(\mathsf{d}(q), \mathsf{d}(r)) = 1$. For any $(p, n) \in \mathbb{P} \times \mathbb{N}$, it follows from part (2) that $c_{p,n}(q + r) = c_{p,n}(q) + c_{p,n}(r)$. Thus,

$$\begin{aligned} c_0(q+r) &= q + r - \sum_{(p,n) \in P \times \mathbb{N}} c_{p,n}(q+r) \frac{1}{p^n} \\ &= q + r - \sum_{(p,n) \in P \times \mathbb{N}} c_{p,n}(q) \frac{1}{p^n} - \sum_{(p,n) \in P \times \mathbb{N}} c_{p,n}(r) \\ &= c_0(q) + c_0(r). \end{aligned}$$

□

For any $q \in M_P$, given the uniqueness of the canonical sum decomposition, we can also write $q = c_0(q) + \sum_{p \in P} c_p(q)$, where $c_p(q) := \sum_{n \in \mathbb{N}} c_{p,n}(q) \frac{1}{p^n}$. Let us now show that the canonical sum decomposition inside the monoids defined in (5.11) behaves well with respect to divisibility.

PROPOSITION 5.17. *Let P be a nonempty finite set of primes, and let M_P be the monoid defined in* (5.11). *For any $r, s \in M_P$, the following statements hold.*

(1) *$c_0(r)$ is the largest integer dividing r in M_P.*

(2) *If* $r \mid_{M_P} s$, *then* $c_0(r) \le c_0(s)$.
(3) *If* $r \mid_{M_P} s$ *and* $c_p(r) > c_p(s)$ *for some* $p \in P$, *then* $c_0(r) < c_0(s)$.

PROOF.

(1) It is clear that $c_0(r)$ is a nonnegative integer dividing r. In addition, if n_r is a nonnegative integer such that $n_r \mid_{M_P} r$, then we can write $r = n_r + r'$ for some $r' \in M_P$ and observe that

$$r = (n_r + c_0(r')) + \sum_{(p,n)\in P\times\mathbb{N}} c_{p,n}(r')\frac{1}{p^n}.$$

Thus, from the uniqueness of the canonical sum decomposition of r, we obtain that $c_0(r) = n_r + c_0(r')$, and so $n_r \le c_0(r)$. Therefore $c_0(r)$ is the largest integer dividing r in M_P.

(2) Write $s = q + r$ for some $q \in M_P$. Then we can write

$$s = c_0(q) + c_0(r) + \sum_{(p,n)\in P\times\mathbb{N}} (c_{p,n}(q) + c_{p,n}(r))\frac{1}{p^n}.$$

As a consequence, $c_0(q) + c_0(r)$ divides s in M_S. Therefore, part (1) guarantees that the inequality $c_0(s) \ge c_0(q) + c_0(r) \ge c_0(r)$ holds.

(3) Now suppose that $r \mid_{M_P} s$ and $c_p(r) > c_p(s)$ for some $p \in P$. Assume, by way of contradiction, that $c_0(r) \ge c_0(s)$ and so that $c_0(r) = c_0(s)$. Set $r' := s - r \in M_P$ and observe that $c_0(r') = 0$ and we can write $c_p(r) + c_p(r') = n_p + r_p$ for some $n_0 \in \mathbb{N}_0$ and $r_p \in \mathbb{N}_0\big[\frac{1}{p}\big] \cap (0,1)$, and so the fact that $c_0(r) = c_0(s) \ge n_p + c_0(r)$ ensures that $n_p = 0$. Thus, $c_p(r) + c_p(r') < 1$ and this implies that $c_p(s) = c_p(r) + c_p(r') \ge c_p(r)$, a contradiction. □

We conclude this section proving that for each finite nonempty set P consisting of primes, M_P is an antimatter monoid and also that M_P is a valuation monoid if and only if $|P| = 1$.

PROPOSITION 5.18. *Let* P *be a finite nonempty set of primes, and let* M_P *be the monoid defined in* (5.11). *Then the following statements hold.*

(1) M_P *is an antimatter.*
(2) M_P *is a valuation monoid if and only if* $|P| = 1$.

PROOF.

(1) As M_P is the internal sum over P of the Puiseux monoids M_p, the fact that M_P is antimatter follows immediately from the fact that, for each $p \in P$, the monoid M_p is antimatter: indeed, none of elements in the generating set $\{\frac{1}{p^n} : n \in \mathbb{N}_0\}$ of M_p is an atom of M_p because $\frac{1}{p^n} = p\frac{1}{p^{n+1}}$ for every $n \in \mathbb{N}_0$.

(2) For the reverse implication, it suffices to observe that if P equals a singleton $\{p\}$ for some $p \in \mathbb{P}$, then M_P is the additive monoid M_p of $\mathbb{N}_0[\frac{1}{p}]$, which is clearly a valuation monoid. For the direct implication, assume that $|P| \ge 2$. Let the pair $(M_P, |_{M_P})$ be the partially ordered with underlying set M_P and order relation given by divisibility inside M_P. Let us argue the following claim, from which we can immediately complete our proof.

CLAIM. For any distinct $p_1, p_2 \in P$ and $(q_1, q_2) \in M_{p_1} \times M_{p_2}$ with $c_0(q_1) = c_0(q_2)$, the following conditions are equivalent.

(a) $\{q_1, q_2\}$ is a chain in the poset $(M_P, |_{M_P})$.

(b) $\{q_1, q_2\}$ intersects $\mathbb{N}_0$.

PROOF OF CLAIM. (a) $\Rightarrow$ (b): Suppose that $\{q_1, q_2\}$ is a chain in the poset $(M_P, |_{M_P})$ and assume, without loss of generality, that $q_1 \mid_{M_P} q_2$. Given the uniqueness of the canonical sum decompositions of q_1 and q_2 inside M_P, the equality $c_{p,n}(q_1) = 0$ holds for all $(p, n) \in P \times \mathbb{N}$ with $p \neq p_1$ while the equality $c_{p,n}(q_2) = 0$ for all $(p, n) \in P \times \mathbb{N}$ with $p \neq p_2$. Hence

$$q_1 = c_0(q_1) + c_{p_1}(q_1) \quad \text{and} \quad q_2 = c_0(q_2) + c_{p_2}(q_2),$$

where $c_{p_1}(q_1) = \sum_{n \in \mathbb{N}} c_{p_1,n}(q_1)\frac{1}{p_1^n} \in M_{p_1}$ and $c_{p_2}(q_2) = \sum_{n \in \mathbb{N}} c_{p_2,n}(q_2)\frac{1}{p_2^n} \in M_{p_2}$. Since $q_1 \mid_{M_P} q_2$ and $c_0(q_1) = c_0(q_2)$, it follows from part (3) of Proposition 5.17 that $c_{p_1}(q_1) \leq c_{p_1}(q_2) = 0$, which enforces the equality $c_{p_1}(q_1) = 0$. Thus, $q_1 = c_0(q_1) \in \mathbb{N}_0$ and so $q_1 \in \{q_1, q_2\} \cap \mathbb{N}_0$.

(b) $\Rightarrow$ (a): Now suppose that $\{q_1, q_2\}$ intersects $\mathbb{N}_0$ and assume, without loss of generality, that $q_1 \in \mathbb{N}_0$. Then $c_0(q_2) = c_0(q_1) = q_1$ by part (1) of Proposition 5.17. Thus, $q_1 = c_0(q_2) \mid_{M_P} q_2$, which implies that $\{q_1, q_2\}$ is a chain in the poset $(M_P, |_{M_P})$. Hence the claim is established.

Finally, we argue that M_P is not a valuation monoid. Take $p_1, p_2 \in P$ with $p_1 \neq p_2$ and then set $q_1 := \frac{1}{p_1}$ and $q_2 := \frac{1}{p_2}$. Note that $(q_1, q_2) \in M_{p_1} \times M_{p_2}$ and $c_0(q_1) = c_0(q_2) = 0$. Thus, by virtue of the established claim, the fact that neither q_1 nor q_2 are integers implies that $q_1 \nmid_{M_P} q_2$ and $q_2 \nmid_{M_P} q_1$. Hence M is not a valuation monoid. $\square$

6. Multiplicatively closed Puiseux monoids

The first study of the atomic structure of a family of multiplicatively closed Puiseux monoids was briefly initiated by the second and third authors in [**34**, Section 5] and continued by the first three authors in [**10**] and by fourth author in [**43**], and all these paper focused on the additive monoids of the rational cyclic semiring $\mathbb{N}_0[q]$. Our purpose in this section is to revisit this class and explore other new classes of Puiseux monoids that are multiplicatively closed.

6.1. Additive monoids of rational cyclic semirings. Let us start by considering, for each $q \in \mathbb{Q}_{>0}$, the cyclic subsemiring of $\mathbb{Q}_{\geq 0}$ generated by q:

$$\mathbb{N}_0[q] := \{f(q) : f(x) \in \mathbb{N}_0[x]\},$$

where $\mathbb{N}_0[x]$ denotes the semiring of polynomials in the indeterminate x over $\mathbb{N}_0$. We let M_q denote the underlying additive monoid of the semiring $\mathbb{N}_0[q]$, which is a Puiseux monoid. Observe that the monoid M_q is additively generated by the powers of q:

$$M_q = \langle q^n : n \in \mathbb{N}_0 \rangle. \tag{6.1}$$

One can readily see that if $\mathsf{n}(q) = 1$, then $M_q = \mathbb{N}_0\left[\frac{1}{d}\right]$, where $d := \mathsf{d}(q)$, and so any nonzero $r \in M_q$ can be written as $r = d(\frac{r}{d})$ and so M_q is an antimatter monoid. In addition, for any $r, s \in M_q$, the divisibility relation $r \mid_{M_q} s$ holds if and only if $r \leq s$, and so M_q is a valuation monoid.

As was the case for various Puiseux monoids we investigated in previous sections, elements inside the monoid M_q have a special sum decomposition (when $q \notin \mathbb{N}$), which we argue as follows.

PROPOSITION 6.1. *For $q \in \mathbb{Q}_{>0} \setminus \mathbb{N}$, let M_q be the underlying additive monoid of the semiring $\mathbb{N}_0[q]$. Then each $r \in \mathbb{N}_0[q]$ can be uniquely written as follows:*

$$r = c_0(r) + \sum_{n \in \mathbb{N}} c_n(r) q^n, \tag{6.2}$$

where $(c_n(r))_{n \geq 0}$ is a sequence with $c_n(r) \in [\![0, \mathsf{d}(q) - 1]\!]$ for every $n \in \mathbb{N}$ such that all but finitely many of whose terms are zero.

PROOF. To argue the existential part of the statement, it is convenient to split our argument into two cases.

CASE 1. $\mathsf{n}(q) = 1$. After setting $d := \mathsf{d}(q)$, we see that $M_q = \langle \frac{1}{d^n} : n \in \mathbb{N}_0 \rangle$. Fix $r \in M_q$ and take the minimum $k \in \mathbb{N}_0$ such that $d^k r \in \mathbb{N}_0$. It follows from the minimality of k that we can represent $d^k r$ in base d as $d^k r = \sum_{n=0}^{k} c_n d^{k-n}$ for some coefficients $c_0, \dots, c_k \in [\![0, d-1]\!]$. From this, we deduce that

$$r = \frac{1}{d^k} \sum_{n=0}^{k} c_n d^{k-n} = c_0 + \sum_{n=1}^{k} c_n \frac{1}{d^n} = c_0 + \sum_{n \in \mathbb{N}} c_n q^n,$$

where the equality $c_n = 0$ holds for every $n > k$. After setting $c_0(r) := c_0$ and $c_n(r) := c_n$, one obtains the sum decomposition of (6.2).

CASE 2. $\mathsf{n}(q) \geq 2$. As $\mathsf{n}(q), \mathsf{d}(q) \geq 2$, it follows from [**34**, Theorem 6.2] that the Puiseux monoid M_q is atomic with set of atoms

$$\mathscr{A}(M_q) = \{q^n : n \in \mathbb{N}_0\}.$$

Fix an element $r \in M_q$, and let us find a sum decomposition of r as that in (6.2). Since M_q is atomic, $\mathsf{Z}(r)$ is a nonempty set. For each factorization $z := \sum_{n \in \mathbb{N}_0} c_n q^n \in \mathsf{Z}(r)$, consider the subset

$$N_z := \{0\} \cup \{n \in \mathbb{N} : c_n \geq \mathsf{d}(q)\}$$

of $\mathbb{N}_0$. Take a factorization $z := \sum_{n \in \mathbb{N}_0} c_n q^n \in \mathsf{Z}(r)$ such that the minimum $m := \min N_z$ is as small as it can possibly be. We claim that $m = 0$. Assume, towards a contradiction, that $m \geq 1$. Since $c_m \geq \mathsf{d}(q)$, we can write $c_m = b_m \mathsf{d}(q) + r_m$ for some $b_m \in \mathbb{N}$ and $r_m \in [\![0, \mathsf{d}(q) - 1]\!]$. Observe that

$$\begin{aligned} c_{m-1} q^{m-1} + c_m q^m &= c_{m-1} q^{m-1} + (b_m \mathsf{d}(q)) q^m + r_m q^m \\ &= (c_{m-1} + b_m \mathsf{n}(q)) q^{m-1} + r_m q^m. \end{aligned}$$

Because $r_m < \mathsf{d}(q)$, after replacing $c_{m-1} q^{m-1} + c_m q^m$ by $(c_{m-1} + b_m \mathsf{n}(q)) q^{m-1} + r_m q^m$ in z, we obtain a new factorization w such that $\min N_w < m$, which contradicts the minimality of m. Hence $m = 0$, which implies that the coefficients in the right-hand side of

$$r = c_0 + \sum_{n \in \mathbb{N}} c_n q^n \tag{6.3}$$

are such that $c_n < \mathsf{d}(q)$ for every $n \in \mathbb{N}$, and so after setting $c_0(r) := c_0$ and $c_n(r) := c_n$, we obtain the sum decomposition of (6.2).

Let us now argue the uniqueness of the sum decomposition in (6.2). Assume, by way of contradiction, that the element $r \in M_q$ has two distinct sum decompositions:

the one shown in (6.2) and also the following

$$r = b_0(r) + \sum_{n\in\mathbb{N}} b_n(r)q^n, \tag{6.4}$$

where $(b_n(r))_{n\geq 0}$ is a sequence with $b_n(r) \in [\![0, \mathsf{d}(q)-1]\!]$ for every $n \in \mathbb{N}$ such that all but finitely many terms of $(b_n(r))_{n\geq 0}$ are zero. After setting $m := \max\{n \in \mathbb{N}_0 : |c_n - b_n| \neq 0\}$ and set equal the right-hand sides of (6.2) and (6.4), we obtain

$$(c_m(r) - b_m(r))\mathsf{n}(q)^m = \sum_{n=0}^{m-1} (b_n(r) - c_n(r))\mathsf{n}(q)^n \mathsf{d}(q)^{m-n}.$$

As $\mathsf{d}(q)$ divides every summand in the right-hand side of the obtained identity, $\mathsf{d}(q)$ must divide the right-hand side of the same identity, whence $\mathsf{d}(q) \mid c_m - b_m$ because $\gcd(\mathsf{d}(q), \mathsf{n}(q)) = 1$. However, this implies that $b_m = c_m$, which is a contradiction. □

We can now use the canonical sum decomposition established in (6.2) for the monoids M_q to argue that it is atomic if and only if it is an LFFM.

THEOREM 6.2. *For $q \in \mathbb{Q}_{>0}$, let M_q be the underlying additive monoid of $\mathbb{N}_0[q]$. Then the following conditions are equivalent.*

(a) *M_q is an LFFM.*
(b) *M_q is atomic.*
(c) *$q \in \mathbb{Q} \setminus \{\frac{1}{n} : n \in \mathbb{N}_{\geq 2}\}$.*

PROOF. (a) ⇒ (b): It follows from the corresponding definitions.

(b) ⇒ (c): If $q = \frac{1}{n}$ for some $n \in \mathbb{N}$ such that $n \geq 2$, then $\frac{1}{n^k} = n\frac{1}{n^{k+1}}$ for every $k \in \mathbb{N}$, and so the set of atoms of M_q is empty. Hence M_q is not atomic.

(c) ⇒ (a): Finally, assume that $q \in \mathbb{Q} \setminus \{\frac{1}{n} : n \in \mathbb{N}_{\geq 2}\}$. In this case, it is well known and not hard to verify that the monoid M_q is atomic. We split the rest of our argument into the following two cases.

CASE 1. $q \geq 1$. In this case, we can argue that M_q is indeed an FFM, whence an LFFM. Observe that if $q \in \mathbb{N}$, then $M_q = \mathbb{N}_0$ and so it is a UFM and, in particular, an FFM. Therefore we can assume that $q \in \mathbb{Q}_{>1} \setminus \mathbb{Z}$. In this case, the set of atoms of M_q is

$$\mathscr{A}(M_q) = \{q^n : n \in \mathbb{N}_0\}. \tag{6.5}$$

Therefore M_q is an increasing positive monoid of an ordered field because the sequence of atoms $(q_n)_{n\geq 0}$ is increasing. Hence M_q is an FFM by virtue of [**28**, Theorem 5.6].

CASE 2. $q < 1$. Assume, towards a contradiction, that M_q is not an LFFM in this case. Using the fact that $q \neq \frac{1}{n}$ for any $n \in \mathbb{N}$ with $n \geq 2$, one can verify that the set of atoms of M_q is given by the equality (6.5) and, therefore, M_q is atomic.

Since M_q is not an LFFM, we can choose $\ell \in \mathbb{N}$ such that $|\mathsf{Z}_\ell(r)| = \infty$ for some $r \in M_q$. We can assume that we have chosen ℓ as small as possible. For each index $m, n \in \mathbb{N}_0$, we let $\mathsf{Z}_{\ell,m}(r, q^n)$ denote the set of factorizations in $\mathsf{Z}_\ell(r)$ having exactly m copies of the atom q^n. Observe that for each $n \in \mathbb{N}_0$, the set $\mathsf{Z}_{\ell,m}(r, q^n)$ is empty for every $m \geq \ell$. Therefore, for each index $n \in \mathbb{N}_0$, the equality

$$\mathsf{Z}_\ell(r) = \bigcup_{m=0}^{\ell} \mathsf{Z}_{\ell,m}(r, q^n)$$

holds. Fix $n \in \mathbb{N}_0$. Since $|\mathsf{Z}_\ell(r)| = \infty$, there is an index $m \in [\![0, \ell]\!]$ such that $|\mathsf{Z}_{\ell,m}(r, q^n)| = \infty$. This implies, in particular, that $r - mq^n$ belongs to M_q. Define the map $f \colon \mathsf{Z}_{\ell,m}(r, q^n) \to \mathsf{Z}_{\ell-m}(r - mq^n)$ as follows:

$$f \colon mq^n + \sum_{k \in \mathbb{N}_0} c_k q^k \mapsto \sum_{k \in \mathbb{N}_0} c_k q^k,$$

for every sequence $(c_k)_{k \ge 0}$ of nonnegative integers such that $c_n = 0$ and $mq^n + \sum_{k \in \mathbb{N}_0} c_k q^k$ is a length-ℓ factorization of r. It is clear that f is an injective function, and so $|\mathsf{Z}_{\ell,m}(r, q^n)| \le |\mathsf{Z}_{\ell-m}(r - mq^n)|$. Thus, $\mathsf{Z}_{\ell-m}(r - mq^n)$ is an infinite set consisting of length-$(\ell - m)$ factorizations of $r - mq^n$. Therefore it follows from the minimality of ℓ that $m = 0$. Thus, for each $n \in \mathbb{N}_0$,

$$(6.6) \qquad |\mathsf{Z}_{\ell,0}(r, q^n)| = \infty \quad \text{and} \quad \Big| \bigcup_{m \in \mathbb{N}} \mathsf{Z}_{\ell,m}(r, q^n) \Big| = \Big| \bigcup_{m=1}^{\ell} \mathsf{Z}_{\ell,m}(r, q^n) \Big| < \infty$$

As $q < 1$, we can choose an index $N \in \mathbb{N}$ large enough so that $q^N < r/\ell$. Now let $\mathsf{Z}_{\ell, \ge N}(r)$ be the set consisting of factorizations in $\mathsf{Z}_\ell(r)$ which do not contain any of the atoms 1, q, $\dots$, q^N. It follows from (6.6) that $\mathsf{Z}_{\ell,m}(r, q^n)$ is a finite set for all $(m, n) \in \mathbb{N} \times [\![0, N-1]\!]$. This implies that

$$\mathsf{Z}_{\ell, \ge N}(r) = \mathsf{Z}_\ell(r) \setminus \Big(\bigcup_{n=0}^{N-1} \bigcup_{m \in \mathbb{N}} \mathsf{Z}_{\ell,m}(r, q^n) \Big),$$

whence the set $\mathsf{Z}_{\ell, \ge N}(r)$ is nonempty. Take a sequence $(c_k)_{k \ge N}$ of nonnegative terms such that $\sum_{k \ge N} c_k q^k$ is a length-ℓ factorization of r in M_q, and notice that

$$r = \sum_{k \ge N} c_k q^k \le \sum_{k \ge N} c_k q^N = q^N \ell,$$

which contradicts our choice of $N \in \mathbb{N}$ such that $q^N < r/\ell$. Hence we conclude that M_q is an LFFM. □

6.2. Internal sum of rational cyclic semirings. Various classes of positive monoids that generalize the class consisting of the additive monoids of rational cyclic semirings have been studied in recent literature (see, for instance, [**1, 2, 42, 43**]). In this final subsection, we study another natural generalization of such class, the Puiseux monoids generated by the nonnegative powers of finitely many positive rationals.

Fix $n \in \mathbb{N}$ and let x_1, $\dots$, x_n be n distinct indeterminates or, more formally, let $X := \{x_1, \dots, x_n\}$ with $|X| = n$ be a subset of a field extension of $\mathbb{Q}$ such that X is algebraically independent over $\mathbb{Q}$. As usual, we let $\mathbb{N}_0[x_1, \dots, x_n]$ denote the polynomial semiring in the indeterminates x_1, $\dots$, x_n over the prototypical semiring $\mathbb{N}_0$. For positive rationals q_1, $\dots$, $q_n \in \mathbb{Q}_{>0}$, set $Q := \{q_1, \dots, q_n\}$. Then

$$\mathbb{N}_0[Q] := \mathbb{N}_0[q_1, \dots, q_n] := \big\{ f(q_1, \dots, q_n) : f \in \mathbb{N}_0[x_1, \dots, x_n] \big\}$$

is the semiring extension of $\mathbb{N}_0$ by Q inside $\mathbb{Q}$, and so the underlying additive monoid of $\mathbb{N}_0[Q]$ is a Puiseux monoid. The monoids we proceed to introduce and study are submonoids of the underlying additive monoid of $\mathbb{N}_0[Q]$.

DEFINITION 6.3. For a nonempty finite set Q consisting of positive rationals, let M_Q be the Puiseux monoid generated by the nonnegative powers of the elements of Q:

$$M_Q = \langle q^n : (q,n) \in Q \times \mathbb{N}_0 \rangle. \tag{6.7}$$

We call M_Q the (Puiseux) monoid *powerly generated* by Q or a *powerly generated* Puiseux monoid.

Throughout the rest of the paper, we will use the following notation: for any $q \in \mathbb{Q}_{>0}$, we write M_q instead of $M_{\{q\}}$. Observe that this is consistent with the notation used in the previous section, where we let M_q denote the underlying additive monoid of the rational cyclic semiring $\mathbb{N}_0[q]$. Thus, for each finite nonempty set Q consisting of positive rationals, the monoid powerly generated by Q is simply the internal sum over Q of the Puiseux monoids M_q:

$$M_Q = \sum_{q \in Q} \langle q^n : n \in \mathbb{N}_0 \rangle = \sum_{q \in Q} M_q.$$

We have already considered a special class of powerly generated monoids: indeed, for any finite nonempty set P consisting of primes, we can set $Q := \{\frac{1}{p} : p \in P\}$ and notice that the monoid M_Q powerly generated by Q is the internal sum over $p \in P$ of the valuation Puiseux monoids $\mathbb{N}_0[\frac{1}{p}]$, which we already considered at the end of Section 5. Although we have seen in Proposition 5.18 that these special powerly generated monoids are antimatter, this is not the case for more general powerly generated monoids. Let us take a look at a powerly generated monoid that is atomic but does not satisfy the ACCP.

EXAMPLE 6.4. Fix two distinct positive integers $n_1, n_2 \in \mathbb{N}$ such that $\gcd(n_1, n_2) = 1$, and take $d_1, d_2 \in \mathbb{N}$ such that $n_1 n_2 < d_2$. Then consider the Puiseux monoid

$$M := \left\langle \left(\frac{n_1}{d_1 d_2}\right)^m, \left(\frac{n_2}{d_1 d_2}\right)^m : m \in \mathbb{N}_0 \right\rangle.$$

Fix $k \in \mathbb{N}$, and consider the numerical monoid $N := \langle n_1^k, n_2^k \rangle$. Now observe that we can bound the Frobenius number $\mathsf{f}(N)$ as follows: $\mathsf{f}(N) < (n_1^k - 1)(n_2^k - 1) < d_2^k$. Therefore there exist coefficients $c_1, c_2 \in \mathbb{N}_0$ such that $c_1 n_1^k + c_2 n_2^k = d_2^k$. This implies that

$$\frac{1}{d_1^k} = \frac{c_1 n_1^k + c_2 n_2^k}{(d_1 d_2)^k} = c_1 \left(\frac{n_1}{d_1 d_2}\right)^k + c_2 \left(\frac{n_2}{d_1 d_2}\right)^k \in M.$$

We have verified that $\frac{1}{d_1^m} \in M$ for every $m \in \mathbb{N}$, from which we obtain that

$$\Big(\frac{1}{d_1^m} + M\Big)_{m \geq 1}$$

is an ascending chain of principal ideals of M that does not stabilize. As a consequence, M is an atomic monoid that does not satisfy the ACCP.

Unlike additive monoids of rational cyclic semirings, for each nonempty finite subset Q of $\mathbb{Q}_{>0}$ with $|Q| \geq 2$, the monoid M_Q powerly generated by Q is not closed under the standard multiplication. We proceed to characterize when M_Q is closed under the standard multiplication.

PROPOSITION 6.5. *For a nonempty finite set Q consisting of positive rationals, the following conditions are equivalent.*

(a) *M_Q is closed under multiplication, which means that $M_Q = \mathbb{N}_0[Q]$.*

(b) $q^i r^j \in M_Q$ *for all* $q, r \in Q$ *and* $i, j \in \mathbb{N}_0$.

PROOF. (a) $\Rightarrow$ (b): This is straightforward.

(b) $\Rightarrow$ (a): Assume now that $q^i r^j \in M_Q$ for all $q, r \in Q$ and $i, j \in \mathbb{N}_0$. If $Q = \{q\}$ for some $q \in \mathbb{Q}_{>0}$, then $M_Q = \mathbb{N}_0[q]$, which is a semiring. Therefore we assume that $|Q| \geq 2$. Let us show that, for any $k \in \mathbb{N}_{\geq 2}$ and $q_1, \ldots, q_k \in Q$, the product $q_1 \cdots q_k$ belongs to M_Q: we proceed by inducting on k. The case $k = 2$, which is our base case, follows from our initial assumption. For the inductive step, fix $k \in \mathbb{N}$ with $k \geq 2$ and suppose that, for any $\ell \in [\![2, k]\!]$, the product of any ℓ elements of Q (repetitions allowed) belongs to M_Q. Take $q_1, \ldots, q_{k+1} \in Q$ and let us verify that $q := q_1 \cdots q_{k+1}$ belongs to M_Q. If $q_1 = \cdots = q_{k+1}$, then $q = q_1^{k+1} \in M_Q$. Otherwise, after relabeling the subindices of $q_1, \ldots, q_{k+1}$, we can take $e \in \mathbb{N}$ with $e < k + 1$ such that $q = (q_1 \cdots q_{k+1-e}) q_{k+1}^e$ and $q_{k+1} \notin \{q_1, \ldots, q_{k+1-e}\}$. As $1 \leq e < k + 1$, our induction hypothesis ensures that $q_1 \cdots q_{k+1-e} \in M_Q$, Therefore we can take a finite-supported sequence $(c_{q,n})_{n \geq 0}$ of nonnegative integer coefficients such that

$$q = q_{k+1}^e \sum_{(q,n) \in Q \times \mathbb{N}_0} c_{q,n} q^n = \sum_{(q,n) \in Q \times \mathbb{N}_0} c_{q,n} q^n q_{k+1}^e,$$

and so we can deduce that $q \in M_Q$ from the fact that $q^n q_{k+1}^e \in M_Q$ for all $(q, n) \in Q \times \mathbb{N}_0$. Hence $c \prod_{q \in Q} q^{e_q} \in M_Q$ for all coefficient $c \in \mathbb{N}_0$ and Q-tuple $(e_q)_{q \in Q}$ with nonnegative integer entries, and so M_Q is closed under the standard multiplication.

Therefore M_Q is a subsemiring of $\mathbb{Q}$ containing both $\mathbb{N}_0$ and Q, and so the inclusion $\mathbb{N}_0[Q] \subseteq M_Q$ follows from the fact that $\mathbb{N}_0[Q]$ is the smallest subsemiring of $\mathbb{Q}$ satisfying the same condition. The other inclusion follows immediately, so $M_Q = \mathbb{N}_0[Q]$. □

Observe that if $k \in Q$ for some $k \in \mathbb{N}_0$, then $k^n \in \mathbb{N}_0$ for all $n \in \mathbb{N}_0$. Thus, $M_Q = \mathbb{N}_0$ when $Q \subset \mathbb{N}$ and $M_Q = M_{Q \setminus \mathbb{N}}$ if $Q \not\subseteq \mathbb{N}$. Therefore we can restrict our attention, without loss of generality, to monoids powerly generated by finite nonempty subsets of $\mathbb{Q}_{>0} \setminus \mathbb{N}$. Using the fact that, for any nonempty finite subset Q of $\mathbb{Q}_{>0}$, the monoid powerly generated by Q is the internal sum of $|Q|$ additive monoids of rational cyclic semirings, we can simultaneously extend the canonical sum decompositions established in Propositions 5.14 and 6.1.

PROPOSITION 6.6. *Let* Q *be a subset of* $\mathbb{Q}_{>0} \setminus \mathbb{N}$ *such that* $2 \leq |Q| < \infty$ *and* $\gcd(\mathsf{d}(Q)) = 1$. *Then each* $r \in M_Q$ *can be uniquely written as follows:*

$$r = c_0(r) + \sum_{(q,n) \in Q \times \mathbb{N}} c_{q,n}(r) q^n, \tag{6.8}$$

where $c_0(r) \in \mathbb{N}_0$ *and, for each* $q \in Q$, *the sequence* $(c_{q,n}(r))_{n \geq 1}$ *is finite-supported with nonnegative integer terms such that* $c_{q,n} \in [\![0, \mathsf{d}(q) - 1]\!]$ *for every* $q \in Q$.

PROOF. Fix a nonzero $r \in M_Q$, and let us argue that r has a unique sum decomposition as the one specified in (6.8). When Q consists of only one element, namely q, then the desired sum decomposition is the canonical sum decomposition of r inside M_q, which we have already established in Proposition 6.1. Thus, for the rest of the proof we assume that $|Q| \geq 2$.

For the existence, first take a Q-tuple $(r_q)_{q \in Q}$ with $r_q \in M_q$ for all $q \in Q$ such that $r = \sum_{q \in Q} r_q$, which is possible because $M_Q = \sum_{q \in Q} M_q$. Then, for each

$q \in Q$, take $c_0(r_q) \in \mathbb{N}_0$ and $(c_{q,n}(r_q))_{n \ge 1}$ such that the right-hand side of equality

$$c_0(r_q) + \sum_{n \in \mathbb{N}} c_{q,n}(r_q) q^n$$

is the canonical sum decomposition of r_q in M_q given in (6.2). Then

$$r = \sum_{q \in Q} r_q = \sum_{q \in Q} \Big(c_0(r_q) + \sum_{n \in \mathbb{N}} c_{q,n}(r_q) q^n \Big) = \sum_{q \in Q} c_0(r_q) + \sum_{(q,n) \in Q \times \mathbb{N}} c_{q,n}(r_q) q^n. \tag{6.9}$$

It is clear that $\sum_{q \in Q} c_0(r_q) \in \mathbb{N}_0$. In addition, for each $q \in Q$, the fact that $c_0(r_q) + \sum_{n \in \mathbb{N}} c_{q,n}(r_q) q^n$ is the canonical sum decomposition of r_q in M_q guarantees that $c_{q,n}(r_q) \in [\![0, \mathsf{d}(q) - 1]\!]$. Thus, we can obtain the desired sum decomposition of r in M_Q from the rightmost part of (6.9) after setting $c_0(r) := \sum_{q \in Q} c_0(r_q)$ and $c_{q,n}(r) := c_{q,n}(r_q)$ for all $(q,n) \in Q \times \mathbb{N}$.

To prove the uniqueness of the sum decomposition given in (6.8), suppose we can write element r inside M_Q in the following two ways:

$$r = b_0(r) + \sum_{(q,n) \in Q \times \mathbb{N}} b_{q,n}(r) q^n \quad \text{and} \quad r = c_0(r) + \sum_{(q,n) \in Q \times \mathbb{N}} c_{q,n}(r) q^n, \tag{6.10}$$

where $b_0(r), c_0(r) \in \mathbb{N}_0$ and, for each $q \in Q$, the sequences $(b_{q,n}(r))_{n \ge 1}$ are finitely supported such that $b_{q,n}(r), c_{q,n}(r) \in [\![0, \mathsf{d}(q) - 1]\!]$ for every $n \in \mathbb{N}$. For each $q \in Q$, the sum $\sum_{n \in \mathbb{N}} (b_{q,n}(r) - c_{q,n}(r)) q^n$ belongs to $\mathbb{N}_0[q]$, and so it can be written as $a_q \frac{1}{\mathsf{d}(q)^{e_q}}$ for some $a_q, e_q \in \mathbb{N}_0$. We can further assume that the pair (a_q, e_q) has been chosen in $\mathbb{N}_0^2$ so that the exponent e_q is as small as it can possibly be. Note that, for each $q \in Q$ such that $a_q \neq 0$, the minimality of e_q implies that $\mathsf{d}(q) \nmid a_q$. Set

$$m := \prod_{q \in Q} \mathsf{d}(q)^{e_q} \quad \text{and} \quad n_q := \frac{m}{\mathsf{d}(q)^{e_q}}$$

for every $q \in Q$. As $\gcd(\mathsf{d}(Q)) = 1$, for each $q \in Q$, we observe that $n_q \in \mathbb{N}$ with $\gcd(\mathsf{d}(q)^{e_q}, n_q) = 1$ and also that, for each $t \in Q$, the divisibility relation $\mathsf{d}(t)^{e_t} \mid n_q$ holds if and only if $t \neq q$. After subtracting the sum decompositions of r given in (6.10), replacing $\sum_{n \in \mathbb{N}} (b_{q,n}(r) - c_{q,n}(r)) q^n$ by $a_q \frac{1}{\mathsf{d}(q)^{e_q}}$, and multiplying the obtained identity by m, one obtains that

$$0 = m(b_0(r) - c_0(r)) + m \sum_{q \in Q} a_q \frac{1}{\mathsf{d}(q)^{e_q}} = m(b_0(r) - c_0(r)) + \sum_{q \in Q} n_q a_q. \tag{6.11}$$

Thus, for each $t \in Q$, the equality $\gcd(\mathsf{d}(t)^{e_t}, n_t) = 1$ holds and so we can use that $\mathsf{d}(t)^{e_t} \mid m$ and also that $\mathsf{d}(t)^{e_t} \mid n_q$ for all $q \in Q$ with $q \neq t$ to infer $\mathsf{d}(t)^{e_t} \mid a_t$, from which $a_t = 0$. This implies that, for each $q \in Q$, the equality $\sum_{n \in \mathbb{N}} b_{q,n}(r) q^n = \sum_{n \in \mathbb{N}} c_{q,n}(r) q^n$. Hence the uniqueness of the canonical sum decomposition inside M_q guarantees that $b_{q,n}(r) = c_{q,n}(r)$ for all pairs $(q,n) \in Q \times \mathbb{N}$ and, therefore, $b_0(r) = c_0(r)$ by virtue of (6.11). Hence we have proved the uniqueness of (6.8), which concludes the proof. □

Now that we have generalized to powerly generated monoids the canonical sum decomposition initially established in Proposition 5.14 for internal sum of p-adic valuation Puiseux monoids, let us also extend the corresponding terminology.

DEFINITION 6.7. Let Q be a subset of $\mathbb{Q}_{>0} \setminus \mathbb{N}$ such that $2 \le |Q| < \infty$ and $\gcd(\mathsf{d}(Q)) = 1$. For each $r \in M_Q$, we call the right-hand side of (6.8) the *canonical sum decomposition* of r inside M_Q.

- We let $c_0 \colon M_Q \to \mathbb{N}_0$ be the function defined via the assignments $c_0 \colon r \mapsto c_0(r)$ for all $r \in M_Q$, where c_0 is as in (6.8).
- For each pair $(q, n) \in Q \times \mathbb{N}$, we let $c_{q,n} \colon M_Q \to \mathbb{N}_0$ be the function defined via the assignments $c_{q,n} \colon r \mapsto c_{q,n}(r)$ for all $r \in M_Q$, where $c_{q,n}(r)$ is as in (6.8).

Keep the notation as in Definition 6.7. First observe that, for any $r \in M_Q$,

$$c_0(r) = \max\{n \in \mathbb{N}_0 : n \mid_{M_Q} r\}. \tag{6.12}$$

Now fix $q \in Q$ and then take $r \in M_q$. If $c_0(r) = 0$, then we can readily infer from the maximality of $c_0(r)$ that any element $s \in M_Q$ such that $s \mid_{M_Q} r$ must satisfy the following two conditions: $c_0(s) = 0$ and $s \in M_q$. Thus, the submonoids M_q of M_Q behave somehow like divisor-closed submonoids. Let us record these observations as the following lemma.

LEMMA 6.8. *Let Q be a subset of $\mathbb{Q}_{>0} \setminus \mathbb{N}$ such that $2 \le |Q| < \infty$ and $\gcd(\mathsf{d}(Q)) = 1$. Then the following statements hold for any $q \in Q$ and $r \in M_q$:*

(1) $c_0(r) = \max\{n \in \mathbb{N}_0 : n \mid_{M_Q} r\}$.
(2) *If $c_0(r) = 0$, then any divisor of r in M_Q must belong to M_q.*

With Lemma 6.8 and the canonical sum decomposition established in Proposition 6.6 at our disposal, we can further investigate the atomic structure of powerly generated monoids. Although powerly generated monoids are not always closed under multiplication, they share various relevant atomic and factorization properties with the additive monoids M_q. The following result, which characterizes when a powerly generated monoid is atomic/antimatter (in terms of its internal summands M_q) generalizes [**34**, Theorem 6.2].

THEOREM 6.9. *Let Q be a subset of $\mathbb{Q}_{>0} \setminus \mathbb{N}$ such that $2 \le |Q| < \infty$ and $\gcd(\mathsf{d}(Q)) = 1$. Then the following statements hold.*

(1) $\big(\bigcup_{q \in Q} \mathscr{A}(M_q)\big) \setminus \{1\} \subseteq \mathscr{A}(M_Q)$.
(2) *M_Q is atomic if and only if M_q is atomic for all $q \in Q$, in which case*
$$\mathscr{A}(M_Q) = \bigcup_{q \in Q} \mathscr{A}(M_q).$$
(3) *M_Q is antimatter if and only if M_q is antimatter for all $q \in Q$, in which case, Q consists of unit fractions.*

PROOF. (1) If M_q is not atomic for any $q \in Q$, then M_q is antimatter for all $q \in Q$ and the desired inclusion trivially holds. Thus, we assume that M_q is atomic for some $q \in Q$. It suffices to fix $a \in Q$ such that M_a is atomic and show that $a^k \in \mathscr{A}(M_Q)$ for every $k \in \mathbb{N}$. Fix $k \in \mathbb{N}$ and take $r, s \in M_Q$ such that $a^k = r + s$. By the uniqueness of the canonical sum decomposition of a^k inside M_Q, we infer that $c_0(a_k) = 0$. Thus $r, s \in M_a$ by virtue of Lemma 6.8. Since $a^k \in \mathscr{A}(M_a)$, either $r = 0$ or $s = 0$. Hence $a^k \in \mathscr{A}(M_Q)$, as desired.

(2) For the direct implication suppose that M_Q is atomic. Assume, towards a contradiction, that there exists $r \in Q$ such that M_r is not atomic: in this case, $r := \frac{1}{d} \in Q$ for some $d \in \mathbb{N}_{\ge 2}$. As $c_0(r) = 0$, it follows from part (2) of Lemma 6.8

that, for each $q \in M \setminus \{r\}$, the only element of M_q that divides r in M_Q is 0. This, along with the fact that $M_Q = \sum_{q \in Q} M_q$, ensures that the only divisors of r in M_Q are those elements dividing r in M_q. Thus, r cannot have any factorization in M_Q because M_q is antimatter, contradicting the initial assumption that M_Q is atomic. For the reverse implication, suppose that M_q is atomic for all $q \in Q$. In this case, the set $\{q^n : n \in \mathbb{N}_0\} \subseteq \mathscr{A}(M_q)$ for all $q \in Q$, and so $\{q^n : (q,n) \in Q \times \mathbb{N}_0\} \subseteq \bigcup_{q \in Q} \mathscr{A}(M_q)$. Hence M_Q is atomic.

Let us argue now that $\mathscr{A}(M_Q) = \bigcup_{q \in Q} \mathscr{A}(M_q)$ under the assumption that M_Q is atomic, which is equivalent to the statement that M_q is atomic for all $q \in Q$. Since $\{q^n : (q,n) \in Q \times \mathbb{N}_0\}$ is a subset of $\bigcup_{q \in Q} \mathscr{A}(M_q)$ that generates the reduced monoid M_Q, the inclusion $\mathscr{A}(M_Q) \subseteq \bigcup_{q \in Q} \mathscr{A}(M_q)$ must hold. In light of part (1), proving the reverse inclusion amounts to argue that $1 \in \mathscr{A}(M_Q)$. To do so, write $1 = r + s$ for some $r, s \in \mathscr{A}(M_Q)$, and then write

$$r = \sum_{q \in Q} r_q \quad \text{and} \quad s = \sum_{q \in Q} s_q,$$

where $r_q, s_q \in M_q$ for all $q \in Q$. As $\gcd(\mathsf{d}(Q)) = 1$, from the equality $1 = \sum_{q \in Q}(r_q + s_q)$, we obtain that $r_q + s_q \in \mathbb{Z}$ for all $q \in Q$, whence we can take $t \in Q$ such that $r_t + s_t = 1$ and $r_q + s_q = 0$ for all $q \in Q \setminus \{t\}$. As M_t is atomic, $1 \in \mathscr{A}(M_t)$, and so $\{r, s\} = \{r_t, s_t\} = \{0, 1\}$. Hence $1 \in \mathscr{A}(M_Q)$.

(3) For the direct implication, suppose that M_Q is antimatter. Assume, towards a contradiction, that M_q is not antimatter for some $q \in Q$. Then M_q is atomic, and the fact that $q \notin \mathbb{N}$ ensures that $\mathscr{A}(M_q) = \{q^n : n \in \mathbb{N}_0\}$. By the uniqueness of the canonical sum decomposition, $c_0(q) = 0$ and so it follows from Lemma 6.8 that the only elements dividing q in M_Q must belong to M_q. Hence q must be an atom of M_Q, contradicting that M_Q is antimatter.

For the reverse implication, suppose that M_q is antimatter for all $q \in Q$. Then, for any $q \in Q$, the fact that M_q is antimatter implies that, for any $n \in \mathbb{N}_0$, the element q^n is not an atom of the monoid M_q, and so that $q^n \notin \mathscr{A}(M_Q$. As a result, none of the elements in the generating set $\{q^n : (q,n) \in Q \times \mathbb{N}_0\}$ of M_Q is an atom. Hence M_Q must be antimatter.

Finally, observe that as Q contains no integers, for each $q \in Q$, the monoid M_q is antimatter if and only if q is a unit fraction. Thus, we conclude that M_Q is antimatter if and only if the set Q consists of unit fractions. □

Inside the class $\{M_q : q \in \mathbb{Q}_{>0}\}$, each monoid that satisfies the ACCP is an FFM: indeed, by virtue of [**17**, Theorem 4.11], being an FFM, being a BFM, and satisfying the ACCP are equivalent conditions in the class $\{\mathbb{N}_0[\alpha] : \alpha \in \mathbb{C} \text{ is algebraic}\}$. Now that we have characterized when powerly generated monoids are atomic, let us characterize when these monoids are FFMs.

THEOREM 6.10. *Let Q be a subset of $\mathbb{Q}_{>0} \setminus \mathbb{N}$ such that $2 \le |Q| < \infty$ and* $\gcd(\mathsf{d}(Q)) = 1$. *Then the monoid powerly generated by Q is an FFM if and only if $q \ge 1$ for all $q \in Q$.*

PROOF. For the direct implication, assume that $q < 1$ for some $q \in Q$. It is well known that, for each rational $r \in (0,1)$, the underlying additive monoid of the rational semiring $\mathbb{N}_0[r]$ does not satisfy the ACCP, whence M_q does not satisfy the ACCP. Let $(r_n + M_q)_{n \ge 1}$ be an ascending chain of principal ideals of M_q that does not stabilize. As M_Q is a reduced monoid, for each $n \in \mathbb{N}$, if the inclusion

$r_n + M_q \subsetneq r_{n+1} + M_q$ is strict, so is the inclusion $r_n + M_Q \subsetneq r_{n+1} + M_Q$. Hence $(r_n + M_Q)_{n \geq 1}$ is an ascending chain of principal ideals of M_Q that does not stabilize, and so M_Q does not satisfy the ACCP. Thus, M_Q is not an FFM.

Conversely, suppose that $q \geq 1$ for every $q \in Q$. Then it follows from [**28**, Proposition 4.5] that M_q is a BFM. Thus, it follows from [**17**, Theorem 4.4] that the monoid M_q is also an FFM. Finally, as M_q is an FFM for all $q \in Q$, the fact that $M_Q = \sum_{q \in Q} M_q$ implies that M_Q is also an FFM, which concludes our proof. □

As the final result of this paper, we provide a necessary condition for a powerly generated monoid to be closed under multiplication.

PROPOSITION 6.11. *Let Q be a subset of $\mathbb{Q}_{>0} \setminus \mathbb{N}$ such that $2 \leq |Q| < \infty$ and* $\gcd(\mathsf{d}(Q)) = 1$. *If the monoid powerly generated by Q is closed under multiplication, then Q cannot contain unit fractions.*

PROOF. Assume that M_Q is closed under multiplication. Suppose, by way of contradiction, that Q contains a unit fraction, namely, r. As M_Q is closed under multiplication, for each $q \in Q \setminus \{r\}$, the product rq belongs to M_Q and so $q \in \mathsf{d}(r)M_Q$, which implies that $q \notin \mathscr{A}(M_Q)$. As the sets $Q \setminus \{r\}$ and $\mathscr{A}(M_Q)$ do not intersect, for each $q \in Q \setminus \{r\}$, the monoid M_q is antimatter and so q must be a unit fraction. Therefore, for any $q \in Q \setminus \{r\}$, the fact that $qr \in M_Q$ allows us to take coefficients $c_q, c_r \in \mathbb{N}_0$ and exponents $e_q, e_r \in \mathbb{N}_0$ such that

$$qr = c_q \frac{1}{\mathsf{d}(q)^{e_q}} + c_r \frac{1}{\mathsf{d}(r)^{e_r}} + N, \tag{6.13}$$

where $N \in \sum_{t \in Q \setminus \{q,r\}} M_t$. Since the coefficients c_q and c_r are nonzero, we can assume that $\mathsf{d}(q) \nmid c_q$ and $\mathsf{d}(r) \nmid c_r$. Now, as $\gcd(\mathsf{d}(Q)) = 1$, from the fact that the denominator of qr is $\mathsf{d}(r)\mathsf{d}(q)$ we obtain that $e_q = e_r = 1$ and $N \in \mathbb{N}_0$. Then the inequality $qr < 1$ guarantees that $N = 0$. After multiplying (6.13) by $\mathsf{d}(q)\mathsf{d}(r)$, we obtain the equality $1 = c_q\mathsf{d}(r) + c_r\mathsf{d}(q)$, which is clearly a contradiction. Hence we conclude that Q cannot contain unit fractions. □

The necessary condition given in Proposition 6.11 cannot be used to characterize powerly generated monoids that are closed under multiplication. The following example illustrates this observation.

EXAMPLE 6.12. Let p_1 and p_2 be two primes such that $p_1 < p_2$. Now set $Q := \{q_1, q_2\}$, where $q_1 := \frac{p_2 - p_1}{p_1}$ and $q_2 := \frac{p_1}{p_2}$. Thus, Q is a non-singleton subset of $\mathbb{Q}_{>0}$ with $\gcd(\mathsf{d}(Q)) = \gcd(p_1, p_2) = 1$, and Q does not contain unit fractions. Now consider the monoid powerly generated by Q:

$$M_Q := \langle q_1^n, q_2^n : n \in \mathbb{N}_0 \rangle.$$

As neither q_1 nor q_2 is a unit fraction, the monoids M_{q_1} and M_{q_2} are both atomic. Hence it follows from part (2) of Theorem 6.9 that M_Q is atomic and also that $1 \in \mathscr{A}(M_Q)$. Since

$$q_1 q_2 + q_2 = \left(\frac{p_2 - p_1}{p_1}\right)\left(\frac{p_1}{p_2}\right) + \frac{p_1}{p_2} = 1,$$

from the fact that 1 is an atom of M_Q we obtain that $q_1 q_2 \notin M_Q$. Hence M_Q is not closed under the standard multiplication.

We have seen in Theorem 6.2 that, for any $q \in \mathbb{Q}_{>0}$, the monoid M_q is an LFFM if and only if q is not a unit fraction. Based on this, we conclude with the following open question.

QUESTION 6.13. Let Q be a subset of $\mathbb{Q}_{>0} \setminus \mathbb{N}$ such that $2 \le |Q| < \infty$ and $\gcd(\mathsf{d}(Q)) = 1$. Can we characterize in terms of Q when the monoid M_Q powerly generated by Q is an LFFM?

Acknowledgments

The authors thank an anonymous referee for her/his helpful suggestions.

References

[1] K. Ajran, J. Bringas, B. Li, E. Singer, and M. Tirador, *Factorization in additive monoids of evaluation polynomial semirings*, Comm. Algebra **51** (2023), no. 10, 4347–4362, DOI 10.1080/00927872.2023.2208672. MR4613278

[2] S. Albizu-Campos, J. Bringas, and H. Polo, *On the atomic structure of exponential Puiseux monoids and semirings*, Comm. Algebra **49** (2021), no. 2, 850–863, DOI 10.1080/00927872.2020.1820514. MR4207942

[3] J. Amos, S. T. Chapman, N. Hine, and J. Paixão, *Sets of lengths do not characterize numerical monoids*, Integers **7** (2007), A50, 8. MR2373112

[4] D. D. Anderson, D. F. Anderson, and M. Zafrullah, *Factorization in integral domains*, J. Pure Appl. Algebra **69** (1990), no. 1, 1–19, DOI 10.1016/0022-4049(90)90074-R. MR1082441

[5] D. D. Anderson and R. O. Quintero, *Some generalizations of GCD-domains*, Factorization in integral domains (Iowa City, IA, 1996), Lecture Notes in Pure and Appl. Math., vol. 189, Dekker, New York, 1997, pp. 189–195. MR1460772

[6] D. F. Anderson and F. Gotti, *Bounded and finite factorization domains*, Rings, monoids and module theory, Springer Proc. Math. Stat., vol. 382, Springer, Singapore, [2021] ©2021, pp. 7–57, DOI 10.1007/978-981-16-8422-7_2. MR4428451

[7] A. Bu, F. Gotti, B. Li, and A. Zhao, *One-dimensional monoid algebras and ascending chains of principal ideals*. Submitted. Preprint on arXiv: `arXiv:2409.00580`.

[8] M. Bras-Amoros and M. Gotti, *Atomicity and density of Puiseux monoids*, Comm. Algebra **49** (2021), no. 4, 1560–1570, DOI 10.1080/00927872.2020.1840574. MR4227645

[9] S. T. Chapman, P. A. García-Sánchez, D. Llena, and J. Marshall, *Elements in a numerical semigroup with factorizations of the same length*, Canad. Math. Bull. **54** (2011), no. 1, 39–43, DOI 10.4153/CMB-2010-068-3. MR2797486

[10] S. T. Chapman, F. Gotti, and M. Gotti, *Factorization invariants of Puiseux monoids generated by geometric sequences*, Comm. Algebra **48** (2020), no. 1, 380–396, DOI 10.1080/00927872.2019.1646269. MR4060036

[11] S. T. Chapman, F. Gotti, and M. Gotti, *When is a Puiseux monoid atomic?*, Amer. Math. Monthly **128** (2021), no. 4, 302–321, DOI 10.1080/00029890.2021.1865064. MR4234728

[12] S. T. Chapman and M. Gotti, *Atomicity of positive monoids*, Quaest. Math. **47** (2024), no. 11, 2177–2193, DOI 10.2989/16073606.2024.2352875. MR4825394

[13] S. T. Chapman, J. Jang, J. Mao, and S. Mao, *On the set of Betti elements of a Puiseux monoid*, Bull. Aust. Math. Soc. **111** (2025), no. 1, 80–92, DOI 10.1017/s0004972724000352. MR4865573

[14] S. T. Chapman and H. Polo, *Arithmetic of additively reduced monoid semidomains*, Semigroup Forum **107** (2023), no. 1, 40–59, DOI 10.1007/s00233-023-10363-0. MR4631078

[15] T. Chen, F. Gotti, T. Lu, and A. Yao, *On the additive structure of algebraic valuations of polynomial semirings II*. Submitted. Preprint on arXiv: `arXiv:2008.13073`.

[16] P. M. Cohn, *Bezout rings and their subrings*, Proc. Cambridge Philos. Soc. **64** (1968), 251–264, DOI 10.1017/s0305004100042791. MR222065

[17] J. Correa-Morris and F. Gotti, *On the additive structure of algebraic valuations of polynomial semirings*, J. Pure Appl. Algebra **226** (2022), no. 11, Paper No. 107104, 20, DOI 10.1016/j.jpaa.2022.107104. MR4414838

[18] J. Coykendall, D. E. Dobbs, and B. Mullins, *On integral domains with no atoms*, Comm. Algebra **27** (1999), no. 12, 5813–5831, DOI 10.1080/00927879908826792. MR1726278

[19] J. Coykendall and F. Gotti, *On the atomicity of monoid algebras*, J. Algebra **539** (2019), 138–151, DOI 10.1016/j.jalgebra.2019.07.033. MR3995239

[20] L. Fuchs, *Infinite abelian groups. Vol. I*, Pure and Applied Mathematics, Vol. 36, Academic Press, New York-London, 1970. MR255673

[21] J. C. Rosales and P. A. García-Sánchez, *Numerical semigroups*, Developments in Mathematics, vol. 20, Springer, New York, 2009, DOI 10.1007/978-1-4419-0160-6. MR2549780

[22] A. Geroldinger and F. Gotti, *On monoid algebras having every nonempty subset of* $\mathbb{N}_{\geq 2}$ *as a length set*, Mediterr. J. Math. **22** (2025), no. 3, Paper No. 69, 19, DOI 10.1007/s00009-025-02835-0. MR4891850

[23] A. Geroldinger and F. Halter-Koch, *Non-unique factorizations*, Pure and Applied Mathematics (Boca Raton), vol. 278, Chapman & Hall/CRC, Boca Raton, FL, 2006. Algebraic, combinatorial and analytic theory, DOI 10.1201/9781420003208. MR2194494

[24] A. Geroldinger and W. A. Schmid, *A realization theorem for sets of lengths in numerical monoids*, Forum Math. **30** (2018), no. 5, 1111–1118, DOI 10.1515/forum-2017-0180. MR3849636

[25] A. Geroldinger and Q. Zhong, *A characterization of length-factorial Krull monoids*, New York J. Math. **27** (2021), 1347–1374. MR4312737

[26] R. Gilmer, *Commutative semigroup rings*, Chicago Lectures in Mathematics, University of Chicago Press, Chicago, IL, 1984. MR741678

[27] V. Gonzalez, F. Gotti, and I. Panpaliya, *On the ascent of almost and quasi-atomicity to monoid semidomains*. Submitted. Preprint on arXiv: `arXiv:2501.04990`.

[28] F. Gotti, *Increasing positive monoids of ordered fields are FF-monoids*, J. Algebra **518** (2019), 40–56, DOI 10.1016/j.jalgebra.2018.10.010. MR3872857

[29] F. Gotti, *Irreducibility and factorizations in monoid rings*, Numerical semigroups, Springer INdAM Ser., vol. 40, Springer, Cham, [2020] ©2020, pp. 129–139. MR4178257

[30] F. Gotti, *On semigroup algebras with rational exponents*, Comm. Algebra **50** (2022), no. 1, 3–18, DOI 10.1080/00927872.2021.1949018. MR4370409

[31] F. Gotti, *On the atomic structure of Puiseux monoids*, J. Algebra Appl. **16** (2017), no. 7, 1750126, 20, DOI 10.1142/S0219498817501262. MR3660409

[32] F. Gotti, *Puiseux monoids and transfer homomorphisms*, J. Algebra **516** (2018), 95–114, DOI 10.1016/j.jalgebra.2018.08.026. MR3863471

[33] F. Gotti, *Systems of sets of lengths of Puiseux monoids*, J. Pure Appl. Algebra **223** (2019), no. 5, 1856–1868, DOI 10.1016/j.jpaa.2018.08.004. MR3906530

[34] F. Gotti and M. Gotti, *Atomicity and boundedness of monotone Puiseux monoids*, Semigroup Forum **96** (2018), no. 3, 536–552, DOI 10.1007/s00233-017-9899-9. MR3808643

[35] F. Gotti and B. Li, *Atomic semigroup rings and the ascending chain condition on principal ideals*, Proc. Amer. Math. Soc. **151** (2023), no. 6, 2291–2302, DOI 10.1090/proc/16295. MR4576298

[36] F. Gotti and C. O'Neill, *The elasticity of Puiseux monoids*, J. Commut. Algebra **12** (2020), no. 3, 319–331, DOI 10.1216/jca.2020.12.319. MR4146363

[37] F. Gotti and H. Rabinovitz, *On the ascent of atomicity to monoid algebras*, J. Algebra **663** (2025), 857–881, DOI 10.1016/j.jalgebra.2024.10.002. MR4811058

[38] A. Grams, *Atomic rings and the ascending chain condition for principal ideals*, Proc. Cambridge Philos. Soc. **75** (1974), 321–329, DOI 10.1017/s0305004100048532. MR340249

[39] P. A. Grillet, *Commutative semigroups*, Advances in Mathematics (Dordrecht), vol. 2, Kluwer Academic Publishers, Dordrecht, 2001, DOI 10.1007/978-1-4757-3389-1. MR2017849

[40] F. Halter-Koch, *Finiteness theorems for factorizations*, Semigroup Forum **44** (1992), no. 1, 112–117, DOI 10.1007/BF02574329. MR1138690

[41] H. Jiang, S. Kanungo, and H. Kim, *A weaker notion of the finite factorization property*, Commun. Korean Math. Soc. **39** (2024), no. 2, 313–329, DOI 10.4134/CKMS.c230178. MR4739845

[42] H. Polo, *Factorization invariants of the additive structure of exponential Puiseux semirings*, J. Algebra Appl. **22** (2023), no. 3, Paper No. 2350077, 22, DOI 10.1142/S0219498823500779. MR4550072

[43] H. Polo, *On the sets of lengths of Puiseux monoids generated by multiple geometric sequences*, Commun. Korean Math. Soc. **35** (2020), no. 4, 1057–1073, DOI 10.4134/CKMS.c200017. MR4168991

[44] M. Sasaki and T. Tamura, *Positive rational semigroups and commutative power joined cancellative semigroups without idempotent*, Czechoslovak Math. J. **21(96)** (1971), 567–576. MR292981

[45] A. Tripathi, *Formulae for the Frobenius number in three variables*, J. Number Theory **170** (2017), 368–389, DOI 10.1016/j.jnt.2016.05.027. MR3541713

MATHEMATICS DEPARTMENT, SAM HOUSTON STATE UNIVERSITY, HUNTSVILLE, TEXAS 77340
Email address: scott.chapman@shsu.edu

MATHEMATICS DEPARTMENT, MASSACHUSETTS INSTITUTE OF TECHNOLOGY (MIT), CAMBRIDGE, MASSACHUSETTS 02139
Email address: fgotti@mit.edu

APPLE INC., ONE APPLE PARK WAY, CUPERTINO, CALIFORNIA 95014
Email address: marlygotti@apple.com

MATHEMATICS DEPARTMENT, UNIVERSITY OF CALIFORNIA IRVINE, IRVINE, CALIFORNIA 92697
Email address: harold.polo@uci.edu

Contemporary Mathematics
Volume **836**, 2026
https://doi.org/10.1090/conm/836/16707

The set of minimal distances of the monoid of plus-minus weighted zero-sum sequences and applications to the characterization problem

Kamil Merito, Oscar Ordaz, and Wolfgang A. Schmid

ABSTRACT. Recently a systematic investigation of monoids of sequences of plus-minus weighted zero-sum sequences had been started, which is among others motivated by applications to monoids of norms of algebraic integers. In the current paper these investigations are continued. The focus is on the set of minimal distances of these monoids, which is an important arithmetical invariant. Applications to the characterization problem are discussed as well.

1. Introduction

There is a wealth of literature on the arithmetic of various types of monoids. A central class are Krull monoids and related to this monoids of zero-sum subsequences over abelian groups.

In a recent paper Boukheche and the authors of the current paper [**4**] considered monoids of weighted zero-sum sequences; while there are many investigations on zero-sum problems with weights, see for example [**1**, **2**, **22**], the arithmetic of these monoids had not been investigated. Moreover, that paper established a connection to certain monoids of norms of algebraic integers. Indeed, these monoids of norms had first been studied under the term normsets by Coykendall [**8**]; see [**9**] for a more recent exposition.

In a subsequent paper Geroldinger, Halter-Koch, and Zhong [**16**] generalized this connection and obtained further results on the arithmetic of monoids of weighted zero-sum sequences in particular on their (local) elasticities. A partcular emphasis was put on monoids of plus-minus weighted zero-sum sequences. For further results see [**10**, **17**].

In the current paper we further pursue the investigations of monoids of weighted zero-sum sequences. In particular, we study their sets of minimal distances, that is, the set formed by the minimal distances of their divisor-closed submonoids, denoted $\Delta^*(H)$; sometimes this set is also called the set of differences of the monoid, but this name is less common and risks to be confused with the set of distances $\Delta(H)$. This set has been studied a lot for Krull monoids, mostly for Krull monoids with finite class group (see [**19**, **29**]); for a result in the case of infinite class groups see [**7**].

2020 *Mathematics Subject Classification.* Primary 11B30, 11B75, 11R04, 20K01.

The work of W. A. Schmid is supported by the project ANR-21-CE39-0009 - BARRACUDA.

©2026 American Mathematical Society

Here, we investigate this set for monoids of plus-minus weighted zero-sum sequences over abelian groups, mainly for finite abelian groups. It turns out that the results depend strongly on the parity of the order of the group. In the case of groups of odd order (and not 1), we obtain strong results on the sets $\Delta^*(H)$; we show that $\max \Delta^*(H) = \exp(G) - 2$ in that case and obtain strong constraints on its other elements that sometimes allow us to obtain a complete description. In the case of groups of even order, we show that $\max \Delta^*(H)$ can be quite large, it can exceed the value of the invariant for the monoid of zero-sum sequences without weights over the same group.

In the final section, we apply these results to characterize certain monoids of plus-minus weighted zero-sum problems by their systems of sets of lengths. In particular, we can show that the exponent of the group is determined by the system of sets of lengths for every group of odd order (excluding a few well-understood cases of very small groups). Then we apply this result to characterize elementary p-groups for p odd and also obtain an alternative charaterization for cyclic groups of odd order (a first characterization was recently obtained in [**10**]).

2. Preliminaries

We recall some key-notions that we use frequently in the current paper. For the most part they are fairly well-established in factorization theory and we refer to [**15**] for a complete exposition.

By $\mathbb{N}$ and $\mathbb{N}_0$ we denote the set of positive and non-negative integers, respectively. For real numbers a and b, we denote by $[a, b] = \{x \in \mathbb{Z} \colon a \le x \le b\}$ the interval of integers.

2.1. Abelian groups. For $n \in \mathbb{N}$, let C_n a cyclic group of order n; usually we use additive notation for abelian groups. It is well-known that for $(G, +, 0)$ a nontrivial finite abelian group there are uniquely determined $1 < n_1 \mid \cdots \mid n_r$ such that $G \cong C_{n_1} \oplus \cdots \oplus C_{n_r}$ where r is the rank of G, denoted by $\mathsf{r}(G)$, and n_r is the exponent of G, denoted by $\exp(G)$. The exponent of a group of order 1 is 1 and its rank is 0. If the exponent is a power of a prime p, then we call G a p-group, and we say that it is an elementary p-group if the exponent is p.

For $m \in \mathbb{N} \setminus \{1\}$, let $\mathsf{r}_m(G) = |\{i \colon m \mid n_i\}|$ denote the m-rank of G; for the most part we use this for m a prime number.

A family $(e_1, \dots, e_s)$ over G is called independent if $\sum_{i=1}^{s} a_i e_i = 0$ with $a_i \in \mathbb{Z}$ implies that $a_i e_i = 0$ for each $i \in [1, s]$. We also use the terminology independent for subsets of G in the obvious way.

2.2. Monoids. Throughout the paper, monoid always means a semigroup with identity that is commutative and satisfies the cancellation law. We usually use multiplicative notation for monoids and denoted the identity element by 1_H or simply 1 if there is no risk of confusion.

An element a of a monoid H is called invertible or a unit if there exists an element $a' \in H$ such $aa' = 1_H$; the set of invertible elements of H is denoted by $H^\times$; it is a subgroup of H. A reduced monoid is a monoid whose only invertible element is 1_H; one calls $H_{\text{red}} = H/H^\times$ the reduced monoid associated to H.

We say that b divides a in H, and we write $b \mid_H a$ if there exists some $c \in H$ such that $a = bc$. If there is no risk of confusion we drop the subscript H. If $bc = a$, then we write $b^{-1}a$ for c.

An element $a \in H$ is called irreducible or an atom (in H) if it has no proper divisors, that is, $a = bc$ with $b, c \in H$ implies that b or c are invertible. The set of irreducible elements of H is denoted by $\mathcal{A}(H)$. An element $p \in H \setminus H^\times$ is called prime if $p \mid bc$ with $b, c \in H$ implies that $p \mid b$ or $p \mid c$.

A submonoid S of H is called a saturated submonoid if for $s, t \in S$ one has $s \mid_S t$ if and only if $s \mid_H t$ (of course one implication is trivial) and it is called a divisor closed submonoid if for $s \in S$ and $a \in H$, one has that $a \mid_H s$ implies that $a \in S$.

For a set P, let $\mathcal{F}(P)$ denote the free commutative monoid over P, as is common in the context of factorization theory we refer to its elements as sequences. To explain the terminology 'sequence' one can note that it is a collection of elements of P where repetitions are possible yet the ordering of elements is disregarded, in that sense they are finite sequences where the ordering is disregarded.

Essentially by definition, for $S \in \mathcal{F}(P)$, there are unique $v_p \in \mathbb{N}_0$, all but finitely many equal to 0, such that $S = \prod_{p \in P} p^{v_p}$; the integer v_p is called the multiplicity of p in S or also the p-adic valuation of S, denoted by $\mathsf{v}_p(S)$. Moreover, there exist up to ordering uniquely determined $p_1, \dots, p_\ell \in P$ (not necessarily distinct) such that $S = p_1 \dots p_\ell$. For a sequence S a divisor T of S in $\mathcal{F}(P)$ is called a subsequence of S. The identity element of the monoid of sequences is called the empty sequence, and is denote it by 1 unless there is a risk of confusion. One calls ℓ the length of S, it is denoted by $|S|$. Moreover, the set $\{p_1, \dots, p_\ell\}$ is called the support of S, denoted $\operatorname{supp}(S)$; its cardinality is bounded above by ℓ yet it is usually strictly smaller as repeated elements are counted only once.

In the current paper we are mainly interested in the case of sequences over abelian groups or subsets of abelian groups. Let G_0 be a subset of an abelian group $(G, +)$. In this case, for a sequence $S = g_1 \dots g_\ell \in \mathcal{F}(G_0)$, we denote by $\sigma(S) = \sum_{i=1}^{\ell} g_i \in G$ its sum and by $\sigma_\pm(S) = \{\sum_{i=1}^{\ell} \epsilon_i g_i \in G \colon \epsilon_i \in \{+1, -1\}\}$ the set of plus-minus weighted sums of S. It is possible to consider other sets of weights, yet in the current paper we deal exclusively with this sets of weights. As an aside, for the case of general weights one takes endomorphisms of G rather than integers as weights; the current set of weights would thus correspond to $\{\mathrm{id}_G, -\mathrm{id}_G\}$.

If $\sigma(S) = 0$, the sequence is called a zero-sum sequence and if $0 \in \sigma_\pm(S)$, the sequence is called a plus-minus weighted zero-sum sequence. The set of all sequences over G_0 that are zero-sum sequences is denoted by $\mathcal{B}(G_0)$, and the set set of all plus-minus weighted zero-sum sequence is denoted by $\mathcal{B}_\pm(G_0)$. It is easy to see that both are submonoids of $\mathcal{F}(G_0)$, and that $\mathcal{B}(G_0)$ is a submonoid of $\mathcal{B}_\pm(G_0)$.

It is important to note though that while $\mathcal{B}(G_0)$ is a saturated submonoid of $\mathcal{F}(G_0)$, it is usually not true that $\mathcal{B}_\pm(G_0)$ is a saturated submonoid of $\mathcal{F}(G_0)$. This is quite relevant because a monoid H that is a saturated submonoid of a factorial monoid is a Krull monoid. Thus, while $\mathcal{B}(G_0)$ is a Krull monoid this usually is not the case for $\mathcal{B}_\pm(G_0)$. Indeed, in case $G_0 = G$, it is known that $\mathcal{B}_\pm(G)$ is Krull if and only if G is an elementary 2-group and it thus actually is equal to $\mathcal{B}(G)$; for all other groups the monoids are not even transfer Krull (see below for a definition and [**4**] for the result).

While Krull monoids are arguably the most intensely studied class of monoids in factorization theory, here we do not go into much detail and refer to, for example, [**13, 20, 27**] for survey articles.

2.3. Factorizations. We continue by recalling some notions from factorization theory. A monoid H is called atomic if each non-invertible element of H can be written as a (finite) product of irreducible elements. The monoid of factorizations of H, denoted by $\mathsf{Z}(H)$, is the monoid $\mathcal{F}(\mathcal{A}(H_{\text{red}}))$.

The homomorphism $\pi_H : \mathsf{Z}(H) \to H_{\text{red}}$, which maps the formal product $a_1 \dots a_k$ to its value, is called the factorization homomorphism. It is surjective if and only if H is atomic. For $a \in H$, the set $\mathsf{Z}_H(a) = \pi_H^{-1}(aH^\times)$ is called the set of factorizations of a in H; if the monoid H is obvious from context we drop H from the notation.

We stress that the monoid of factorizations formalizes the idea that one identifies factorizations of elements of H that differ only by the ordering of the terms or multiplication by units.

For $z \in \mathsf{Z}(H)$, one calls $|z|$, which is defined as $\mathsf{Z}(H)$ is a free monoid, the length of the factorization; informally it is the number of irreducible elements in the factorization where multiplicities are taken into account. Moreover, one calls $\mathsf{L}_H(a) = \{|z| \colon z \in \mathsf{Z}_H(a)\}$ the set of lengths of a in H. Finally, the system of set of lengths of H is defined as $\mathcal{L}(H) = \{\mathsf{L}_H(a) \colon a \in H\}$.

The monoid is called a bounded factorization monoid, BF-monoid for short, if $\mathsf{L}_H(a)$ is finite for each $a \in H$. Similarly, monoids for which even the sets of factorizations are finite are called finite factorization monoids. We note that H is factorial if and only if $|\mathsf{Z}_H(a)| = 1$ for each $a \in H$. One says that H is half-factorial if $|\mathsf{L}_H(a)| = 1$ for each $a \in H$.

We end the section by briefly recalling the concept of transfer homorphism, see for example, [**12**, Section 1.3]. For a generalization to not necessarily commutative monoids see [**3**].

Let H and $\mathcal{B}$ be monoids. A monoid homomorphism $\Theta : H \to \mathcal{B}$ is a transfer homomorphism if it satisfies the following:

T1 $\mathcal{B} = \Theta(H)\mathcal{B}^\times$ and $\Theta^{-1}(\mathcal{B}^\times) = H^\times$.

T2 If $u \in H$ and $b, c \in \mathcal{B}$ with $\Theta(u) = bc$, then there exist $v, w \in H$ such that $u = vw$, $\Theta(v) \simeq b$ and $\Theta(w) \simeq c$.

A monoid (not necessarily commutative) is called transfer Krull if it admits a tranfer homomorphism to a Krull monoid.

The relevance of the notion of transfer homomorphism is due to the fact that it preserves many arithmetical properties. In particular, if $\Theta : H \to \mathcal{B}$ is a transfer homomorphism, then $\mathsf{L}_H(a) = \mathsf{L}_{\mathcal{B}}(\Theta(a))$ for each $a \in H$, and in particular $\mathcal{L}(H) = \mathcal{L}(\mathcal{B})$. There are various arithmetical invariants that are derived from sets of lengths and that are thus also preserved. We recall some of them in the next section.

3. Structure of sets of lengths

In the current section we recall a few notions and results on sets of lengths, focusing on the context at hand. For a more complete presentation see, e.g., [**13**, **15**, **20**].

Let H be a BF-monoid. To avoid trivial corner cases we assume that $H \neq H^\times$. For $k \in \mathbb{N}$ one sets $\rho_k(H) = \sup\{\max \mathsf{L}_H(a) \colon k \in \mathsf{L}_H(a),\, a \in H\} \in \mathbb{N}_0 \cup \{\infty\}$, that is, $\rho_k(H)$ is (in case it is finite) the largest number of factors that can appear in the factorization into irreducibles of an element that is the product of k irreducible elements. Obviously, $\rho_1(H) = 1$ and $\rho_k(H) \geq k$ for each k. The value $\rho_k(H)$ is sometimes called the k-th local elasticity of H. This terminology is derived from that of the elasticity of a monoid, denoted $\rho(H)$. If $A \subseteq \mathbb{N}$ is a finite subset, we call $\rho(A) = \max A / \min A \in \mathbb{Q}_{\geq 1}$ the elasticity of A, and we set $\rho(\{0\}) = 1$. The elasticity of an element $a \in H$, denoted $\rho(a)$, is just $\rho(\mathsf{L}_H(a))$. Finally, the elasticity of H, denoted $\rho(H)$, is defined as $\sup\{\rho(a) \colon a \in H\} \in \mathbb{R}_{\geq 1} \cup \{\infty\}$. It is not difficult to see that $\rho(H) = \lim_{k\to\infty} \rho_k(H)/k$.

Another important notion to investigate sets of lengths are sets of distances. For $A \subseteq \mathbb{Z}$ a finite subset, we denote by $\Delta(A)$ the set of (successive) distances of A, that is, for $A = \{a_0, \ldots, a_k\}$ with $a_0 < \cdots < a_k$ one has $\Delta(A) = \{a_1 - a_0, a_2 - a_1, \ldots, a_k - a_{k-1}\} = \{a_{i+1} - a_i \colon i \in [1, k]\}$. Thus, for each $a \in A$, and a not the maximum of A, we have $d \in \Delta(A)$ if d is the smallest $d \in \mathbb{N}$ such that $a + d \in A$.

For $a \in H$, we set $\Delta(a) = \Delta(\mathsf{L}(a))$ and

$$\Delta(H) = \bigcup_{a \in H} \Delta(\mathsf{L}(a)).$$

We observe that H is half-factorial if and only if $\rho(H) = 1$ if and only if $\Delta(H) = \emptyset$. It is well-known that if $\Delta(H) \neq \emptyset$, then $\min \Delta(H) = \gcd \Delta(H)$.

Moreover, we set

$$\Delta^*(H) = \{\min \Delta(S) \colon S \subseteq H \text{ is a divisor closed submonoid with } \Delta(S) \neq \emptyset\}$$

the set of minimal distances of H. The relevance of the notion $\Delta^*(H)$ is due to the fact that it is a natural choice in the structure theorem for sets of lengths, which we recall below.

Specifically, for H a finitely generated monoid and not half-factorial, there is some $M \in \mathbb{N}_0$ such that each set of lengths L of H is an almost arithmetical multiprogression with bound M and difference $d \in \Delta^*(H)$, that is,

$$L = y + (L_1 \cup L^* \cup (\max L^* + L_2)) \subseteq y + \mathcal{D} + d\mathbb{Z}$$

with $y \in \mathbb{N}_0$, $\{0, d\} \subseteq \mathcal{D} \subseteq [0, d]$, $-L_1, L_2 \subseteq [1, M]$, $\min L^* = 0$ and $L^* = [0, \max L^*] \cap \mathcal{D} + d\mathbb{Z}$.

The set $\Delta^*(H)$ is not the only choice for which this result holds, indeed it obviously would hold for any superset of that set and in the other direction, if $d, d' \in \Delta^*(H)$ with $d' \mid d$, then possibly altering the constant M slightly, the result would also hold for $\Delta^*(H) \setminus \{d'\}$.

However, the set $\Delta^*(H)$ is a natural choice in that it is a set that naturally comes out of a proof of this result. Very roughly speaking, for an element that has 'many' irreducible factors there ought to be some irreducibles that occur with a 'high' multiplicity and they will induce a repeating pattern corresponding to the minimal distance of the divisor-closed submonoid they generate.

Or, looked at it from another perspective, considering an element that is a product of multiple copies of all irreducible elements from a divisor-closed submonoid, one can see that its set of lengths will be to an almost arithmetic progression whose difference is the minimal distance of the submonoid.

Indeed, if one denotes by $\Delta_1(H)$ the set of all $d \in \mathbb{N}$ such that $\mathcal{L}(H)$ contains almost arithmetical progression with difference d of arbitrarily large length, then it is known that $\Delta^*(H) \subseteq \Delta_1(H)$ and each $d_1 \in \Delta_1(H)$ divides some $d \in \Delta^*(H)$. Recall that an almost arithmetic progression is an almost arithmetic multiprogression with $\mathcal{D} = \{0, d\}$, that is an arithmetic progression where some elements at the start and at the end might be missing.

While we mentioned the result for finitely generated monoids, this is not the only class of monoids for which the result holds; one way to extend the result, yet not the only way, is to use tranfer homomorphisms to finitely generated monoids.

We recall a few well-known facts that are useful in the remainder of the paper. Let H be a BF-monoid with $\Delta(H) \neq \emptyset$. We already recalled that $\min \Delta(H) = \gcd \Delta(H)$, and thus in particular, for every choice of $a \in H$ and $l, l' \in \mathsf{L}(a)$, one has $\min \Delta(H) \mid l' - l$.

Moreover, if S is a divisor-closed submonoid of H, then $\Delta(S) \subseteq \Delta(H)$ and thus $\min \Delta(H) \le \min \Delta(S)$ and indeed $\min \Delta(H) \mid \min \Delta(S)$.

The results recalled hitherto apply to all finitely generated monoids and it is not hard to see that monoids of plus-minus weighted zero-sum sequences over subsets of finite abelian groups are finitely generated, and thus the results apply. More precisely, for G a finite abelian group and $H = \mathcal{B}_{\pm}(G)$ one has that $\{|A| \colon A \in \mathcal{A}(H)\}$ is bounded above by the Davenport constant of G, denoted $\mathsf{D}(G)$. Moreover, $\max\{|A| \colon A \in \mathcal{A}(H)\}$ is called the Davenport constant of the monoid H, denoted $\mathsf{D}(H)$; this definition is used for every monoid embedded into a free monoid, not just monoids of (weighted) zero-sum sequences.

We recall that the Davenport constant of a finite abelian group G, denoted $\mathsf{D}(G)$, is defined as the smallest integer ℓ such that each sequence of lenght at least ℓ admits a subsequence that is a zero-sum subsequence. It is well-known and easy to see that it is equal to $\mathsf{D}(\mathcal{B}(G))$ the Davenport constant of the monoid $\mathcal{B}(G)$. It is well known that for $G \cong C_{n_1} \oplus \cdots \oplus C_{n_r}$ with $1 < n_1 \mid \cdots \mid n_r$ one has $|G| \ge \mathsf{D}(G) \ge \sum_{i=1}^{r}(n_i - 1) + 1$, and the latter quantity is denoted by $\mathsf{D}^*(G)$. Indeed, for p-groups and groups of rank at most two, one knows that $\mathsf{D}(G) = \sum_{i=1}^{r}(n_i - 1) + 1$, yet this is not true in general.

We point out that, as is the case without weights, the situation is very different for infinite abelian groups. Indeed, Geroldinger and Kainrath [**17**] recently showed that for G an infinite abelian group every finite subset of $\mathbb{N}_{\ge 2}$ is a set of lengths of some element of $H = \mathcal{B}_{\pm}(G)$, just like for $\mathcal{B}(G)$. Note that since these monoids are still BF-monoids and the only sets of lengths that contain elements less than 2 are $\{0\}$ and $\{1\}$, this yields indeed a complete description of the system of sets of lengths of $\mathcal{B}_{\pm}(G)$ for infinite G. We recall that by a result of Kainrath [**21**] that was already known for $\mathcal{B}(G)$.

However, while $\Delta^*(\mathcal{B}(G)) = \mathbb{N}$ (see [**7**]), it is not always true that $\Delta^*(\mathcal{B}_{\pm}(G)) = \mathbb{N}$; this was already noted in [**17**] and we provide further results on that problem.

We close this section with observations and results specific to monoids of plus-minus weighted zero-sum sequences. Since we want to study $\Delta^*(H)$, which is defined using divisor-closed submonoids of H, it is crucial to know the divisor-closed submonoids. Let G be an abelian group. While it is immediate that for $G_0 \subseteq G$, the monoid $\mathcal{B}_{\pm}(G_0)$ is a divisor-closed submonoid, there might in principle be other divisor-closed submonoids. However, it turns out that this is not the case. We include a short proof for the particular case; the result if obtained for arbitrary sets of weights for finite abelian groups in [**10**].

LEMMA 3.1. *Let G be an abelian group. Then $\mathcal{B}_{\pm}(G_0)$ with $G_0 \subseteq G$ a subset are all the divisor-closed submonoids of $\mathcal{B}_{\pm}(G)$.*

PROOF. It is clear that $\mathcal{B}_{\pm}(G_0)$ for $G_0 \subseteq G$ is a divisor-closed submonoid of $\mathcal{B}_{\pm}(G)$. Assume that $H \subseteq \mathcal{B}_{\pm}(G)$ is divisor-closed. We need to show that there exists some $G_0 \subseteq G$ such that $H = \mathcal{B}_{\pm}(G_0)$.

Set $G_0 = \bigcup_{B \in H} \operatorname{supp}(B)$. Then $H \subseteq \mathcal{B}_{\pm}(G_0)$ and we need to show the converse inclusion. Let $S \in \mathcal{B}_{\pm}(G_0)$, say, $S = g_1 \dots g_k$ with $g_i \in G_0$ and let $\epsilon_i \in \{+1, -1\}$ such that $\sum_{i=1}^{k} \epsilon_i g_i = 0$. By the definition of G_0, for each g_i there exists some $T_i \in H$ such that g_i occurs in T_i.

Now S is a subsequence of $T_1 \dots T_k$. It remains to show that S divides $T_1 \dots T_k$ in $\mathcal{B}_{\pm}(G)$, that is, $S^{-1}T_1 \dots T_k \in \mathcal{B}_{\pm}(G)$.

Let $T_i' = g_i^{-1}T_i$. Then $S^{-1}T_1 \dots T_k = T_1' \dots T_k'$. Since each T_i is a plus-minus weighted zero-sum sequence we know that g_i or $-g_i$ is in $-\sigma_{\pm}(T_i')$, yet flipping all the weights we see that indeed both are contained in this set. Therefore for each i we get that $\epsilon_i g_i \in \sigma_{\pm}(T_i')$, which implies that $\sum_{i=1}^{k} \epsilon_i g_i \in \sigma_{\pm}(T_1' \dots T_k')$. Yet $\sum_{i=1}^{k} \epsilon_i g_i$ is 0 by assumption and $T_1' \dots T_k'$ is indeed a plus-minus weighted zero-sum sequence. □

4. Minimal distances

In the current section we carry out our investigation of sets of distances, in particular their minima.

We first recall a result essentially in [**16**].

LEMMA 4.1. *Let G be a finite abelian group and let $H = \mathcal{B}_{\pm}(G)$. Then $\Delta(H) = \emptyset$ if and only if $|G| \le 2$. If $|G| \ge 3$, then $1 \in \Delta(H)$ and in particular* $\min \Delta^*(H) = \min \Delta(H) = 1$.

PROOF. The first assertion is Lemma 6.1 in [**16**], that $\min \Delta(H) = 1$ is implicit in its proof, and the assertion on $\min \Delta^*(H)$ follows as H is a divisor-closed submonoid of itself. □

We now give a simple yet powerful lemma; the restriction that $0 \notin G_0$ is no actual restriction; since 0 is a prime element of $\mathcal{B}_{\pm}(G)$, it is easy to see that $\mathcal{B}_{\pm}(G_0)$ and $\mathcal{B}_{\pm}(G_0 \cup \{0\})$ have the same set of distances.

LEMMA 4.2. *Let G be an abelian group and let $G_0 \subseteq G \setminus \{0\}$ be non-empty. Let $H = \mathcal{B}_{\pm}(G_0)$. For each $A \in \mathcal{A}(H)$, we have that $\{2, |A|\} \subseteq \mathsf{L}(A^2)$. In particular, if $\Delta(H) \neq \emptyset$, then* $\min \Delta(H) \mid |A| - 2$. *Moreover,*

$$\min \Delta(H) \mid \gcd\{|A| - 2 \colon A \in \mathcal{A}(H)\} = \gcd\{|A| - |A'| \colon A, A' \in \mathcal{A}(H)\}.$$

PROOF. Let $A = g_1 \dots g_l \in \mathcal{A}(H)$. Note that since $g_i \neq 0$, we have $l \ge 2$ and $g_i^2 \in \mathcal{A}(H)$ for each $i \in [1, l]$. Thus A^2 and $g_1^2 \dots g_l^2$ are factorizations of the same element, the former has length 2 and the latter length $l = |A|$. While $|A| - 2$ might not be in $\Delta(H)$ it is at least a sum of elements from $\Delta(H)$, and since $\min \Delta(H) = \gcd \Delta(H)$ it follows that $\min \Delta(H) \mid |A| - 2$.

Furthermore, as $\min \Delta(H) \mid |A| - 2$ for each $A \in \mathcal{A}(H)$, it divides also $\gcd\{|A| - 2 \colon A \in \mathcal{A}(H)\}$. Finally $\gcd\{|A| - 2 \colon A \in \mathcal{A}(H)\} = \gcd\{|A| - |A'| \colon A, A' \in \mathcal{A}(H)\}$ holds as there is some A' of length 2 and by elementary properties of the greatest common divisor. □

This argument is clearly specific to the case of plus-minus weighted zero-sum sequences. It is however possible to view it as a particular instance of a more general argument that would also encompass a related argument involving cross numbers for the the classical case, which we do not develop here.

A natural question is to what extent $\min \Delta(H)$ and $\gcd\{|A| - 2 \colon A \in \mathcal{A}(H)\}$ can differ. It turns out that they are tightly linked. Before turning to this question we give a few additional general results. In particular, we characterize when $\Delta(H)$ is empty.

LEMMA 4.3. *Let G be an abelian group and let $G_0 \subseteq G$. Let $H = \mathcal{B}_{\pm}(G_0)$.*
(1) *The monoid H is half-factorial if and only if* $\mathsf{D}(H) \le 2$.
(2) *If $\Delta(H) \ne \emptyset$ and $\mathsf{D}(H)$ is finite, we have* $\min \Delta(H) \mid \mathsf{D}(H) - 2$.

PROOF. By Lemma 4.2 it is immediate that $\mathsf{D}(H) > 2$ implies that $\Delta(H)$ is non-empty. To see the converse it suffices to note that the only element of $\mathcal{A}(H)$ of length 1, is the prime element 0.

The second part also follows by Lemma 4.2 as there is an $A \in \mathcal{A}(H)$ with $|A| = \mathsf{D}(H)$. □

This allows us to establish the following bound.

COROLLARY 4.4. *Let G be an abelian group and let $G_0 \subseteq G$. Let $H = \mathcal{B}_{\pm}(G_0)$. We have* $\max \Delta^*(H) \le \mathsf{D}(H) - 2$.

PROOF. If $\mathsf{D}(H)$ is infinite, the claim is trivial. Thus assume that $\mathsf{D}(H)$ is finite. By Lemma 4.3 we know that $\min \Delta(H) \mid \mathsf{D}(H) - 2$ and the same holds true for each divisor-closed submonoid H' of H. It suffices to observe that $\mathsf{D}(H') \le \mathsf{D}(H)$. □

LEMMA 4.5. *Let G be an abelian group and let $G_0 \subseteq G \setminus \{0\}$ be non-empty. Let $H = \mathcal{B}_{\pm}(G_0)$. Then*

$$\min \Delta(H) \mid \gcd\{|A| - 2 \colon A \in \mathcal{A}(H)\} \mid 2 \min \Delta(H).$$

In particular, if $\gcd\{|A| - 2 \colon A \in \mathcal{A}(H)\}$ *is odd, then* $\min \Delta(H) = \gcd\{|A| - 2 \colon A \in \mathcal{A}(H)\}$.

PROOF. The first divisibility relation is just Lemma 4.2. We need to show that $d = \gcd\{|A| - 2 \colon A \in \mathcal{A}(H)\} \mid 2 \min \Delta(H)$. Let $B \in H$ such that B has a factorization of length k and of length $l = k + \min \Delta(H)$, say $B = A_1 \dots A_k = C_1 \dots C_l$ with $A_i, C_j \in \mathcal{A}(H)$.

As $\sum_{i=1}^{k} |A_i| = \sum_{j=1}^{l} |C_j|$, we have $\sum_{i=1}^{k} |A_i| \equiv \sum_{j=1}^{l} |C_j| \pmod d$. Now, essentially by the definition of d, the length of each $A \in \mathcal{A}(H)$ is congruent to 2 modulo d. Therefore $2k \equiv 2l \pmod d$, which implies $2 \min \Delta(H) \equiv 0 \pmod d$ as claimed. Of course in case d is odd, this further yields $\Delta(H) \equiv 0 \pmod d$. □

It might be interesting to observe in case $\gcd\{|A| - 2 \colon A \in \mathcal{A}(H)\}$ is even, there actually can be a discrepancy. A simple example can be obtained by considering elementary 2-groups. Recall that in this case, plus-minus weighted zero-sum sequences are just classical zero-sum sequences.

EXAMPLE 4.6. Let $G = C_2^4 = \langle e_1, e_2, e_3, e_4 \rangle$. Let $G_0 = e_1 + \langle e_2, e_3, e_4 \rangle$ and $H = \mathcal{B}_{\pm}(G_0)$. Then $\min \Delta(H) = 1$ while $\gcd\{|A| - 2 \colon A \in \mathcal{A}(H)\} = 2$.

To see that this is indeed the case, it suffices to note on the one hand that $|A|$ is necessarily even for $A \in \mathcal{A}(H)$, which is clear considering the projection to $\langle e_1 \rangle$. On the other hand, considering $A_0 = (e_1 + e_2 + e_3 + e_4)(e_1 + e_2)(e_1 + e_3)(e_1 + e_4)$, $A_1 = (e_1 + e_2 + e_3 + e_4)(e_1 + e_2)(e_1 + e_3)e_1$ and $A_2 = (e_1 + e_2 + e_3 + e_4)(e_1 + e_2 + e_3)(e_1 + e_4)e_1$, one has $A_1 A_2 = A_0 \cdot e_1^2 \cdot (e_1 + e_2 + e_3 + e_4)^2$, which yields a distance of 1.

Furthermore, this is not an isolated phenomenon and not limited to the minimal distance 1. For instance [**25**, Theorem 5.6] yields examples of other minimal distances for elementary 2-groups.

We recall that if $G_1 \subseteq G_0$, then $\min \Delta(\mathcal{B}_{\pm}(G_0)) \mid \min \Delta(\mathcal{B}_{\pm}(G_1))$.

In view of this studying minimal distances for small sets is of particular relevance. The following lemma gives a complete answer for singletons. The result depends on the parity of the order of the element.

LEMMA 4.7. *Let G be an abelian group. Let $g \in G \setminus \{0\}$ and let $H = \mathcal{B}_{\pm}(\{g\})$.*

(1) *If the order of g is even or infinite, then H is isomorphic to $(\mathbb{N}_0, +)$. In particular, $\Delta(H) = \emptyset$.*
(2) *If the order of g is odd, then H is isomorphic to the numerical monoid $\langle 2, \mathrm{ord}(g)\rangle \subseteq (\mathbb{N}_0, +)$. In particular, $\Delta(H) = \{\mathrm{ord}(g) - 2\}$.*

PROOF. We determine the irreducible elements in the respective monoids. In case the order of g is infinite, it is easy to see that g^2 is the only plus-minus weighted minimal zero-sum sequence over $\{g\}$. This is also the case when the order of g is even; just note that $g^{\mathrm{ord}(g)}$ is not a plus-minus weighted minimal zero-sum sequence if $\mathrm{ord}(g)$ is even and not 2.

Thus in these two cases, the monoid has a unique irreducible element and thus is isomorphic to $(\mathbb{N}_0, +)$. By contrast, if $\mathrm{ord}(g)$ is odd, then there are two plus-minus weighted minimal zero-sum sequences, namely g^2 and $g^{\mathrm{ord}(g)}$; while the latter does have proper subsequences that are also plus-minus weighted zero-sum sequences it cannot be decomposed into two nonempty plus-minus weighted zero-sum sequences.

The isomorphism to $\langle 2, \mathrm{ord}(g)\rangle$ is obtained by considering the multiplicity of g, equivalently the length of the sequence. The result on the set of distances follows noting that the only minimal relation among the irreducibles is $(g^2)^{\mathrm{ord}(g)} = (g^{\mathrm{ord}(g)})^2$ or by invoking results on distances in numerical monoids (see [**5**]). □

LEMMA 4.8. *Let G be an abelian group and let $G_0 \subseteq G$. Let $H = \mathcal{B}_{\pm}(G_0)$. We have $\{\mathrm{ord}(g) - 2 \colon g \in G_0, \mathrm{ord}(g) \geq 3 \text{ odd}\} \subseteq \Delta(H)$.*

PROOF. Let $g \in G_0$ be a nonzero element of odd order. Let $S = \mathcal{B}_{\pm}(\{g\})$. Then $S \subseteq H$ is a divisorclosed submonoid and thus $\Delta(S) \subseteq \Delta(H)$. By Lemma 4.7 we know that $\Delta(S) = \{\mathrm{ord}(g) - 2\}$ and the claim follows. □

4.1. Groups of odd order. In the following result we obtain quite precise information on $\Delta^*(H)$ for $H = \mathcal{B}_{\pm}(G)$ where G is a group of odd order.

THEOREM 4.9. *Let G be a finite abelian group exponent n and let $H = \mathcal{B}_{\pm}(G)$. Assume that n is odd and at least 3. Let $D_1 = \{d - 2 \colon d \mid n,\ d \geq 3\}$ and let $D_2 = \{d' \colon d' \mid d,\ d \in D_1\}$. Then $D_1 \subseteq \Delta^*(H) \subseteq D_2$. In particular, $\max \Delta^*(H) = n - 2$.*

PROOF. To show that $D_1 \subseteq \Delta^*(H)$ it suffices to note that for each $d \mid n$ there exists an element $g \in G$ of order d and to invoke Lemma 4.8. To see that $\Delta^*(H) \subseteq D_2$, let $S \subseteq H$ be a divisor-closed submonoid, that is, $S = \mathcal{B}_{\pm}(G_0)$ for some $G_0 \subseteq G$, see Lemma 3.1. If $\Delta(S) \neq \emptyset$, which is the only case that is relevant, then G_0 contains a nonzero element g whose order, denote it d, divides n and thus is odd. By Lemma 4.7 we have $\Delta(\mathcal{B}_{\pm}(\{g\})) = \{d - 2\}$. Since $\min \Delta(S) \mid \min \Delta(\mathcal{B}_{\pm}(\{g\})) = d - 2$, we have $\min \Delta(S) \in D_2$, establishing the claim. □

The description is not always precise. In the following results we pursue the investigations of this. First, we note that in certain cases the description in fact is complete.

COROLLARY 4.10. *Let $n \in \mathbb{N}$ such that n and $n-2$ are prime, that is n is the larger one of a couple of twin primes. For $H = \mathcal{B}_{\pm}(C_n)$, we have $\Delta^*(H) = \{1, n-2\}$.*

PROOF. Since n is prime each non-zero element C_n has order n. Since $n-2$ is prime as well, using the notation of Theorem 4.9, we have $D_2 = \{1, n-2\}$. Since by Lemma 4.1 we also have $1 \in \Delta^*(H)$, the claim is established. □

However, if $n-2$ is not prime, we do not know which divisors of $n-2$ are contained in $\Delta^*(H)$. We show that at least sometimes such divisors occur as a minimal distance.

LEMMA 4.11. *Let $n \in \mathbb{N}$ be odd and of the form $n = a^2+1$ for $a \in \mathbb{N}$. For $H = \mathcal{B}_{\pm}(C_n)$, we have $a-1 \in \Delta^*(H)$.*

PROOF. Let $C_n = \langle e \rangle$ and let $G_0 = \{e, ae\}$. We show that for $H' = \mathcal{B}_{\pm}(G_0)$ we have $\min \Delta(H') = a-1$. By Lemma 4.5 it suffices to study the length of the elements of $\mathcal{A}(H')$.

We first recall that there are two elements of $\mathcal{A}(H')$ containing only e, namely e^2 and e^n. Likewise there are two elements containing only ae, namely $(ae)^2$ and $(ae)^n$; note that the order of ae is n, since a and n are co-prime.

Now assume that $e^v(ae)^w$, with $v, w \in \mathbb{N}$ is in $\mathcal{A}(H')$. We observe that due to the minimality in a plus-minus weighted zero-sum of this sequence all the weights of e are equal and likewise all the weights of ae are equal. This means that either $ve + wae = 0$ or $ve + w(-a)e = 0$. In the former case $v = n - wa$ for w from 1 to a (note that $n - a^2 = 1$) yields indeed minimal plus-minus weighted zero-sum sequences, while for $w > a$ the sequence would be divisible by $e(ae)^a$ and hence not minimal. In the latter case, for $v = a-k$ for k from 1 to $a-1$ implies that $w = 1 + ka$ and these are indeed minimal plus-minus weighted zero-sum sequences, while other other choices of v and w do not yield minimal plus-minus weighted zero-sum sequences.

The length of these sequences are $(n - wa) + w = n - (a-1)w$ for w from 1 to a and $(a-k) + (1+ka) = a + 1 + k(a-1)$ for k from 1 to $a-1$. It follows that $\gcd\{|A| - 2 \colon A \in \mathcal{A}(H')\} = a-1$; observe that $a-1$ divides $n-2 = a^2-1$. □

The problem of determining the set $\Delta^*(H)$ entirely for $H = \mathcal{B}_{\pm}(C_n)$ might be difficult. It certainly is in the case without weights (see [**6**, **24**]). It seems plausible that results on the minimal distances in the case without weights can be employed to make further progress, yet we do not pursue this line of investigation in this paper.

For groups of higher rank, we show that for elementary p-groups, the problem for groups of large rank can be reduced almost entirely to the problem for cyclic groups. Again, the exclusion of 0 in the lemma below is not actual restriction.

LEMMA 4.12. *Let $G = C_p^r$ be an elementary p-group of rank r for some odd prime p. Let $G_0 \subseteq G \setminus \{0\}$ and $H' = \mathcal{B}_{\pm}(G_0)$. If $\min \Delta(H') > 1$, then $G_0 = G_1 \uplus \cdots \uplus G_k$ where $\langle G_i \rangle$ is cyclic for each $i \in [1,k]$ and $\langle G_0 \rangle = \oplus_{i=1}^k \langle G_i \rangle$.*

PROOF. Let $G_0 \subseteq G$, and let $e_1, \dots, e_k$ be a basis $\langle G_0 \rangle$. To show the assertion it suffices to show that if G_0 contains some element $e_0 = \sum_{i=1}^k a_i e_i$ with $a_i \in [0, p-1]$ and at least two of the a_i non-zero, then $\min \Delta(H') = 1$.

Without loss we can assume that all the a_i are non-zero and $k \geq 2$. Moreover we can assume that $a_i < p/2$, since we could replace e_i by $(-e_i)$ without affecting the problem.

Let $U = e_0 e_1^{p-a_1} e_2^{p-a_2} e_3^{a_3} \dots e_k^{a_k}$ and $V = e_0^2 e_1^{p-2a_1} e_2^{p-2a_2} e_3^{2a_3} \dots e_k^{2a_k}$. These are both minimal plus-minus weighted zero-sum sequences. We have $U^2 = V e_1^p e_2^p$ and thus the minimal distance is 1. □

This lemma shows that $\Delta^*(\mathcal{B}_\pm(C_p^r))$ and $\Delta^*(\mathcal{B}_\pm(C_p))$ are very closely linked. Before stating this relation in detail, we note that the argument in the proof of the preceding lemma needs neither the orders of the e_i to be prime nor them to be all equal. We record this as a lemma as it might be useful for future investigations.

LEMMA 4.13. *Let G be a finite abelian group and let $e_1, \dots, e_k$ be independent elements of odd order, say* $\operatorname{ord}(e_i) = n_i \geq 3$. *Let $e_0 = a_1 e_1 + \dots + a_k e_k$ with $a_i \in [0, n_i - 1]$. Put $G_0 = \{e_0, e_1, \dots, e_k\}$ and $H = \mathcal{B}_\pm(G_0)$. If at least two a_i are non-zero, then* $\min \Delta(H) = 1$.

PROOF. Without loss we can assume that all the a_i are non-zero. Suppose $k \geq 2$. Moreover we can assume that $a_i < n_i/2$ for each i, since we could replace e_i by $(-e_i)$ without affecting the problem.

Now let $U = e_0 e_1^{n_1-a_1} e_2^{n_2-a_2} e_3^{a_3} \dots e_k^{a_k}$ and $V = e_0^2 e_1^{n_1-2a_1} e_2^{n_2-2a_2} e_3^{2a_3} \dots e_k^{2a_k}$. These are both minimal plus-minus weighted zero-sum sequences. We have $U^2 = V e_1^{n_1} e_2^{n_2}$ and thus the minimal distance is 1. □

We now give the description for elementary p-groups.

THEOREM 4.14. *Let p be an odd prime and $r \geq 1$. We have $\Delta^*(\mathcal{B}_\pm(C_p^r)) = \{\gcd(d_1, \dots, d_r) \colon d_i \in \Delta^*(\mathcal{B}_\pm(C_p))\}$.*

PROOF. By Lemma 4.12 each subset G_0 that yields an element of $\Delta^*(\mathcal{B}_\pm(C_p^r))$ other than 1 can be written as $G_0 \setminus \{0\} = G_1 \uplus \dots \uplus G_k$ where $\langle G_i \rangle$ is cyclic for each $i \in [1, k]$ and $\langle G_0 \rangle = \oplus_{i=1}^k \langle G_i \rangle$. Now, the minimal distance of $\mathcal{B}_\pm(G_0)$ is the greatest common divisor of the minimal distances of $\mathcal{B}_\pm(G_i)$ for $i \in [1, k]$, which shows that each element of $\Delta^*(\mathcal{B}_\pm(C_p^r))$ is of the claimed form. Conversely, given $d_1, \dots, d_r \in \Delta^*(\mathcal{B}_\pm(C_p))$, there are $G_1, \dots, G_r \subseteq C_p^r$ such that $\langle G_i \rangle$ is cyclic, $\min \Delta(\mathcal{B}_\pm(G_i)) = d_i$ and $C_p^r = \langle G_1 \rangle \oplus \dots \oplus \langle G_r \rangle$. Then $\min \mathcal{B}_\pm(G_1 \cup \dots \cup G_r) = \gcd(d_1, \dots, d_r)$. □

4.2. Groups of even order. In the current subsection we focus on groups of even order and start with a result that allows us to construct sets with relatively large minimal distances.

LEMMA 4.15. *Let G be a finite abelian group and let $e_1, \dots, e_r$ be independent elements of even order, say* $\operatorname{ord}(e_i) = 2m_i$. *Assume that $m_1 + \dots + m_r \geq 2$. Let $e_0 = m_1 e_1 + \dots + m_r e_r$, $G_0 = \{e_0, e_1, \dots, e_r\}$ and $H = \mathcal{B}_\pm(G_0)$. Then $\Delta(H) = \{m_1 + \dots + m_r - 1\}$.*

PROOF. We determine $\mathcal{A}(H)$. Clearly $e_i^2 \in \mathcal{A}(H)$ for each $0 \leq i \leq r$. Let $A \in \mathcal{A}(H)$. If $e_0 \nmid A$, then the distinct elements in A are independent, and A being minimal implies that A is equal to e_i^k for some $1 \leq i \leq r$. Since the order of e_i is even, it follows that $k = 2$, compare Lemma 4.7. If $e_0^2 \mid A$, then we claim that $e_0^2 = A$. To see this it suffices to note that the order of e_0 is 2, and thus, 0 is the only plus-minus weighted sum of e_0^2, which implies that $A e_0^{-2}$ has also a plus-minus weighted zero-sum. The only way how this does not contradict the minimality of A is that Ae_0^{-2} is empty.

Thus, it remains to consider the case that A contains e_0 exactly once. We show that in this case A is equal to $U = e_0 \prod_{i=1}^{r} e_i^{m_i}$. We note that $(m_i e_i) e_i^k$ is a minimal plus-minus weighted zero-sum sequence if and only if $k = m_i$, we refer to Lemma 6.6 in [**4**] for a detailed argument. The claim now follows from the independence of the e_i.

Thus we established that $\mathcal{A}(H) = \{U, e_0^2, e_1^2, \dots, e_r^2\}$. Now $U^2 = e_0^2 \prod_{i=1}^{r} (e_i^2)^{m_i}$ is the only relevant relation between these atoms and the claim follows. □

The argument also shows that $\mathsf{c}(H) = m_1 + \cdots + m_r + 1$ where $\mathsf{c}(H)$ denotes the catenary degree of H. It is interesting to note that this value coincides with the value that one expects for many types of groups for the catenary degree of the monoid of classical zero-sum sequences over $\langle e_1, \dots, e_r \rangle$ (see for example [**14**]).

We note that if $e_1, \dots, e_r$ is an independent generating set of elements of even order of G, say $\mathrm{ord}(e_i) = 2m_i$, with $\mathrm{ord}(e_i) \mid \mathrm{ord}(e_{i+1})$, then $\mathsf{D}^*(\mathcal{B}_{\pm}(G)) = m_1 + \cdots + m_r + 1$. This is a lower bound for $\mathsf{D}(\mathcal{B}_{\pm}(G))$ that was obtained in [**4**].

Thus, for a group with $\mathsf{r}_2(G) = \mathsf{r}(G)$ we have $\mathsf{D}^*(\mathcal{B}_{\pm}(G)) - 2 \in \Delta^*(\mathcal{B}_{\pm}(G))$. We recall that we always have $\max \Delta^*(H) \le \mathsf{D}(H) - 2$. Thus, if we know that $\mathsf{D}^*(\mathcal{B}_{\pm}(G)) = \mathsf{D}(\mathcal{B}_{\pm}(G))$, we get the exact value of the maximum. However, result of this type are rare. Yet, it was established for C_{2m} in [**4**] and for $C_2 \oplus C_4$, in [**10**]. The result on elementary 2-groups is already in [**17**].

PROPOSITION 4.16. *Let G be a finite abelian group and let $H = \mathcal{B}_{\pm}(G)$.*
(1) *If G is cyclic of order at least 3, then* $\max \Delta^*(H) = \mathsf{D}(H) - 2$.
(2) *If G is an elementary 2-group of order at least 4, then* $\Delta^*(H) = [1, \mathsf{D}(H) - 2]$.
(3) *If G is a group of order at least 4 exponent a divisor of 4, then* $[1, \mathsf{D}^*(H) - 2] \subseteq \Delta^*(H) \subseteq [1, \mathsf{D}(H) - 2]$.
(4) *If $G = C_n \oplus C_{2n}$ for some odd $n \ge 3$, then* $\mathsf{D}^*(H) - 2 \le \max \Delta^*(H) \le \mathsf{D}(H) - 2$.

PROOF. We recall that we always have $\max \Delta^*(H) \le \mathsf{D}(H) - 2$.

Let $G = \langle e_1 \rangle$ be cyclic of order $n \ge 3$. If n is odd, then $\mathsf{D}(H) - 2 = n - 2$, and $\Delta(\{e\}) = n - 2$. Thus the claim follows.

In all other cases except for 4, we have $\mathsf{r}_2(G) = \mathsf{r}(G)$ and we have $\mathsf{D}^*(\mathcal{B}_{\pm}(G)) - 2 \in \Delta^*(\mathcal{B}_{\pm}(G))$. If $n = 2m$ is even and $G = C_n$, then $\mathsf{D}(H) - 2 = m - 1$ and $\Delta(\{e, me\}) = m - 1$.

If $G = \oplus_{i=1}^{r} \langle e_i \rangle$ is an elementary 2-group of rank r, then $\mathsf{D}^*(H) - 2 = r - 1$ and thus $r - 1$ is an element of $\Delta^*(H)$. Since $H_s = \mathcal{B}_{\pm}(\oplus_{i=1}^{s} \langle e_i \rangle)$ is a divisor-closed submonoid of H for each $s \le r$ we have that $\Delta(H_s) \subseteq \Delta(H)$. Since $\mathsf{D}^*(\mathcal{B}_{\pm}(\oplus_{i=1}^{s} \langle e_i \rangle)) = s + 1$, the claim follows using the same argument as for $s = r$.

Suppose $G = \oplus_{i=1}^{r} \langle e_i \rangle$ with $\mathrm{ord}(e_i) \mid \mathrm{ord}(e_{i+1})$ is a group of exponent dividing 4 and rank r. As above we have $\mathsf{D}^*(\mathcal{B}_{\pm}(G)) - 2 \in \Delta^*(\mathcal{B}_{\pm}(G))$. To complete the argument it suffices to show that G has a subgroup G' with $\mathsf{D}^*(\mathcal{B}_{\pm}(G')) = D$ for each $D \in [3, \mathsf{D}^*(\mathcal{B}_{\pm}(G))]$. We set $G_1 = \oplus_{i=1}^{r-1} \langle e_i \rangle \oplus \langle 2e_r \rangle$. Then $\mathsf{D}^*(\mathcal{B}_{\pm}(G_1)) = \mathsf{D}^*(\mathcal{B}_{\pm}(G)) - 1$. Now the argument follows by a direct induction.

We show the fourth part. Let $C_n \oplus C_{2n} = \langle e_1 \rangle \oplus \langle e_2 \rangle$ with $\mathrm{ord}(e_1) = n$ and $\mathrm{ord}(e_2) = 2n$. We consider $G_0 = \{e_1 + e_2, e_2\}$. We determine $\mathcal{A}(\mathcal{B}_{\pm}(G_0))$. Clearly we have $V_1 = (e_1 + e_2)^2$ and $V_2 = e_2^2$ and no other elements with only one of the two elements. In addition we find $U = (e_1 + e_2)^n e_2^n$. We have $U^2 = V_1^n V_2^n$ and indeed $\mathsf{L}(U^2) = \{2, 2n\}$, and we find that $\Delta(\mathcal{B}_{\pm}(G_0)) = \{2n-2\}$. We recall that $\mathsf{D}^*(H) = n - 1 + (2n)/2 + 1 = 2n$, establishing the claim. □

In the following result we characterize the groups for which $\max \Delta^*(H)$ is small.

PROPOSITION 4.17. *Let G be a finite abelian group and let $H = \mathcal{B}_{\pm}(G)$.*

(1) *Then* $\max \Delta^*(H) = 1$ *if and only if* $\exp(G) = 3$ *or* $G = C_2^2$ *or* $G = C_4$.
(2) *Then* $\max \Delta^*(H) = 2$ *if and only* $G = C_2^3$, $G = C_2 \oplus C_4$, *or* $G = C_6$.

PROOF. If $|G| \le 2$, then $\Delta^*(H) = \emptyset$ by Lemma 4.1 and thus $\max \Delta^*(H) \ne 1$. Thus assume $|G| \ge 3$ and let $n = \exp(G)$.

Supose n is odd. Then, by Proposition 4.16 $\max \Delta^*(H) = n - 2$. Therefore, $\max \Delta^*(H) = 1$ if and only if and $n = 3$, and $\max \Delta^*(H) = 2$ is impossible.

Now, suppose that n is even. Then, by Proposition 4.16, $\Delta^*(H) = [1, r-1]$, where r denotes the rank of G. Thus $\max \Delta^*(H) = 1$ if and only if $G = C_2^2$, and $\max \Delta^*(H) = 2$ if and only if $G = C_2^3$.

If $n \ge 4$, by Lemma 4.15, we have $n/2 - 1 \in \Delta^*(H)$. Thus, $\max \Delta^*(H) = 1$ implies that $n = 4$, and $\max \Delta^*(H) = 2$ implies $n \le 6$. If $n = 4$, then Lemma 4.15 shows that $\max \Delta^*(H) \ge \mathsf{r}_2(G) + \mathsf{r}_4(G) - 1$. Therefore, $\max \Delta^*(H) = 1$, implies $G = C_4$, by Proposition 4.16 indeed $\Delta^*(\mathcal{B}_{\pm}(C_4)) = \{1\}$, the claim is established. Moreover $\max \Delta^*(H) = 2$, implies $G = C_2 \oplus C_4$. Since $\mathsf{D}(H) = 4$ in that case, indeed $\max \Delta^*(\mathcal{B}_{\pm}(C_2 \oplus C_4)) = 2$. For $n = 6$ we see that for $G = C_6$ the maximum is indeed $2 = \mathsf{D}(H) - 2$ while in all other cases, using Lemma 4.15 and Proposition 4.16(4), we see that the maximum is strictly greater than 2. □

We end the section with a result that might seem peculiar at first, but is highly useful in the context of the subsequent section.

THEOREM 4.18. *Let G be a finite abelian group with $|G| \ge 5$. The following statements are equivalent.*

(1) $|G|$ *is even.*
(2) $\Delta^*(H)$ *contains an even element.*
(3) $\Delta_1(H)$ *contains an even element.*

PROOF. The equivalence of the second and third statement is clear by the inclusions between $\Delta^*(H)$ and $\Delta_1(H)$ and the divors of $\Delta^*(H)$ recalled in Section 3. We show that the first and second statement are equivalent.

Assume that $\Delta^*(H)$ contains an even element. Then $|G|$ cannot be odd, as this would contradict Theorem 4.9 as all elements of D_2 are odd. It remains to show that if $|G|$ is even, then $\Delta^*(H)$ contains an even element. If G is not a 2-group, then G contains an element of order $2m$ for some odd $m \ge 2$. By Lemma 4.15 we know that $m - 1 \in \Delta^*(H)$, and this is even.

Thus assume that G is a 2-group. If the rank of G is at least 3, then by Proposition 4.16 $\Delta^*(H)$ contains 2, and if the rank is 2 then by Proposition 4.16 it contains $1 + (\exp(G)/2 - 1) = \exp(G)/2$, which is even as $\exp(G) \ne 2$ by the assumption $|G| \ge 5$.

It remains to study the case of cyclic 2-groups. We note that for e an element of order 8, the monoid $\mathcal{B}_{\pm}(\{e, 3e\})$ has minimal distance 2; note that the irreducibles are $e^2, (3e)^2, e^3(3e), e(3e)^3$. □

4.3. Groups of infinite order. In this section, we gather a few results on groups of infinite order, which basically reduce the problem of determining the set of minimal distances to the one for finite groups.

THEOREM 4.19. *Let G be a an abelian group.*

(1) *If G is not a torsion group, then $\Delta^*(\mathcal{B}_\pm(G)) = \mathbb{N}$.*
(2) *If G is a torsion group, then*

$$\Delta^*(\mathcal{B}_\pm(G)) = \bigcup_{G' \subseteq G \text{ finite subgroup}} \Delta^*(\mathcal{B}_\pm(G')).$$

PROOF. To prove the first claim, let $g \in G$ be an element of infinite order. We assert that for each $n \in \mathbb{N}$, one has $\Delta(\mathcal{B}_\pm(\{g, (n+1)g\})) = n$. One checks that $\mathcal{A}(\mathcal{B}_\pm(\{g, (n+1)g\})) = \{g^2, ((n+1)g)^2, g^{n+1}((n+1)g)\}$. Noting that $(g^{n+1}((n+1)g))^2 = g^2((n+1)g)^2$ is the only relevant relation, the claim follows.

We now turn to the second claim. It is clear that

$$\Delta^*(\mathcal{B}_\pm(G)) \supset \bigcup_{G' \subseteq G \text{ finite subgroup}} \Delta^*(\mathcal{B}_\pm(G')).$$

Let $d \in \Delta^*(\mathcal{B}_\pm(G))$. We need to show that there is a finite subgroup G' of G such that $d \in \Delta^*(\mathcal{B}_\pm(G'))$. Let $G_0 \subseteq G$ such that $d = \min \Delta(\mathcal{B}_\pm(G_0))$. Let $B \in \mathcal{B}_\pm(G_0)$ such that $d \in \Delta(B)$. Then $d = \min \Delta(\mathcal{B}_\pm(\operatorname{supp}(B)))$ holds as well. Let $G' = \langle \operatorname{supp}(B) \rangle$, which is a finite subgroup of G as $\operatorname{supp}(B)$ is finite and the group is a torsion group. Since $\operatorname{supp}(B) \subseteq G'$, it follows that $d \in \Delta^*(\mathcal{B}_\pm(G'))$. □

COROLLARY 4.20. *Let G be an elementary p-group of infinite rank. Let $H = \mathcal{B}_\pm(G)$.*

(1) *If $p = 2$, then $\Delta^*(H) = \mathbb{N}$.*
(2) *If p is odd, then $\Delta^*(H) = \{\gcd(d_1, \dots, d_s) \colon s \in \mathbb{N},\, d_i \in \Delta^*(\mathcal{B}_\pm(C_p))\}$.*

PROOF. The result is a direct consequence of Theorem 4.19 and the results on finite elementary p-groups obtained in Theorem 4.14 and Proposition 4.16. □

COROLLARY 4.21. *Let G be an abelian torsion group without an element of even order. Let D_1 denote the set of all $d \in \mathbb{N}$ such that G contains an element of order $d+2$ and let $D_2 \subseteq \mathbb{N}$ denote the set of all d' that divide an element of D_1. Then $D_1 \subseteq \Delta^*(\mathcal{B}(G)) \subseteq D_2$.*

PROOF. The result is a direct consequence of Theorem 4.19 and Theorem 4.9. □

5. Applications to the characterization problem

It is obvious that isomorphic abelian groups G and G' yield isomorphic monoids of zero-sum sequences $\mathcal{B}(G)$ and $\mathcal{B}(G')$, and hence also the systems of sets of lengths are equal $\mathcal{L}(\mathcal{B}(G)) = \mathcal{L}(\mathcal{B}(G'))$. The same hold true if we consider plus-minus weighted zero-sum sequences (or in fact weighted zero-sum sequences for any choice of weights).

A natural question to ask is to what extent these implications can be reversed. For the first implication this is called the Isomophism Problem, that is it is the problem of determinig if isomorphy of the monoids of (weighted) zero-sum sequences implies isomorphy of the underlying groups, the problem is completely answered for abelian groups in the case without weights and for groups that are the direct sum of cyclic groups in the case with plus-minus weights (see [**10**] and the references there).

For the question involving length this is called the Characterization Problem. In its classical form it can be stated as asking whether for G and G' finite abelian groups such that $\mathcal{L}(\mathcal{B}(G)) = \mathcal{L}(\mathcal{B}(G'))$, it follows that G and G' are isomorphic. The history of this problem goes back more than fifty years and initially arose in the context of obtaining arithmetic characterizations of the class group of a ring of algebraic integers, whence the name (see [**23**]). We mention that the condition that the group is finite is crucial. All infinite abelian groups yield the same system of sets of lengths, which differs from that of every finite group. This is known as Kainrath's Theorem and we discussed it in Section 3.

It is well-known and easy to see that C_1 and C_2 both give rise to half-factorial monoids of zero-sum sequences, indeed the monoids are even free. This simple example apart, there is one more pair of non-isomorphic groups known that yield the same system of set of lengths for their monoids of zero-sum sequences, namely C_3 and C_2^2. The standing conjecture is that these are the only cases where this arises and that for each finite abelian group G that is not isomorphic to one of those groups, a condition that ensures this is to assume that $\mathsf{D}(\mathcal{B}(G)) \geq 4$, it is indeed true that $\mathcal{L}(\mathcal{B}(G)) = \mathcal{L}(\mathcal{B}(G'))$ implies that G and G' are isomorphic. We refer to [**26**] for an overview of older results on the subject and to [**18**, **28**] for more recent contributions.

A first investigation of the Characterization Problem for plus-minus weighted zero-sum problems, and more broadly of the problem of describing the system of sets of lengths for plus-minus weighted zero-sum problems has been undertaken in [**10**]. Very recently, as already mentioned in Section 3, a result like Kainrath's Theorem has also been obtained for the monoids of plus-minus weighted zero-sum sequences [**17**].

As in the case without weights C_1 and C_2 both give rise to half-factorial monoids of zero-sum sequences, indeed even free monoids. For other groups of small order in [**10**] the authors proved the following result.

THEOREM 5.1. $\mathcal{L}(\mathcal{B}_{\pm}(C_2^2)) = \mathcal{L}(\mathcal{B}_{\pm}(C_3)) = \mathcal{L}(\mathcal{B}_{\pm}(C_4)) = \{y+2k+[0,k] : y, k \in \mathbb{N}_0\}$ *Conversely, if* $\mathcal{L}(\mathcal{B}_{\pm}(G)) = \{y + 2k + [0,k] : y, k \in \mathbb{N}_0\}$ *for some finite abelian group* G*, then* G *is isomorphic to one of the three groups* C_2^2, C_3, C_4.

Note that the sets $y + 2k + [0,k]$ are just the sets $\{m, m+1, m+2, \dots, n\}$ with $n/m \leq 3/2$. Moreover the authors settled the Characterization Problem for cyclic groups of odd order. *In view of this result, we assume for our subsequent investigations that all groups* G *and* G' *have order at least* 5.

In the current section, we try to make further progress on this problem using our results on Δ^*. It is in most cases quite difficult to describe $\mathcal{L}(H)$ completely. In order to solve, the characterization problem one often proceeds like this. One establishes that if $\mathcal{L}(H) = \mathcal{L}(H')$, then certain arithmetic invariants of H and H' are equal. Using results on these arithmetic invariants, one obtains information on H and H'.

For invariants that are essentially defined directly via the system of sets of lengths, such as ρ, ρ_k, Δ, and Δ_1 it is clear that equality of $\mathcal{L}(H)$ and $\mathcal{L}(H')$ implies the equality of the respective invariants for H and H'. Since for $H = \mathcal{B}_{\pm}(G)$ and $H' = \mathcal{B}_{\pm}(G')$, it is known that $\rho_2(H) = \mathsf{D}(H)$ and $\rho_2(H') = \mathsf{D}(H')$ (see [**4**]), it follows that the equality of $\mathcal{L}(H)$ and $\mathcal{L}(H')$ also implies that $\mathsf{D}(H) = \mathsf{D}(H')$. Since we use the result, we also recall that there it was proved that in case $|G|$ is odd, one has $\mathsf{D}(H) = \mathsf{D}(G)$.

Likewise the fact that

$$\Delta^*(H) \subseteq \Delta_1(H) \subseteq \{d' \mid d\colon d \in \Delta^*(H)\},$$

implies that in case $\mathcal{L}(H) = \mathcal{L}(H')$ also the sets $\Delta^*(H)$ and $\Delta^*(H')$ are closely linked: $\max \Delta^*(H) = \max \Delta^*(H')$ and more precisely, if M denotes this maximum, the sets $\Delta^*(H) \cap \mathbb{N}_{>M/2}$ and $\Delta^*(H') \cap \mathbb{N}_{>M/2}$ are equal.

Then, one uses results that give the values of these arithmetical invariants in terms of algebraic invariants of the groups, to establish conditions on the groups, and ideally their isomorphy.

THEOREM 5.2. *Let G be a finite abelian group of odd order and let G' be a finite abelian group. We set $H = \mathcal{B}_{\pm}(G)$ and $H' = \mathcal{B}_{\pm}(G')$. If $\mathcal{L}(H) = \mathcal{L}(H')$, then $\exp(G) = \exp(G')$ and $\mathsf{D}(G) = \mathsf{D}(G')$.*

PROOF. We know that $\rho(H) = \rho(H')$ and $\Delta_1(H) = \Delta_1(H')$. By Theorem 4.9 we know that $\max \Delta_1(H) = \exp(G) - 2$.

By Theorem 4.18 we know that $\Delta_1(H)$ only contains odd elements. Thus, $\Delta_1(H')$ only contains odd elements as well, and G' is a group of odd order too. Thus, $\max \Delta_1(H') = \exp(G') - 2$ as well and $\exp(G) = \exp(G')$.

To obtain the result on the Davenport constants it suffices to recall that $\rho_2(H) = \mathsf{D}(H)$ and $\rho_2(H') = \mathsf{D}(H')$, and since $|G|$ and $|G'|$ are odd we have $\mathsf{D}(H) = \mathsf{D}(G)$ and $\mathsf{D}(H') = \mathsf{D}(G')$. $\square$

We insist that indeed the classical Davenport constants of G and G' are equal as for groups of odd order or equivalently odd exponent, as recalled above, they are equal to $\mathsf{D}(\mathcal{B}_{\pm}(G))$ and $\mathsf{D}(\mathcal{B}_{\pm}(G'))$ respectively. Below we freely use the results on the Davenport constant recalled in Section 3.

This result allows us to prove that elementary p-groups for p an odd prime are characterized by the system of sets of length of for plus-minus weighted zero-sum sequences. Since the same reasoning allows us to characterize certain other p-groups and groups of odd exponent, we start with a technical proposition.

PROPOSITION 5.3. *Let G be a finite abelian group of exponent n odd and Davenport constant D such that there is no group non-isomorphic to G that also has exponent n and Davenport constant D. If $\mathcal{L}(\mathcal{B}_{\pm}(G)) = \mathcal{L}(\mathcal{B}_{\pm}(G'))$, then $G \cong G'$.*

PROOF. By Theorem 5.2 we know that the exponents of G and G' are equal and that the Davenport constants of G and G' are equal. The claim is thus obviously true by the assumption on G. $\square$

We first note that this result yields an alternative proof for the characterization of cyclic groups of odd order already obtained in [**10**].

COROLLARY 5.4. *Let G be a cyclic group whose order is odd. If $\mathcal{L}(\mathcal{B}_{\pm}(G)) = \mathcal{L}(\mathcal{B}_{\pm}(G'))$, then $G \cong G'$.*

PROOF. A cyclic group fulfills the condition of Proposition 5.3 as cyclic groups are the only groups for which the exponent equals the Davenport constant. $\square$

We now establish the result for elementary p-groups of odd order.

COROLLARY 5.5. *Let G be an elementary p-group for an odd prime p. If $\mathcal{L}(\mathcal{B}_{\pm}(G)) = \mathcal{L}(\mathcal{B}_{\pm}(G'))$, then $G \cong G'$.*

PROOF. For a group G with $\exp(G)$ prime, the Davenport constant is given by $1 + \mathsf{r}(G)(\exp(G) - 1)$ and thus elementary p-groups fulfill the condition of Proposition 5.3. □

We briefly discuss the problem for elementary 2-groups at the end of the section. Yet, before this we establish two results that can be obtained via our method.

COROLLARY 5.6. *Let $G = C_p \oplus C_n$ where n is odd and p is the smallest prime divisor of n. If $\mathcal{L}(\mathcal{B}_{\pm}(G)) = \mathcal{L}(\mathcal{B}_{\pm}(G'))$, then $G \cong G'$.*

PROOF. It suffices to show that the groups fulfill the condition of Proposition 5.3. First recall that the Davenport constant of G is $n + p - 1$, as the group has rank two. Now, if G' is a group with exponent n, then either G' is cyclic or of the form $H \oplus C_n$ where $1 \neq \exp(H) \mid n$. In the former case the Davenport of G' is n and thus not equal to that of G in the latter case the Davenport constant is at least $n + \mathsf{D}(H) - 1 \geq n + \exp(H) - 1$ and thus strictly greater than $n + p - 1$ unless $\exp(H) = p$ and $\mathsf{D}(H) = \exp(H)$. Thus we get that H is cyclic of order p, implying the claim. □

To obtain a result for p-groups, we establish a simple lemma on the values of the Davenport constant of p-groups.

LEMMA 5.7. *Let K be a p-group with Davenport constant D.*

(1) *If $D < p^2$ and K' is another p-group with Davenport constant D, then K and K' are isomorphic.*
(2) *If $D \geq p^2$ then there exists a p-group K' with Davenport constant D such that K and K' are not isomorphic.*

PROOF. To see the first point, it suffices to note that $D < p^2$ implies $\exp(K) < p^2$ and $\exp(K') < p^2$. Thus, both have exponent p and the claim follows directly.

For the second point, we distinguish two cases. Assume that the exponent of K is p, say $K \simeq C_p^r$. By the assumption on the Davenport constant we get that $r \geq p + 1$. Yet then $K' = C_p^{r-p-1} \oplus C_{p^2}$ has the same Davenport constant.

Now, assume that the exponent of K is not prime, say, it is p^k with $k > 1$. Let K_0 be a group such that $K \simeq K_0 \oplus C_{p^k}$. Then set $K' = K_0 \oplus C_p^{(p^k-1)/(p-1)}$, which has the same Davenport constant. □

COROLLARY 5.8. *Let G be a p-group for p odd and assume that $\mathsf{D}(G) < \exp(G) + p^2 - 1$. If $\mathcal{L}(\mathcal{B}_{\pm}(G)) = \mathcal{L}(\mathcal{B}_{\pm}(G'))$, then $G \cong G'$.*

PROOF. To proof the corollary, we show that a group G that fulfills the conditions in this corollary, indeed fulfills the conditions in Proposition 5.3, which directly yields the result. That is, we need to show that for such a G there are no other groups (up to isomorphism) with the same exponent and Davenport constant. Indeed, let K be a group such that $G \simeq K \oplus C_{\exp(G)}$ fulfills the conditions of the corollary. Then $\mathsf{D}(G) = \exp(G) + \mathsf{D}(K) - 1$, as G is a p-group. It follows that $\mathsf{D}(K) < p^2$, and thus by Lemma 5.7 the group K, which is a p-group, is determined (up to isomorphism) by its Davenport constant. □

We did not make progress on the characterization problem for groups of even order as our results on $\Delta^*(\mathcal{B}_\pm(G))$ are weaker in that case. Moreover, even in a case where we have good knowledge of $\Delta^*(\mathcal{B}_\pm(G))$, such as elementary 2-groups it is not clear how to exploit it, and characterizing elementary 2-groups might actually be difficult. It is worth noting that in [**10**] the characterization problem was solved for C_2^3 and $C_2 \oplus C_4$, yet needs a detailed analysis of the respective systems of sets of lengths.

Acknowledgment

The authors thank the referee for numerous helpful remarks.

References

[1] S. D. Adhikari, D. J. Grynkiewicz, and Z.-W. Sun, *On weighted zero-sum sequences*, Adv. in Appl. Math. **48** (2012), no. 3, 506–527, DOI 10.1016/j.aam.2011.11.007. MR2899967

[2] S. D. Adhikari and P. Rath, *Davenport constant with weights and some related questions*, Integers **6** (2006), A30, 6. MR2264845

[3] N. R. Baeth and D. Smertnig, *Factorization theory: from commutative to noncommutative settings*, J. Algebra **441** (2015), 475–551, DOI 10.1016/j.jalgebra.2015.06.007. MR3391936

[4] S. Boukheche, K. Merito, O. Ordaz, and W. A. Schmid, *Monoids of sequences over finite abelian groups defined via zero-sums with respect to a given set of weights and applications to factorizations of norms of algebraic integers*, Comm. Algebra **50** (2022), no. 10, 4195–4217, DOI 10.1080/00927872.2022.2058009. MR4447457

[5] C. Bowles, S. T. Chapman, N. Kaplan, and D. Reiser, *On delta sets of numerical monoids*, J. Algebra Appl. **5** (2006), no. 5, 695–718, DOI 10.1142/S0219498806001958. MR2269412

[6] S. Chang, S. T. Chapman, and W. W. Smith, *On minimum delta set values in block monoids over cyclic groups*, Ramanujan J. **14** (2007), no. 1, 155–171, DOI 10.1007/s11139-006-9000-x. MR2299046

[7] S. T. Chapman, W. A. Schmid, and W. W. Smith, *On minimal distances in Krull monoids with infinite class group*, Bull. Lond. Math. Soc. **40** (2008), no. 4, 613–618, DOI 10.1112/blms/bdn040. MR2441133

[8] J. Coykendall, *Normsets and determination of unique factorization in rings of algebraic integers*, Proc. Amer. Math. Soc. **124** (1996), no. 6, 1727–1732, DOI 10.1090/S0002-9939-96-03261-3. MR1317034

[9] J. Coykendall and R. Hasenauer, *Norms, normsets, and factorization*, Recent progress in ring and factorization theory, Springer Proc. Math. Stat., vol. 477, Springer, Cham, [2025] ©2025, pp. 161–193, DOI 10.1007/978-3-031-75326-8_6. MR4928634

[10] F. Fabsits, A. Geroldinger, A. Reinhart, and Q. Zhong, *On monoids of plus-minus weighted zero-sum sequences: the isomorphism problem and the characterization problem*, J. Commut. Algebra **16** (2024), no. 1, 1–23, DOI 10.1216/jca.2024.16.1. MR4690592

[11] W. Gao and A. Geroldinger, *Zero-sum problems in finite abelian groups: a survey*, Expo. Math. **24** (2006), no. 4, 337–369, DOI 10.1016/j.exmath.2006.07.002. MR2313123

[12] A. Geroldinger, *Additive group theory and non-unique factorizations*, Combinatorial number theory and additive group theory, Adv. Courses Math. CRM Barcelona, Birkhäuser Verlag, Basel, 2009, pp. 1–86, DOI 10.1007/978-3-7643-8962-8. MR2522037

[13] A. Geroldinger, *Sets of lengths*, Amer. Math. Monthly **123** (2016), no. 10, 960–988, DOI 10.4169/amer.math.monthly.123.10.960. MR3593640

[14] A. Geroldinger, D. J. Grynkiewicz, and W. A. Schmid, *The catenary degree of Krull monoids I* (English, with English and French summaries), J. Théor. Nombres Bordeaux **23** (2011), no. 1, 137–169, DOI 10.5802/jtnb.754. MR2780623

[15] A. Geroldinger and F. Halter-Koch, *Non-unique factorizations*, Pure and Applied Mathematics (Boca Raton), vol. 278, Chapman & Hall/CRC, Boca Raton, FL, 2006. Algebraic, combinatorial and analytic theory, DOI 10.1201/9781420003208. MR2194494

[16] A. Geroldinger, F. Halter-Koch, and Q. Zhong, *On monoids of weighted zero-sum sequences and applications to norm monoids in Galois number fields and binary quadratic forms*, Acta Math. Hungar. **168** (2022), no. 1, 144–185, DOI 10.1007/s10474-022-01270-x. MR4517537

[17] F. Fabsits, A. Geroldinger, A. Reinhart, and Q. Zhong, *On monoids of plus-minus weighted zero-sum sequences: the isomorphism problem and the characterization problem*, J. Commut. Algebra **16** (2024), no. 1, 1–23, DOI 10.1216/jca.2024.16.1. MR4690592

[18] A. Geroldinger and W. A. Schmid, *A characterization of class groups via sets of lengths*, J. Korean Math. Soc. **56** (2019), no. 4, 869–915, DOI 10.4134/JKMS.j180467. MR3978236

[19] A. Geroldinger and Q. Zhong, *The set of minimal distances in Krull monoids*, Acta Arith. **173** (2016), no. 2, 97–120, DOI 10.4064/aa7906-1-2016. MR3503061

[20] A. Geroldinger and Q. Zhong, *Factorization theory in commutative monoids*, Semigroup Forum **100** (2020), no. 1, 22–51, DOI 10.1007/s00233-019-10079-0. MR4061567

[21] F. Kainrath, *Factorization in Krull monoids with infinite class group*, Colloq. Math. **80** (1999), no. 1, 23–30, DOI 10.4064/cm-80-1-23-30. MR1684567

[22] L. E. Marchan, O. Ordaz, and W. A. Schmid, *Remarks on the plus-minus weighted Davenport constant*, Int. J. Number Theory **10** (2014), no. 5, 1219–1239, DOI 10.1142/S1793042114500225. MR3231411

[23] W. Narkiewicz, *Elementary and analytic theory of algebraic numbers*, 3rd ed., Springer Monographs in Mathematics, Springer-Verlag, Berlin, 2004, DOI 10.1007/978-3-662-07001-7. MR2078267

[24] A. Plagne and W. A. Schmid, *On congruence half-factorial Krull monoids with cyclic class group*, J. Comb. Algebra **3** (2019), no. 4, 331–400, DOI 10.4171/jca/34. MR4046034

[25] M. Radziejewski and W. A. Schmid, *Weakly half-factorial sets in finite abelian groups*, Forum Math. **19** (2007), no. 4, 727–747, DOI 10.1515/FORUM.2007.028. MR2336971

[26] W. A. Schmid, *Characterization of class groups of Krull monoids via their systems of sets of lengths: a status report*, Number theory and applications, Hindustan Book Agency, New Delhi, 2009, pp. 189–212. MR2547500

[27] W. A. Schmid, *Some recent results and open problems on sets of lengths of Krull monoids with finite class group*, Multiplicative ideal theory and factorization theory, Springer Proc. Math. Stat., vol. 170, Springer, [Cham], 2016, pp. 323–352, DOI 10.1007/978-3-319-38855-7_14. MR3565815

[28] Q. Zhong, *A characterization of finite abelian groups via sets of lengths in transfer Krull monoids*, Comm. Algebra **46** (2018), no. 9, 4021–4041, DOI 10.1080/00927872.2018.1430811. MR3820613

[29] Q. Zhong, *Sets of minimal distances and characterizations of class groups of Krull monoids*, Ramanujan J. **45** (2018), no. 3, 719–737, DOI 10.1007/s11139-016-9873-2. MR3776434

Laboratoire Analyse, Géométrie et Applications, LAGA, Université Sorbonne Paris Nord, CNRS, UMR 7539, F-93430, Villetaneuse, France; and Laboratoire Analyse, Géométrie et Applications (LAGA, UMR 7539), COMUE Université Paris Lumières, Université Paris 8, CNRS, 93526 Saint-Denis cedex, France

Email address: kamilmerito@hotmail.com

Escuela de Matemáticas y Laboratorio MoST, Centro ISYS, Facultad de Ciencias, Universidad Central de Venezuela, Ap. 47567, Caracas 1041–A, Venezuela

Email address: oscarordaz55@gmail.com

Laboratoire Analyse, Géométrie et Applications, LAGA, Université Sorbonne Paris Nord, CNRS, UMR 7539, F-93430, Villetaneuse, France; and Laboratoire Analyse, Géométrie et Applications (LAGA, UMR 7539), COMUE Université Paris Lumières, Université Paris 8, CNRS, 93526 Saint-Denis cedex, France

Email address: wolfgang.schmid@univ-paris8.fr, schmid@math.univ-paris13.fr

Contemporary Mathematics
Volume **836**, 2026
https://doi.org/10.1090/conm/836/16708

On counterexamples to Mordell's Pellian Equation Conjecture and the AAC Conjecture: A non-computer-based approach

Andreas Reinhart

ABSTRACT. In this note, we discuss recently discovered counterexamples to Mordell's Pellian Equation Conjecture and the AAC Conjecture. We provide a verification of the counterexample to Mordell's Pellian Equation Conjecture that can be checked with marginal computer assistance.

1. Introduction and main result

We fix notation which is needed to formulate the Ankeny–Artin–Chowla Conjecture and Mordell's Pellian Equation Conjecture.

Let $\mathbb{P}$, $\mathbb{N}$, $\mathbb{N}_0$, $\mathbb{Z}$, $\mathbb{Q}$ denote the sets of prime numbers, positive integers, non-negative integers, integers and rational numbers, respectively. For each $p \in \mathbb{P}$ and each $a \in \mathbb{Z} \setminus \{0\}$, let $\mathrm{v}_p(a)$ be the p-adic exponent of a (i.e., the largest $k \in \mathbb{N}_0$ with $p^k \mid a$). Let $d \in \mathbb{N}_{\geq 2}$ be squarefree, let $K = \mathbb{Q}\left(\sqrt{d}\right)$, let $\mathcal{O}_K$ be the ring of algebraic integers of K, let $\mathcal{O}_d = \mathbb{Z} + d\mathcal{O}_K$ and let $\varepsilon > 1$ be the fundamental unit of $\mathcal{O}_K$. By $\mathcal{O}_K^\times$ (respectively $\mathcal{O}_d^\times$) we denote the unit group of $\mathcal{O}_K$ (respectively, of $\mathcal{O}_d$). Let $\mathrm{N} : K \to \mathbb{Q}$ defined by $\mathrm{N}\left(a + b\sqrt{d}\right) = a^2 - db^2$ for each $a, b \in \mathbb{Q}$ be the norm map. Moreover, for each nonzero ideal I of $\mathcal{O}_K$, let $\mathrm{N}(I) = |\mathcal{O}_K/I|$ be the ideal norm of I. We set

$$\omega = \begin{cases} \sqrt{d} & \text{if } d \equiv 2, 3 \mod 4, \\ \frac{1+\sqrt{d}}{2} & \text{if } d \equiv 1 \mod 4. \end{cases}$$

Observe that $\mathcal{O}_K = \mathbb{Z}[\omega] = \mathbb{Z} \oplus \omega\mathbb{Z}$. Let $p \in \mathbb{P}$. Then p is called *ramified* in $\mathcal{O}_K$ if $p\mathcal{O}_K$ is the square of a maximal ideal of $\mathcal{O}_K$. Moreover, we say that p *splits* in $\mathcal{O}_K$ if $p\mathcal{O}_K$ is the product of two distinct maximal ideals of $\mathcal{O}_K$.

We recall two (long standing) conjectures.

CONJECTURE 1.1 (Ankeny–Artin–Chowla Conjecture). *Let $x, y \in \mathbb{N}_0$ be such that $\varepsilon = x + y\omega$. If $d \in \mathbb{P}$ and $d \equiv 1 \mod 4$, then $d \nmid y$.*

CONJECTURE 1.2 (Mordell's Pellian Equation Conjecture). *Let $x, y \in \mathbb{N}_0$ be such that $\varepsilon = x + y\omega$. If $d \in \mathbb{P}$ and $d \equiv 3 \mod 4$, then $d \nmid y$.*

2020 *Mathematics Subject Classification.* Primary 11R11, 11R27.
Key words and phrases. Fundamental unit, Pell equation, quadratic number field.
This work was supported by the Austrian Science Fund FWF, Project Number P36852-N.

©2026 American Mathematical Society

We briefly discuss the history and the significance of both conjectures. The Ankeny-Artin-Chowla Conjecture was first mentioned in 1952 (see [**2**, page 480]) and it was subsequently investigated in a multitude of papers [**1,3,7,8,11,13,19, 21**]. Mordell's Pellian Equation Conjecture was stated by L. J. Mordell in 1961 (see [**14**, page 283]) and independently by A. A. Kiselev and I. Sh. Slavutskiĭ in 1959 (see [**10**]). It was investigated in various papers and by many authors (for instance, see [**3,5,12,19,21**]).

Both conjectures were recently disproved by counterexamples which were found by a long numerical computer search [**17,18**]. For the methods, algorithms and programs that were used to find (and check) these counterexamples, we refer to [**17**, Section 3] and [**18**, Section 2]. Note that the algorithms (that were used in [**17,18**] to check the counterexamples) require a multitude of computational steps and make (at least to some extent) use of continued fraction expansions. For instance, the fundamental unit of $\mathcal{O}_K$ for d in Theorem 1.3 is larger than $10^{1880000}$ and the period length of the continued fraction expansion of $\sqrt{d}$ is 3650856.

The purpose of this note is to demonstrate how to verify the given counterexamples with manageable effort and only minor computer assistance. We do this in detail for the counterexample to Mordell's Pellian Equation Conjecture. The verification of the counterexample to the AAC Conjecture can be done along similar lines. Also note that we do not use continued fraction expansions, and thus we get additional independent verifications of the counterexamples.

Note that both the Ankeny–Artin–Chowla Conjecture and Mordell's Pellian Equation Conjecture have ties with some problems in (commutative) algebra. Let $\mathcal{O}$ be an order in K. Then $\mathcal{O}$ is said to have *torsionfree cancellation* if for all torsionfree finitely generated $\mathcal{O}$-modules A, B, C with $A \oplus B \cong A \oplus C$, it follows that $B \cong C$. Moreover, $\mathcal{O}$ is said to have *mixed cancellation* if for all finitely generated $\mathcal{O}$-modules A, B, C with $A \oplus B \cong A \oplus C$, we have that $B \cong C$. Let $d \in \mathbb{P}$ and let $x, y \in \mathbb{N}_0$ be such that $\varepsilon = x + y\omega$. It is well-known that $\mathcal{O}_d$ has torsionfree cancellation if and only if $d \nmid y$ [**22**, Theorem 4.9]. In particular, it follows from [**22**, Theorem 4.9] and [**9**, Theorem 5.12] that if $d \geq 7$, then $\mathcal{O}_d$ has torsionfree cancellation, but not mixed cancellation if and only $d \nmid y$.

We also want to mention a problem from factorization theory that is connected to the condition $d \mid y$. It was recently studied in [**16,17**]. For a profound introduction to factorization theory see [**6**]. Let $\mathcal{O}$ be an order in K. A nonzero nonunit of $\mathcal{O}$ that cannot be written as a product of two nonunits of $\mathcal{O}$ is called an *atom* of $\mathcal{O}$. It is well-known that every nonzero nonunit of $\mathcal{O}$ is a finite product of atoms of $\mathcal{O}$ (see [**6**]). Let $x \in \mathcal{O}$ be a nonzero nonunit. If $\ell \in \mathbb{N}$ is such that x is a product of ℓ atoms of $\mathcal{O}$, then ℓ is called a *length* of x. Let $\mathsf{L}(x)$ be the set of lengths of x. Set $L = \min\{r - s \mid y \in \mathcal{O}$ is a nonzero nonunit, $r, s \in \mathsf{L}(y), s < r\}$ (note that we use $\min \emptyset = 0$ here for convenience). It is known (see [**16**]) that $L \leq 2$. We say that $\mathcal{O}$ is *unusual* if $L = 2$. The condition $d \mid y$ determines in many cases whether K has an unusual order (see [**17**, Proposition 2.4 and Theorem 2.5]).

For a survey on Conjectures 1.1 and 1.2 and further discussion of their importance, we refer to [**20**]. For a more detailed study of the property $d \mid y$ for arbitrary squarefree $d \in \mathbb{N}_{\geq 2}$, see [**17**]. Next, we present the main result of this note.

THEOREM 1.3. *The number $d = 39028039587479$ is a counterexample to Mordell's Pellian Equation Conjecture.*

The proof of Theorem 1.3 in Section 3 was created with computer assistance (and with the help of the programs Mathematica 12.0.0 and Pari/GP 2.15.2). We used the command bnfunits(bnfinit(X^2-d,1)) in Pari/GP (together with some elementary algebraic manipulations) to obtain the values in the table (with the entries a_i, b_i, c_i, d_i, e_i and f_i) below.

Observe that (very dedicated) arithmeticians may be able to verify this proof without computer assistance. We define an arithmetical operation as *hard* if it is a multiplication of integers a and b with $|a| \geq 10^3$, $|b| \geq 10^3$ and $|ab| \geq 10^{10}$. All other arithmetical operations in this note (additions, subtractions and multiplications of integers) are defined to be *easy*. We define additions and subtractions where (at least) one of the members is a single digit integer as *trivial*. A rough estimate yields that there are approximately 550 hard operations and about 2200 (nontrivial) easy operations in the proof of Theorem 1.3 below. If we assume that (the verification of) a hard operation takes about 15 minutes (on average) and (the verification of) an easy (but nontrivial) operation takes about 3 minutes (on average), then a verification of the proof below is doable in about 250 hours (without computers). If all arithmetical operations are done with computers, then it should only take a modest amount of time (to check all the details).

2. Preliminaries

We recall a variant of a result of H. C. Pocklington (cf. [**4**, Theorem 4] and [**15**]) that will be used to show that d is prime. For the readers' convenience we include a proof. This proof goes along the lines of the proof of [**4**, Theorem 4 and Corollary 1].

PROPOSITION 2.1. *Let $a, b, c \in \mathbb{N}$ be such that $a^2 > c > 1$, $a \mid c-1$, $b^{c-1} \equiv 1 \mod c$ and for each $p \in \mathbb{P}$ with $p \mid a$, $\gcd\left(b^{\frac{c-1}{p}} - 1, c\right) = 1$. Then $c \in \mathbb{P}$.*

PROOF.

CLAIM. For each $q \in \mathbb{P}$ with $q \mid c$, it follows that $q > a$.

PROOF. Let $q \in \mathbb{P}$ be such that $q \mid c$. Then $q \nmid b$ (since $b^{c-1} \equiv 1 \mod c$). Set $e = \min\{k \in \mathbb{N} \mid b^k \equiv 1 \mod q\}$. We obtain that $e \mid q-1$. It remains to show that $\mathrm{v}_p(a) \leq \mathrm{v}_p(e)$ for each $p \in \mathbb{P}$ with $p \mid a$. (Then $a \mid e \mid q-1$, and hence $q = 1 + a\ell$ for some $\ell \in \mathbb{N}$, which implies that $q > a$.) Let $p \in \mathbb{P}$ be such that $p \mid a$. Since $b^{c-1} \equiv 1 \mod c$, we have that $b^{c-1} \equiv 1 \mod q$, and thus $e \mid c-1$. Moreover, since $\gcd\left(b^{\frac{c-1}{p}} - 1, c\right) = 1$, we infer that $q \nmid b^{\frac{c-1}{p}} - 1$. Therefore, $b^{\frac{c-1}{p}} \not\equiv 1 \mod q$, and hence $e \nmid \frac{c-1}{p}$. In particular, we obtain that $p \nmid \frac{c-1}{e}$, and thus $\mathrm{v}_p(c-1) = \mathrm{v}_p\left(\frac{c-1}{e}\right) + \mathrm{v}_p(e) = \mathrm{v}_p(e)$. Since $a \mid c-1$, we have that $\mathrm{v}_p(a) \leq \mathrm{v}_p(e)$. □(Claim)

Assume to the contrary that $c \notin \mathbb{P}$. Then there are some $r, s \in \mathbb{P}$ such that $rs \mid c$. We infer by the claim that $c \geq rs > a^2 > c$, a contradiction. □

Next we mention a simple (well-known) lemma that will be used to compute integer powers modulo d.

LEMMA 2.2. *Let $a, n \in \mathbb{N}$, let $m \in \mathbb{N}_0$ and let $(g_i)_{i=0}^m$ be the sequence of integers with $g_i \in \{0,1\}$ for each $i \in [0,m]$, $g_m = 1$ and $n = \sum_{j=0}^m g_j 2^j$ (i.e., $(g_i)_{i=0}^m$ is the* *binary representation of* n). *Furthermore, let* $(h_i)_{i=0}^{m+1}$ *be defined recursively by* $h_0 = 1$ *and* $h_{r+1} = a^{g_{m-r}} h_r^2$ *for each* $r \in [0, m]$. *Then* $h_{m+1} = a^n$.

PROOF. It is straightforward to show by induction that $h_{r+1} = a^{\sum_{j=0}^{r} g_{m-j} 2^{r-j}}$ for each $r \in [0, m]$. We obtain that $h_{m+1} = a^{\sum_{j=0}^{m} g_{m-j} 2^{m-j}} = a^{\sum_{j=0}^{m} g_j 2^j} = a^n$. □

3. Proof of Theorem 1.3

In what follows, we will use the sieve of Eratosthenes and well-known facts about $\mathcal{O}_K$ (e.g. that $\mathcal{O}_K$ is a Dedekind domain) without further mention. Besides that, we will indicate the binary representation of a positive integer by using the subscript "$_2$".

Let $d = 39028039587479$ and let $x, y \in \mathbb{N}_0$ be such that $\varepsilon = x + y\omega$. We need to show that $d \equiv 3 \mod 4$, $d \in \mathbb{P}$ and $d \mid y$. Since $79 = 19 \cdot 4 + 3$, we have that $d \equiv 79 \equiv 3 \mod 4$. Observe that $\omega = \sqrt{d}$, $x, y \in \mathbb{N}$ and $\varepsilon = x + y\sqrt{d}$.

CLAIM 1. $d \in \mathbb{P}$.

Set $p = 3617$, $q = 4021$ and $a = pq$. By Proposition 2.1, it is sufficient to show that $p, q \in \mathbb{P}$, $a^2 > d$, $a \mid d - 1$, $2^{d-1} \equiv 1 \mod d$ and

$$\gcd\left(2^{\frac{d-1}{p}} - 1, d\right) = \gcd\left(2^{\frac{d-1}{q}} - 1, d\right) = 1.$$

Note that $\{t \in \mathbb{P} \mid t^2 \leq p\} = \{2, 3, 5, 7, 11, 13, 17, 19, 23, 29, 31, 37, 41, 43\} \cup \{47, 53, 59\}$ and

$$\begin{aligned} p &= 1808 \cdot 2 + 1 = 1205 \cdot 3 + 2 = 723 \cdot 5 + 2 = 516 \cdot 7 + 5 = 328 \cdot 11 + 9 \\ &= 278 \cdot 13 + 3 = 212 \cdot 17 + 13 = 190 \cdot 19 + 7 = 157 \cdot 23 + 6 = 124 \cdot 29 + 21 \\ &= 116 \cdot 31 + 21 = 97 \cdot 37 + 28 = 88 \cdot 41 + 9 = 84 \cdot 43 + 5 \\ &= 76 \cdot 47 + 45 = 68 \cdot 53 + 13 = 61 \cdot 59 + 18. \end{aligned}$$

This implies that $p \in \mathbb{P}$.

Moreover, $\{t \in \mathbb{P} \mid t^2 \leq q\} = \{2, 3, 5, 7, 11, 13, 17, 19, 23, 29, 31, 37, 41, 43\} \cup \{47, 53, 59, 61\}$ and

$$\begin{aligned} q &= 2010 \cdot 2 + 1 = 1340 \cdot 3 + 1 = 804 \cdot 5 + 1 = 574 \cdot 7 + 3 = 365 \cdot 11 + 6 \\ &= 309 \cdot 13 + 4 = 236 \cdot 17 + 9 = 211 \cdot 19 + 12 = 174 \cdot 23 + 19 = 138 \cdot 29 + 19 \\ &= 129 \cdot 31 + 22 = 108 \cdot 37 + 25 = 98 \cdot 41 + 3 = 93 \cdot 43 + 22 = 85 \cdot 47 + 26 \\ &= 75 \cdot 53 + 46 = 68 \cdot 59 + 9 = 65 \cdot 61 + 56. \end{aligned}$$

Consequently, $q \in \mathbb{P}$. Observe that $a^2 > (3000 \cdot 4000)^2 > 10^{14} > d$ and $d = 1 + 10790168534p = 1 + 9706053118q = 1 + 2683454a$. In particular, we obtain that $a \mid d - 1$.

Next, we show that $2^{d-1} \equiv 1 \mod d$ by applying Lemma 2.2. Each item in the list of equations below consists of two parts. The first part is used to derive the binary representation of $d - 1$, while the second part is the outcome of determining the representatives of the integers h_i (in Lemma 2.2) modulo d. Also note that the second part contains an additional factor 2 if and only if the remainder (modulo 2) in the first part is 1. (Keep in mind that the computation of the binary representation of $d - 1$ is technically done from the bottom to the top.)

- $2 \cdot 0 + 1 = 1$, $2 \cdot 1^2 = 2$.
- $2 \cdot 1 = 2$, $2^2 = 4$.
- $2 \cdot 2 = 4$, $4^2 = 16$.
- $2 \cdot 4 = 8$, $16^2 = 256$.

- $2 \cdot 8 + 1 = 17$, $2 \cdot 256^2 = 131072$.
- $2 \cdot 17 + 1 = 35$, $2 \cdot 131072^2 = 34359738368$.
- $2 \cdot 35 = 70$, $34359738368^2 = 30249831d + 18934861837375$.
- $2 \cdot 70 + 1 = 141$, $2 \cdot 18934861837375^2 = 18372892750447d + 8878320928137$.
- $2 \cdot 141 + 1 = 283$, $2 \cdot 8878320928137^2 = 4039382112766d + 4872136924624$.
- $2 \cdot 283 + 1 = 567$, $2 \cdot 4872136924624^2 = 1216444303284d + 37954390101716$.
- $2 \cdot 567 + 1 = 1135$, $2 \cdot 37954390101716^2 = 73820552772801d + 14864000930633$.
- $2 \cdot 1135 + 1 = 2271$, $2 \cdot 14864000930633^2 = 11322040563715d + 5306939836893$.
- $2 \cdot 2271 + 1 = 4543$, $2 \cdot 5306939836893^2 = 1443250069954d + 7254835280932$.
- $2 \cdot 4543 = 9086$, $7254835280932^2 = 1348585158511d + 1741096904855$.
- $2 \cdot 9086 + 1 = 18173$, $2 \cdot 1741096904855^2 = 155345667583d + 18644662148793$.
- $2 \cdot 18173 + 1 = 36347$, $2 \cdot 18644662148793^2 = 17814034746144d + 34365896782722$.
- $2 \cdot 36347 + 1 = 72695$, $2 \cdot 34365896782722^2 = 60521352041448d + 19084525628976$.
- $2 \cdot 72695 = 145390$, $19084525628976^2 = 9332242211824d + 32132914656880$.
- $2 \cdot 145390 + 1 = 290781$, $2 \cdot 32132914656880^2 = 52911917444994d + 20520829038674$.
- $2 \cdot 290781 + 1 = 581563$, $2 \cdot 20520829038674^2 = 21579583749811d + 29933591140083$.
- $2 \cdot 581563 = 1163126$, $29933591140083^2 = 22958362449471d + 9944907473280$.
- $2 \cdot 1163126 = 2326252$, $9944907473280^2 = 2534105881245d + 7176351027045$.
- $2 \cdot 2326252 = 4652504$, $7176351027045^2 = 1319564462056d + 17827533235201$.
- $2 \cdot 4652504 + 1 = 9305009$, $2 \cdot 17827533235201^2 = 16286800188352d + 30177590176194$.
- $2 \cdot 9305009 + 1 = 18610019$, $2 \cdot 30177590176194^2 = 46668341964810d + 29055194037282$.
- $2 \cdot 18610019 = 37220038$, $29055194037282^2 = 21630712417718d + 35791355394602$.
- $2 \cdot 37220038 + 1 = 74440077$, $2 \cdot 35791355394602^2 = 65646193584044d + 32203984891732$.
- $2 \cdot 74440077 + 1 = 148880155$, $2 \cdot 32203984891732^2 = 53146233009337d + 18209288628225$.
- $2 \cdot 148880155 + 1 = 297760311$, $2 \cdot 18209288628225^2 = 16991793379874d + 22534432303604$.
- $2 \cdot 297760311 + 1 = 595520623$, $2 \cdot 22534432303604^2 = 26022349296203d + 5977961735395$.
- $2 \cdot 595520623 + 1 = 1191041247$, $2 \cdot 5977961735395^2 = 1831300105645d + 26065072393095$.
- $2 \cdot 1191041247 = 2382082494$, $26065072393095^2 = 17407689600562d + 2873437115827$.
- $2 \cdot 2382082494 = 4764164988$, $2873437115827^2 = 211556638403d + 25924124537892$.
- $2 \cdot 4764164988 + 1 = 9528329977$, $2 \cdot 25924124537892^2 = 34439866319688d + 247631620776$.
- $2 \cdot 9528329977 = 19056659954$, $247631620776^2 = 1571214446d + 9241789320542$.
- $2 \cdot 19056659954 + 1 = 38113319909$, $2 \cdot 9241789320542^2 = 4376887527432d + 9150470123600$.
- $2 \cdot 38113319909 + 1 = 76226639819$, $2 \cdot 9150470123600^2 = 4290817800121d + 10455437635041$.
- $2 \cdot 76226639819 = 152453279638$, $10455437635041^2 = 2800965082942d + 24962411388463$.
- $2 \cdot 152453279638 + 1 = 304906559277$, $2 \cdot 24962411388463^2 = 31932015490051d + 31490910333309$.
- $2 \cdot 304906559277 = 609813118554$, $31490910333309^2 = 25409358094908d + 8143824432549$.
- $2 \cdot 609813118554 = 1219626237108$, $8143824432549^2 = 1699339169714d + 30354920226395$.
- $2 \cdot 1219626237108 + 1 = 2439252474217$, $2 \cdot 30354920226395^2 = 47218419971389d + 29633566753719$.
- $2 \cdot 2439252474217 = 4878504948434$, $29633566753719^2 = 22500445521451d + 15832914818932$.
- $2 \cdot 4878504948434 + 1 = 9757009896869$, $2 \cdot 15832914818932^2 = 12846209766784d + 7728534743712$.
- $2 \cdot 9757009896869 + 1 = 19514019793739$, $2 \cdot 7728534743712^2 = 3060889038553d + 1$.
- $2 \cdot 19514019793739 = 39028039587478$, $1^2 = 1$.

This shows that $d-1 = 1000110111111011101100011011111001011010010110_2 = 39028039587478$. By Lemma 2.2, we obtain that $2^{d-1} \equiv 1 \mod d$. Now we show that $2^{\frac{d-1}{p}} \equiv 10285064380914 \mod d$ by applying Lemma 2.2 again. This is done along the lines of the proof of $2^{d-1} \equiv 1 \mod d$ above.

- $2 \cdot 0 + 1 = 1$, $2 \cdot 1^2 = 2$.
- $2 \cdot 1 = 2$, $2^2 = 4$.
- $2 \cdot 2 + 1 = 5$, $2 \cdot 4^2 = 32$.
- $2 \cdot 5 = 10$, $32^2 = 1024$.
- $2 \cdot 10 = 20$, $1024^2 = 1048576$.
- $2 \cdot 20 = 40$, $1048576^2 = 1099511627776$.
- $2 \cdot 40 = 80$, $1099511627776^2 = 30975827440d + 31390886082416$.
- $2 \cdot 80 = 160$, $31390886082416^2 = 25248199485668d + 897251646084$.
- $2 \cdot 160 + 1 = 321$, $2 \cdot 897251646084^2 = 41255493481d + 23445371345713$.
- $2 \cdot 321 + 1 = 643$, $2 \cdot 23445371345713^2 = 28168744489781d + 6603555904639$.
- $2 \cdot 643 = 1286$, $6603555904639^2 = 1117323622877d + 23588814563238$.
- $2 \cdot 1286 = 2572$, $23588814563238^2 = 14257241162513d + 36673782069917$.
- $2 \cdot 2572 + 1 = 5145$, $2 \cdot 36673782069917^2 = 68923077127515d + 7453272389093$.
- $2 \cdot 5145 = 10290$, $7453272389093^2 = 1423368170504d + 28404889843233$.
- $2 \cdot 10290 = 20580$, $28404889843233^2 = 20673284529132d + 20389277954061$.
- $2 \cdot 20580 + 1 = 41161$, $2 \cdot 20389277954061^2 = 21303793881634d + 11174538322756$.
- $2 \cdot 41161 = 82322$, $11174538322756^2 = 3199502410231d + 12534158337887$.
- $2 \cdot 82322 = 164644$, $12534158337887^2 = 4025442397307d + 8237359105716$.
- $2 \cdot 164644 + 1 = 329289$, $2 \cdot 8237359105716^2 = 3477196690058d + 10027987161530$.
- $2 \cdot 329289 + 1 = 658579$, $2 \cdot 10027987161530^2 = 5153245080958d + 10764345756918$.
- $2 \cdot 658579 + 1 = 1317159$, $2 \cdot 10764345756918^2 = 5937840629415d + 7258376622663$.
- $2 \cdot 1317159 = 2634318$, $7258376622663^2 = 1349902064087d + 13688184444896$.
- $2 \cdot 2634318 = 5268636$, $13688184444896^2 = 4800814885347d + 11643098680603$.
- $2 \cdot 5268636 + 1 = 10537273$, $2 \cdot 11643098680603^2 = 6946889893478d + 1481444325256$.
- $2 \cdot 10537273 + 1 = 21074547$, $2 \cdot 1481444325256^2 = 112466693794d + 1299469525746$.

- $2 \cdot 21074547 + 1 = 42149095$, $2 \cdot 1299469525746^2 = 86533736574d + 22516898956086$.
- $2 \cdot 42149095 + 1 = 84298191$, $2 \cdot 22516898956086^2 = 25981870673373d + 12486243382125$.
- $2 \cdot 84298191 + 1 = 168596383$, $2 \cdot 12486243382125^2 = 7989449403329d + 36029689713659$.
- $2 \cdot 168596383 = 337192766$, $36029689713659^2 = 33261689661681d + 18099097476082$.
- $2 \cdot 337192766 + 1 = 674385533$, $2 \cdot 18099097476082^2 = 16786768329189d + 8643489516917$.
- $2 \cdot 674385533 = 1348771066$, $8643489516917^2 = 1914262458958d + 16355540998007$.
- $2 \cdot 1348771066 + 1 = 2697542133$, $2 \cdot 16355540998007^2 = 13708283796212d + 16782573114550$.
- $2 \cdot 2697542133 + 1 = 5395084267$, $2 \cdot 16782573114550^2 = 14433456731225d + 18430065073225$.
- $2 \cdot 5395084267 = 10790168534$, $18430065073225^2 = 8703160655609d + 10285064380914$.

Note that

$$\frac{d-1}{p} = 10790168534 = 1010000011001001001110011111010110_2.$$

We infer by Lemma 2.2 that

$$2^{\frac{d-1}{p}} - 1 \equiv 10285064380913 \mod d.$$

Since $7336389398826 \cdot 10285064380913 = 1933359658541d - 1$, we have that

$$\gcd\left(2^{\frac{d-1}{p}} - 1, d\right) = 1.$$

Finally, we prove that $2^{\frac{d-1}{q}} \equiv 15901499388071 \mod d$ by applying Lemma 2.2. We proceed along the lines of the proof of $2^{d-1} \equiv 1 \mod d$ above.

- $2 \cdot 0 + 1 = 1$, $2 \cdot 1^2 = 2$.
- $2 \cdot 1 = 2$, $2^2 = 4$.
- $2 \cdot 2 = 4$, $4^2 = 16$.
- $2 \cdot 4 + 1 = 9$, $2 \cdot 16^2 = 512$.
- $2 \cdot 9 = 18$, $512^2 = 262144$.
- $2 \cdot 18 = 36$, $262144^2 = 68719476736$.
- $2 \cdot 36 = 72$, $68719476736^2 = 120999325d + 36711407762021$.
- $2 \cdot 72 = 144$, $36711407762021^2 = 34532286892056d + 31998527837617$.
- $2 \cdot 144 + 1 = 289$, $2 \cdot 31998527837617^2 = 52470264691605d + 38620486063583$.
- $2 \cdot 289 = 578$, $38620486063583^2 = 38217188450990d + 25764109643679$.
- $2 \cdot 578 + 1 = 1157$, $2 \cdot 25764109643679^2 = 34016022979769d + 6589566597731$.
- $2 \cdot 1157 = 2314$, $6589566597731^2 = 1112594647461d + 6956899607542$.
- $2 \cdot 2314 = 4628$, $6956899607542^2 = 1240094369611d + 30110629581095$.
- $2 \cdot 4628 = 9256$, $30110629581095^2 = 23230734194007d + 5999347360672$.
- $2 \cdot 9256 = 18512$, $5999347360672^2 = 922213084091d + 7444474594995$.
- $2 \cdot 18512 + 1 = 37025$, $2 \cdot 7444474594995^2 = 2840019769443d + 7552148495853$.
- $2 \cdot 37025 + 1 = 74051$, $2 \cdot 7552148495853^2 = 2922767707846d + 10516484734984$.
- $2 \cdot 74051 = 148102$, $10516484734984^2 = 2833769063220d + 28051642057876$.
- $2 \cdot 148102 + 1 = 296205$, $2 \cdot 28051642057876^2 = 40324578454903d + 38284942303215$.
- $2 \cdot 296205 = 592410$, $38284942303215^2 = 37555993656179d + 37268879953484$.
- $2 \cdot 592410 = 1184820$, $37268879953484^2 = 35589013121550d + 38577918665806$.
- $2 \cdot 1184820 + 1 = 2369641$, $2 \cdot 38577918665806^2 = 76265978220592d + 10176084091704$.
- $2 \cdot 2369641 + 1 = 4739283$, $2 \cdot 10176084091704^2 = 5306578989668d + 15426040080260$.
- $2 \cdot 4739283 + 1 = 9478567$, $2 \cdot 15426040080260^2 = 12194448661681d + 9515608643001$.
- $2 \cdot 9478567 = 18957134$, $9515608643001^2 = 2320044993389d + 36985824109670$.
- $2 \cdot 18957134 + 1 = 37914269$, $2 \cdot 36985824109670^2 = 70100942785264d + 9512534908344$.
- $2 \cdot 37914269 + 1 = 75828539$, $2 \cdot 9512534908344^2 = 4637092784516d + 22708882969508$.
- $2 \cdot 75828539 + 1 = 151657079$, $2 \cdot 22708882969508^2 = 26426813704896d + 7292632926944$.
- $2 \cdot 151657079 + 1 = 303314159$, $2 \cdot 7292632926944^2 = 2725348009752d + 37465133263064$.
- $2 \cdot 303314159 + 1 = 606628319$, $2 \cdot 37465133263064^2 = 71929629325755d + 32577343114547$.
- $2 \cdot 606628319 + 1 = 1213256639$, $2 \cdot 32577343114547^2 = 54385682479598d + 31854972276976$.
- $2 \cdot 1213256639 + 1 = 2426513279$, $2 \cdot 31854972276976^2 = 52000524212466d + 35754130095938$.
- $2 \cdot 2426513279 + 1 = 4853026559$, $2 \cdot 35754130095938^2 = 65509712116177d + 10262965651905$.
- $2 \cdot 4853026559 = 9706053118$, $10262965651905^2 = 2698789513526d + 15901499388071$.

Observe that

$$\frac{d-1}{q} = 9706053118 = 1001000010100001101001110111111110_2$$

and $2^{\frac{d-1}{q}} - 1 \equiv 15901499388070 \mod d$ by Lemma 2.2. Since $11826010015564 \cdot 15901499388070 = 4818363746001d + 1$, it follows that

$$\gcd\left(2^{\frac{d-1}{q}} - 1, d\right) = 1.$$

This shows that $d \in \mathbb{P}$. □(Claim 1)

Finally, we show that $d \mid y$. Set $r = 57$ and $s = 56$, and let

$$(a_i)_{i=1}^{r}, (b_i)_{i=1}^{r}, (c_i)_{i=1}^{s}, (d_i)_{i=1}^{s}, (e_i)_{i=1}^{s}, (f_i)_{i=1}^{s}$$

be defined as follows.

TABLE 1. Important precomputed numbers

i	a_i	b_i	c_i	d_i	e_i	f_i
1	−3033477	3410579	2	84702667	1	19514019793739
2	−2271496	3206358	5	146634276	2	7805607917495
3	−1787484	82862	23	112959051	7	1696871286410
4	−1758518	4049619	29	57435327	14	1345794468527
5	−1699348	2129966	37	54701496	5	1054811880742
6	−1427442	8014362	47	60589478	3	830383821010
7	−832653	717212	53	37470722	22	736378105415
8	−748505	9346906	61	9465381	6	639803927663
9	−674823	15717595	79	40989357	9	494025817562
10	−554788	3962534	83	26886549	11	470217344426
11	−521509	17200418	89	16628268	8	438517298735
12	−482142	13587629	97	33227623	5	402350923582
13	−445057	8567013	107	27266087	25	364748033522
14	−425248	2262519	109	16545406	22	358055409055
15	−322867	3536609	113	31075882	50	345380881283
16	−280973	3410579	137	14887445	36	284876201359
17	−252892	4558274	139	18716004	49	280777263202
18	−226498	2007139	157	18190318	57	248586239390
19	−215202	3644517	163	23253236	3	239435825690
20	−207973	4049619	167	4568510	23	233700835850
21	−206198	5460335	179	16995512	63	218033740690
22	−193242	2284255	193	16028724	94	202217821651
23	−126128	8014362	199	21271274	15	196120801946
24	−121640	5480460	223	1909615	52	175013630425
25	−54422	13915416	227	19110033	90	171929689777
26	−48408	8365791	229	9530079	41	170428120462
27	−40763	864323	239	4315464	40	163297236761
28	−34641	7529543	263	13737288	63	148395587770
29	−22980	14887445	269	6324658	125	145085648966
30	3843	1909615	271	16934137	49	144014906218
31	29419	6660471	277	5275486	9	140895449774
32	37368	3205421	283	6324658	96	137908267061
33	38302	2075306	307	8891039	68	127127164765
34	80167	4486315	311	8094883	49	125492088698
35	83134	7103618	337	8014362	90	115810206467
36	102633	6324658	349	21686733	103	111828193630
37	115435	9578599	353	17200418	128	110561018615
38	116792	11851993	359	6177547	132	108713202145
39	156001	4568510	383	35556425	23	101900886650
40	197807	3013477	409	11851993	182	95423079595
41	211873	4414221	421	4049619	1	92703181918
42	219482	1158859	433	1909615	54	90134040611
43	229892	6177547	439	3410579	13	88902140290
44	264088	6019577	443	13587629	158	88099412105
45	269512	147111	449	5020616	202	86922137075
46	315998	4315464	467	20082812	8	83571819245
47	339243	1212808	557	26469383	149	70068293654
48	471496	2548	563	15717595	214	69321562241
49	499231	1109106	569	5278544	205	68590579166
50	596696	1909615	593	6589566	75	65814569278
51	1124049	5020616	601	4414221	280	64938501679
52	1154432	1909615	631	13309440	139	61851092818
53	1214853	23621570	641	17283280	265	60886177094
54	1403789	16545406	647	13915416	74	60321544949
55	1947578	5128375	683	11066152	192	57142078405
56	1991126	10145918	691	9578599	38	56480520385
57	2126573	6589566				

For each $j \in [1, s]$, set

$$P_j = c_j \mathcal{O}_K + \left(e_j + \sqrt{d}\right) \mathcal{O}_K \quad \text{and} \quad \overline{P}_j = c_j \mathcal{O}_K + \left(e_j - \sqrt{d}\right) \mathcal{O}_K.$$

Then $P_1 = \overline{P}_1$, $c_1 f_1 + e_1^2 = d$, $c_1 \in \mathbb{P}$ and $\gcd(c_1, e_1) = 1$. Observe that $c_j f_j + e_j^2 = d$, $c_j \in \mathbb{P}$ and $\gcd(c_j, 2e_j) = 1$ for each $j \in [2, s]$. In particular, $P_j, \overline{P}_j \in$

$\max(\mathcal{O}_K)$, $c_j\mathcal{O}_K = P_j\overline{P}_j$ and $\mathrm{N}(P_j) = \mathrm{N}(\overline{P}_j) = c_j$ for each $j \in [1,s]$. Set $z = \prod_{i=1}^r \left(a_i + \sqrt{d}\right)^{b_i}$, set $n = \prod_{j=1}^s c_j^{d_j}$ and set $\eta = \frac{z}{n}$. Clearly, $\eta \in K$ and $\eta > 0$.

CLAIM 2. $\eta \in \mathcal{O}_K^\times$.

Observe the following.

$$\begin{aligned} d &= 1561121583499c_2^2 + 2^2 = 312224316694c_2^3 + 27^2 = 12488972039c_2^5 + 1402^2 \\ &= 73777012451c_3^2 + 30^2 = 46406705350c_4^2 + 623^2 = 17667740470c_6^2 + 943^2 \\ &= 13893926015c_7^2 + 1188^2 = 5665268302c_{10}^2 + 2501^2 = 3056458675c_{15}^2 + 10898^2 \\ &= 1399382135c_{20}^2 + 25908^2 = 1218065438c_{21}^2 + 2211^2 = 1047746695c_{22}^2 + 22868^2 \\ &= 320417518c_{36}^2 + 29419^2 = 125690710c_{47}^2 + 180617^2, \end{aligned}$$

$2 = e_2$, $27 = 5c_2 + e_2$, $1402 = 280c_2 + e_2$, $30 = c_3 + e_3$, $623 = 21c_4 + e_4$, $943 = 20c_6 + e_6$, $1188 = 22c_7 + e_7$, $2501 = 30c_{10} + e_{10}$, $10898 = 96c_{15} + e_{15}$, $25908 = 155c_{20} + e_{20}$, $2211 = 12c_{21} + e_{21}$, $22868 = 118c_{22} + e_{22}$, $29419 = 84c_{36} + e_{36}$, $180617 = 324c_{47} + e_{47}$.

Let $j \in [2,s]$, $k, m \in \mathbb{N}_0$, $\ell \in \mathbb{N}$ be such that $d = kc_j^\ell + (mc_j + e_j)^2$. Set

$$I = c_j^\ell\mathcal{O}_K + \left(mc_j + e_j + \sqrt{d}\right)\mathcal{O}_K \quad \text{and} \quad \overline{I} = c_j^\ell\mathcal{O}_K + \left(mc_j + e_j - \sqrt{d}\right)\mathcal{O}_K.$$

Clearly, I and $\overline{I}$ are ideals of $\mathcal{O}_K$ such that $I \subseteq P_j$ and $\overline{I} \subseteq \overline{P}_j$. Observe that $I\overline{I} = c_j^\ell J$ for some ideal J of $\mathcal{O}_K$ with $c_j^\ell, 2(mc_j + e_j) \in J$. Since $\gcd\left(c_j^\ell, 2(mc_j + e_j)\right) = 1$, we obtain that $J = \mathcal{O}_K$, $I \not\subseteq \overline{P}_j$ and $\overline{I} \not\subseteq P_j$. Since $I\overline{I} = c_j^\ell\mathcal{O}_K = P_j^\ell\overline{P}_j^\ell$, we infer that $P_j^\ell = I$ and $\overline{P}_j^\ell = \overline{I}$. Using this, we get the following equations (and in analogy the equations for the conjugates $\overline{P}_j^\ell$).

Equations, part A.

$P_2^2 = c_2^2\mathcal{O}_K + \left(2 + \sqrt{d}\right)\mathcal{O}_K$, $P_2^3 = c_2^3\mathcal{O}_K + \left(27 + \sqrt{d}\right)\mathcal{O}_K$,
$P_2^5 = c_2^5\mathcal{O}_K + \left(1402 + \sqrt{d}\right)\mathcal{O}_K$, $P_3^2 = c_3^2\mathcal{O}_K + \left(30 + \sqrt{d}\right)\mathcal{O}_K$,
$P_4^2 = c_4^2\mathcal{O}_K + \left(623 + \sqrt{d}\right)\mathcal{O}_K$, $P_6^2 = c_6^2\mathcal{O}_K + \left(943 + \sqrt{d}\right)\mathcal{O}_K$,
$P_7^2 = c_7^2\mathcal{O}_K + \left(1188 + \sqrt{d}\right)\mathcal{O}_K$, $P_{10}^2 = c_{10}^2\mathcal{O}_K + \left(2501 + \sqrt{d}\right)\mathcal{O}_K$,
$P_{15}^2 = c_{15}^2\mathcal{O}_K + \left(10898 + \sqrt{d}\right)\mathcal{O}_K$, $P_{20}^2 = c_{20}^2\mathcal{O}_K + \left(25908 + \sqrt{d}\right)\mathcal{O}_K$,
$P_{21}^2 = c_{21}^2\mathcal{O}_K + \left(2211 + \sqrt{d}\right)\mathcal{O}_K$, $P_{22}^2 = c_{22}^2\mathcal{O}_K + \left(22868 + \sqrt{d}\right)\mathcal{O}_K$,
$P_{36}^2 = c_{36}^2\mathcal{O}_K + \left(29419 + \sqrt{d}\right)\mathcal{O}_K$, $P_{47}^2 = c_{47}^2\mathcal{O}_K + \left(180617 + \sqrt{d}\right)\mathcal{O}_K$.

The following equations can easily be derived (by using elementary arithmetic).

Equations, part B.

$a_1 = -1516739c_1 + e_1 = -121339c_2^2 - 2 = -131890c_3 - e_3 = -81986c_5 + e_5 = -64542c_6 - e_6 = -57235c_7 - e_7 = -6910c_{43} + e_{43} = -4732c_{53} - e_{53}$,
$a_2 = -4294c_3^2 + 30 = -42858c_7 - e_7 = -28753c_9 - e_9 = -16580c_{16} - e_{16} = -9504c_{27} - e_{27} = -4864c_{46} - e_{46}$,
$a_3 = -77717c_3 + e_3 = -20084c_{11} - e_{11} = -15818c_{15} - e_{15} = -4667c_{39} - e_{39} = -2833c_{52} + e_{52} = -2789c_{53} + e_{53}$,
$a_4 = -351704c_2 + e_2 = -76457c_3 - e_3 = -33180c_7 + e_7 = -18129c_{12} - e_{12} = -4982c_{37} + e_{37} = -4300c_{40} + e_{40} = -4177c_{41} - e_{41}$,
$a_5 = -13595c_2^3 + 27 = -73885c_3 + e_3 = -17519c_{12} - e_{12} = -4437c_{39} + e_{39} = -3018c_{48} - e_{48} = -2828c_{51} + e_{51}$,
$a_6 = -285488c_2 - e_2 = -62063c_3 + e_3 = -18069c_9 + e_9 = -13096c_{14} + e_{14} = -7173c_{23} - e_{23} = -4236c_{35} + e_{35} = -2563c_{47} + e_{47}$,
$a_7 = -416327c_1 + e_1 = -166531c_2 + e_2 = -36202c_3 - e_3 = -22504c_5 - e_5 = -8584c_{12} - e_{12} = -3006c_{31} + e_{31} = -2319c_{38} - e_{38} = -1783c_{46} + e_{46}$,
$a_8 = -374253c_1 + e_1 = -1415c_3^2 + 30 = -25811c_4 + e_4 = -20230c_5 + e_5 = -9018c_{10} - e_{10} = -1186c_{52} - e_{52} = -1157c_{54} + e_{54}$,
$a_9 = -337412c_1 + e_1 = -26993c_2^2 + 2 = -14358c_6 + e_6 = -4140c_{19} - e_{19} = -1762c_{39} + e_{39} = -1445c_{46} - e_{46} = -1199c_{48} + e_{48}$,

$a_{10} = -110958c_2 + e_2 = -5090c_{14} + e_{14} = -1572c_{37} + e_{37} = -1188c_{46} + e_{46} = -879c_{52} - e_{52} = -812c_{55} - e_{55}$,
$a_{11} = -260755c_1 + e_1 = -22674c_3 - e_3 = -11096c_6 + e_6 = -2297c_{25} - e_{25} = -1494c_{36} - e_{36} = -1477c_{37} - e_{37} = -814c_{53} + e_{53}$,
$a_{12} = -96428c_2 - e_2 = -20963c_3 + e_3 = -13031c_5 + e_5 = -3469c_{17} + e_{17} = -1833c_{28} - e_{28} = -1088c_{44} - e_{44} = -856c_{48} - e_{48}$,
$a_{13} = -222529c_1 + e_1 = -89011c_2 - e_2 = -19350c_3 - e_3 = -5362c_{10} - e_{10} = -3939c_{15} + e_{15} = -2486c_{21} - e_{21} = -1961c_{25} + e_{25} = -1005c_{44} + e_{44}$,
$a_{14} = -17010c_2^2 + 2 = -5383c_9 + e_9 = -2137c_{23} + e_{23} = -1569c_{30} - e_{30} = -747c_{49} - e_{49} = -663c_{53} - e_{53}$,
$a_{15} = -161434c_1 + e_1 = -64573c_2 - e_2 = -14038c_3 + e_3 = -8726c_5 - e_5 = -5293c_8 + e_8 = -1038c_{34} - e_{34} = -915c_{37} + e_{37} = -473c_{55} + e_{55}$,
$a_{16} = -140487c_1 + e_1 = -2248c_2^3 + 27 = -7594c_5 + e_5 = -1790c_{18} + e_{18} = -1412c_{23} + e_{23} = -915c_{33} - e_{33} = -640c_{43} - e_{43}$,
$a_{17} = -50578c_2 - e_2 = -478c_3^2 - 30 = -933c_{30} - e_{30} = -913c_{31} + e_{31} = -813c_{34} - e_{34} = -401c_{52} + e_{52}$,
$a_{18} = -9060c_2^2 + 2 = -7c_{21}^2 - 2211 = -738c_{33} + e_{33} = -642c_{37} + e_{37} = -504c_{45} - e_{45}$,
$a_{19} = -8608c_2^2 - 2 = -4060c_7 - e_7 = -3528c_8 + e_8 = -2011c_{13} - e_{13} = -16c_{15}^2 - 10898 = -610c_{37} + e_{37}$,
$a_{20} = -103987c_1 + e_1 = -67c_2^5 + 1402 = -9042c_3 - e_3 = -7171c_4 - e_4 = -2144c_{12} - e_{12} = -908c_{26} - e_{26} = -494c_{41} + e_{41}$,
$a_{21} = -8248c_2^2 + 2 = -1313c_{18} - e_{18} = -1265c_{19} - e_{19} = -767c_{29} + e_{29} = -574c_{38} - e_{38} = -327c_{52} + e_{52}$,
$a_{22} = -38648c_2 - e_2 = -6664c_4 + e_4 = -3168c_8 + e_8 = -6c_{20}^2 - 25908 = -735c_{28} + e_{28} = -322c_{51} + e_{51}$,
$a_{23} = -25226c_2 + e_2 = -3409c_5 + e_5 = -1179c_{13} + e_{13} = -803c_{18} - e_{18} = -4c_{22}^2 + 22868 = -374c_{35} - e_{35}$,
$a_{24} = -230c_3^2 + 30 = -4194c_4 - e_4 = -1994c_8 - e_8 = -531c_{26} - e_{26} = -396c_{33} - e_{33} = -205c_{50} - e_{50}$,
$a_{25} = -10884c_2 - e_2 = -1471c_5 + e_5 = -689c_9 + e_9 = -561c_{12} - e_{12} = -347c_{18} + e_{18} = -201c_{30} + e_{30} = -84c_{54} - e_{54}$,
$a_{26} = -9682c_2 + e_2 = -2105c_3 + e_3 = -544c_{11} + e_{11} = -499c_{12} - e_{12} = -297c_{19} + e_{19} = -139c_{36} + e_{36} = -70c_{56} - e_{56}$,
$a_{27} = -20382c_1 + e_1 = -8153c_2 + e_2 = -77c_3^2 - 30 = -260c_{18} + e_{18} = -152c_{29} + e_{29} = -133c_{33} + e_{33} = -72c_{49} + e_{49}$,
$a_{28} = -17321c_1 + e_1 = -1195c_4 + e_4 = -307c_{15} + e_{15} = -179c_{22} - e_{22} = -174c_{23} - e_{23} = -153c_{25} + e_{25} = -51c_{55} + e_{55}$,
$a_{29} = -489c_6 + e_6 = -434c_7 + e_7 = -291c_9 + e_9 = -277c_{10} + e_{10} = -215c_{13} + e_{13} = -168c_{16} + e_{16} = -141c_{19} + e_{19}$,
$a_{30} = 1921c_1 + e_1 = 769c_2 - e_2 = 133c_4 - e_4 = 104c_5 - e_5 = 28c_{17} - e_{17} = 17c_{24} + e_{24} = 14c_{30} + e_{30} = 9c_{42} - e_{42}$,
$a_{31} = 14709c_1 + e_1 = 626c_6 - e_6 = 215c_{16} - e_{16} = 212c_{17} - e_{17} = 164c_{21} + e_{21} = 29419$,
$a_{32} = 7474c_2 - e_2 = 1625c_3 - e_3 = 795c_6 + e_6 = 349c_{13} + e_{13} = 163c_{26} + e_{26} = 80c_{46} + e_{46} = 59c_{52} + e_{52}$,
$a_{33} = 1532c_2^2 + 2 = 1665c_3 + e_3 = 815c_6 - e_6 = 628c_8 - e_8 = 235c_{19} - e_{19} = 123c_{34} + e_{34} = 82c_{46} + e_{46}$,
$a_{34} = 40083c_1 + e_1 = 16033c_2 + e_2 = 1513c_7 - e_7 = 12c_{10}^2 - 2501 = 709c_{15} + e_{15} = 296c_{30} - e_{30} = 230c_{36} - e_{36}$,
$a_{35} = 2247c_5 - e_5 = 934c_{11} + e_{11} = 857c_{12} + e_{12} = 217c_{39} + e_{39} = 178c_{46} + e_{46} = 122c_{55} - e_{55}$,
$a_{36} = 51316c_1 + e_1 = 20527c_2 - e_2 = 4462c_3 + e_3 = 2774c_5 - e_5 = 448c_{26} + e_{26} = 390c_{28} + e_{28} = 382c_{29} - e_{29} = 363c_{32} - e_{32}$,
$a_{37} = 57717c_1 + e_1 = 138c_4^2 - 623 = 3120c_5 - e_5 = 2456c_6 + e_6 = 1190c_{12} + e_{12} = 580c_{23} + e_{23} = 167c_{56} + e_{56}$,
$a_{38} = 23358c_2 + e_2 = 2485c_6 - e_6 = 1407c_{10} + e_{10} = 10c_{15}^2 - 10898 = 305c_{39} - e_{39} = 286c_{40} - e_{40}$,
$a_{39} = 78000c_1 + e_1 = 2943c_7 + e_7 = 1431c_{14} + e_{14} = 994c_{18} - e_{18} = 934c_{20} + e_{20} = 784c_{23} - e_{23} = 241c_{54} + e_{54}$,
$a_{40} = 98903c_1 + e_1 = 39561c_2 + e_2 = 8600c_3 + e_3 = 5346c_5 + e_5 = 2504c_9 - e_9 = 871c_{25} + e_{25} = 441c_{45} - e_{45} = 348c_{49} - e_{49}$,
$a_{41} = 105936c_1 + e_1 = 8475c_2^2 - 2 = 4508c_6 - e_6 = 1184c_{21} - e_{21} = 782c_{30} - e_{30} = 372c_{49} + e_{49} = 353c_{51} - e_{51}$,
$a_{42} = 43896c_2 + e_2 = 9543c_3 - e_3 = 2466c_{11} + e_{11} = 2051c_{13} + e_{13} = 1103c_{23} - e_{23} = 573c_{39} + e_{39} = 470c_{46} - e_{46}$,
$a_{43} = 45978c_2 + e_2 = 9995c_3 + e_3 = 4338c_7 - e_7 = 2034c_{15} + e_{15} = 812c_{32} + e_{32} = 640c_{38} + e_{38} = 413c_{47} - e_{47}$,
$a_{44} = 52818c_2 - e_2 = 9106c_4 + e_4 = 3343c_9 - e_9 = 1475c_{21} + e_{21} = 1327c_{23} + e_{23} = 860c_{33} + e_{33} = 849c_{34} + e_{34}$,
$a_{45} = 53902c_2 + e_2 = 9294c_4 - e_4 = 3247c_{10} + e_{10} = 1717c_{18} - e_{18} = 1025c_{28} - e_{28} = 973c_{31} - e_{31} = 952c_{32} + e_{32}$,
$a_{46} = 12640c_2^2 - 2 = 375c_4^2 + 623 = 2796c_{15} + e_{15} = 1765c_{21} + e_{21} = 1322c_{27} + e_{27} = 825c_{39} + e_{39}$,
$a_{47} = 169621c_1 + e_1 = 67849c_2 - e_2 = 14750c_3 - e_3 = 154c_6^2 - 943 = 1252c_{30} - e_{30} = 829c_{40} + e_{40} = 491c_{56} - e_{56}$,
$a_{48} = 16258c_4 + e_4 = 12743c_5 + e_5 = 1793c_{28} - e_{28} = 1231c_{39} + e_{39} = 829c_{49} - e_{49} = 747c_{52} + e_{52}$,
$a_{49} = 249615c_1 + e_1 = 21706c_3 - e_3 = 10622c_6 - e_6 = 2089c_{27} - e_{27} = 1842c_{30} + e_{30} = 1069c_{46} + e_{46} = 842c_{50} - e_{50}$,
$a_{50} = 25943c_3 + e_3 = 212c_7^2 + 1188 = 9782c_8 - e_8 = 7553c_9 + e_9 = 2676c_{24} - e_{24} = 1071c_{47} + e_{47}$,
$a_{51} = 562024c_1 + e_1 = 48872c_3 - e_3 = 23916c_6 - e_6 = 8205c_{16} - e_{16} = 2537c_{44} + e_{44} = 2503c_{45} + e_{45} = 1754c_{53} - e_{53}$,
$a_{52} = 230886c_2 + e_2 = 50193c_3 - e_3 = 31201c_5 - e_5 = 5982c_{22} - e_{22} = 5086c_{25} - e_{25} = 2666c_{42} + e_{42} = 2472c_{46} + e_{46}$,

$a_{53} = 607426c_1+e_1 = 242971c_2-e_2 = 52820c_3-e_3 = 41891c_4+e_4 = 32834c_5-e_5 = 25848c_6-e_6 = 15378c_9 - e_9 = 11354c_{13} - e_{13} = 3172c_{39} - e_{39}$,
$a_{54} = 701894c_1 + e_1 = 61034c_3 + e_3 = 48407c_4 - e_4 = 26487c_7 - e_7 = 15773c_{11} - e_{11} = 14472c_{12} + e_{12} = 12879c_{14} - e_{14} = 2520c_{47} + e_{47}$,
$a_{55} = 389516c_2 - e_2 = 84677c_3 + e_3 = 24653c_9 - e_9 = 14011c_{17} + e_{17} = 7405c_{28} + e_{28} = 7031c_{31} - e_{31} = 5085c_{39} + e_{39}$,
$a_{56} = 3764c_3^2 - 30 = 68660c_4 - e_4 = 37568c_7 + e_7 = 14325c_{17} - e_{17} = 7c_{47}^2 - 180617$,
$a_{57} = 1063286c_1 + e_1 = 85063c_2^2 - 2 = 92460c_3 - e_3 = 11019c_{22} - e_{22} = 5199c_{40} + e_{40} = 3586c_{50} + e_{50} = 3318c_{53} - e_{53}$.

It is straightforward to prove that the equations below are satisfied. Note that these equations will be used later to show that the ideal norms of certain nonzero ideals of $\mathcal{O}_K$ coincide.

Equations, part C.

$$
\begin{aligned}
d =& a_1^2 + c_1c_2^2c_3c_5c_6c_7c_{43}c_{53} = a_2^2 + c_3^2c_7c_9c_{16}c_{27}c_{46} = a_3^2 + c_3c_{11}c_{15}c_{39}c_{52}c_{53} \\
=& a_4^2 + c_2c_3c_7c_{12}c_{37}c_{40}c_{41} = a_5^2 + c_2^3c_3c_{12}c_{39}c_{48}c_{51} = a_6^2 + c_2c_3c_9c_{14}c_{23}c_{35}c_{47} \\
=& a_7^2 + c_1c_2c_3c_5c_{12}c_{31}c_{38}c_{46} = a_8^2 + c_1c_3^2c_4c_5c_{10}c_{52}c_{54} = a_9^2 + c_1c_2^2c_6c_{19}c_{39}c_{46}c_{48} \\
=& a_{10}^2 + c_2c_{14}c_{37}c_{46}c_{52}c_{55} = a_{11}^2 + c_1c_3c_6c_{25}c_{36}c_{37}c_{53} = a_{12}^2 + c_2c_3c_5c_{17}c_{28}c_{44}c_{48} \\
=& a_{13}^2 + c_1c_2c_3c_{10}c_{15}c_{21}c_{25}c_{44} = a_{14}^2 + c_2^2c_9c_{23}c_{30}c_{49}c_{53} \\
=& a_{15}^2 + c_1c_2c_3c_5c_8c_{34}c_{37}c_{55} = a_{16}^2 + c_1c_2^3c_5c_{18}c_{23}c_{33}c_{43} = a_{17}^2 + c_2c_3^2c_{30}c_{31}c_{34}c_{52} \\
=& a_{18}^2 + c_2^2c_{21}^2c_{33}c_{37}c_{45} = a_{19}^2 + c_2^2c_7c_8c_{13}c_{15}^2c_{37} = a_{20}^2 + c_1c_2^5c_3c_4c_{12}c_{26}c_{41} \\
=& a_{21}^2 + c_2^2c_{18}c_{19}c_{29}c_{38}c_{52} = a_{22}^2 + c_2c_4c_8c_{20}^2c_{28}c_{51} = a_{23}^2 + c_2c_5c_{13}c_{18}c_{22}^2c_{35} \\
=& a_{24}^2 + c_3^2c_4c_8c_{26}c_{33}c_{50} = a_{25}^2 + c_2c_5c_9c_{12}c_{18}c_{30}c_{54} = a_{26}^2 + c_2c_3c_{11}c_{12}c_{19}c_{36}c_{56} \\
=& a_{27}^2 + c_1c_2c_3^2c_{18}c_{29}c_{33}c_{49} = a_{28}^2 + c_1c_4c_{15}c_{22}c_{23}c_{25}c_{55} = a_{29}^2 + c_6c_7c_9c_{10}c_{13}c_{16}c_{19} \\
=& a_{30}^2 + c_1c_2c_4c_5c_{17}c_{24}c_{30}c_{42} = a_{31}^2 + c_1c_6c_{16}c_{17}c_{21}c_{36}^2 = a_{32}^2 + c_2c_3c_6c_{13}c_{26}c_{46}c_{52} \\
=& a_{33}^2 + c_2^2c_3c_6c_8c_{19}c_{34}c_{46} = a_{34}^2 + c_1c_2c_7c_{10}^2c_{15}c_{30}c_{36} = a_{35}^2 + c_5c_{11}c_{12}c_{39}c_{46}c_{55} \\
=& a_{36}^2 + c_1c_2c_3c_5c_{26}c_{28}c_{29}c_{32} = a_{37}^2 + c_1c_4^2c_5c_6c_{12}c_{23}c_{56} = a_{38}^2 + c_2c_6c_{10}c_{15}^2c_{39}c_{40} \\
=& a_{39}^2 + c_1c_7c_{14}c_{18}c_{20}c_{23}c_{54} = a_{40}^2 + c_1c_2c_3c_5c_9c_{25}c_{45}c_{49} = a_{41}^2 + c_1c_2^2c_6c_{21}c_{30}c_{49}c_{51} \\
=& a_{42}^2 + c_2c_3c_{11}c_{13}c_{23}c_{39}c_{46} = a_{43}^2 + c_2c_3c_7c_{15}c_{32}c_{38}c_{47} = a_{44}^2 + c_2c_4c_9c_{21}c_{23}c_{33}c_{34} \\
=& a_{45}^2 + c_2c_4c_{10}c_{18}c_{28}c_{31}c_{32} = a_{46}^2 + c_2^2c_4^2c_{15}c_{21}c_{27}c_{39} = a_{47}^2 + c_1c_2c_3c_6^2c_{30}c_{40}c_{56} \\
=& a_{48}^2 + c_4c_5c_{28}c_{39}c_{49}c_{52} = a_{49}^2 + c_1c_3c_6c_{27}c_{30}c_{46}c_{50} = a_{50}^2 + c_3c_7^2c_8c_9c_{24}c_{47} \\
=& a_{51}^2 + c_1c_3c_6c_{16}c_{44}c_{45}c_{53} = a_{52}^2 + c_2c_3c_5c_{22}c_{25}c_{42}c_{46} \\
=& a_{53}^2 + c_1c_2c_3c_4c_5c_6c_9c_{13}c_{39} = a_{54}^2 + c_1c_3c_4c_7c_{11}c_{12}c_{14}c_{47} \\
=& a_{55}^2 + c_2c_3c_9c_{17}c_{28}c_{31}c_{39} = a_{56}^2 + c_3^2c_4c_7c_{17}c_{47}^2 = a_{57}^2 + c_1c_2^2c_3c_{22}c_{40}c_{50}c_{53}.
\end{aligned}
$$

Observe that 2 and d are precisely the ramified primes of $\mathcal{O}_K$ (since $d \in \mathbb{P}$ and $4d$ is the discriminant of $\mathcal{O}_K$). Therefore, c_j splits in $\mathcal{O}_K$ for each $j \in [2, s]$. We infer that all positive powers of the maximal ideals P_1, P_j and $\overline{P}_j$ are pairwise comaximal for each $j \in [2, s]$. Next, we discuss an elementary fact that will subsequently be used. Let I and J be nonzero ideals of $\mathcal{O}_K$ such that $I \subseteq J$ and $\mathrm{N}(I) = \mathrm{N}(J)$. It follows by the theorem of Lagrange that $|\mathcal{O}_K/J| = \mathrm{N}(J) = \mathrm{N}(I) = |\mathcal{O}_K/I| = |\mathcal{O}_K/J||J/I|$, and thus $|J/I| = 1$. This implies that $I = J$.

By combining the equations in the parts A, B and C above, we obtain the next list of equations. Note that the equations in parts A and B together show that each left hand side is contained in each right hand side. The equations in part C prove that the ideal norms on each side coincide, and hence both sides coincide. We will make these statements clearer, by giving a detailed proof for the first equality below. (The remaining equations can be proved in analogy.)

Equations, part D.

$$
\begin{gathered}
\left(a_1+\sqrt{d}\right)\mathcal{O}_K=P_1\overline{P}_2^2\overline{P}_3P_5\overline{P}_6\overline{P}_7P_{43}\overline{P}_{53},\ \left(a_2+\sqrt{d}\right)\mathcal{O}_K=P_3^2\overline{P}_7\overline{P}_9\overline{P}_{16}\overline{P}_{27}\overline{P}_{46},\\
\left(a_3+\sqrt{d}\right)\mathcal{O}_K=P_3\overline{P}_{11}\overline{P}_{15}\overline{P}_{39}P_{52}P_{53},\ \left(a_4+\sqrt{d}\right)\mathcal{O}_K=P_2\overline{P}_3P_7\overline{P}_{12}P_{37}P_{40}\overline{P}_{41},\\
\left(a_5+\sqrt{d}\right)\mathcal{O}_K=P_2^3P_3\overline{P}_{12}P_{39}\overline{P}_{48}P_{51},\ \left(a_6+\sqrt{d}\right)\mathcal{O}_K=\overline{P}_2P_3P_9P_{14}\overline{P}_{23}P_{35}P_{47},\\
\left(a_7+\sqrt{d}\right)\mathcal{O}_K=P_1P_2\overline{P}_3\overline{P}_5\overline{P}_{12}P_{31}\overline{P}_{38}P_{46},\ \left(a_8+\sqrt{d}\right)\mathcal{O}_K=P_1P_3^2P_4P_5\overline{P}_{10}\overline{P}_{52}P_{54},\\
\left(a_9+\sqrt{d}\right)\mathcal{O}_K=P_1P_2^2P_6\overline{P}_{19}P_{39}\overline{P}_{46}P_{48},\ \left(a_{10}+\sqrt{d}\right)\mathcal{O}_K=P_2P_{14}P_{37}P_{46}\overline{P}_{52}\overline{P}_{55},\\
\left(a_{11}+\sqrt{d}\right)\mathcal{O}_K=P_1\overline{P}_3P_6\overline{P}_{25}\overline{P}_{36}\overline{P}_{37}P_{53},\ \left(a_{12}+\sqrt{d}\right)\mathcal{O}_K=\overline{P}_2P_3P_5P_{17}\overline{P}_{28}\overline{P}_{44}\overline{P}_{48},\\
\left(a_{13}+\sqrt{d}\right)\mathcal{O}_K=P_1\overline{P}_2\overline{P}_3\overline{P}_{10}P_{15}\overline{P}_{21}P_{25}P_{44},\ \left(a_{14}+\sqrt{d}\right)\mathcal{O}_K=P_2^2P_9P_{23}\overline{P}_{30}\overline{P}_{49}\overline{P}_{53},\\
\left(a_{15}+\sqrt{d}\right)\mathcal{O}_K=P_1\overline{P}_2P_3\overline{P}_5P_8\overline{P}_{34}P_{37}P_{55},\ \left(a_{16}+\sqrt{d}\right)\mathcal{O}_K=P_1P_2^3P_5P_{18}P_{23}\overline{P}_{33}\overline{P}_{43},\\
\left(a_{17}+\sqrt{d}\right)\mathcal{O}_K=\overline{P}_2\overline{P}_3^2\overline{P}_{30}P_{31}\overline{P}_{34}P_{52},\ \left(a_{18}+\sqrt{d}\right)\mathcal{O}_K=P_2^2\overline{P}_{21}^2P_{33}P_{37}\overline{P}_{45},\\
\left(a_{19}+\sqrt{d}\right)\mathcal{O}_K=\overline{P}_2^2\overline{P}_7P_8\overline{P}_{13}\overline{P}_{15}^2P_{37},\ \left(a_{20}+\sqrt{d}\right)\mathcal{O}_K=P_1P_2^5\overline{P}_3\overline{P}_4\overline{P}_{12}\overline{P}_{26}P_{41},\\
\left(a_{21}+\sqrt{d}\right)\mathcal{O}_K=P_2^2\overline{P}_{18}\overline{P}_{19}P_{29}\overline{P}_{38}P_{52},\ \left(a_{22}+\sqrt{d}\right)\mathcal{O}_K=\overline{P}_2P_4P_8\overline{P}_{20}^2P_{28}P_{51},\\
\left(a_{23}+\sqrt{d}\right)\mathcal{O}_K=P_2P_5P_{13}\overline{P}_{18}P_{22}^2\overline{P}_{35},\ \left(a_{24}+\sqrt{d}\right)\mathcal{O}_K=P_3^2\overline{P}_4\overline{P}_8\overline{P}_{26}\overline{P}_{33}\overline{P}_{50},\\
\left(a_{25}+\sqrt{d}\right)\mathcal{O}_K=\overline{P}_2P_5P_9\overline{P}_{12}P_{18}P_{30}\overline{P}_{54},\ \left(a_{26}+\sqrt{d}\right)\mathcal{O}_K=P_2P_3P_{11}\overline{P}_{12}P_{19}P_{36}\overline{P}_{56},\\
\left(a_{27}+\sqrt{d}\right)\mathcal{O}_K=P_1P_2\overline{P}_3^2P_{18}P_{29}P_{33}P_{49},\ \left(a_{28}+\sqrt{d}\right)\mathcal{O}_K=P_1P_4P_{15}\overline{P}_{22}\overline{P}_{23}P_{25}P_{55},\\
\left(a_{29}+\sqrt{d}\right)\mathcal{O}_K=P_6P_7P_9P_{10}P_{13}P_{16}P_{19},\ \left(a_{30}+\sqrt{d}\right)\mathcal{O}_K=P_1\overline{P}_2\overline{P}_4\overline{P}_5\overline{P}_{17}P_{24}P_{30}\overline{P}_{42},\\
\left(a_{31}+\sqrt{d}\right)\mathcal{O}_K=P_1\overline{P}_6\overline{P}_{16}\overline{P}_{17}P_{21}P_{36}^2,\ \left(a_{32}+\sqrt{d}\right)\mathcal{O}_K=\overline{P}_2\overline{P}_3P_6P_{13}P_{26}P_{46}P_{52},\\
\left(a_{33}+\sqrt{d}\right)\mathcal{O}_K=P_2^2P_3\overline{P}_6\overline{P}_8\overline{P}_{19}P_{34}P_{46},\ \left(a_{34}+\sqrt{d}\right)\mathcal{O}_K=P_1P_2\overline{P}_7\overline{P}_{10}^2P_{15}\overline{P}_{30}\overline{P}_{36},\\
\left(a_{35}+\sqrt{d}\right)\mathcal{O}_K=\overline{P}_5P_{11}P_{12}P_{39}P_{46}\overline{P}_{55},\ \left(a_{36}+\sqrt{d}\right)\mathcal{O}_K=P_1\overline{P}_2P_3\overline{P}_5P_{26}P_{28}\overline{P}_{29}\overline{P}_{32},\\
\left(a_{37}+\sqrt{d}\right)\mathcal{O}_K=P_1\overline{P}_4^2\overline{P}_5P_6P_{12}P_{23}P_{56},\ \left(a_{38}+\sqrt{d}\right)\mathcal{O}_K=P_2\overline{P}_6P_{10}\overline{P}_{15}^2\overline{P}_{39}\overline{P}_{40},\\
\left(a_{39}+\sqrt{d}\right)\mathcal{O}_K=P_1P_7P_{14}\overline{P}_{18}P_{20}\overline{P}_{23}P_{54},\ \left(a_{40}+\sqrt{d}\right)\mathcal{O}_K=P_1P_2P_3P_5\overline{P}_9P_{25}\overline{P}_{45}\overline{P}_{49},\\
\left(a_{41}+\sqrt{d}\right)\mathcal{O}_K=P_1\overline{P}_2^2\overline{P}_6\overline{P}_{21}\overline{P}_{30}P_{49}\overline{P}_{51},\ \left(a_{42}+\sqrt{d}\right)\mathcal{O}_K=P_2\overline{P}_3P_{11}P_{13}\overline{P}_{23}P_{39}\overline{P}_{46},\\
\left(a_{43}+\sqrt{d}\right)\mathcal{O}_K=P_2P_3\overline{P}_7P_{15}P_{32}P_{38}\overline{P}_{47},\ \left(a_{44}+\sqrt{d}\right)\mathcal{O}_K=\overline{P}_2P_4\overline{P}_9P_{21}P_{23}P_{33}P_{34},\\
\left(a_{45}+\sqrt{d}\right)\mathcal{O}_K=P_2\overline{P}_4P_{10}\overline{P}_{18}\overline{P}_{28}\overline{P}_{31}P_{32},\ \left(a_{46}+\sqrt{d}\right)\mathcal{O}_K=\overline{P}_2^2P_4^2P_{15}P_{21}P_{27}P_{39},\\
\left(a_{47}+\sqrt{d}\right)\mathcal{O}_K=P_1\overline{P}_2\overline{P}_3\overline{P}_6^2\overline{P}_{30}P_{40}\overline{P}_{56},\ \left(a_{48}+\sqrt{d}\right)\mathcal{O}_K=P_4P_5\overline{P}_{28}P_{39}\overline{P}_{49}P_{52},\\
\left(a_{49}+\sqrt{d}\right)\mathcal{O}_K=P_1\overline{P}_3\overline{P}_6\overline{P}_{27}P_{30}P_{46}\overline{P}_{50},\ \left(a_{50}+\sqrt{d}\right)\mathcal{O}_K=P_3P_7^2\overline{P}_8P_9\overline{P}_{24}P_{47},\\
\left(a_{51}+\sqrt{d}\right)\mathcal{O}_K=P_1\overline{P}_3\overline{P}_6\overline{P}_{16}P_{44}P_{45}\overline{P}_{53},\ \left(a_{52}+\sqrt{d}\right)\mathcal{O}_K=P_2\overline{P}_3\overline{P}_5\overline{P}_{22}\overline{P}_{25}P_{42}P_{46},\\
\left(a_{53}+\sqrt{d}\right)\mathcal{O}_K=P_1\overline{P}_2\overline{P}_3P_4\overline{P}_5\overline{P}_6\overline{P}_9\overline{P}_{13}\overline{P}_{39},\ \left(a_{54}+\sqrt{d}\right)\mathcal{O}_K=P_1P_3\overline{P}_4\overline{P}_7\overline{P}_{11}P_{12}\overline{P}_{14}P_{47},\\
\left(a_{55}+\sqrt{d}\right)\mathcal{O}_K=\overline{P}_2P_3\overline{P}_9P_{17}P_{28}\overline{P}_{31}P_{39},\ \left(a_{56}+\sqrt{d}\right)\mathcal{O}_K=\overline{P}_3^2\overline{P}_4P_7\overline{P}_{17}\overline{P}_{47}^2,\\
\left(a_{57}+\sqrt{d}\right)\mathcal{O}_K=P_1\overline{P}_2^2\overline{P}_3\overline{P}_{22}P_{40}P_{50}\overline{P}_{53}.
\end{gathered}
$$

Now, we show that the first equation in part D holds. By the equations in part B, we have that

$$a_1+\sqrt{d}=-1516739c_1+e_1+\sqrt{d}\in c_1\mathcal{O}_K+\left(e_1+\sqrt{d}\right)\mathcal{O}_K=P_1 \text{ and}$$

$$a_1+\sqrt{d}=-131890c_3-\left(e_3-\sqrt{d}\right)\in c_3\mathcal{O}_K+\left(e_3-\sqrt{d}\right)\mathcal{O}_K=\overline{P}_3.$$

Furthermore, we have that $a_1+\sqrt{d}=-121339c_2^2-\left(2-\sqrt{d}\right)\in c_2^2\mathcal{O}_K+\left(2-\sqrt{d}\right)\mathcal{O}_K=\overline{P}_2^2$ by the equations in parts A and B. Along the same lines it can be proved that $a_1+\sqrt{d}\in P_5\cap\overline{P}_6\cap\overline{P}_7\cap P_{43}\cap\overline{P}_{53}$.

This implies that $a_1+\sqrt{d}\in P_1\cap\overline{P}_2^2\cap\overline{P}_3\cap P_5\cap\overline{P}_6\cap\overline{P}_7\cap P_{43}\cap\overline{P}_{53}$. Therefore, $\left(a_1+\sqrt{d}\right)\mathcal{O}_K\subseteq P_1\cap\overline{P}_2^2\cap\overline{P}_3\cap P_5\cap\overline{P}_6\cap\overline{P}_7\cap P_{43}\cap\overline{P}_{53}=P_1\overline{P}_2^2\overline{P}_3P_5\overline{P}_6\overline{P}_7P_{43}\overline{P}_{53}$

(since the powers of the maximal ideals in the aforementioned intersection are pairwise comaximal).

We infer by the equations in part C that

$$\begin{aligned}\mathrm{N}\left(\left(a_1+\sqrt{d}\right)\mathcal{O}_K\right) =&\left|\mathrm{N}\left(a_1+\sqrt{d}\right)\right| = d-a_1^2 = c_1c_2^2c_3c_5c_6c_7c_{43}c_{53}\\ =&\mathrm{N}\left(P_1\right)\mathrm{N}\left(\overline{P}_2\right)^2\mathrm{N}\left(\overline{P}_3\right)\mathrm{N}\left(P_5\right)\mathrm{N}\left(\overline{P}_6\right)\mathrm{N}\left(\overline{P}_7\right)\mathrm{N}\left(P_{43}\right)\mathrm{N}\left(\overline{P}_{53}\right)\\ =&\mathrm{N}\left(P_1\overline{P}_2^2\overline{P}_3P_5\overline{P}_6\overline{P}_7P_{43}\overline{P}_{53}\right).\end{aligned}$$

Therefore, $\left(a_1+\sqrt{d}\right)\mathcal{O}_K = P_1\overline{P}_2^2\overline{P}_3P_5\overline{P}_6\overline{P}_7P_{43}\overline{P}_{53}$.

We want to emphasize that the equations in part D show that the maximal ideals P_j and $\overline{P}_j$ for $j\in[1,s]$ are the only maximal ideals of $\mathcal{O}_K$ that can occur in the factorization of $\left(a_i+\sqrt{d}\right)\mathcal{O}_K$ into maximal ideals for each $i\in[1,r]$.

Next, we determine the factorization of $z\mathcal{O}_K$ into maximal ideals of $\mathcal{O}_K$. Let I be a nonzero ideal of $\mathcal{O}_K$ and let P be maximal ideal of $\mathcal{O}_K$. In what follows, let $\mathrm{v}_P(I)$ be the unique nonnegative integer k such that $I=P^kJ$ for some nonzero ideal J of $\mathcal{O}_K$ with $J\not\subseteq P$. Note that the equations in part E below consist of two equalities. Each of the first equalities is an immediate consequence of the equations in part D, while each of the second equalities is (a consequence of) elementary arithmetic.

Equations, part E.

$\mathrm{v}_{P_1}(z\mathcal{O}_K) = b_1+b_7+b_8+b_9+b_{11}+b_{13}+b_{15}+b_{16}+b_{20}+b_{27}+b_{28}+b_{30}+b_{31}+b_{34}+b_{36}+b_{37}+b_{39}+b_{40}+b_{41}+b_{47}+b_{49}+b_{51}+b_{53}+b_{54}+b_{57} = 2d_1$,
$\mathrm{v}_{P_2}(z\mathcal{O}_K) = b_4+3b_5+b_7+2b_9+b_{10}+2b_{14}+3b_{16}+2b_{18}+5b_{20}+2b_{21}+b_{23}+b_{26}+b_{27}+2b_{33}+b_{34}+b_{38}+b_{40}+b_{42}+b_{43}+b_{45}+b_{52} = d_2$,
$\mathrm{v}_{\overline{P}_2}(z\mathcal{O}_K) = 2b_1+b_6+b_{12}+b_{13}+b_{15}+b_{17}+2b_{19}+b_{22}+b_{25}+b_{30}+b_{32}+b_{36}+2b_{41}+b_{44}+2b_{46}+b_{47}+b_{53}+b_{55}+2b_{57} = d_2$,
$\mathrm{v}_{P_3}(z\mathcal{O}_K) = 2b_2+b_3+b_5+b_6+2b_8+b_{12}+b_{15}+2b_{24}+b_{26}+b_{33}+b_{36}+b_{40}+b_{43}+b_{50}+b_{54}+b_{55} = d_3$,
$\mathrm{v}_{\overline{P}_3}(z\mathcal{O}_K) = b_1+b_4+b_7+b_{11}+b_{13}+2b_{17}+b_{20}+2b_{27}+b_{32}+b_{42}+b_{47}+b_{49}+b_{51}+b_{52}+b_{53}+2b_{56}+b_{57} = d_3$,
$\mathrm{v}_{P_4}(z\mathcal{O}_K) = b_8+b_{22}+b_{28}+b_{44}+2b_{46}+b_{48}+b_{53} = d_4$,
$\mathrm{v}_{\overline{P}_4}(z\mathcal{O}_K) = b_{20}+b_{24}+b_{30}+2b_{37}+b_{45}+b_{54}+b_{56} = d_4$,
$\mathrm{v}_{P_5}(z\mathcal{O}_K) = b_1+b_8+b_{12}+b_{16}+b_{23}+b_{25}+b_{40}+b_{48} = d_5$,
$\mathrm{v}_{\overline{P}_5}(z\mathcal{O}_K) = b_7+b_{15}+b_{30}+b_{35}+b_{36}+b_{37}+b_{52}+b_{53} = d_5$,
$\mathrm{v}_{P_6}(z\mathcal{O}_K) = b_9+b_{11}+b_{29}+b_{32}+b_{37} = d_6$,
$\mathrm{v}_{\overline{P}_6}(z\mathcal{O}_K) = b_1+b_{31}+b_{33}+b_{38}+b_{41}+2b_{47}+b_{49}+b_{51}+b_{53} = d_6$,
$\mathrm{v}_{P_7}(z\mathcal{O}_K) = b_4+b_{29}+b_{39}+2b_{50}+b_{56} = d_7$, $\mathrm{v}_{\overline{P}_7}(z\mathcal{O}_K) = b_1+b_2+b_{19}+b_{34}+b_{43}+b_{54} = d_7$,
$\mathrm{v}_{P_8}(z\mathcal{O}_K) = b_{15}+b_{19}+b_{22} = d_8$, $\mathrm{v}_{\overline{P}_8}(z\mathcal{O}_K) = b_{24}+b_{33}+b_{50} = d_8$,
$\mathrm{v}_{P_9}(z\mathcal{O}_K) = b_6+b_{14}+b_{25}+b_{29}+b_{50} = d_9$, $\mathrm{v}_{\overline{P}_9}(z\mathcal{O}_K) = b_2+b_{40}+b_{44}+b_{53}+b_{55} = d_9$,
$\mathrm{v}_{P_{10}}(z\mathcal{O}_K) = b_{29}+b_{38}+b_{45} = d_{10}$, $\mathrm{v}_{\overline{P}_{10}}(z\mathcal{O}_K) = b_8+b_{13}+2b_{34} = d_{10}$,
$\mathrm{v}_{P_{11}}(z\mathcal{O}_K) = b_{26}+b_{35}+b_{42} = d_{11}$, $\mathrm{v}_{\overline{P}_{11}}(z\mathcal{O}_K) = b_3+b_{54} = d_{11}$,
$\mathrm{v}_{P_{12}}(z\mathcal{O}_K) = b_{35}+b_{37}+b_{54} = d_{12}$, $\mathrm{v}_{\overline{P}_{12}}(z\mathcal{O}_K) = b_4+b_5+b_7+b_{20}+b_{25}+b_{26} = d_{12}$,
$\mathrm{v}_{P_{13}}(z\mathcal{O}_K) = b_{23}+b_{29}+b_{32}+b_{42} = d_{13}$, $\mathrm{v}_{\overline{P}_{13}}(z\mathcal{O}_K) = b_{19}+b_{53} = d_{13}$,
$\mathrm{v}_{P_{14}}(z\mathcal{O}_K) = b_6+b_{10}+b_{39} = d_{14}$, $\mathrm{v}_{\overline{P}_{14}}(z\mathcal{O}_K) = b_{54} = d_{14}$,
$\mathrm{v}_{P_{15}}(z\mathcal{O}_K) = b_{13}+b_{28}+b_{34}+b_{43}+b_{46} = d_{15}$, $\mathrm{v}_{\overline{P}_{15}}(z\mathcal{O}_K) = b_3+2b_{19}+2b_{38} = d_{15}$,
$\mathrm{v}_{P_{16}}(z\mathcal{O}_K) = b_{29} = d_{16}$, $\mathrm{v}_{\overline{P}_{16}}(z\mathcal{O}_K) = b_2+b_{31}+b_{51} = d_{16}$,
$\mathrm{v}_{P_{17}}(z\mathcal{O}_K) = b_{12}+b_{55} = d_{17}$, $\mathrm{v}_{\overline{P}_{17}}(z\mathcal{O}_K) = b_{30}+b_{31}+b_{56} = d_{17}$,
$\mathrm{v}_{P_{18}}(z\mathcal{O}_K) = b_{16}+b_{25}+b_{27} = d_{18}$, $\mathrm{v}_{\overline{P}_{18}}(z\mathcal{O}_K) = b_{21}+b_{23}+b_{39}+b_{45} = d_{18}$,
$\mathrm{v}_{P_{19}}(z\mathcal{O}_K) = b_{26}+b_{29} = d_{19}$, $\mathrm{v}_{\overline{P}_{19}}(z\mathcal{O}_K) = b_9+b_{21}+b_{33} = d_{19}$,
$\mathrm{v}_{P_{20}}(z\mathcal{O}_K) = b_{39} = d_{20}$, $\mathrm{v}_{\overline{P}_{20}}(z\mathcal{O}_K) = 2b_{22} = d_{20}$,
$\mathrm{v}_{P_{21}}(z\mathcal{O}_K) = b_{31}+b_{44}+b_{46} = d_{21}$, $\mathrm{v}_{\overline{P}_{21}}(z\mathcal{O}_K) = b_{13}+2b_{18}+b_{41} = d_{21}$,
$\mathrm{v}_{P_{22}}(z\mathcal{O}_K) = 2b_{23} = d_{22}$, $\mathrm{v}_{\overline{P}_{22}}(z\mathcal{O}_K) = b_{28}+b_{52}+b_{57} = d_{22}$,
$\mathrm{v}_{P_{23}}(z\mathcal{O}_K) = b_{14}+b_{16}+b_{37}+b_{44} = d_{23}$, $\mathrm{v}_{\overline{P}_{23}}(z\mathcal{O}_K) = b_6+b_{28}+b_{39}+b_{42} = d_{23}$,
$\mathrm{v}_{P_{24}}(z\mathcal{O}_K) = b_{30} = d_{24}$, $\mathrm{v}_{\overline{P}_{24}}(z\mathcal{O}_K) = b_{50} = d_{24}$,
$\mathrm{v}_{P_{25}}(z\mathcal{O}_K) = b_{13}+b_{28}+b_{40} = d_{25}$, $\mathrm{v}_{\overline{P}_{25}}(z\mathcal{O}_K) = b_{11}+b_{52} = d_{25}$,
$\mathrm{v}_{P_{26}}(z\mathcal{O}_K) = b_{32}+b_{36} = d_{26}$, $\mathrm{v}_{\overline{P}_{26}}(z\mathcal{O}_K) = b_{20}+b_{24} = d_{26}$,

$\mathrm{v}_{P_{27}}(z\mathcal{O}_K) = b_{46} = d_{27}$, $\mathrm{v}_{\overline{P}_{27}}(z\mathcal{O}_K) = b_2 + b_{49} = d_{27}$,
$\mathrm{v}_{P_{28}}(z\mathcal{O}_K) = b_{22} + b_{36} + b_{55} = d_{28}$, $\mathrm{v}_{\overline{P}_{28}}(z\mathcal{O}_K) = b_{12} + b_{45} + b_{48} = d_{28}$,
$\mathrm{v}_{P_{29}}(z\mathcal{O}_K) = b_{21} + b_{27} = d_{29}$, $\mathrm{v}_{\overline{P}_{29}}(z\mathcal{O}_K) = b_{36} = d_{29}$,
$\mathrm{v}_{P_{30}}(z\mathcal{O}_K) = b_{25} + b_{30} + b_{49} = d_{30}$, $\mathrm{v}_{\overline{P}_{30}}(z\mathcal{O}_K) = b_{14} + b_{17} + b_{34} + b_{41} + b_{47} = d_{30}$,
$\mathrm{v}_{P_{31}}(z\mathcal{O}_K) = b_7 + b_{17} = d_{31}$, $\mathrm{v}_{\overline{P}_{31}}(z\mathcal{O}_K) = b_{45} + b_{55} = d_{31}$,
$\mathrm{v}_{P_{32}}(z\mathcal{O}_K) = b_{43} + b_{45} = d_{32}$, $\mathrm{v}_{\overline{P}_{32}}(z\mathcal{O}_K) = b_{36} = d_{32}$,
$\mathrm{v}_{P_{33}}(z\mathcal{O}_K) = b_{18} + b_{27} + b_{44} = d_{33}$, $\mathrm{v}_{\overline{P}_{33}}(z\mathcal{O}_K) = b_{16} + b_{24} = d_{33}$,
$\mathrm{v}_{P_{34}}(z\mathcal{O}_K) = b_{33} + b_{44} = d_{34}$, $\mathrm{v}_{\overline{P}_{34}}(z\mathcal{O}_K) = b_{15} + b_{17} = d_{34}$,
$\mathrm{v}_{P_{35}}(z\mathcal{O}_K) = b_6 = d_{35}$, $\mathrm{v}_{\overline{P}_{35}}(z\mathcal{O}_K) = b_{23} = d_{35}$,
$\mathrm{v}_{P_{36}}(z\mathcal{O}_K) = b_{26} + 2b_{31} = d_{36}$, $\mathrm{v}_{\overline{P}_{36}}(z\mathcal{O}_K) = b_{11} + b_{34} = d_{36}$,
$\mathrm{v}_{P_{37}}(z\mathcal{O}_K) = b_4 + b_{10} + b_{15} + b_{18} + b_{19} = d_{37}$, $\mathrm{v}_{\overline{P}_{37}}(z\mathcal{O}_K) = b_{11} = d_{37}$,
$\mathrm{v}_{P_{38}}(z\mathcal{O}_K) = b_{43} = d_{38}$, $\mathrm{v}_{\overline{P}_{38}}(z\mathcal{O}_K) = b_7 + b_{21} = d_{38}$,
$\mathrm{v}_{P_{39}}(z\mathcal{O}_K) = b_5 + b_9 + b_{35} + b_{42} + b_{46} + b_{48} + b_{55} = d_{39}$, $\mathrm{v}_{\overline{P}_{39}}(z\mathcal{O}_K) = b_3 + b_{38} + b_{53} = d_{39}$,
$\mathrm{v}_{P_{40}}(z\mathcal{O}_K) = b_4 + b_{47} + b_{57} = d_{40}$, $\mathrm{v}_{\overline{P}_{40}}(z\mathcal{O}_K) = b_{38} = d_{40}$,
$\mathrm{v}_{P_{41}}(z\mathcal{O}_K) = b_{20} = d_{41}$, $\mathrm{v}_{\overline{P}_{41}}(z\mathcal{O}_K) = b_4 = d_{41}$,
$\mathrm{v}_{P_{42}}(z\mathcal{O}_K) = b_{52} = d_{42}$, $\mathrm{v}_{\overline{P}_{42}}(z\mathcal{O}_K) = b_{30} = d_{42}$,
$\mathrm{v}_{P_{43}}(z\mathcal{O}_K) = b_1 = d_{43}$, $\mathrm{v}_{\overline{P}_{43}}(z\mathcal{O}_K) = b_{16} = d_{43}$,
$\mathrm{v}_{P_{44}}(z\mathcal{O}_K) = b_{13} + b_{51} = d_{44}$, $\mathrm{v}_{\overline{P}_{44}}(z\mathcal{O}_K) = b_{12} = d_{44}$,
$\mathrm{v}_{P_{45}}(z\mathcal{O}_K) = b_{51} = d_{45}$, $\mathrm{v}_{\overline{P}_{45}}(z\mathcal{O}_K) = b_{18} + b_{40} = d_{45}$,
$\mathrm{v}_{P_{46}}(z\mathcal{O}_K) = b_7 + b_{10} + b_{32} + b_{33} + b_{35} + b_{49} + b_{52} = d_{46}$, $\mathrm{v}_{\overline{P}_{46}}(z\mathcal{O}_K) = b_2 + b_9 + b_{42} = d_{46}$,
$\mathrm{v}_{P_{47}}(z\mathcal{O}_K) = b_6 + b_{50} + b_{54} = d_{47}$, $\mathrm{v}_{\overline{P}_{47}}(z\mathcal{O}_K) = b_{43} + 2b_{56} = d_{47}$,
$\mathrm{v}_{P_{48}}(z\mathcal{O}_K) = b_9 = d_{48}$, $\mathrm{v}_{\overline{P}_{48}}(z\mathcal{O}_K) = b_5 + b_{12} = d_{48}$,
$\mathrm{v}_{P_{49}}(z\mathcal{O}_K) = b_{27} + b_{41} = d_{49}$, $\mathrm{v}_{\overline{P}_{49}}(z\mathcal{O}_K) = b_{14} + b_{40} + b_{48} = d_{49}$,
$\mathrm{v}_{P_{50}}(z\mathcal{O}_K) = b_{57} = d_{50}$, $\mathrm{v}_{\overline{P}_{50}}(z\mathcal{O}_K) = b_{24} + b_{49} = d_{50}$,
$\mathrm{v}_{P_{51}}(z\mathcal{O}_K) = b_5 + b_{22} = d_{51}$, $\mathrm{v}_{\overline{P}_{51}}(z\mathcal{O}_K) = b_{41} = d_{51}$,
$\mathrm{v}_{P_{52}}(z\mathcal{O}_K) = b_3 + b_{17} + b_{21} + b_{32} + b_{48} = d_{52}$, $\mathrm{v}_{\overline{P}_{52}}(z\mathcal{O}_K) = b_8 + b_{10} = d_{52}$,
$\mathrm{v}_{P_{53}}(z\mathcal{O}_K) = b_3 + b_{11} = d_{53}$, $\mathrm{v}_{\overline{P}_{53}}(z\mathcal{O}_K) = b_1 + b_{14} + b_{51} + b_{57} = d_{53}$,
$\mathrm{v}_{P_{54}}(z\mathcal{O}_K) = b_8 + b_{39} = d_{54}$, $\mathrm{v}_{\overline{P}_{54}}(z\mathcal{O}_K) = b_{25} = d_{54}$,
$\mathrm{v}_{P_{55}}(z\mathcal{O}_K) = b_{15} + b_{28} = d_{55}$, $\mathrm{v}_{\overline{P}_{55}}(z\mathcal{O}_K) = b_{10} + b_{35} = d_{55}$,
$\mathrm{v}_{P_{56}}(z\mathcal{O}_K) = b_{37} = d_{56}$, $\mathrm{v}_{\overline{P}_{56}}(z\mathcal{O}_K) = b_{26} + b_{47} = d_{56}$.

The equations in part E show that $\mathrm{v}_{P_1}(z\mathcal{O}_K) = 2d_1 = \mathrm{v}_{P_1}(n\mathcal{O}_K)$ and $\mathrm{v}_{P_j}(z\mathcal{O}_K) = \mathrm{v}_{\overline{P}_j}(z\mathcal{O}_K) = d_j = \mathrm{v}_{\overline{P}_j}(n\mathcal{O}_K) = \mathrm{v}_{P_j}(n\mathcal{O}_K)$ for each $j \in [2, s]$, and thus $\eta\mathcal{O}_K = \mathcal{O}_K$. We infer that $\eta \in \mathcal{O}_K^\times$. □(Claim 2)

CLAIM 3. $\eta \in \mathcal{O}_d^\times$.

Observe that $d > c_j$ and $c_j \in \mathbb{P}$ for each $j \in [1, s]$. It follows that $d \nmid n$ (since $d \in \mathbb{P}$ by Claim 1). Since $z \in \mathcal{O}_K$, there are some $g, h \in \mathbb{Z}$ such that $z = g + h\sqrt{d}$. Moreover, since $\eta \in \mathcal{O}_K$ (by Claim 2) and $\{1, \sqrt{d}\}$ is a $\mathbb{Z}$-basis of $\mathcal{O}_K$, we obtain that $n \mid g$, $n \mid h$ and $\eta = \frac{g}{n} + \frac{h}{n}\sqrt{d}$. Clearly, $\mathcal{O}_K^\times \cap \mathcal{O}_d = \mathcal{O}_d^\times$. Since $d \nmid n$, $d \in \mathbb{P}$ and $\eta \in \mathcal{O}_K^\times$ (by Claim 2), we infer that $\eta \in \mathcal{O}_d^\times$ if and only if $\eta \in \mathcal{O}_d$ if and only if $d \mid \frac{h}{n}$ if and only if $d \mid h$ if and only if $z \in \mathcal{O}_d$. Consequently, it remains to show that $z \in \mathcal{O}_d$.

Set $z' = \prod_{i=1}^r \left(a_i + b_i\sqrt{d}\right)$. Since $\left(u + \sqrt{d}\right)^v \equiv u^{v-1}\left(u + v\sqrt{d}\right) \mod d\mathcal{O}_K$ for all $u \in \mathbb{Z}$ and $v \in \mathbb{N}$, there is some $t \in \mathbb{Z}$ such that $z \equiv tz' \mod d\mathcal{O}_K$. For this reason, it is sufficient to show that $z' \in \mathcal{O}_d$. Next we determine $\prod_{i=1}^k \left(a_i + b_i\sqrt{d}\right)$ mod $d\mathcal{O}_K$ step-by-step for each $k \in [2, r]$. We use that $\left(v + w\sqrt{d}\right)\left(v' + w'\sqrt{d}\right) \equiv vv' + (wv' + vw')\sqrt{d} \mod d\mathcal{O}_K$ for each $v, v', w, w' \in \mathbb{Z}$. Note that the items below consist of two equations that give us the reductions of u_k and v_k modulo d (for each $k \in [2, r]$), where $u_k, v_k \in \mathbb{Z}$ are such that $\prod_{i=1}^k \left(a_i + b_i\sqrt{d}\right) = u_k + v_k\sqrt{d}$.

- $a_1a_2 = 6890530871592$,
  $b_1a_2 + a_1b_2 = -d + 21554509784529$.
- $6890530871592a_3 = -315587d + 28244816980645$,
  $21554509784529a_3 + 6890530871592b_3 = -972567d + 5178868509861$.

- $28244816980645a_4 = -1272650d + 15513835265240$,
  $5178868509861a_4 + 28244816980645b_4 = 2697384d + 4467486423321$.
- $15513835265240a_5 = -675500d + 35811027000980$,
  $4467486423321a_5 + 15513835265240b_5 = 652149d + 30537129644761$.
- $35811027000980a_6 = -1309781d + 20714594938939$,
  $30537129644761a_6 + 35811027000980b_6 = 6236863d + 16497568105021$.
- $20714594938939a_7 = -441941d + 21223637681572$,
  $16497568105021a_7 + 20714594938939b_7 = 28697d + 18837954382492$.
- $21223637681572a_8 = -407041d + 13338881989779$,
  $18837954382492a_8 + 21223637681572b_8 = 4721606d + 17458165870498$.
- $13338881989779a_9 = -230639d + 3661427934964$,
  $17458165870498a_9 + 13338881989779b_9 = 5070046d + 17002573835617$.
- $3661427934964a_{10} = -52048d + 15123266299360$,
  $17002573835617a_{10} + 3661427934964b_{10} = 130052d + 34143299535672$.
- $15123266299360a_{11} = -202084d + 22867483171996$,
  $34143299535672a_{11} + 15123266299360b_{11} = 6208881d + 10414810162433$.
- $22867483171996a_{12} = -282499d + 8083910734589$,
  $10414810162433a_{12} + 22867483171996b_{12} = 7832662d + 17492146626900$.
- $8083910734589a_{13} = -92186d + 37797607362521$,
  $17492146626900a_{13} + 8083910734589b_{13} = 1575020d + 23141664104777$.
- $37797607362521a_{14} = -411842d + 28944089196110$,
  $23141664104777a_{14} + 37797607362521b_{14} = 1939038d + 6609389335501$.
- $28944089196110a_{15} = -239446d + 16720583049264$,
  $6609389335501a_{15} + 28944089196110b_{15} = 2568152d + 34718516820815$.
- $16720583049264a_{16} = -120376d + 6912281518232$,
  $34718516820815a_{16} + 16720583049264b_{16} = 1211229d + 10227378309170$.
- $6912281518232a_{17} = -44790d + 5195414457466$,
  $10227378309170a_{17} + 6912281518232b_{17} = 741048d + 289652693936$.
- $5195414457466a_{18} = -30152d + 22465854532740$,
  $289652693936a_{18} + 5195414457466b_{18} = 265509d + 17460040771835$.
- $22465854532740a_{19} = -123878d + 18660863010082$,
  $17460040771835a_{19} + 22465854532740b_{19} = 2001631d + 19162392372661$.
- $18660863010082a_{20} = -99441d + 31621822715453$,
  $19162392372661a_{20} + 18660863010082b_{20} = 1834171d + 26774899887696$.
- $31621822715453a_{21} = -167069d + 18945559551357$,
  $26774899887696a_{21} + 31621822715453b_{21} = 4282685d + 14809237415832$.
- $18945559551357a_{22} = -93807d + 25490759313159$,
  $14809237415832a_{22} + 18945559551357b_{22} = 1035530d + 16642252647821$.
- $25490759313159a_{23} = -82380d + 31410566401668$,
  $16642252647821a_{23} + 25490759313159b_{23} = 5180714d + 7665156542464$.
- $31410566401668a_{24} = -97899d + 24750475711101$,
  $7665156542464a_{24} + 31410566401668b_{24} = 4386896d + 12345706811136$.
- $24750475711101a_{25} = -34513d + 4341133124105$,
  $12345706811136a_{25} + 24750475711101b_{25} = 8807546d + 33705248273090$.
- $4341133124105a_{26} = -5385d + 20420906899575$,
  $33705248273090a_{26} + 4341133124105b_{26} = 888730d + 19138455539665$.
- $20420906899575a_{27} = -21329d + 11628413963866$,
  $19138455539665a_{27} + 20420906899575b_{27} = 432256d + 14371072675706$.
- $11628413963866a_{28} = -10322d + 27536499676132$,
  $14371072675706a_{28} + 11628413963866b_{28} = 2230673d + 20483449818325$.
- $27536499676132a_{29} = -16214d + 11871313871146$,
  $20483449818325a_{29} + 27536499676132b_{29} = 10491877d + 23841147446157$.
- $11871313871146a_{30} = 1168d + 36708968638606$,
  $23841147446157a_{30} + 11871313871146b_{30} = 583202d + 29824187122383$.
- $36708968638606a_{31} = 27670d + 35292993605984$,
  $29824187122383a_{31} + 36708968638606b_{31} = 6287182d + 30828612734725$.
- $35292993605984a_{32} = 33791d + 32099367907223$,
  $30828612734725a_{32} + 35292993605984b_{32} = 2928174d + 15667131309718$.
- $32099367907223a_{33} = 31502d + 8686497691888$,
  $15667131309718a_{33} + 32099367907223b_{33} = 1722251d + 13069916858845$.
- $8686497691888a_{34} = 17842d + 32178145784978$,
  $13069916858845a_{34} + 8686497691888b_{34} = 1025368d + 38021671372563$.
- $32178145784978a_{35} = 68542d + 38082283375434$,
  $38021671372563a_{35} + 32178145784978b_{35} = 5937837d + 11732682959923$.
- $38082283375434a_{36} = 100145d + 35965182833267$,
  $11732682959923a_{36} + 38082283375434b_{36} = 6202247d + 37211608563518$.
- $35965182833267a_{37} = 106375d + 33169240097520$,
  $37211608563518a_{37} + 35965182833267b_{37} = 8936948d + 26020836879171$.
- $33169240097520a_{38} = 99259d + 17708055977779$,
  $26020836879171a_{38} + 33169240097520b_{38} = 10150666d + 32544641385778$.
- $17708055977779a_{39} = 70781d + 30770548150680$,
  $32544641385778a_{39} + 17708055977779b_{39} = 2202939d + 36915064492287$.
- $30770548150680a_{40} = 155955d + 11904176271315$,
  $36915064492287a_{40} + 30770548150680b_{40} = 2562988d + 165258901717$.
- $11904176271315a_{41} = 64624d + 25508831080099$,
  $165258901717a_{41} + 11904176271315b_{41} = 1347305d + 5907415462461$.
- $25508831080099a_{42} = 143454d + 872140076252$,
  $5907415462461a_{42} + 25508831080099b_{42} = 790654d + 34225185690977$.
- $872140076252a_{43} = 5137d + 10987048845161$,
  $34225185690977a_{43} + 872140076252b_{43} = 339647d + 26138731918415$.
- $10987048845161a_{44} = 74345d + 8152289751913$,
  $26138731918415a_{44} + 10987048845161b_{44} = 1871482d + 38377823723539$.

- $8152289751913a_{45} = 56296d + 17399000858672,$
  $38377823723539a_{45} + 8152289751913b_{45} = 295750d + 32817075202061.$
- $17399000858672a_{46} = 140874d + 13424492118010,$
  $32817075202061a_{46} + 17399000858672b_{46} = 2189576d + 33163475056782.$
- $13424492118010a_{47} = 116689d + 22068166729399,$
  $33163475056782a_{47} + 13424492118010b_{47} = 705436d + 24070914532262.$
- $22068166729399a_{48} = 266604d + 20874064459588,$
  $24070914532262a_{48} + 22068166729399b_{48} = 292240d + 15318085049644.$
- $20874064459588a_{49} = 267012d + 25167892634080,$
  $15318085049644a_{49} + 20874064459588b_{49} = 789145d + 30753673516637.$
- $25167892634080a_{50} = 384789d + 20538358542749,$
  $30753673516637a_{50} + 25167892634080b_{50} = 1701637d + 23065592907429.$
- $20538358542749a_{51} = 591526d + 21236595367747,$
  $23065592907429a_{51} + 20538358542749b_{51} = 3306393d + 31259701520158.$
- $21236595367747a_{52} = 628169d + 663951808753,$
  $31259701520158a_{52} + 21236595367747b_{52} = 1963739d + 37377022803680.$
- $663951808753a_{53} = 20667d + 11352564579816,$
  $37377022803680a_{53} + 663951808753b_{53} = 1565314d + 35652369557844.$
- $11352564579816a_{54} = 408337d + 12677902910401,$
  $35652369557844a_{54} + 11352564579816b_{54} = 6095135d + 24252482581547.$
- $12677902910401a_{55} = 632652d + 37493335194470,$
  $24252482581547a_{55} + 12677902910401b_{55} = 2876153d + 28715685383254.$
- $37493335194470a_{56} = 1912828d + 27624385992608,$
  $28715685383254a_{56} + 37493335194470b_{56} = 11211960d + 33470792628624.$
- $27624385992608a_{57} = 1505206d + 34038147456710,$
  $33470792628624a_{57} + 27624385992608b_{57} = 6487920d.$

Therefore, $z' \equiv 34038147456710 \mod d\mathcal{O}_K$, and thus $z' \in \mathcal{O}_d$. □(Claim 3)

CLAIM 4. $\eta \neq 1$.

First, we show that $z \equiv \sqrt{d} \mod 3\mathcal{O}_K$. We have that $d \equiv 2 \mod 3\mathcal{O}_K$ and $\left(0+\sqrt{d}\right)^2 \equiv \left(1+\sqrt{d}\right)^4 \equiv \left(2+\sqrt{d}\right)^4 \equiv 2 \mod 3\mathcal{O}_K$. Moreover, $\left(0+\sqrt{d}\right)^4 \equiv \left(1+\sqrt{d}\right)^8 \equiv \left(2+\sqrt{d}\right)^8 \equiv 1 \mod 3\mathcal{O}_K$. Using this, we infer that

$$
\begin{aligned}
z \equiv & \left(0+\sqrt{d}\right)^{3}\left(2+\sqrt{d}\right)^{6}\left(0+\sqrt{d}\right)^{2}\left(1+\sqrt{d}\right)^{3}\left(2+\sqrt{d}\right)^{6}\left(0+\sqrt{d}\right)^{2}\left(0+\sqrt{d}\right)^{0} \\
& \left(1+\sqrt{d}\right)^{2}\left(0+\sqrt{d}\right)^{3}\left(2+\sqrt{d}\right)^{6}\left(2+\sqrt{d}\right)^{2}\left(0+\sqrt{d}\right)^{1}\left(2+\sqrt{d}\right)^{5}\left(2+\sqrt{d}\right)^{7} \\
& \left(2+\sqrt{d}\right)^{1}\left(1+\sqrt{d}\right)^{3}\left(2+\sqrt{d}\right)^{2}\left(2+\sqrt{d}\right)^{3}\left(0+\sqrt{d}\right)^{1}\left(2+\sqrt{d}\right)^{3}\left(1+\sqrt{d}\right)^{7} \\
& \left(0+\sqrt{d}\right)^{3}\left(1+\sqrt{d}\right)^{2}\left(1+\sqrt{d}\right)^{4}\left(1+\sqrt{d}\right)^{0}\left(0+\sqrt{d}\right)^{3}\left(1+\sqrt{d}\right)^{3}\left(0+\sqrt{d}\right)^{3} \\
& \left(0+\sqrt{d}\right)^{1}\left(0+\sqrt{d}\right)^{3}\left(1+\sqrt{d}\right)^{7}\left(0+\sqrt{d}\right)^{1}\left(1+\sqrt{d}\right)^{2}\left(1+\sqrt{d}\right)^{3}\left(1+\sqrt{d}\right)^{2} \\
& \left(0+\sqrt{d}\right)^{2}\left(1+\sqrt{d}\right)^{7}\left(2+\sqrt{d}\right)^{1}\left(1+\sqrt{d}\right)^{6}\left(2+\sqrt{d}\right)^{5}\left(1+\sqrt{d}\right)^{5}\left(2+\sqrt{d}\right)^{3} \\
& \left(2+\sqrt{d}\right)^{3}\left(1+\sqrt{d}\right)^{1}\left(1+\sqrt{d}\right)^{7}\left(2+\sqrt{d}\right)^{0}\left(0+\sqrt{d}\right)^{0}\left(1+\sqrt{d}\right)^{4}\left(1+\sqrt{d}\right)^{2} \\
& \left(2+\sqrt{d}\right)^{7}\left(0+\sqrt{d}\right)^{0}\left(2+\sqrt{d}\right)^{7}\left(0+\sqrt{d}\right)^{2}\left(2+\sqrt{d}\right)^{6}\left(2+\sqrt{d}\right)^{7}\left(2+\sqrt{d}\right)^{6} \\
& \left(2+\sqrt{d}\right)^{6} \\
\equiv & \left(0+\sqrt{d}\right)^{2}\left(1+\sqrt{d}\right)^{6}\left(2+\sqrt{d}\right)^{4} \\
\equiv & 2^3\left(1+\sqrt{d}\right)^{2} \\
\equiv & \sqrt{d} \mod 3\mathcal{O}_K.
\end{aligned}
$$

Since $z \equiv \sqrt{d} \mod 3\mathcal{O}_K$, it follows that $z \notin \mathbb{Z}$, and hence $\eta \neq 1$. □(Claim 4)

It remains to show that $d \mid y$ (where y was defined at the beginning of Section 3). Observe that for each $j \in [1,r]$, $1 \le -3033477 + \sqrt{d} \le a_j + \sqrt{d} \le 2126573 + \sqrt{d} \le 2^{100}$. Moreover, $\max\{b_i \mid i \in [1,r]\} = 23621570 \le 10^8$ and $r \le 100$, and thus

$$\eta \le \prod_{i=1}^{r} \left(a_i + \sqrt{d}\right)^{b_i} \le \prod_{i=1}^{r} \left(2^{100}\right)^{10^8} = 2^{10^{10} r} \le 2^{10^{12}} < 2^d \le \left(1 + \sqrt{d}\right)^d \le \varepsilon^d.$$

Also note that for each $j \in [1,s]$, $c_j \le 2^{10}$. Furthermore, $\max\{d_i \mid i \in [1,s]\} = 146634276 \le 10^9$ and $s \le 100$, and hence

$$\eta^{-1} \le \prod_{j=1}^{s} c_j^{d_j} \le \prod_{j=1}^{s} \left(2^{10}\right)^{10^9} = 2^{10^{10} s} \le 2^{10^{12}} < 2^d \le \left(1 + \sqrt{d}\right)^d \le \varepsilon^d.$$

It follows from Claims 2 and 4 that $\eta \in \mathcal{O}_K^\times \setminus \{1\}$. Since $\varepsilon^{-d} < \eta < \varepsilon^d$, there is some $k \in \mathbb{Z} \setminus \{0\}$ such that $-d < k < d$ and $\eta = \varepsilon^k$. Set $\operatorname{ord}\left(\varepsilon\mathcal{O}_d^\times\right) = \min\{m \in \mathbb{N} \mid \left(\varepsilon\mathcal{O}_d^\times\right)^m = \mathcal{O}_d^\times\}$. Observe that $\operatorname{ord}\left(\varepsilon\mathcal{O}_d^\times\right) = \left(\mathcal{O}_K^\times : \mathcal{O}_d^\times\right) \mid d$ (because $\mathcal{O}_K^\times/\mathcal{O}_d^\times$ is generated by $\varepsilon\mathcal{O}_d^\times$ and $d \in \mathbb{P}$ is ramified in $\mathcal{O}_K$). Since $\left(\varepsilon\mathcal{O}_d^\times\right)^k = \eta\mathcal{O}_d^\times = \mathcal{O}_d^\times$ by Claim 3, we have that $\operatorname{ord}\left(\varepsilon\mathcal{O}_d^\times\right) \mid k$, and thus $\left(\mathcal{O}_K^\times : \mathcal{O}_d^\times\right) = 1$ (since $d \in \mathbb{P}$). We infer that $\varepsilon \in \mathcal{O}_d$, and hence $d \mid y$. □(Theorem 1.3)

Acknowledgment

We want to thank A. Geroldinger for helpful remarks and comments that improved this note.

References

[1] T. Agoh, *Congruences related to the Ankeny-Artin-Chowla conjecture*, Integers **16** (2016), Paper No. A12, 30. MR3475253

[2] N. C. Ankeny, E. Artin, and S. Chowla, *The class-number of real quadratic number fields*, Ann. of Math. (2) **56** (1952), 479–493, DOI 10.2307/1969656. MR49948

[3] Y. Benmerieme and A. Movahhedi, *Ankeny-Artin-Chowla and Mordell conjectures in terms of p-rationality*, J. Number Theory **257** (2024), 202–214, DOI 10.1016/j.jnt.2023.10.013. MR4673311

[4] J. Brillhart, D. H. Lehmer, and J. L. Selfridge, *New primality criteria and factorizations of* $2^m \pm 1$, Math. Comp. **29** (1975), 620–647, DOI 10.2307/2005583. MR384673

[5] D. Chakraborty and A. Saikia, *On a conjecture of Mordell*, Rocky Mountain J. Math. **49** (2019), no. 8, 2545–2556, DOI 10.1216/RMJ-2019-49-8-2545. MR4058336

[6] A. Geroldinger and F. Halter-Koch, *Non-unique factorizations*, Pure and Applied Mathematics (Boca Raton), vol. 278, Chapman & Hall/CRC, Boca Raton, FL, 2006. Algebraic, combinatorial and analytic theory, DOI 10.1201/9781420003208. MR2194494

[7] J. Harrington and L. Jones, *A new condition equivalent to the Ankeny-Artin-Chowla conjecture*, J. Number Theory **192** (2018), 240–250, DOI 10.1016/j.jnt.2018.04.011. MR3841554

[8] R. Hashimoto, *Ankeny-Artin-Chowla conjecture and continued fraction expansion*, J. Number Theory **90** (2001), no. 1, 143–153, DOI 10.1006/jnth.2001.2652. MR1850878

[9] W. Hassler, *Direct-sum cancellation for modules over real quadratic orders*, J. Pure Appl. Algebra **208** (2007), no. 2, 575–589, DOI 10.1016/j.jpaa.2006.03.001. MR2277697

[10] A. A. Kiselev and I. Š. Slavutskiĭ, *On the number of classes of ideals of a quadratic field and its rings* (Russian), Dokl. Akad. Nauk SSSR **126** (1959), 1191–1194. MR122808

[11] H. W. Lu, *Hecke operators and Pellian equation conjecture. II*, Chinese Ann. Math. Ser. B **13** (1992), no. 1, 46–53. A Chinese summary appears in Chinese Ann. Math. Ser. A **13** (1992), no. 1, 132. MR1166874

[12] H. W. Lu, *Hecke operator and Pellian equation conjecture. III*, Systems Sci. Math. Sci. **4** (1991), no. 2, 97–103. MR1119285

[13] L. J. Mordell, *On a Pellian equation conjecture*, Acta Arith. **6** (1960), 137–144, DOI 10.4064/aa-6-2-137-144. MR118699

[14] L. J. Mordell, *On a Pellian equation conjecture. II*, J. London Math. Soc. **36** (1961), 282–288, DOI 10.1112/jlms/s1-36.1.282. MR126411

[15] H. C. Pocklington, *The determination of the prime or composite nature of large numbers by Fermat's theorem*, Proc. Cambridge Philos. Soc. **18** (1914), 29–30.

[16] A. Reinhart, *On orders in quadratic number fields with unusual sets of distances*, Acta Arith. **211** (2023), no. 1, 61–92, DOI 10.4064/aa230515-4-10. MR4666668

[17] A. Reinhart, *A counterexample to the Pellian equation conjecture of Mordell*, Acta Arith. **215** (2024), no. 1, 85–95, DOI 10.4064/aa240214-3-4. MR4772273

[18] A. Reinhart, *A counterexample to the Conjecture of Ankeny, Artin, and Chowla*, Preprint, arXiv:2410.21864, (2024).

[19] S. V. Sidorov and P. A. Shcherbakov, *On the period length modulo D of sequences of numerators and denominators of convergents for the square root of a non-square D*, Mathematical modeling and supercomputer technologies, Commun. Comput. Inf. Sci., vol. 1914, Springer, Cham, [2024] ©2024, pp. 28–43, DOI 10.1007/978-3-031-52470-7_3. MR4952580

[20] I. Sh. Slavutskiĭ, *A real quadratic field and the Ankeny-Artin-Chowla conjecture* (Russian, with Russian summary), Zap. Nauchn. Sem. S.-Peterburg. Otdel. Mat. Inst. Steklov. (POMI) **286** (2002), no. Anal. Teor. Chisel i Teor. Funkts. 18, 159–168, 230–231, DOI 10.1023/B:JOTH.0000035243.24060.83; English transl., J. Math. Sci. (N.Y.) **122** (2004), no. 6, 3673–3678. MR1937375

[21] A. J. Stephens and H. C. Williams, *Some computational results on a problem concerning powerful numbers*, Math. Comp. **50** (1988), no. 182, 619–632, DOI 10.2307/2008629. MR929558

[22] R. Wiegand, *Cancellation over commutative rings of dimension one and two*, J. Algebra **88** (1984), no. 2, 438–459, DOI 10.1016/0021-8693(84)90077-2. MR747527

Institut für Mathematik und Wissenschaftliches Rechnen, Karl-Franzens-Universität Graz, NAWI Graz, Heinrichstrasse 36, 8010 Graz, Austria

Email address: andreas.reinhart@uni-graz.at

Contemporary Mathematics
Volume **836**, 2026
https://doi.org/10.1090/conm/836/16709

A generalized Davenport constant of the second kind

Jared Kettinger

ABSTRACT. In this paper, we explore a ring invariant which is closely related to the Davenport constant of a group. In particular, we will calculate this invariant for a certain class of rings of integers and their orders and use it to understand factorization properties of the latter. To this end, we also examine the well-behaved class of Galois-invariant orders.

1. Introduction and notation

The Davenport constant was first introduced by H. Davenport at a conference on group theory and number theory in 1966 ([**8**]). Denoted $D(G)$ for a finite abelian group G, let us recall the following two equivalent definitions of this constant.

DEFINITION 1.1. $\max\{n \mid \exists$ a zero-sum sequence of length n with no non-empty proper zero-sum subsequence$\}$.

DEFINITION 1.2. $\min\{n \mid$ any G-sequence of length n must have a non-empty zero-sum subsequence$\}$.

While this constant has combinatorial interest in its own right, having been studied in this capacity by the likes of Erdös ([**2**]) and Olson ([**24**]), it was first posited due to its connection with factorization. In particular, if G is the class group of the ring of integers $\mathcal{O}_K$, and $\alpha \in \mathcal{O}_K$ an irreducible element, $D(G)$ is the maximum number of prime ideals (counting multiplicity) in the factorization of the principal ideal $\alpha\mathcal{O}_K$. While rings of integers fail to be unique factorization domains in general, they are Dedekind, meaning all nonzero proper ideals factor uniquely into prime ideals. Thus, understanding the prime factorization of a principal ideal $\beta\mathcal{O}_K$ should help us understand the factorization of the element $\beta \in \mathcal{O}_K$. The Davenport constant plays a central role in elucidating this connection.

Now, for a given atomic domain, we would like a way to characterize its factorization. With this in mind, we recall the following definition of elasticity.

DEFINITION 1.3. Let R be an atomic domain. The *elasticity* of a nonzero, nonunit $r \in R$ is defined as

$$\rho(r) = \sup\left\{\frac{n}{m} \mid r = \alpha_1\alpha_2\cdots\alpha_n = \beta_1\beta_2\cdots\beta_m\right\}$$

2020 *Mathematics Subject Classification.* Primary 13A05, 11R04; Secondary 11R65.
Key words and phrases. Factorization theory, Davenport constant, orders.
The author would like to thank his advisor Dr. James Coykendall.

©2026 American Mathematical Society

where $\alpha_i, \beta_j \in Irr(R)$ for all i, j.

Similarly, the *elasticity* of the domain R is defined as

$$\rho(R) = \sup\left\{\frac{n}{m} \mid \alpha_1\alpha_2\cdots\alpha_n = \beta_1\beta_2\cdots\beta_m\right\}$$

where $\alpha_i, \beta_j \in Irr(R)$ for all i, j.

Elasticity essentially serves as a measure of the failure of unique factorization in a domain—in particular, unique factorization length. It has long been known that the class number of a ring of integers also serves as such a measure. For example, a ring of integers is a unique factorization domain (UFD) if and only if it has class number $h = 1$. Carlitz ([**3**]) further showed that a ring of integers R is a half-factorial domain (HFD), meaning $\rho(R) = 1$, if and only if $h \leq 2$. The direct relationship between the Davenport constant and factorization is best illustrated by the following result due to Narkiewicz ([**22**]) and Valenza ([**26**]).

THEOREM 1.4 ([**22**]). *Let R be a ring of integers with class number $h > 1$. Then*

$$\rho(R) = \frac{D(Cl(R))}{2}.$$

Note that if $h = 1$, R is a UFD and thus $\rho(R) = 1$.

Now, a natural progression from this result would be to study the factorization of orders $\mathcal{O}$ in the rings of integers $\mathcal{O}_K$. However, any proper order of a ring of integers fails to be Dedekind. In particular, only those prime ideals ideals prime to the conductor $\mathfrak{f}_{\mathcal{O}} = \{\alpha \in \mathcal{O}_K | \alpha\mathcal{O}_K \subseteq \mathcal{O}\}$ are invertible in $\mathcal{O}$ ([**6**]), so the Davenport constant of the generalized class group can only help us directly compute the elasticity of certain elements. Hence, we need to develop some more sophisticated tools to study factorization in these orders entirely. We would like to find a way to use our knowledge of factorization in the ring of integers above to give us insight into factorization in the order below. Just as Davenport leverages the factorization of the ideal $\beta\mathcal{O}_K$ to understand the factorization of the element $\beta \in \mathcal{O}_K$, we seek to leverage the factorization of the element β in $\mathcal{O}_K$ to understand its factorization in an order $\mathcal{O} \subseteq \mathcal{O}_K$. More generally, we might ask: how can we use our knowledge of factorization in a ring R to gain knowledge about factorization in a given subring?

Let us consider the simplest case where $\mathcal{O} \subseteq R$ is an integral extension of atomic domains with R a UFD and $U(R) = U(\mathcal{O})$. Then, for any nonzero, nonunit $x \in \mathcal{O}$, we may consider it also as an element of R where it enjoys a unique factorization into prime elements:

$$x = p_1p_2\cdots p_n.$$

Now, if $x = \pi_1\pi_2\cdots\pi_k$ is an irreducible factorization of x in $\mathcal{O}$, by uniqueness, each

$$\pi_i = \prod_{j\in A\subseteq[1,n]} p_j$$

up to a unit multiple. Hence, if we can determine how elements (in particular, prime elements) of R combine to produce elements of $\mathcal{O}$, we can completely determine the factorization properties of $\mathcal{O}$.

EXAMPLE 1.5. Ignoring for the moment the discrepancy in their unit groups, consider the Gaussian integers $\mathbb{Z}[i]$ and the order $\mathbb{Z}[7i] \subseteq \mathbb{Z}[i]$. $\mathbb{Z}[i]$ is a unique factorization domain, so the element 490 factors *uniquely* into primes as $490 =$

$7 \cdot 7(1+i)(1-i)(1+2i)(1-2i)$. Now, we want to group these elements to form irreducibles of $\mathbb{Z}[7i]$. One possible grouping is $[7][7][(1+i)(1-i)][(1+2i)(1-2i)] = 7 \cdot 7 \cdot 2 \cdot 5$. Another option is $[7(1+i)(1-2i)][7(1-i)(1+2i)] = (21-7i)(21+7i)$. The fact that these elements are all irreducible in $\mathbb{Z}[7i]$ follows from the fact that the prime factorization in $\mathbb{Z}[i]$ is unique, and there is no way to non-trivially partition the primes we have grouped together to give a finer factorization in $\mathbb{Z}[7i]$. For example, looking at $[7(1+i)(1-2i)]$, $(1+i)(1-2i) = 3-i \notin \mathbb{Z}[7i]$. Also, despite $7(1+i)$ and $7(1-2i)$ being in $\mathbb{Z}[7i]$, this leaves us with a factor which is not, namely $1+i$ or $1-2i$. Thus, we have irreducible factorizations of lengths 2 and 4, so $\rho(490) \geq \frac{4}{2} = 2$ in $\mathbb{Z}[7i]$.

With this example in mind, we present the following definitions.

DEFINITION 1.6. Let R be an atomic monoid and $\mathcal{O} \subseteq R$ a non-empty subset. We say an R-product $\pi_1\pi_2\cdots\pi_n$ of elements $\pi_i \in \operatorname{Irr}(R)$ is an *$\mathcal{O}$-product* if $\pi_1\pi_2\cdots\pi_n \in \mathcal{O}$.

DEFINITION 1.7. Let R be an atomic monoid and $\mathcal{O} \subseteq R$ a non-empty subset.

$$D_{\mathcal{O}}(R) = \min\{n \mid \text{any } R\text{-product of length } n \text{ has an } \mathcal{O}\text{-subproduct}\}$$

If no such n exists, we say $D_{\mathcal{O}}(R) = \infty$. In the case that $\mathcal{O}$ is a domain and R its integral closure, we will use $\bar{D}(\mathcal{O})$ to denote $D_{\mathcal{O}}(R)$.

We will refer to this as the generalized Davenport constant of R over $\mathcal{O}$. In this paper, we will focus almost exclusively on the case when R is a domain. Now, the use of $\mathcal{O}$ to denote the subset is suggestive in two ways. First, it hints that our primary objects of study will be orders of a ring of integers – in which case we will make extensive use of the $\bar{D}(\mathcal{O})$ notation. Second, it is reminiscent of the fact that the subset $\mathcal{O}$ is taking the place of 0 in the definition of the Davenport constant.

Returning to finite abelian groups, we can make a similar generalization. For an abelian group G and subgroup $H \leq G$, we call a G-sequence $g_1g_2\ldots g_n$ an H-sum sequence if $g_1+g_2+\cdots+g_n \in H$. Then we may allow $D_H(G)$ to denote the minimum n such that any G-sequence of length n must have an H-sum subsequence. This coincides with the study of $\mathcal{B}_H(G)$, the monoid of H-sum sequences, introduced by Halter-Koch in [**16**]. In our notation, we have the nice symmetry $D_H(G) = D(G/H)$ which can be found in [**1**]. We will see that *quotient rings* also play an important role in understanding $D_{\mathcal{O}}(R)$.

Now, our definition of $D_{\mathcal{O}}(R)$ clearly echoes the second definition of $D(G)$: $\min\{n \mid \text{any } G\text{-sequence of length } n \text{ must have a non-empty zero-sum subsequence}\}$. Thus, it is natural to ask if $D_{\mathcal{O}}(R)$ is equal to $\sup\{n \mid \exists \text{ an } \mathcal{O}\text{-product of length } n \text{ with no non-empty proper } \mathcal{O}\text{-subproduct}\}$ in parallel with $D(G)$. The following example demonstrates that these values are distinct.

EXAMPLE 1.8. Consider the ideal $5\mathbb{Z}[i] = (5) \subseteq \mathbb{Z}[i]$ which factors into prime ideals as $(2+i)(2-i)$. Hence, the element $(2+i)^n \notin (5)$ for any $n \in \mathbb{N}$. Thus, we have $D_{5\mathbb{Z}[i]}(\mathbb{Z}[i]) = \infty$. However, given any product $\pi_1\pi_2\cdots\pi_n \in 5\mathbb{Z}[i]$ with $\pi_i \in \operatorname{Irr}(R)$, as $\mathbb{Z}[i]$ is a UFD and 5 divides the product, we may assume without loss of generality that $\pi_1 = 2+i$ and $\pi_2 = 2-i$, so $\pi_1\pi_2 = 5 \in 5\mathbb{Z}[i]$. Note, this is technically up to a $U(\mathbb{Z}[i])$ multiple of 5, but these all lie in $5\mathbb{Z}[i]$. Therefore, the longest $5\mathbb{Z}[i]$-product with no proper $5\mathbb{Z}[i]$-subproduct has length 2.

This example is also significant as it distinguishes this generalized Davenport constant from the omega invariant. First introduced by Geroldinger in [**9**], for a integral domain R and nonzero nonunit $a \in R$, $\omega_R(a)$ denotes the smallest $n \in \mathbb{N}$ such that for any $b_1, b_2, \ldots, b_{n+1} \in R$ such that a divides $\prod_{i=1}^{n+1} b_i$, there exists a non-empty, proper subset $S \subset \{1, 2, \ldots, n+1\}$ such that a divides $\prod_{i\in S} b_i$. If no such n exists, we write $\omega_R(a) = \infty$. Subsequently, for an atomic domain R, we define $\omega(R) = \sup\{\omega(a) \mid a \text{ is an irreducible element of } R\}$. For an atomic domain R and a proper ideal $I \subseteq R$, we have the related definition: $\omega(I)$ is the smallest $n \in \mathbb{N}$ such that for any $a_1, a_2, \ldots, a_{n+1} \in R$ with $\prod_{i=1}^{n+1} a_i \in I$ there exists a non-empty, proper subset $S \subset \{1, 2, \ldots, n+1\}$ such that $\prod_{i\in S} a_i \in I$. If no such n exists, we write $\omega(I) = \infty$. Now, recalling Example 1.8, we see that $D_{5\mathbb{Z}[i]}(\mathbb{Z}[i]) = \infty$ while $\omega(5\mathbb{Z}[i]) = 2$.

In many ways, the omega invariant is reminiscent of the first definition of the Davenport constant: $\max\{n \mid \exists$ a zero-sum sequence of length n with no non-empty proper zero-sum subsequence$\}$. However, the two invariants cannot be reconciled because, unlike in the group setting, given an R-product $a_1a_2\cdots a_n$ with no $\mathcal{O}$-subproduct, there is no guarantee that there exists $a_{n+1} \in R$ such that $a_1a_2\cdots a_{n+1}$ is an $\mathcal{O}$-product with no proper $\mathcal{O}$-subproduct.

Now, the omega invariant has long been studied in connection with factorization—see for example [**11**] and [**13**]. Concurrently with this work, Choi made extensive use of the omega invariant to study the factorization properties of orders of a PID ([**4**]). As we will see, the distinctions between $\omega(R)$ and $D_{\mathcal{O}}(R)$ result in distinct techniques and constructions. As with the omega invariant, some of the definitions and constructions developed here have ideal-theoretic generalizations. These are explored in part by Moles and the current author in [**19**] to determine properties of arbitrary orders of rings of integers with prime conductor.

Before moving on, let us remark on a few other generalizations of the Davenport constant. The large and small Davenport constants, defined for a finite group as $D(G) = \max\{n \mid \exists$ a zero-sum sequence with no proper non-empty zero-sum subsequence$\}$ and $d(G) = \max\{m \mid \exists$ a G-sequence of length m with no non-empty zero-sum subsequence$\}$, respectively, were introduced and studied in [**10**] and [**14**]. These definitions similarly attempt to account for the discrepancy between the two definitions of $D(G)$ when G is no longer assumed to be abelian. In many ways, the omega invariant represents a generalization of the large Davenport constant as $D_{\mathcal{O}}(R)$ represents a generalization of the small davenport constant. For a more detailed exploration of these concepts, the interested reader is encouraged to see [**7**]. Finally, the term "generalized Davenport constant" is sometimes used in reference to the kth Davenport constant $D_k(G) = \min\{l \mid$ any G-sequence of length l has at least k disjoint non-empty zero-sum subsequences$\}$ first defined in [**15**].

Returning to the definition in question, when $\mathcal{O}$ is a subring, we will find that $D_{\mathcal{O}}(R)$ and $D(G)$ are not only similarly defined, but in many cases $D_{\mathcal{O}}(R)$ is intimately connected with the Davenport constant of a group related to $\mathcal{O}$. As previously mentioned, we will primarily be concerned with orders contained in rings of integers. Throughout the rest of the paper, we will use $\mathcal{O}$ to denote an order of a ring of integers $\mathcal{O}_K$ and $\mathfrak{f}$ to denote the conductor of this extension. In Section 2, we will study how $D_{\mathcal{O}}(\mathcal{O}_K) = \bar{D}(\mathcal{O})$ relates to the elasticity of $\mathcal{O}$. In Section 3, we will apply results and techniques developed in Section 2 to prove some stronger results for a class of orders in quadratic rings of integers.

2. Conditions for finite elasticity of orders in rings of integers

Intuitively, if $\bar{D}(R)$ is finite and $\bar{R}$ is well-behaved, we would expect the same of R to some extent. In this section, we explore some conditions under which $\bar{D}(R)$ gives us information about the finiteness of $\rho(R)$ and vice versa. To aid in this process, we will investigate a class of well-behaved orders in rings of integers. The following lemma will play an integral part in exploring the relationship between these two constants.

LEMMA 2.1. *Assume $\mathcal{O}_K$ has class number 1. If $\alpha_1, \alpha_2, \ldots, \alpha_m$ are irreducibles of $\mathcal{O}_K$ and $r \in \mathfrak{f}$ an irreducible of $\mathcal{O}_K$, then $r\alpha_1\alpha_2\cdots\alpha_m$ is irreducible in $\mathcal{O}$ if and only if $\alpha_1\alpha_2\cdots\alpha_m$ has no $\mathcal{O}$-subproduct up to multiplication by units of $\mathcal{O}_K$.*

PROOF. ($\Rightarrow$) Proceeding by contraposition, we will assume without loss of generality that there exists $u \in U(\mathcal{O}_K)$ such that $u\alpha_1\alpha_2\cdots\alpha_t \in \mathcal{O}$ for some $1 \leq t \leq m$. Thus, $r \in \mathfrak{f}$ implies $r\alpha_1\alpha_2\cdots\alpha_m = (u\alpha_1\alpha_2\cdots\alpha_t)(r(u^{-1}\alpha_{t+1}\cdots\alpha_m))$ is a non-trivial factorization in $\mathcal{O}$.

($\Leftarrow$) We once again proceed by contraposition and assume $r\alpha_1\alpha_2\cdots\alpha_m$ is not irreducible in $\mathcal{O}$. Then, there exist nonzero, nonunits $x, y \in \mathcal{O}$ such that $r\alpha_1\alpha_2\cdots\alpha_m = xy$. The result now follows from the fact that $\mathcal{O}_K$ is a unique factorization domain. □

This lemma will prove vital to finding elements of maximal elasticity in orders of number fields. The requirement that $\alpha_1\cdots\alpha_m$ has no $\mathcal{O}$-subproduct *up to units of $\mathcal{O}_K$* is a subtle but important condition. We will explore this idea further in Section 3.

Before moving on, let us recall some important theory of orders in number fields. Any order $\mathcal{O}$ properly contained in the ring of integers $\mathcal{O}_K$ fails to be Dedekind as it fails to be integrally closed. However, ideals relatively prime to the conductor $\mathfrak{f}_{\mathcal{O}} = \{\alpha \in \mathcal{O}_K \mid \alpha\mathcal{O}_K \subseteq \mathcal{O}\}$ are still invertible. In many ways, these ideals and their elements behave similarly to those in the ring of integers above. We will leverage this fact often throughout the rest of the paper. Notably, we can form a *generalized* class group $Cl(\mathcal{O})$ by taking the fractional ideals relatively prime to $\mathfrak{f}$ modulo the principal ideals relatively prime to $\mathfrak{f}$ ([**6**]). This leads us to the following theorem.

THEOREM 2.2 ([**5**]). *Let $\mathbb{Z}[\omega]$ be the ring of integers of the number field $\mathbb{Q}(\sqrt{d})$, then for $n > 1$,*

$$|Cl(\mathbb{Z}[n\omega])| = h(d)\psi_n(n)/u$$

where $h(d)$ is the class number of

$$\mathbb{Z}[\omega], u = (\mathbb{Z}[\omega]^* : \mathbb{Z}[n\omega]^*), \quad \text{and} \quad \psi_d(n) = n\prod_{q|n}(1 - (d/q)/q).$$

Significantly, we note that this implies $|Cl(\mathbb{Z}[n\omega])|$ is finite. What is more, by [**23**, Theorem 12.12], the class group of an order in any number field is always finite. We are now prepared to explore one direction of the relationship between $\bar{D}(\mathcal{O})$ and $\rho(\mathcal{O})$.

THEOREM 2.3. *Let $\mathcal{O}_K$ be a ring of integers with class number 1, and $\mathcal{O} \subseteq \mathcal{O}_K$ an order with conductor prime in $\mathcal{O}_K$. If $\bar{D}(\mathcal{O})$ is finite, then $\rho(\mathcal{O})$ is finite.*

PROOF. Assume for the purpose of contradiction that $\rho(\mathcal{O}) = \infty$. Then, for all $n \in \mathbb{N}$, there exist $\alpha_1, \alpha_2, \ldots, \alpha_k, \delta_1, \delta_2, \ldots, \delta_N \in \mathrm{Irr}(\mathcal{O})$ where $N \geq nk$ such that

$$\alpha_1\alpha_2\cdots\alpha_k = \delta_1\delta_2\cdots\delta_N.$$

As $\mathcal{O}_K$ is the integral closure of $\mathcal{O}$, each δ_j remains a nonunit in $\mathcal{O}_K$, and thus admits an irreducible factorization in the overring. In fact, as $\mathcal{O}_K$ is a unique factorization domain, each δ_j factors into a product of primes in R. Let $p_1p_2\cdots p_s$ be the prime factorization of $\delta_1\delta_2\cdots\delta_N$ in $\mathcal{O}_K$. By the above argument, we must have $s \geq N \geq nk$. As this factorization is unique, it must be the case that some α_i factors into a product of at least n primes in $\mathcal{O}_K$. Without loss of generality, $\alpha_1 = p_1p_2\cdots p_m$ where $m \geq n$. Thus, we have shown that an irreducible of $\mathcal{O}$ can factor into an arbitrarily long product of primes in $\mathcal{O}_K$.

Now, $\mathcal{O}_K$ is a Dedekind UFD and thus a PID. As the conductor of $\mathcal{O}$ is a prime ideal of $\mathcal{O}_K$, it is of the form (p) where $p \in \mathcal{O}_K$ is prime. Assume for the purpose of contradiction that p does not divide α_1 in $\mathcal{O}_K$. Then, $(\alpha_1) \subseteq \mathcal{O}$ is relatively prime to the conductor and hence has a prime factorization corresponding to $(\alpha_1) = (p_1)(p_2)\cdots(p_m) \subseteq R$ (**[6]**). Say

$$(\alpha_1) = \mathfrak{P}_1\mathfrak{P}_2\cdots\mathfrak{P}_m \subseteq \mathcal{O}$$

where the $\mathfrak{P}_i$'s are not necessarily distinct elements of $Spec(\mathcal{O})$. As m may be arbitrarily large and $D(Cl(\mathcal{O}))$ is finite, we may assume $m > D(Cl(\mathcal{O}))$. However, α_1 irreducible in $\mathcal{O}$ implies $[\mathfrak{P}_1][\mathfrak{P}_2]\ldots[\mathfrak{P}_m]$ is a minimal zero-sum sequence in $Cl(\mathcal{O})$ which implies $D(Cl(\mathcal{O})) \geq m$, contradicting our previous statement. Hence, without loss of generality, we may assume $p_1 = p$.

Finally, as p is in the conductor of $\mathcal{O}$ and $\alpha_1 = pp_2p_3\cdots p_m$ is irreducible in $\mathcal{O}$, by Lemma 2.1, $p_2p_3\cdots p_m$ has no $\mathcal{O}$-subproduct. However, as m is arbitrarily large, this implies $\bar{D}(\mathcal{O}) = \infty$, so we must have $\rho(\mathcal{O}) < \infty$. □

The following result follows directly from the preceding proof.

PORISM 2.4. Let $\mathcal{O}_K$ be a ring of integers with class number 1, and $\mathcal{O} \subseteq \mathcal{O}_K$ an order with conductor prime in $\mathcal{O}_K$. Then

$$\rho(\mathcal{O}) \leq \max\{D(Cl(\mathcal{O})), \bar{D}(\mathcal{O})\}.$$

Already, we are seeing the interconnectedness of these three constants. We should also be encouraged by the symmetry between this corollary and Theorem 1.4. Now, the following example demonstrates the significance of the prime conductor assumption.

EXAMPLE 2.5. Let $R = \mathbb{Z}[i]$ and consider the orders $\mathbb{Z}[3i]$ and $\mathbb{Z}[5i]$. It is easy to see that these orders have conductors (3) and (5) respectively. Now, $3 \equiv 3 \bmod 4$ implies it remains prime in $\mathbb{Z}[i]$ while $5 \equiv 1 \bmod 4$ implies that 5 splits. In fact, as an element, $5 = (2-i)(2+i)$. Theorem 2.3 thus implies that $\rho(\mathbb{Z}[3i]) < \infty$. In Section 3, we will show that $\rho(\mathbb{Z}[3i]) = \frac{3}{2}$. Contrastingly, note that $(2-i)^n, (2+i)^n \notin \mathbb{Z}[5i]$ for any $n \in \mathbb{N}$. This is easiest to see by considering the coefficients modulo 5. Thus, we have $(5(2-i)^n)\cdot(5(2+i)^n) = 5^{n+2}$ as irreducible factorizations in $\mathbb{Z}[5i]$ of lengths 2 and $n+2$. Thus, $\rho(\mathbb{Z}[5i]) \geq \frac{n+2}{2}$ for all $n \in \mathbb{N}$, so $\rho(\mathbb{Z}[5i]) = \infty$.

Once again, we note the theme of using products with no $\mathcal{O}$-subproduct to form elements of large elasticity. We can also see that the structure of the conductor plays a vital role in determining factorization properties of the order. This relationship is made most explicit by Halter-Koch in [**17**]. In general, if there are multiple prime ideals of $\mathcal{O}_K$ containing $\mathfrak{f}\mathcal{O}$, we can form an element of arbitrarily large elasticity in $\mathcal{O}$. Similarly, Choi ([**4**]) shows that factorization is well-behaved for orders of PIDs with primary conductor. We will explore the explicit calculation for some such orders in Section 3.

We have now seen how our generalized Davenport constant can be used to form elements of maximal elasticity and circumstances under which finiteness of $\bar{D}(\mathcal{O})$ implies finiteness of $\rho(\mathcal{O})$. We would naturally like to know if and when the converse of Theorem 2.3 is true. Namely, when does finite elasticity imply $\bar{D}(\mathcal{O})$ is finite? Equivalently, when does $\bar{D}(\mathcal{O}) = \infty$ imply $\rho(\mathcal{O}) = \infty$? What properties of quadratic orders are we taking advantage of to create these elements of large elasticity? In effort to explore these questions, we present the following definition.

DEFINITION 2.6. An order $\mathcal{O}$ of a Galois number field K is called *Galois-invariant* if it is closed under Galois conjugation. That is, for any $\alpha \in \mathcal{O}$ and $\sigma \in \text{Gal}(K/\mathbb{Q})$, $\sigma(\alpha) \in \mathcal{O}$.

This term was first formally used in in 2015 by Lee and Louboutin in [**20**] where the authors considered the family of Galois number fields generated by a totally real cubic algebraic unit. More recently, Geroldinger et al. made connection between Galois-invariance and factorization theory in [**12**]–investigating the factorization properties of the normsets of Galois-invariant orders. It appears that work on Galois-invariant orders qua algebraic objects remains relatively sparse in the literature. We will soon see the relevance of this definition to our exploration of $\bar{D}(\mathcal{O})$, but for now, let us consider a few classes of Galois-invariant orders.

PROPOSITION 2.7. *Let K be a Galois number field. The following orders are Galois-invariant.*

(1) *The ring of integers $\mathcal{O}_K$.*
(2) *If K is quadratic, any order of K.*
(3) *Any order generated over $\mathbb{Z}$ by a normal basis of K.*

First, for any $\beta \in \mathcal{O}_K$, $\sigma(\beta)$ is also an algebraic integer contained in K, so we must have $\sigma(\beta) \in \mathcal{O}_K$. All quadratic number fields are of the form $K = \mathbb{Q}(\sqrt{d})$ for some square-free $d \in \mathbb{Z}$. It is well known that every order of K is of the form $\mathbb{Z}[n\omega]$ where $n \in \mathbb{N}$ and

$$\omega = \begin{cases} \sqrt{d} & d \equiv 2, 3 \bmod 4 \\ \frac{1+\sqrt{d}}{2} & d \equiv 1 \bmod 4. \end{cases}$$

Now, the only elements of $\text{Gal}(K/\mathbb{Q})$ are the identity and complex conjugation, so in order to show $\mathbb{Z}[n\omega]$ is Galois-invariant, it suffices to show $n\bar{\omega} \in \mathbb{Z}[n\omega]$. To this end, we note that $tr(n\omega) = n\omega + n\bar{\omega} = m \in \mathbb{Z}$ which implies $n\bar{\omega} = m - n\omega \in \mathbb{Z}[n\omega]$ as desired.

Finally, recall that a normal basis of a K (over $\mathbb{Q}$) is a set of the form $\{\sigma(\alpha) | \sigma \in \text{Gal}(K/\mathbb{Q})\}$ for some $\alpha \in K$ which forms a basis of K as a vector space over $\mathbb{Q}$. Essentially, it is a basis formed by Galois conjugates. Hence, it is clear that any order of the form $\mathbb{Z}[\alpha_1, \alpha_2, \ldots, \alpha_n]$ where $\{\alpha_1, \alpha_2, \ldots, \alpha_n\}$ forms a normal (integral) basis, is Galois-invariant. Recall that the normal basis theorem guarantees any Galois number field admits a normal basis. Also, every order contains an integral basis of the extension $[K : \mathbb{Q}]$. Unfortunately, the existence of a normal *integral* basis is not guaranteed in general, but in the case K does admit such a basis, this implies the existence of many Galois-invariant orders in the form of (3).

EXAMPLE 2.8. For any rational prime p, the p-th cyclotomic number field $\mathbb{Q}(\zeta_p)$ has normal integral basis $\{\zeta_p, \zeta_p^2, \ldots, \zeta_p^{p-1}\}$ ([**21**]). Then, if the greatest common divisor of $\{n_1, n_2, \ldots, n_{p-1}\}$, with $n_i \in \mathbb{N}$, is a member of the set, the order $\mathbb{Z}[n_1\zeta_p, n_2\zeta_p^2, \ldots, n_{p-1}\zeta_p^{p-1}]$ is Galois-invariant.

This leads us to a natural question: are all Galois-invariant orders generated by a normal basis? The following example shows that this is not the case.

EXAMPLE 2.9. Consider the Gaussian integers $\mathbb{Z}[i] \subseteq \mathbb{Q}(i)$. Both 1 and 2 imply that $\mathbb{Z}[i]$ is Galois-invariant. However, it is well known that the Gaussian field does not admit a normal integral basis. Additionally, recalling that an order of a number field K is a $\mathbb{Z}$-module of rank $[K : \mathbb{Q}]$, one can observe directly that $\mathbb{Z}[i]$ cannot be expressed in the form $\langle a + bi, a - bi\rangle_{\mathbb{Z}}$.

This tells us that there are "non-obvious" examples of Galois-invariant orders. We might even begin to wonder if all orders in Galois number fields are Galois-invariant. This turns out not to be the case.

EXAMPLE 2.10. Consider the Galois number field $\mathbb{Q}(\sqrt[6]{3}i) = \mathbb{Q}(\alpha)$ and the order $\mathbb{Z}[\alpha] = \langle 1, \alpha, \alpha^2, \ldots, \alpha^5\rangle_{\mathbb{Z}}$. Observe that this order does not contain the conjugate $\frac{\alpha}{2} + \frac{\alpha^4}{2} = \alpha \cdot \omega$ (where $\omega = \frac{1}{2} - \frac{\sqrt{3}}{2}i$ is a 6th root of unity) and thus is not Galois-invariant. Additionally, we note that $\mathbb{Z}[\alpha] = \mathbb{Z}[\alpha, 2 \cdot \sigma_1(\alpha), 2 \cdot \sigma_2(\alpha), \ldots, 2 \cdot \sigma_5(\alpha)]$, which motivates this example. Many non-invariant orders in number fields of degree 3 or greater can be constructed in this manner.

With this concept in hand, we are now prepared to return to the question of when finite elasticity implies $\bar{D}(\mathcal{O})$ is finite.

THEOREM 2.11. *Let K be a Galois number field with class number 1, and $\mathcal{O}$ a Galois-invariant order of K with prime conductor such that $U(\mathcal{O}) = U(\mathcal{O}_K)$. Then if $\rho(\mathcal{O})$ is finite, $\bar{D}(\mathcal{O})$ is finite. In particular, $\bar{D}(\mathcal{O}) \leq n(\rho(\mathcal{O}) - 1) + 1$ where $n = [K : \mathbb{Q}]$.*

PROOF. Let $\alpha_1\alpha_2 \cdots \alpha_k$ be an $\mathcal{O}_K$-product with no $\mathcal{O}$-subproduct. Let us write

$$Gal(K/\mathbb{Q}) = \{\sigma_1, \sigma_2, \ldots, \sigma_n\}.$$

Now, $\mathcal{O}$ Galois-invariant implies $\sigma_i(\alpha_1)\sigma_i(\alpha_2) \cdots \sigma_i(\alpha_k)$ has no $\mathcal{O}$-subproduct for any $\sigma_i \in Gal(K/\mathbb{Q})$. This is because $\sigma_i(\alpha_1)\sigma_i(\alpha_2) \cdots \sigma_i(\alpha_t) = \sigma_i(\alpha_1\alpha_2 \cdots \alpha_t) \in \mathcal{O}$ implies $(\sigma_i^{-1} \circ \sigma_i)(\alpha_1\alpha_2 \cdots \alpha_t) = \alpha_1\alpha_2 \cdots \alpha_t \in \mathcal{O}$. Note that as $\mathcal{O}_K$ is a PID, every prime ideal is generated by a prime element, so we may choose $r \in \mathfrak{f}$ prime in $\mathcal{O}_K$. Thus, as $U(\mathcal{O}) = U(\mathcal{O}_K)$, by Lemma 2.1, $r\sigma_i(\alpha_1)\sigma_i(\alpha_2) \cdots \sigma_i(\alpha_k)$ is irreducible in $\mathcal{O}$ for all $1 \leq i \leq n$. Now, consider the following element of $\mathcal{O}$

$$\prod_{i=1}^{n}(r\sigma_i(\alpha_1)\sigma_i(\alpha_2) \cdots \sigma_i(\alpha_k)) = r^n \prod_{j=1}^{k}\prod_{i=1}^{n} \sigma_i(\alpha_j) = r^n \prod_{j=1}^{k} N(\alpha_j).$$

Now, each $N(\alpha_j) \in \mathbb{Z} \subseteq \mathcal{O}$, and $\alpha_j \notin U(\mathcal{O}_K)$ implies $N(\alpha_j) \notin U(\mathcal{O}_K) = U(\mathcal{O})$ for all $1 \leq j \leq k$. Thus, we have $\rho(\mathcal{O}) \geq \frac{n+k}{n}$, so k is bounded above. Now, if $\rho(\mathcal{O}) = 1$, we must have $k = 0$. Otherwise we can rearrange to achieve $k \leq n(\rho(\mathcal{O}) - 1)$. This establishes the result. □

EXAMPLE 2.12. Consider the ring of integers $\mathbb{Z}[\frac{1+\sqrt{-7}}{2}] = \mathbb{Z}[\omega]$ and the order $\mathbb{Z}[5\omega]$. We will soon show that $\rho(\mathbb{Z}[5\omega]) = \frac{7}{2}$. Thus, Theorem 2.11 implies $\bar{D}(\mathbb{Z}[5\omega]) \leq 2\left(\frac{7}{2} - 1\right) + 1 = 6$. Observe that the product ω^5 has no $\mathbb{Z}[5\omega]$ subproduct. Hence, we must have $\bar{D}(\mathbb{Z}[5\omega]) = 6$.

In the following section, we consider some classes of orders for which Theorems 2.3 and 2.11 both hold. Namely, orders in which $\bar{D}(\mathcal{O})$ is finite if and only if $\rho(\mathcal{O})$ is finite. We will also develop some techniques for computing both constants in such orders.

3. Elasticity of orders in quadratic number fields

The results in the section are due to work done concurrently with but separate from Choi [**4**]. In particular, an equivalent of Theorem 3.6 can also be found in [**4**] using a very distinct approach from this paper. This section also represents an extension of the theory developed in [**18**].

LEMMA 3.1 ([**25**]). *Let $\mathbb{Z}[\omega]$ be a quadratic ring of integers with class number 1, and $p \in \mathbb{Z}[\omega]$ a prime element. Then, for any $r \in \mathbb{Z}[\omega]$ relatively prime to p, there exists a prime element $q \in \mathbb{Z}[\omega]$ such that $q \equiv r$ mod p.*

In fact, much more than this is true. The prime elements of $\mathbb{Z}[\omega]$ are asymptotically uniformly distributed among the invertible classes of $\mathbb{Z}[\omega]/(p)$, but for our purposes, the existence of a single element in each of these classes will be sufficient.

We are now ready to calculate $\bar{D}(\mathcal{O})$ for a large class of orders. Recall once again that all orders of a quadratic number field are of the form $\mathcal{O} = \mathbb{Z}[n\omega]$ for any $n \in \mathbb{Z}$. Hence, the conductor $\mathfrak{f} = n\mathcal{O}_K$ is always a principal ideal of the ring of integers. This is a unique and desirable property of which we will make extensive use. The following theorem makes explicit the connection between $\mathfrak{f}_{\mathcal{O}}$ and $\bar{D}(\mathcal{O})$ hinted at previously.

THEOREM 3.2. *Let $\mathbb{Q}(\sqrt{d})$ be a quadratic number field with class number 1 and $d < -3$. If p is a rational prime such that $\left(\frac{d}{p}\right) = -1$, then $\bar{D}(\mathbb{Z}[p\omega]) = p + 1$.*

Note that the assumption $d < -3$ ensures $U(\mathbb{Z}[p\omega]) = U(\mathbb{Z}[\omega])$, and $\left(\frac{d}{p}\right) = -1$ if and only if p is inert in $\mathbb{Z}[\omega]$.

PROOF. First, we note that as $(p) \subseteq \mathbb{Z}[\omega]$ is prime, $\mathbb{Z}[\omega]/(p)$ is a finite field of order p^2. If $\hat{\omega}$ is a root of $x^2 + d$, the defining polynomial of $\mathbb{Q}(\sqrt{d})$, over the field $\mathbb{F}_p$, then the map $[a + b\omega] \overset{\phi}{\mapsto} \hat{a} + \hat{b}\hat{\omega}$, where the coefficients are reduced mod p, is a field isomorphism from $\mathbb{Z}[\omega]/(p)$ to $\mathbb{F}_p[\hat{\omega}]$.

Now, considering $\mathbb{Z}[\omega]/(p)$, the elements of $\mathbb{Z}[p\omega]$ comprise the cosets of (p) of the form $[a + 0 \cdot \omega]$, which have isomorphic image $\mathbb{F}_p$ under the map defined above. Thus, we can identify $\mathbb{Z}[p\omega] \subseteq \mathbb{Z}[\omega]$ with $\mathbb{F}_p \subseteq \mathbb{F}_p[\hat{\omega}]$. Reducing mod p, we see that a $\mathbb{Z}[\omega]$-product $\pi_1 \cdots \pi_N$ has no $\mathbb{Z}[p\omega]$-subproduct if and only if $\hat{\pi}_1 \cdots \hat{\pi}_n$ has no subproduct in $\mathbb{F}_p$. In other words, $\bar{D}(\mathbb{Z}[p\omega])$ is less than or equal to the minimum N such that any product of length N in $\mathbb{F}_p[\hat{\omega}]$ must have a subproduct in $\mathbb{F}_p$. Also, because $\mathbb{F}_p[\hat{\omega}]$ has no nonzero zero divisors, this is equivalent to finding the minimum N such that any product of length N in $\mathbb{F}_p[\hat{\omega}]^*$ must have a subproduct in $\mathbb{F}_p^*$.

Now, $\mathbb{F}_p$ and $\mathbb{F}_p[\hat{\omega}]$ are finite fields, so $\mathbb{F}_p^*$ and $\mathbb{F}_p[\hat{\omega}]^*$ are finite cyclic groups with $\mathbb{F}_p^* \subseteq \mathbb{F}_p[\hat{\omega}]^*$. Hence, $G = \mathbb{F}_p[\hat{\omega}]^*/\mathbb{F}_p^*$ is a finite abelian group of order $\frac{p^2-1}{p-1} = \frac{(p+1)(p-1)}{p-1} = p+1$. Thus, finding the finding the minimum N such that any product of length N in $\mathbb{F}_p[\hat{\omega}]^*$ must have a subproduct in $\mathbb{F}_p^*$ is equivalent to finding the minimum N such that any G-sequence has a zero-sum subsequence. Notice, this is just $D(G)$. As G is a quotient group of a cyclic group, it is also cyclic, so $D(G) = |G| = p+1$. Thus, we have $\bar{D}(\mathbb{Z}[p\omega]) \leq p+1$.

Once again, as $p \in \mathbb{Z}[\omega]$ is prime, every nonzero class of $\mathbb{Z}[\omega]/(p)$ is invertible. Let $r+(p)$ be a class whose image under the map ϕ mod $\mathbb{F}_p^*$ generates G. This must obviously be a nonzero class. Hence, by Lemma 3.1, we may choose $\gamma \in r+(p)$ prime, and by construction, γ^p will have no $\mathbb{Z}[p\omega]$-subsequence. Therefore, we conclude $\bar{D}(\mathbb{Z}[p\omega]) = p+1$. □

Now, we not only know $\bar{D}(\mathbb{Z}[p\omega])$, but we have a procedure for finding a product achieving it.

EXAMPLE 3.3. Returning to a previous example, consider the ring of integers $\mathbb{Z}[\frac{1+\sqrt{-7}}{2}] = \mathbb{Z}[\omega]$ and the order $\mathbb{Z}[5\omega]$. Theorem 3.2 reaffirms that $\bar{D}(\mathbb{Z}[5\omega]) = 6$. It also motivates our choice of ω^5 as our maximal $\mathbb{Z}[\omega]$-product with no $\mathbb{Z}[5\omega]$-subproduct. In this case, as $|\mathbb{F}_5[\hat{\omega}]^*/\mathbb{F}^*| = 6$, we can see that one third of the non-zero equivalence classes mod p will work for our purposes. To see that $\omega = \frac{1+\sqrt{-7}}{2}$ works, by Lagrange's theorem, we need only check that ω^n mod $p \notin \mathbb{F}_p$ for $n \leq 3$, and in fact $\omega^2 = -2+\omega$ while $\omega^3 = -2-\omega$. As a final check of our theory, note that indeed $\omega^6 = 2+5\omega \equiv 2$ mod p.

Now, before we can use this result to find the elasticity of $\mathbb{Z}[p\omega]$, we must develop some machinery unique to quadratic rings of integers.

LEMMA 3.4. *Let $\mathbb{Z}[\omega]$ be a quadratic ring of integers and p a rational prime. If $\pi_1 \in \mathbb{Z}[p\omega]$ and $\pi_2 \in \mathbb{Z}[\omega] \setminus \mathbb{Z}[p\omega]$, then $\pi_1\pi_2 \in \mathbb{Z}[p\omega]$ if and only if $\pi_1 \in p\mathbb{Z}[\omega]$.*

PROOF. Let $\pi_1 = a+pb\omega$ and $\pi_2 = c+d\omega$. Then $\pi_1\pi_2 = ac+px+(ad+py)\omega \in \mathbb{Z}[p\omega]$ if and only if $p|ad$ if and only if $p|a$ or $p|d$. Equivalently, $\pi_1\pi_2 \in \mathbb{Z}[p\omega]$ if and only if $a+pb\omega = \pi_1 \in p\mathbb{Z}[\omega]$ or $c+d\omega = \pi_2 \in \mathbb{Z}[p\omega]$. The result then follows from the fact that $\pi_2 \notin \mathbb{Z}[p\omega]$. □

As we know, for any order $\mathcal{O} \subseteq \mathcal{O}_k$, the elements of the conductor are those $\alpha \in \mathcal{O}_K$ such that $\alpha\mathcal{O}_K \subseteq \mathcal{O}$. That is, the elements of the conductor "pull" elements of the ring of integers down into the order. Lemma 3.4 tells us that for $\mathbb{Z}[p\omega] \subseteq \mathbb{Z}[\omega]$ as above, elements of $\mathcal{O}$ outside the conductor *never* do this. We will now leverage this fact to demonstrate a nice property of factorizations in $\mathbb{Z}[p\omega]$.

THEOREM 3.5. *Let $\mathbb{Z}[\omega]$ be a quadratic ring of integers with class number 1 and $\beta \in \mathbb{Z}[p\omega]$ with prime factorization $p^n\pi_1\cdots\pi_k$ such that $(\pi_i, p) = 1$ in $\mathbb{Z}[\omega]$. For a fixed j, assume $\pi_j \in \mathbb{Z}[p\omega]$. Then, if $\alpha_1\cdots\alpha_m$ is an irreducible factorization of β in $\mathbb{Z}[p\omega]$, $\pi_j = \alpha_t$ for some $1 \leq t \leq m$.*

PROOF. Without loss of generality we will prove the statement for $\pi_1 \in \mathbb{Z}[p\omega]$. Assume $\pi_1 \in \mathbb{Z}[p\omega]$. Note, $\pi_1 \notin p\mathbb{Z}[\omega]$ as $(\pi_1, p) = 1$.

Let $\beta = \alpha_1\alpha_2\cdots\alpha_n$ be an irreducible factorization of β in $\mathbb{Z}[p\omega]$. As π_1 is prime and $\pi_1|\beta$, without loss of generality $\pi_1|\alpha_1$ as an element of $\mathbb{Z}[\omega]$. Thus, by the uniqueness of the prime factorization of β in $\mathbb{Z}[\omega]$, $\alpha_1 = \pi_1\cdot(p^r\pi_{i_1}\cdots\pi_{i_l})u$ where

$u \in U(\mathbb{Z}[\omega]) = U(\mathbb{Z}[p\omega]) = \{\pm 1\}$. If $r \geq 1$, then $(p^r \pi_{i_1} \cdots \pi_{i_l})u \in \mathbb{Z}[p\omega]$, and thus α_1 is not irreducible. Hence, we assume $r = 0$. So we consider $\alpha_1 = \pi_1 \cdot (\pi_{i_1} \cdots \pi_{i_l})u$. If $(\pi_{i_1} \cdots \pi_{i_l})u \notin \mathbb{Z}[p\omega]$, by Lemma 3.4, $\pi_1 \notin p\mathbb{Z}[i]$ implies $\pi_1 \cdot (\pi_{i_1} \cdots \pi_{i_l})u = \alpha_1 \notin \mathbb{Z}[pi]$, a contradiction.

So, $(\pi_{i_1} \cdots \pi_{i_l})u \in \mathbb{Z}[p\omega]$, but $\alpha_1 = \pi_1 \cdot (\pi_{i_1} \cdots \pi_{i_l})u$ is irreducible in $\mathbb{Z}[p\omega]$, so $(\pi_{i_1} \cdots \pi_{i_l})u \in U(\mathbb{Z}[p\omega])$. Thus, $\pi_1 = \alpha_1$ up to a unit, and π_1 appears in every irreducible factorization of β in $\mathbb{Z}[p\omega]$. □

Now, to determine the elasticity of $\mathbb{Z}[p\omega]$, we need only consider elements in the conductor $p\mathbb{Z}[\omega]$. This is because ideals relatively prime to the conductor retain unique factorization into prime ideals. In particular, principal ideals generated by elements relatively prime to the conductor. In our case, the elements outside the conductor. Thus, these elements behave in the order as they would in a ring of integers. That is, their elasticity is bounded above by $\frac{D(Cl(\mathbb{Z}[p\omega]))}{2} \leq \frac{|Cl(\mathbb{Z}[p\omega])|}{2} = \frac{(p+1)\cdot|Cl(\mathbb{Z}[\omega])|}{2} = \frac{p+1}{2}$ when $\left(\frac{d}{p}\right) = -1$ (Theorem 2.2). As we will soon see, elements in the conductor will achieve an upper bound which exceeds this value. Theorem 3.5 allows us to narrow our search even further.

Take any $\gamma \in p\mathbb{Z}[\omega]$ with prime factorization $\gamma = p^r \pi_1 \pi_2 \cdots \pi_m$ in $\mathbb{Z}[\omega]$, where $r \geq 1$ and $(p, \pi_i) = 1$. If $\pi_1 \in \mathbb{Z}[p\omega]$, let $\delta = p^r \pi_2 \cdots \pi_m \in p\mathbb{Z}[\omega]$. We may write $\rho(\delta) = \frac{n_1}{n_2}$ where n_1 and n_2 are the lengths of the longest and shortest irreducible factorizations of β in $\mathbb{Z}[p\omega]$ respectively. Note, the existence of n_1 is guaranteed by [**17**]. Theorem 3.5, together with the fact that $\mathbb{Z}[p\omega]$ is a domain, implies that every irreducible factorization of γ is of the form $\pi_1 \delta_1 \cdots \delta_t$ where $\delta_1 \cdots \delta_t$ is an irreducible factorization of δ. Thus, $\rho(\gamma) = \frac{n_1+1}{n_2+1} \leq \frac{n_1}{n_2} = \rho(\delta)$.

So for any element $p^r \pi_1 \pi_2 \cdots \pi_m$ in the conductor with $\pi_i \in \mathbb{Z}[p\omega]$, we can find an element $p^r \hat{\pi}_1 \hat{\pi}_2 \cdots \hat{\pi}_k$ in the conductor with greater elasticity and $\hat{\pi}_i \notin \mathbb{Z}[p\omega]$ for all $1 \leq i \leq k$. Therefore, to determine $\rho(\mathbb{Z}[p\omega])$, we need only consider elements of the latter form. We are now prepared to prove the ultimate result of this section.

THEOREM 3.6. *Let $\mathbb{Q}(\sqrt{d})$ be a quadratic number field with class number* 1 *and $d < -3$. If p is a rational prime such that $\left(\frac{d}{p}\right) = -1$, $\rho(\mathbb{Z}[p\omega]) = 1 + \frac{p}{2}$.*

PROOF. Let $\alpha \in \mathbb{Z}[\omega]$ with prime factorization $\alpha = p^k \pi_1 \pi_2 \cdots \pi_m$ in $\mathbb{Z}[\omega]$ where $(\pi_i, p) = 1$. Recall that for $\beta \notin p\mathbb{Z}[\omega] = \mathfrak{f}$, we have

$$\rho(\beta) \leq \frac{D(Cl(\mathbb{Z}[p\omega]))}{2} \leq \frac{|Cl(\mathbb{Z}[p\omega])|}{2} = \frac{p+1}{2} < 1 + \frac{p}{2}.$$

Thus, we may assume $\alpha \in p\mathbb{Z}[\omega]$ ($k \geq 1$). Furthermore, as noted above, Theorem 3.5 implies that we may also assume $\pi_i \notin \mathbb{Z}[p\omega]$ for all $1 \leq i \leq k$. Hence, the finest possible irreducible factorization of α in $\mathbb{Z}[p\omega]$ would have the form $p^k(\pi_1\pi_2)\cdots(\pi_{m-1}\pi_m)$ if m is even, and $p^k(\pi_1\pi_2)\cdots(\pi_{m-2}\pi_{m-1}\pi_m)$ if m is odd. That is, a factorization of length $k + \lfloor \frac{m}{2} \rfloor$.

CASE 1. $m < kp$. We know that every irreducible factor of α can contain at most one factor of p. Thus, every irreducible factorization has length at least k, so

$$\rho(\alpha) \leq \frac{k + \lfloor \frac{m}{2} \rfloor}{k} \leq \frac{k + \frac{m}{2}}{k} < \frac{k + \frac{kp}{2}}{k} = 1 + \frac{p}{2}.$$

CASE 2. $m \geq kp$. Now, any irreducible factorization of α must take the form

$$(p\pi_{1,1} \cdots \pi_{1,n_1}) \cdots (p\pi_{k,1} \cdots \pi_{k,n_k})(\pi_{k+1,1} \cdots \pi_{k+1,n_{k+1}}) \cdots (\pi_{t,1} \cdots \pi_{t,n_t})$$

By Lemma 2.1 and Theorem 3.2, $n_i \leq p$ for $1 \leq i \leq k$. Also, for all $k+1 \leq i \leq t$, $\beta = (\pi_{j,1} \cdots \pi_{j,n_j})(\bar{\pi}_{j,1} \cdots \bar{\pi}_{j,n_j}) \in \mathbb{Z}[p\omega] \setminus p\mathbb{Z}[\omega]$ where $\bar{\pi}$ is the complex conjugate of π. Once again, this implies $\rho(\beta) \leq \frac{p+1}{2}$. Using the factorizations $(\pi_{j,1} \cdots \pi_{j,n_j})(\bar{\pi}_{j,1} \cdots \bar{\pi}_{j,n_j}) = (\pi_{j,1}\bar{\pi}_{j,1}) \cdots (\pi_{j,n_j}\bar{\pi}_{j,n_j})$, we find $\frac{n_j}{2} \leq \rho(\beta) \leq \frac{p+1}{2}$ which implies $n_j \leq p+1$ for $k+1 \leq j \leq t$. This is sufficient to determine that the length of any given factorization is at least $k + \lceil \frac{m-kp}{p+1} \rceil$. Hence,

$$\rho(\alpha) \leq \frac{k + \lfloor \frac{m}{2} \rfloor}{k + \lceil \frac{m-kp}{p+1} \rceil} \leq \frac{k + \frac{m}{2}}{k + \frac{m-kp}{p+1}} = \frac{2k+m}{2k + \frac{2m-2kp}{p+1}} = (p+1)\frac{2k+m}{2k+2m}$$

which is maximized when m is minimized. So $m \geq kp$ implies

$$\rho(\alpha) \leq (p+1)\frac{2k+kp}{2k+2kp} = \frac{p+1}{2} \cdot \frac{2+p}{1+p} = 1 + \frac{p}{2}.$$

Hence, we have determined $\rho(\mathbb{Z}[p\omega]) \leq 1 + \frac{p}{2}$.

Finally, let $\gamma_1\gamma_2 \cdots \gamma_p$ be a $\mathbb{Z}[\omega]$-product with no $\mathbb{Z}[p\omega]$-subproduct. Note, this is guaranteed by Theorem 3.2. By Lemma 2.1, $p\gamma_1\gamma_2 \cdots \gamma_p$ is irreducible in $\mathbb{Z}[p\omega]$. If we let $\delta = p^2\gamma_1\gamma_2 \cdots \gamma_p\bar{\gamma}_1\bar{\gamma}_2 \cdots \bar{\gamma}_p$, the factorizations

$$(p\gamma_1\gamma_2 \cdots \gamma_p)(p\bar{\gamma}_1\bar{\gamma}_2 \cdots \bar{\gamma}_p) = p^2(\gamma_1\bar{\gamma}_1)(\gamma_2\bar{\gamma}_2) \cdots (\gamma_p\bar{\gamma}_p)$$

show us that $\rho(\delta) \geq \frac{2+p}{2} = 1 + \frac{p}{2}$, and consequently $\rho(\mathbb{Z}[p\omega]) \geq 1 + \frac{p}{2}$ – establishing the result. □

First, we note that $\rho(\mathbb{Z}[p\omega]) = 1 + \frac{p}{2} = \frac{\bar{D}(\mathcal{O})+1}{2}$. This is remarkably similar to the formula for elasticity of rings of integers given in Theorem 1.4—further motivating our approach. In particular, consider our construction of an element obtaining the upper bound and that of Narkiewicz [**22**]. From the proof, we can also see that $\mathbb{Z}[p\omega]$ has *accepted* elasticity. That is, there exists $x \in \mathbb{Z}[p\omega]$ such that $\rho(x) = \rho(\mathbb{Z}[p\omega])$. On top of this, we have a procedure for finding such an element. The orders which have accepted elasticity in rings of integers were first characterized by Halter-Koch [**17**] in 1995. He also gave an upper bound on said elasticities. This was the same year in which Narkiewicz [**22**] used the Davenport constant to give an explicit formula for the elasticity of rings of integers. However, despite all this activity, we have not produced results on calculating elasticity of proper orders over the last 30 years. The generalized Davenport constant gave us a natural way to use the information in the rings of integers to answer this question and led us to a very constructive solution. It will be interesting to see where else this concept can be applied.

EXAMPLE 3.7. Theorem 3.6 affirms the previous claim that $\rho(\mathbb{Z}[5 \cdot \frac{1+\sqrt{-7}}{2}]) = \rho(\mathbb{Z}[5\omega]) = 1 + \frac{5}{2} = \frac{7}{2}$. Recalling that ω^5 has no $\mathbb{Z}[5\omega]$-subproduct, we see that $(5\omega^5)(5\bar{\omega}^5) = 5^2(\omega\bar{\omega})^5 = 5^2 2^5$ are irreducible factorizations of the element 800, so we have $\rho(800) = \frac{7}{2}$ as desired.

As noted, we assumed $d < -3$ to ensure the unit groups of the order and its corresponding ring of integers were the same. Of course, this only precludes two of the imaginary quadratic unique factorization domains. It turns out, while our analysis must change slightly, the approach we have taken is viable for these rings as well.

EXAMPLE 3.8. Consider the Gaussian integers $\mathbb{Z}[i]$ and the order $\mathbb{Z}[19i]$ which have units $\{\pm 1, \pm i\}$ and $\{\pm 1\}$ respectively. Note that the prime $2 + 3i$ generates

$\mathbb{F}_{19}[i]^*$, so as an element of $\mathbb{Z}[i]$, $(2+3i)^{19}$ has no $\mathcal{O}$-subproduct. However, $19 \cdot (2+3i)^{19}$ is not irreducible in $\mathbb{Z}[19i]$ as Lemma 2.1 might suggest at first glance. This is because, considering the coefficients mod 19, $(2+3i)^{10} \in i\mathbb{F}_{19}$, so $19 \cdot (2+3i)^{19}$ factors non-trivially in $\mathbb{Z}[19i]$ as $[i(2+3i)^{10}] \cdot [19(2+3i)^9(-i)]$. This happens because $\pm i$ are units in $\mathbb{Z}[i]$ but not $\mathbb{Z}[19i]$. The same can be said for any primitive element γ because γ^{10} will always be in $i\mathbb{F}_{19}^*$ as this subgroup contains all elements of order 2. Regardless, with some minor modifications to the proof above, one can show that $\rho(\mathbb{Z}[pi]) = \frac{p+3}{4}$ for $p \equiv 3 \bmod 4$.

Finally, this approach continues to bear fruit as we move to consider the case where the ring of integers $\mathcal{O}_K$ is a non-UFD. While we lose unique factorization for elements, we retain it for ideals. This would suggest that a more comprehensive analysis would require a more ideal-theoretic approach. However, as the following example illustrates, the generalized Davenport constant remains a vital tool for constructing elements of large elasticity without unique factorization.

EXAMPLE 3.9. Consider the ring of integers $\mathbb{Z}[\sqrt{-14}]$ and the order $\mathbb{Z}[11 \cdot \sqrt{-14}]$ ($p = 11$). It is easy to verify that the class group of $\mathbb{Z}[\sqrt{-14}]$ is cyclic of order 4. Thus, we have $D(Cl(\mathbb{Z}[\sqrt{-14}])) = 4$. In keeping with the theory of [**22**], we would like to find a maximal zero-sum sequence in $Cl(\mathbb{Z}[\sqrt{-14}])$. Allowing $(19, 9+\sqrt{-14}) = \mathcal{P}$, we see that $[\mathcal{P}]^4$ is one such sequence. Now, $\mathcal{P}^4 = (325 + 42\sqrt{-14})$, and we note also that the image of $325+42\sqrt{-14}$ under the map defined in Theorem 3.2 is a generator of $\mathbb{F}_{11}[\hat{\omega}]^*$. Notably, this implies that $(325+42\sqrt{-14})^{11}$ has no $\mathbb{Z}[11 \cdot \sqrt{-14}]$-subproduct. Therefore, the element $[11 \cdot (325+42\sqrt{-14})^{11}][11 \cdot (325-42\sqrt{-14})^{11}] = 11^2 \cdot 19^{44}$ has elasticity at least $\frac{2+44}{2} = 1 + 11 \cdot \frac{4}{2} = 1 + 11 \cdot \frac{D(Cl(\mathbb{Z}[\sqrt{-14}]))}{2} = 1 + 11 \cdot \rho(\mathbb{Z}[\sqrt{-14}])$ in $\mathbb{Z}[11 \cdot \sqrt{-14}]$.

Using ideal-theoretic versions of some of the techniques developed in this paper, it has been subsequently shown by Moles and the author of the current paper that the elasticity of $\mathbb{Z}[11 \cdot \sqrt{-14}]$ is $\frac{D(Cl(\mathbb{Z}[11 \cdot \sqrt{-14}]))}{2}$. However, the construction of an element of large elasticity in Example 3.9 remains integral as it is fundamentally different from those which require knowledge of the structure of $Cl(\mathbb{Z}[11 \cdot \sqrt{-14}])$. In fact, it allows not only for the exact calculation of $\rho(\mathbb{Z}[11 \cdot \sqrt{-14}])$, but in so doing also determines the structure of $Cl(\mathbb{Z}[11 \cdot \sqrt{-14}])$. For further details, see [**19**].

References

[1] N. Baeth and J. Hoffmeier, *Atoms of the relative block monoid*, Involve **2** (2009), no. 1, 29–36, DOI 10.2140/involve.2009.2.29. MR2501343

[2] J. D. Bovey, P. Erdős, and I. Niven, *Conditions for a zero sum modulo n*, Canad. Math. Bull. **18** (1975), no. 1, 27–29, DOI 10.4153/CMB-1975-004-4. MR392901

[3] L. Carlitz, *A characterization of algebraic number fields with class number two*, Proc. Amer. Math. Soc. **11** (1960), 391–392, DOI 10.2307/2034782. MR111741

[4] H. S. Choi, *Class group and factorization in orders of a PID*, J. Number Theory **265** (2024), 226–269, DOI 10.1016/j.jnt.2024.06.008. MR4779820

[5] H. Cohn, *Advanced number theory*, Dover Books on Advanced Mathematics, Dover Publications, Inc., New York, 1980. Reprint of *A second course in number theory*, 1962. MR594936

[6] D. A. Cox, *Primes of the form $x^2 + ny^2$—Fermat, class field theory, and complex multiplication*, 3rd ed., AMS Chelsea Publishing, Providence, RI, [2022] ©2022. With contributions by Roger Lipsett. MR4502401

[7] K. Cziszter, M. Domokos, and A. Geroldinger, *The interplay of invariant theory with multiplicative ideal theory and with arithmetic combinatorics*, Multiplicative ideal theory and factorization theory, Springer Proc. Math. Stat., vol. 170, Springer, [Cham], 2016, pp. 43–95, DOI 10.1007/978-3-319-38855-7_3. MR3565804

[8] H. Davenport, *Midwestern conference on group theory and number theory*, Ohio State University (1966).

[9] A. Geroldinger, *Chains of factorizations in weakly Krull domains*, Colloq. Math. **72** (1997), no. 1, 53–81, DOI 10.4064/cm-72-1-53-81. MR1425546

[10] A. Geroldinger and D. J. Grynkiewicz, *The large Davenport constant I: Groups with a cyclic, index 2 subgroup*, J. Pure Appl. Algebra **217** (2013), no. 5, 863–885, DOI 10.1016/j.jpaa.2012.09.004. MR3003311

[11] A. Geroldinger and F. Halter-Koch, *Non-unique factorizations: Algebraic, combinatorial and analytic theory*, Pure and Applied Mathematics (Boca Raton), vol. 278, Chapman & Hall/CRC, Boca Raton, FL, 2006, DOI 10.1201/9781420003208. MR2194494

[12] A. Geroldinger, F. Halter-Koch, and Q. Zhong, *On monoids of weighted zero-sum sequences and applications to norm monoids in Galois number fields and binary quadratic forms*, Acta Math. Hungar. **168** (2022), no. 1, 144–185, DOI 10.1007/s10474-022-01270-x. MR4517537

[13] A. Geroldinger and F. Kainrath, *On the arithmetic of tame monoids with applications to Krull monoids and Mori domains*, J. Pure Appl. Algebra **214** (2010), no. 12, 2199–2218, DOI 10.1016/j.jpaa.2010.02.023. MR2660909

[14] D. J. Grynkiewicz, *The large Davenport constant II: General upper bounds*, J. Pure Appl. Algebra **217** (2013), no. 12, 2221–2246, DOI 10.1016/j.jpaa.2013.03.002. MR3057306

[15] F. Halter-Koch, *A generalization of Davenport's constant and its arithmetical applications*, Colloq. Math. **63** (1992), no. 2, 203–210, DOI 10.4064/cm-63-2-203-210. MR1180633

[16] F. Halter-Koch, *Relative block semigroups and their arithmetical applications*, Comment. Math. Univ. Carolin. **33** (1992), no. 3, 373–381. MR1209280

[17] F. Halter-Koch, *Elasticity of factorizations in atomic monoids and integral domains* (English, with English and French summaries), J. Théor. Nombres Bordeaux **7** (1995), no. 2, 367–385, DOI 10.5802/jtnb.147. MR1378586

[18] J. Kettinger, *Elasticity of orders in quadratic rings of integers*, Master's thesis, Clemson University, 2024.

[19] J. Kettinger and G. Moles, *Elasticity of orders with prime conductor*, J. Number Theory **279** (2026), 579–593, DOI 10.1016/j.jnt.2025.07.013. MR4948501

[20] J. H. Lee and S. R. Louboutin, *Determination of the orders generated by a cyclic cubic unit that are Galois invariant*, J. Number Theory **148** (2015), 33–39, DOI 10.1016/j.jnt.2014.09.031. MR3283165

[21] W. Narkiewicz, *Elementary and analytic theory of algebraic numbers*, Monografie Matematyczne [Mathematical Monographs], Tom 57, PWN—Polish Scientific Publishers, Warsaw, 1974. MR347767

[22] W. Narkiewicz, *A note on elasticity of factorizations*, J. Number Theory **51** (1995), no. 1, 46–47, DOI 10.1006/jnth.1995.1034. MR1321723

[23] J. Neukirch, *Algebraic number theory*, Grundlehren der mathematischen Wissenschaften [Fundamental Principles of Mathematical Sciences], vol. 322, Springer-Verlag, Berlin, 1999. Translated from the 1992 German original and with a note by Norbert Schappacher; With a foreword by G. Harder, DOI 10.1007/978-3-662-03983-0. MR1697859

[24] J. E. Olson, *A combinatorial problem on finite Abelian groups. I*, J. Number Theory **1** (1969), 8–10, DOI 10.1016/0022-314X(69)90021-3. MR237641

[25] B. D. Sittinger, *Quotients of primes in an algebraic number ring*, arXiv preprint, `arXiv:1607.08319`, (2016).

[26] R. J. Valenza, *Elasticity of factorization in number fields*, J. Number Theory **36** (1990), no. 2, 212–218, DOI 10.1016/0022-314X(90)90074-2. MR1072466

Department of Mathematical and Statistical Sciences, Clemson University
Email address: `jkettin@clemson.edu`

Contemporary Mathematics
Volume **836**, 2026
https://doi.org/10.1090/conm/836/16710

Additive subgroups of a module that are saturated with respect to a subset of the ring

Jesse Elliott and Neil Epstein

Dedicated to K. Alan Loper on the occasion of his retirement

Abstract. Let T be a subset of a ring A, and let M be an A-module. We study the additive subgroups F of M such that, for all $x \in M$, if $tx \in F$ for some $t \in T$, then $x \in F$. We call any such subset F of M a *T-factroid of M*, which is a kind of dual to the notion of a T-submodule of M. We connect the notion with the zero-divisors on M, various classes of primary and prime ideals of A, Euclidean domains, and the recent concepts of unit-additive commutative rings and of Egyptian fractions with respect to a multiplicative subset of a commutative ring. We also introduce a common generalization of local rings and unit-additive rings, called *sublocalizable* rings, and relate them to T-factroids.

1. Introduction

Throughout this paper, A denotes a ring, M denotes a (left) A-module, and $W \subseteq A$ denotes a multiplicatively closed set. All rings are unital unless otherwise stated.

We introduce here a new type of algebraic structure that is motivated on several grounds. Specifically, for a given subset T of A, we consider the additive subgroups F of M such that, for all $x \in M$, if $tx \in F$ for some $t \in T$, then $x \in F$. We call any such subset F of M a *T-factroid of M*. Its obvious dual notion is that of a *T-submodule of M*, which is an additive subgroup F of M such that $tx \in F$ whenever $x \in F$ and $t \in T$. Since a T-submodule of M is equivalently a $\mathbb{Z}\langle T\rangle$-submodule of M by restriction of scalars, where $\mathbb{Z}\langle T\rangle$ is the subring of A generated by T, the notion of a T-submodule can be reduced to that of a B-submodule for the subrings B of A. The notion of a T-factroid, however, appears to be new, and it is thus motivated abstractly as a dual to the notion of a T-submodule.

However, the initial motivation for our exploration of T-factroids was our discovery of their intimate connection the study of Egyptian domains, the reciprocal complement of an integral domain, and their respective generalizations to commutative rings with zerodivisors. Recall [**GLO24**] that for an integral domain A with

2020 *Mathematics Subject Classification.* Primary 13A99, 16D10; Secondary 16U99, 13P05, 13F07.

Jesse Elliott passed away June 29, 2025, after acceptance but before publication of this article. Jesse was a crucially valued member of the commutative ring theory community, typically finding the right generalization of a concept before anyone else, and will be greatly missed.

©2026 American Mathematical Society

fraction field K, an element $\alpha \in K$ is *A-Egyptian* if there exist distinct nonzero elements $d_1, \ldots, d_n$ of A such that $\alpha = \sum_{i=1}^{n} \frac{1}{d_i}$, so-called because of Egyptian fractions in number theory [**BE22**] (and ultimately in medieval European [**DG66**] and ancient Egyptian [**Rei14**] mathematics). It follows from [**GLO24**, Theorem 1] and [**Eps25**, Remark 2.2] that the A-Egyptian elements of K form a subring $R(A)$ of K, called the *reciprocal complement* of A. An integral domain A is said to be *Egyptian* if every element of A, or, equivalently, of its quotient field K, is A-Egyptian [**GLO24**]. Clearly, then, a domain A is Egyptian if and only if its reciprocal complement is as large as it could be, namely K. Although some reciprocal complements are easy to understand (e.g. classically $R(\mathbb{Z}) = \mathbb{Q}$ (see [**vAvL63**] and [**DG66**]), and the reciprocal complement of a Euclidean domain is a DVR or a field [**Eps25**, Main Theorem]), some are not (e.g. the reciprocal complement of $k[x,y]$, for k a field, is objectively very strange [**EGL24**]). For more on Egyptian domains and reciprocal complements, see also [**Eps24a, Gue25, Eps23, Eps24b**].

It is trivial to see that $R(A)$, for any integral domain A, is an $(A \setminus \{0\})$-factroid of its quotient field K. In fact, the connection between the $R(A)$ and the $(A \setminus \{0\})$-factroids is deeper than that. Note first that the T-factroids of M are closed under arbitrary intersections (see Proposition 2.10), so one can define the *T-factroid of M generated by a set S*, denoted $[S]_M^T$, to be the smallest T-factroid of M containing S. One then has the following:

THEOREM 1.1 (see Theorem 7.5). *Let A be an integral domain, K its fraction field, $0 \neq \alpha \in K$, and $T = A \setminus \{0\}$. Then the following are equivalent:*

(a) $\alpha \in R(A)$.
(b) *There exist $f, g \in A$ with $\alpha = f/g$ and $f \in [\{g\}]_A^T$.*
(c) *For any $f, g \in A$ with $\alpha = f/g$, there is some nonzero $h \in A$ with $hf \in [\{hg\}]_A^T$.*

Of course, Theorem 1.1 begs to be generalized to the T-factroids of a general A-module M for any arbitrary subset T of A. In Section 7, this is achieved in the case where A is any commutative ring.

In Section 8, we use factroids to analyze Euclidean domains. Namely, we classify the factroids of Euclidean domains in terms of the values of their Euclidean functions, and show that one version of the classification holds if and only if the domain is of the form $k[x]$.

A commutative ring is said to be *unit-additive* if a sum of units is either a unit or nilpotent (see [**ES25**]). We define the following more general class of rings: a commutative ring A is *sublocalizable* if a sum of units is either a unit or an element of its Jacobson radical $J(A)$. Such a ring always has a distinguished local subring consisting of its units along with $J(A)$, which we call its *sublocalization*. We show that the sublocalization of a sublocalizable ring is a particularly nice factroid of A, namely, the subring $\mathbb{Z}\langle A^\times\rangle$ of A generated by the group of units of A. In Section 9, we provide strong connections between T-factroids, unit-additive rings, and sublocalizable rings.

Examples of T-factroids are ubiquitous in both commutative and noncommutative algebra. Indeed, for any additive subgroup F of M, it is clear that

$$W_M(F) := W(F) := \{a \in A \mid (F :_M a) \subseteq F\}$$

is a subset T of A for which F is a T-factroid of M. In fact, by its definition it is the largest such subset, and one can easily check that it is also a multiplicative submonoid of A. In many instances, one can express $W_M(F)$ in terms of known constructs, and vice versa. For example, if F is an A-submodule of M, then $W_M(F)$ is just the saturated multiplicative monoid consisting of all elements of A that are regular on the A-module M/F (see Proposition 3.10). Thus, for example $W_{{}_AA}(\{0\})$ is just the monoid of all left-regular elements of A, where ${}_AA$ denotes A as a left A-module. From the theory of primary decomposition, then, it follows that, if M is a finitely generated module over a commutative Noetherian ring A, then $W_M(0)$ is equal to the complement of the union of the associated primes of M (see Proposition 3.13). Moreover, if A is an arbitrary commutative ring, then one has $W_A(\sqrt{I}) = A \smallsetminus \bigcup \mathrm{Min}(I)$ for any ideal I of A, where $\mathrm{Min}(I)$ is the set of all prime ideals of A that are minimal over I (see Proposition 4.8). Thus, I is prime if and only if $W_A(I) = A \smallsetminus I$. In particular, $W(\sqrt{0})$ is the saturated multiplicative set A° of all elements of A that do not lie in any minimal prime of A, a set that figures centrally in the definition of tight closure over a commutative Noetherian ring of prime characteristic [**HH90**]. Also, it is easy to check that, again if A is commutative, then any intersection $\bigcap \mathfrak{q}_i$ of primary ideals $\mathfrak{q}_i$ of A is an $(A \smallsetminus \bigcup \sqrt{\mathfrak{q}_i})$-factroid of A (see Proposition 4.7).

Further instances of T-factroids abound. For example, if $A = k[x]$ for some field k, then the $(A \smallsetminus \{0\})$-factroids of A besides $\{0\}$ and A are precisely the k-spans of $\{1, x, x^2, \dots, x^m\}$ for all nonnegative integers m. (See Theorem 8.3 for a generalization of this fact to more general Euclidean domains.) However, if $A = k[x_1, \dots, x_n]$ for $n \geq 2$, then there are more $(A \smallsetminus \{0\})$-factroids than just the k-spans of $\{1, x_i, x_i^2, \dots, x_i^m\}$ for $m \geq 0$ and $1 \leq i \leq n$. For instance, if $f \in A$ such that $f + c$ is an irreducible polynomial for all $c \in k$, then $\mathrm{span}_k\{1, f\}$ is an $(A \smallsetminus \{0\})$-factroid (see Example 4.17). The problem of determining the $(A \smallsetminus \{0\})$-factroids of A remains open, even when $n = 2$.

Our paper is structured as follows. Section 2 introduces most of the main definitions and basic results. Section 3 develops duality and interaction of factroid theory with module theory; at the end of the section, we include an extension of the theory to the context of abelian group endomorphisms. In Section 4, we provide the reader with examples of factroids of rings for various familiar classes and constructions of commutative rings. In Section 5, we explore what it means for a set to *generate* a W-factroid of a module or a ring. In Section 6, we develop the theory of W-*regular* T-factroids, the necessary theoretical foundation for Section 7, where we make the connection with Egyptian domains and related topics. In Section 8, we show that the factroids of a Euclidean domain have a particularly elegant description in terms of Euclidean functions. Section 9 is where we discuss unit-additive and sublocalizable rings. In Section 10, we discuss how the objects and operations of factroid theory work when passing along ring and module homomorphisms. Finally, in Section 11, we show that in certain common graded settings, finiteness conditions apply that can make it easier to determine explicitly the membership in factroids generated by a given set.

We thus hope to have shown the use and appeal of T-factroids in commutative and noncommutative algebra.

2. T-factroids of a module

In this section, we define and prove some basic facts about the *T-factroids of M*. The no(ta)tions developed here will be crucial for all that follows.

DEFINITION 2.1. Let A be a ring, I a (two-sided) ideal of A, and M a (left) A-module. Let $T \subseteq A$ and $S, U \subseteq M$. Throughout the paper, we use the following notation and conventions:

- $Z(A) :=$ the center of A.
- $A^\times :=$ the group of units of A.
- $\operatorname{reg}(A) :=$ the multiplicative monoid of all left-regular elements, i.e., all left nonzerodivisors, of A.
- $\langle T \rangle$ denotes the multiplicative submonoid of A generated by T.
- $\mathbb{Z}\langle T \rangle$ denotes the subring of A generated by T.
  - Thus, $\mathbb{Z}\langle T \rangle = \mathbb{Z}\langle \langle T \rangle \rangle$ is the set of all sums of elements of $\langle T \rangle$.
- $\widetilde{T} = \{a \in A : ba \in T \text{ for some } b \in A\}$ denotes the *(left) saturation* of T.
- One says that T is *(left) saturated* if $\widetilde{T} = T$, that is, if $ba \in T$ implies $a \in T$ for any $a, b \in A$.
- T^{sm} denotes the smallest saturated multiplicative submonoid of A containing T.
  - T^{sm} exists because saturated sets and multiplicative submonoids are closed under arbitrary intersections.
  - Note that if A is commutative, then $T^{\mathrm{sm}} = \widetilde{\langle T \rangle}$.
- If $T \subseteq A^\times$, then $T^{-1} := \{t^{-1} \mid t \in T\} \subseteq A^\times$.
- $(S :_M T) := \{x \in M \mid tx \in S \text{ for all } t \in T\}$.
  - When $T = \{t\}$ is singleton, write $(S :_M t) := (S :_M T)$.
- $(S :_A U) := \{a \in A \mid au \in S \text{ for all } u \in U\}$.
  - When $U = \{u\}$ is a singleton, write $(S :_A u) := (S :_A U)$.
- $\operatorname{reg}_M(A) := \{a \in A \mid (0 :_M a) = 0\}$ denotes the set of all *nonzerodisivors on* M (also known as *M-regular elements*).
  - Note that $\operatorname{reg}(A) = \operatorname{reg}_{{}_AA}(A)$, where ${}_AA$ is the left A-module A.
- When A is commutative, $A^\circ :=$ the complement of the union of the minimal (i.e., height zero) prime ideals of A.
  - Note that A° is a saturated multiplicative submonoid of A.
  - Note that $\operatorname{reg}(A) \subseteq A^\circ$, with equality if A is reduced.
- $J(A) :=$ the Jacobson radical of A.
- $\sqrt{I} :=$ the radical of I.
- The set $\mathbb{N}$ of all natural numbers includes 0, by convention.

DEFINITION 2.2. Let $T \subseteq A$.

We say that $S \subseteq M$ is *T-saturated (in M)* if $(S :_M t) \subseteq S$ for all $t \in T$. In other words, S is T-saturated if, for all $x \in M$ such that $tx \in S$ for some $t \in T$, one has $x \in S$. In particular, a subset T of A is a saturated set in the sense of Definition 2.1 if and only if it is A-saturated in (the left module) A.

We say that $F \subseteq M$ is a *T-factroid of M* if F is a T-saturated additive subgroup of M. In other words, F is a *T-factroid of M* if F is an additive subgroup of M such that, for all $x \in M$ such that $tx \in F$ for some $t \in T$, one has $x \in F$. By a *(left) T-factroid of A* we mean a T-factroid of the left A-module ${}_AA$.

Colloquially, a T-factroid of an A-module is an additive subgroup of the module that absorbs colons from T.

DEFINITION 2.3. Let $S \subseteq M$ and $T \subseteq A$. The *T-saturation of S* is the set

$$\mathrm{Sat}_M^T(S) := \bigcup_{w \in \langle T \rangle} (S :_M w) = \{x \in M \mid wx \in S \text{ for some } w \in \langle T \rangle\}.$$

It is the smallest T-saturated subset of M containing S.

EXAMPLE 2.4. One has $\widetilde{T} = \mathrm{Sat}_A^A(T)$ for any subset T of A.

REMARK 2.5. If $U \subseteq T$ are subsets of A, then any T-factroid of M is a U-factroid of M. If $T \subseteq \{\pm 1\}$, then a T-factroid of M is the same as an additive subgroup of M. Suppose that $-1 \in T$. Then to check that $F \subseteq M$ is a T-factroid, in addition to showing that $(F :_M t) \subseteq F$ for all $t \in T$, one need only show that F is nonempty and additively closed. This is because for any $x \in F$, we have $-x \in (F :_M -1)$.

EXAMPLE 2.6. It is clear that M is a T-factroid of M. However, the trivial A-submodule $\{0\}$ of M is a T-factroid of M if and only if every element of T is a nonzerodivisor on M. Moreover, if $0 \in T$, or more generally if $T \cap \mathrm{ann}_A M \neq \varnothing$, then M is the only T-factroid of M.

REMARK 2.7. Let $T \subseteq A$. Suppose that F is a T-factroid of M and B is a subring of A. Then F is a $(T \cap B)$-factroid of the B-module M obtained by restriction of scalars. Thus, if $T \subseteq B$, then F is also a T-factroid of the B-module M.

REMARK 2.8. Let $T \subseteq A$. A T-factroid of M is equivalently an additive subgroup F of M such that $\ker(M \xrightarrow{t\cdot} M \to M/F) \subseteq \ker(M \to M/F)$ for all $t \in T$, where $M \to M/F$ is the canonical projection of abelian groups.

PROPOSITION 2.9. *Let $T \subseteq A$. A T-saturated subset of M is equivalently a $\langle T \rangle$-saturated subset of M. Similarly, a T-factroid of M is equivalently a $\langle T \cup \{\pm 1\} \rangle$-factroid of M.*

PROOF. Let S be any subset of M. Clearly, for any $a, b \in A$, if $(S :_M a) \subseteq S$ and $(S :_M b) \subseteq S$, then $(S :_M ab) = ((S :_M a) :_M b)) \subseteq S$. Moreover, one has $(S :_M 1) = S$, and $(S :_M -1) = S$ if $S = -S$. □

Thus, one may always assume without loss of generality that T is multiplicatively closed, or even a submonoid of A containing -1.

PROPOSITION 2.10. *Let $T \subseteq A$. The intersection of any collection of T-saturated subsets of M is a T-saturated subset of M. Likewise, the intersection of any collection of T-factroids of M is a T-factroid of M.*

PROOF. Let $\{S_\alpha \mid \alpha \in \Lambda\}$ be a collection of T-saturated subsets of M, and let $S = \bigcap_{\alpha \in \Lambda} S_\alpha$. Let $t \in T$ and $x \in M$ with $tx \in S$. Then one has $tx \in S_\alpha$, whence $x \in S_\alpha$, for all $\alpha \in \Lambda$, and therefore $x \in S$. Moreover, if each S_α is an additive subgroup of M, then so is S. □

In light of Proposition 2.10, we can make the following definition.

DEFINITION 2.11. Let $T \subseteq A$. For any subset S of M, the *T-factroid of M generated by S* is the smallest T-factroid M containing S, denoted $[S]_M^T$. In particular,

$$[S]_M^T = \bigcap \{F \mid F \text{ is a } T\text{-factroid of } M \text{ and } S \subseteq F\}.$$

When M is understood, we abbreviate $[S]_M^T$ to $[S]^T$. A T-factroid generated by a singleton is *principal*. In this case, when $S = \{x\}$ for some $x \in M$, we write $[x]_M^T$ or $[x]^T$ in place of $[\{x\}]_M^T$.

By Proposition 2.9, we have the following.

COROLLARY 2.12. *For all $S \subseteq M$ and all $T \subseteq A$, $[S]_M^T = [S]_M^{\langle T\rangle} = [S]_M^{\langle T\cup\{\pm 1\}\rangle}$.*

LEMMA 2.13. *Let $S \subseteq M$ and $T \subseteq A$. One has*

$$[S]_M^T \supseteq \mathrm{Sat}_M^T(S) = \bigcup_{w\in\langle T\rangle} (S :_M w) = \{x \in M \mid wx \in S \text{ for some } w \in \langle T\rangle\}.$$

PROOF. This follows because any T-factroid of M is T-saturated in M. □

PROPOSITION 2.14. *Let $T \subseteq A$. The T-factroid $[\varnothing]_M^T = [0]_M^T$ of M is the smallest T-factroid of M. Moreover, one has*

$$[\varnothing]_M^T \supseteq \mathrm{Sat}_M^T(\{0\}) = \bigcup_{w\in\langle T\rangle} (0 :_M w) = \{x \in M \mid wx = 0 \text{ for some } w \in \langle T\rangle\}.$$

Equality holds if $T \subseteq Z(A)$, or more generally if $st = ts$ for every $t, s \in T$.

PROOF. By Lemma 2.13, we need only prove the last statement of the proposition. We may suppose without loss of generality that $W := T$ is a multiplicative submonoid of A. It suffices to show that the T-saturation F of $\{0\}$ in M is an additive subgroup of M. To see this, let $x, y \in F$, so that $wx = 0$ and $vy = 0$ for some $w, v \in W$. Then $vw(x - y) = vwx - wvy = 0$, since $vw = wv$. Finally, since $vw \in W$, it follows that $x - y \in F$. □

COROLLARY 2.15. *If the ring A is commutative, then for any multiplicative set W,*

$$[\varnothing]_M^W = \ker(M \to W^{-1}M),$$

where $M \to W^{-1}M$ is the localization map.

EXAMPLE 2.16. If the elements of $T \subseteq A$ do not commute with one another, then the union in Proposition 2.14 need not be additively closed.

For instance, let $A = k\langle x, y\rangle$ (a ring freely generated over a field k by two noncommuting indeterminates), $W = T = \langle x, y\rangle$, and $M = (A \oplus A)/(xA \oplus yA)$. Then in M, we have that $x(1,0) = (\bar{x}, 0) = 0$ and $y(0,1) = (0, \bar{y}) = 0$, but $\mathrm{ann}_A(1,1) = 0$. Thus, $(1,0), (0,1) \in \bigcup_{w\in W}(0 :_M w)$, but their sum is not.

DEFINITION 2.17. Given a subset S of M, we let

$$W_M(S) := W(S) := \{a \in A \mid (S :_M a) \subseteq S\}.$$

PROPOSITION 2.18. *Let S be any subset of M. Then $W(S)$ is a multiplicative submonoid of A. Moreover, $W(S)$ is the largest subset T of A for which S is T-saturated in M; in other words, for any subset T of A, S is T-saturated in M if and only if $T \subseteq W(S)$.*

PROOF. This is clear from the definitions and from Proposition 2.9. □

COROLLARY 2.19. *Let F be any additive subgroup of M. Then $W(F)$ is a multiplicative submonoid of A containing -1. Moreover, $W(F)$ is the largest subset T of A for which F is a T-factroid of M; in other words, for any subset T of A, F is a T-factroid of M if and only if $T \subseteq W(F)$.*

REMARK 2.20. An additive subgroup F of M is proper if and only if $0 \notin W(F)$, if and only if $W(F) \subsetneq A$. Moreover, $W(F)$ is saturated if and only if, for all $T \subseteq A$, F is a T-factroid of M if and only if F is a T^{sm}-factroid of M.

The fact that the intersection of any collection of T-factroids of M is a T-factroid of M for any subset T of A is equivalent to the inclusion $W_M(\bigcap_{\lambda \in \Lambda} F_\lambda) \supseteq \bigcap_{\lambda \in \Lambda} W_M(F_\lambda)$ for any collection $\{F_\lambda : \lambda \in \Lambda\}$ of additive subgroups of M. Indeed, each F_λ is a T-factroid of M if and only if $T \subseteq W(F_\lambda)$ for all λ, if and only if $T \subseteq \bigcap_{\lambda \in \Lambda} W(F_\lambda)$; and $\bigcap_{\lambda \in \Lambda} F_\lambda$ is a T-factroid of M if and only if $T \subseteq W(\bigcap_{\lambda \in \Lambda} F_\lambda)$.

PROPOSITION 2.21. *One has* $W_M(\{0\}) = \mathrm{reg}_M(A)$. *Equivalently, the saturated submonoid* $\mathrm{reg}_M(A)$ *of* A *is the largest subset* T *of* A *such that* $\{0\}$ *is a* T*-factroid of* M.

PROOF. This is immediate from the definitions. □

COROLLARY 2.22. *One has* $W_A((0)) = \mathrm{reg}(A)$. *Equivalently, the saturated submonoid* $\mathrm{reg}(A)$ *of* A *is the largest subset* T *of* A *such that the zero ideal* (0) *of* A *is a* T*-factroid of* A.

3. The duality between factroids and submodules

If A is an integral domain, so that $\mathrm{reg}(A) = A^\circ = A \setminus \{0\}$, then the notion of a $\mathrm{reg}(A)$-factroid of M is in a sense dual to the notion of an A-submodule of M. Indeed, it is clear that an additive subgroup L of M is an A-submodule of M if and only if $L \subseteq (L :_M a)$ for every $a \in \mathrm{reg}(A)$, while the notion of a $\mathrm{reg}(A)$-factroid of M has the given inclusion reversed.

We can make the above duality more explicit and general as follows:

DEFINITION 3.1. Let $T \subseteq A$. We say that a *T-submodule of M* is an additive subgroup F of M such that $TF \subseteq F$, or equivalently such that $F \subseteq (F :_M t)$ for all $t \in T$.

In this section, we explore the duality between T-submodules and T-factroids. The correspondence is especially strong when T consists of units. We also give special attention to the theory of regular elements and primary decomposition. We conclude the section with an extension of factroid theory to the setting of endomorphisms of abelian groups.

REMARK 3.2. If B is a subring of A, then a B-submodule of M by restriction of scalars is equivalently a B-submodule of M in the sense of Definition 3.1.

DEFINITION 3.3. For any additive subgroup F of M, we let

$$A(F) := (F :_A F) = \{a \in F : aF \subseteq F\} = \{a \in A : F \subseteq (F :_M a)\}.$$

PROPOSITION 3.4. *Let F be an additive subgroup of M. Then the following statements hold:*

(1) *$A(F)$ is the largest subset T of A for which F is a T-submodule of M.*
(2) *For any subset T of A, F is a T-submodule of M if and only if $T \subseteq A(F)$.*
(3) *$A(F)$ is a subring of A. Indeed, it is the largest subring B of A for which F is a B-submodule of M.*
(4) *F is an A-submodule of M if and only if $A(F) = A$.*
(5) *For any subset T of A, F is a T-submodule of M if and only if F is a $\mathbb{Z}\langle T\rangle$-submodule of M.*

PROOF. All statements follow immediate from the definitions. □

REMARK 3.5. Let $\mathrm{Mod}_{\mathbb{Z}}(M)$ denote the (complete) poset of all additive subgroups of M. Then $C_a : F \mapsto (F :_M a)$ is an operation on the set $\mathrm{Mod}_{\mathbb{Z}}(M)$, for any $a \in A$. Moreover, the association $a \mapsto C_a$ defines a map $A \to \mathrm{End}_{\mathbf{Set}}(\mathrm{Mod}_{\mathbb{Z}}(M))$ with $C_1 = \mathrm{id}$ and $C_b \circ C_a = C_{ab}$ for all $a, b \in A$. For all $F \in \mathrm{Mod}_{\mathbb{Z}}(M)$, one has $W(F) = \{a \in A : C_a(F) \subseteq F\}$ and $A(F) = \{a \in A : F \subseteq C_a(F)\}$.

PROPOSITION 3.6. *Let T be a subset of $A^\times$. Then $\mathbb{Z}\langle T^{-1}\rangle = [1]_A^T$, and a T-factroid of M is equivalently a T^{-1}-submodule of M, or equivalently a $\mathbb{Z}\langle T^{-1}\rangle$-submodule of M. Conversely, $\mathbb{Z}\langle T\rangle = [1]_A^{T^{-1}}$, and a T-submodule of M is equivalently a $\mathbb{Z}\langle T\rangle$-submodule of M, or equivalently a T^{-1}-factroid of M.*

PROOF. This follows from the fact that when $w \in A^\times$, $(F :_M w) = w^{-1}F$, and therefore $(F :_M w) \subseteq F$ if and only if $w^{-1}F \subseteq F$, for any subset F of M. □

COROLLARY 3.7. *Suppose that W is a subgroup of $A^\times$. Then a W-factroid of M is equivalently a W-submodule of M, or equivalently a B-submodule of M, where $B = \mathbb{Z}\langle W\rangle = [1]_A^W$ is the subring of A consisting of all finite sums of elements of W.*

COROLLARY 3.8. *Suppose that A is commutative, and let $V \subseteq W$ be multiplicative subsets of A. Let $\frac{V}{1} = \{v/1 : v \in V\}$ denote the image of the set V in $W^{-1}A$, and let $\frac{1}{V} = \left(\frac{V}{1}\right)^{-1}$ denote the multiplicative set of reciprocals of the elements of $\frac{V}{1}$ in $W^{-1}A$. Then a $\frac{V}{1}$-factroid of $W^{-1}M$ is equivalently a $\frac{1}{V}$-submodule of $W^{-1}M$, or equivalently a $\mathbb{Z}\langle\frac{1}{V}\rangle$-submodule of $W^{-1}M$. Conversely, a $\frac{V}{1}$-submodule of $W^{-1}M$ is equivalently a $\mathbb{Z}\langle\frac{V}{1}\rangle$-submodule of $W^{-1}M$, or equivalently a $\frac{1}{V}$-factroid of $W^{-1}M$. Moreover, one has $\mathbb{Z}\langle\frac{1}{V}\rangle = [1]_{W^{-1}M}^{\frac{V}{1}}$ and $\mathbb{Z}\langle\frac{V}{1}\rangle = [1]_{W^{-1}M}^{\frac{1}{V}}$.*

PROPOSITION 3.9. *Let F be any additive subgroup of M. Then the units of A in $W(F)$ are precisely the inverses of the units of A in $A(F)$, that is,*

$$W(F) \cap A^\times = (A(F) \cap A^\times)^{-1}.$$

Thus, the units of A in $W(F)$ are completely determined by the ring $A(F)$. Moreover, the multiplicative monoid

$$W(F) \cap A(F) = \{a \in A \mid (F :_M a) = F\}$$

is the largest subset T of A for which F is both a T-submodule and T-factroid of M.

PROOF. This follows readily from the fact that $(F :_M w) = w^{-1}F$ for all units w of A. □

We next look at when A-submodules are T-factroids.

PROPOSITION 3.10. *Let L be an A-submodule of M. Then $W_M(L) = \mathrm{reg}_{M/L}(A)$; therefore $W_M(L)$ is saturated. Moreover, for any subset T of A, the following are equivalent:*

(a) *L is a T-factroid of M.*
(b) *$T \subseteq \mathrm{reg}_{M/L}(A)$.*
(c) *L is a T^{sm}-factroid of M.*

Thus, if A is commutative, then $W_M(L) = \ker(M/L \to W^{-1}(M/L))$, where W is the multiplicative set $\langle T\rangle$ (or $T^{\mathrm{sm}} = \widetilde{\langle T\rangle}$) and $M/L \to W^{-1}(M/L)$ is the localization map.

PROOF. We have the following equivalences for any $t \in A$:

$$\begin{aligned} t \in W_M(L) &\Leftrightarrow (L :_M t) \subseteq L \\ &\Leftrightarrow (0 :_{M/L} t) = 0 \\ &\Leftrightarrow t \in \mathrm{reg}_{M/L}(A). \end{aligned}$$

The proposition follows. □

Since every additive subgroup F of M is an $A(F)$-submodule of M, so that M/F is an $A(F)$-module, one can relativize Proposition 3.10 to characterize the additive subgroups of M that are both T-submodules of M and T-factroids F of M in terms of the $A(F)$-module M/F, as follows.

COROLLARY 3.11. *Let T be a subset of A. An additive subgroup F of M is both a T-submodule F of M and T-factroid of M if and only if $T \subseteq A(F)$ and the elements of T are all nonzerodivisors on the $A(F)$-module M/F.*

Recall that a proper submodule L of a module M over a commutative ring A is *primary* if for any $r \in A$, either $(L :_M r) = L$ or $r^n M \subseteq L$ for some $n \in \mathbb{N}$.

PROPOSITION 3.12. *Suppose that A is commutative. Let L be proper A-submodule of M, and let $T \subseteq A$. If L is a T-factroid of M, then $\langle T \rangle \cap \mathrm{ann}_A(M/L) = \varnothing$. The converse holds if L is a primary submodule of M.*

PROOF. Without loss of generality, we may assume that $W := T$ is multiplicatively closed. Suppose there is some $r \in W \cap \mathrm{ann}_A(M/L)$. Let $x \in M \setminus L$. Then $rx \in L$ but $x \notin L$, so L is not a W-factroid of M.

On the other hand, suppose $I \cap W = \varnothing$, where $I = \mathrm{ann}_A(M/L)$, and $L \subseteq M$ is primary. Let $r \in W$ and $x \in M$ with $rx \in L$. Since L is primary in M, either $r \in \sqrt{I}$ or $x \in L$. But if $r \in \sqrt{I}$, then there is some n with $r^n \in I \cap W$, contrary to hypothesis. Hence, $x \in L$. Thus, L is a W-factroid of M. □

One can identify even more precisely what is happening when primary decompositions are present.

PROPOSITION 3.13. *Let A be a commutative Noetherian ring and $L \subseteq M$ finite A-modules. Then $W(L) = A \setminus \bigcup \mathrm{Ass}(M/L)$. More generally, if $L \subseteq M$ admits a primary decomposition, then $W(L)$ is the complement of the union of the primes belonging to L in M, in the sense of* [**AM69**, Exercises 4.22-23].

PROOF. In either situation, the zerodivisors on M/L are precisely the elements of the given union. Then Proposition 3.10 finishes the proof. □

REMARK 3.14. Proposition 3.13 does not always hold when $M = A$ and $L = (0)$ does not admit a primary decomposition, even if the set $\mathrm{Ass}(M)$ is replaced by the set of strong Krull or weak Bourbaki primes of M. See [**ES14**, Example 6.4].

REMARK 3.15. Although the elementary aspects of this subject apply to noncommutative rings, most of the examples and applications we found were for commutative rings. This leaves the subject ripe for further study over noncommutative rings.

REMARK 3.16. Let G be an abelian group, and let $E \subseteq \mathrm{End}(G) = \mathrm{End}_{\mathbb{Z}}(G)$. Define an *$E$-factroid of G* to be a subgroup F of G such that $\varphi^{-1}(F) \subseteq F$ for all $\varphi \in E$. Given a subgroup F of G, let $E(F)$ denote the set of all $\varphi \in \mathrm{End}(G)$ such that $\varphi^{-1}(F) \subseteq F$. Then $E(F)$ is a submonoid of the multiplicative monoid of the endomorphism ring $\mathrm{End}(G)$, and a subgroup F of G is an E-factroid of G if and only if E is a subset of $E(F)$. Moreover, $E(F)$ is the set of all endomorphisms φ of G such that $\ker(G \xrightarrow{\varphi} G \to G/F) \subseteq \ker(G \to G/F)$, where $G \to G/F$ denotes the canonical projection.

Suppose, for instance, that $G = M$ is an A-module. Then the canonical ring homomorphism $i : A \to \mathrm{End}_{\mathbb{Z}}(M)$ restricts to a monoid homomorphism $W(F) \to E(F)$, and F is a T-factroid of M, where T is a subset of A, if and only if F is an $i(T)$-factroid of M, if and only if $i(T) \subseteq i(W(F))$. In other words, one has $i(W(F)) = E(F)$. Note also that $i(Z(A)) \subseteq \mathrm{End}_A(M)$, so, if A is commutative, then any T-factroid of M for $T \subseteq A$ is an E-factroid of M for some $E \subseteq \mathrm{End}_A(M)$. Thus in this case, one may avoid group endomorphisms of M that are not A-linear.

The dual notion to that of an E-factroid of G is the following: say that an *E-submodule of G* is a subgroup F of G such that $\varphi(F) \subseteq F$ (or, equivalently, $F \subseteq \varphi^{-1}(F)$) for all $\varphi \in E$. This extends the notion of a T-submodule of an A-module in the obvious way. For any subgroup F of G, the set $S(F)$ of all $\varphi \in \mathrm{End}(G)$ such that $\varphi(F) \subseteq F$ is the largest subset (or subring) E of $\mathrm{End}(G)$ such that F is an E-submodule of G, and in fact F is an $S(F)$-submodule of G in the module-over-a-ring sense. Moreover, $S(F)$ is equivalently the set of all $\varphi \in \mathrm{End}(G)$ such that $\ker(G \to G/F) \subseteq \ker(G \xrightarrow{\varphi} G \to G/F)$, or, equivalently, such that $\mathrm{im}(F \to G \xrightarrow{\varphi} G) \subseteq \mathrm{im}(F \to G)$, where $F \to G$ is the canonical inclusion. Furthermore, one has $i(A(F)) = S(F)$ for any additive subgroup F of an A-module M, where, again, i is the canoncial ring homomomorphism $i : A \to \mathrm{End}_{\mathbb{Z}}(M)$.

The definitions and equivalences noted above situate the theory of T-factroids and T-submodules of an A-module developed in this paper in the broader context of the category of abelian groups, and, even more broadly, in the context of any abelian category. A possible future direction is to undertake a study of the theory in these more general categorical contexts.

4. W-factroids of a ring

Recall that by a *W-factroid of A* we mean a W-factroid of the left A-module ${}_AA$, that is, an additive subgroup F of A such that $a \in F$ whenever $w \in W$, $a \in A$, and $wa \in A$. In this section, we investigate the W-factroids of A for various examples and classes of rings A and multiplicative subsets W.

EXAMPLE 4.1. We determine the W-factroids of $\mathbb{Z}$ for the various multiplicative subsets W of $\mathbb{Z}$.

For an arbitrary nontrivial subgroup $n\mathbb{Z}$ of $\mathbb{Z}$, one has

$$W(n\mathbb{Z}) = \{g \in \mathbb{Z} \mid (n\mathbb{Z} :_{\mathbb{Z}} g) \subseteq n\mathbb{Z}\} = \{g \in \mathbb{Z} : (g, n) = 1\},$$

where also

$$W(0\mathbb{Z}) = \mathbb{Z} \setminus \{0\}.$$

Thus, a nontrivial subgroup $n\mathbb{Z}$ of $\mathbb{Z}$ is a W-factroid of $\mathbb{Z}$ if and only if all of the elements of W are relatively prime to n. In particular, if $W = \mathbb{Z} \setminus \{0\}$, then the only W-factroids of $\mathbb{Z}$ are $\{0\}$ and $\mathbb{Z}$. Note that all of the multiplicative subsets $W(n\mathbb{Z})$ of $\mathbb{Z}$ are saturated, as follows also from Proposition 3.10. Of course $A(n\mathbb{Z}) = \mathbb{Z}$ for all n, since there are no other subrings of $\mathbb{Z}$.

EXAMPLE 4.2. We determine the W-factroids of $\mathbb{Q}$ for its various multiplicative subsets W.

Let $A = M = \mathbb{Q}$. One has $A(\mathbb{Z}) = \mathbb{Z}$, and therefore, by Proposition 3.9,

$$W(\mathbb{Z}) = W(\mathbb{Z}) \cap \mathbb{Q}^\times = (A(\mathbb{Z}) \cap \mathbb{Q}^\times)^{-1} = \{1/n : n \in \mathbb{Z} \smallsetminus \{0\}\}$$

is largest submonoid W of $\mathbb{Q}$ for which $\mathbb{Z}$ is a W-factroid of $\mathbb{Q}$. Let F be any proper subgroup of $\mathbb{Q}$. Then

$$A(F) = (F :_{\mathbb{Q}} F) = \mathbb{Z}[1/p : (1/p)F \subseteq F,\ p \text{ prime}]$$

and therefore

$$\begin{aligned} W(F) &= (A(F) \cap \mathbb{Q}^\times)^{-1} \\ &= \langle p \text{ prime} : (1/p)F \subseteq F \rangle \cdot \{1/n : n \in \mathbb{Z} \smallsetminus \{0\}\} \\ &= \langle p \text{ prime} : (1/p)F \subseteq F \rangle \cdot W(\mathbb{Z}), \end{aligned}$$

where $\langle - \rangle$ denotes "submonoid generated by," and where $\cdot$ denotes the monoid compositum of pointwise products. In particular, if P is any set of prime numbers and $F = \mathbb{Z}[1/p : p \in P]$, then $A(F) = F$ and $W(F) = \langle P \rangle \cdot W(Z)$. This, along with the characterization of the additive subgroups of $\mathbb{Q}$ [**BZ51**], effectively classifies all W-factroids of $\mathbb{Q}$.

EXAMPLE 4.3. One can replace $\mathbb{Z}$ in Example 4.2 with any PID A and $\mathbb{Q}$ with its quotient field K and use the method employed in the example to characterize all of the A-submodules of K that are W-factroids of K for the various multiplicative subsets W of K.

DEFINITION 4.4. A *factroid of A* is a $\text{reg}(A)$-factroid of A, that is, it is an additive subgroup F of A such that $a \in F$ for any $a \in A$ such that $ba \in F$ for some left nonzerodivisor b of A.

EXAMPLE 4.5. Since $W_A((0)) = \text{reg}(A)$, a factroid of A is equivalently a $W_A((0))$-factroid of A. In particular, the ideals (0) and A are both factroids of A.

EXAMPLE 4.6. Suppose that A is commutative. By Proposition 3.12, any primary ideal of A not intersecting W is a W-factroid of A. Thus, since any ideal that is maximal among the ideals not intersecting W is prime, any such ideal is a W-factroid of A.

An immediate consequence of the above is that any primary ideal of A containing only zerodivisors is a factroid of A. Since the minimal prime ideals of A contain only zerodivisors, it follows that any minimal prime ideal of A is a factroid of A. In fact, since any minimal prime does not meet A°, any minimal prime is an A°-factroid of A. More generally, the intersection of any collection of minimal primes of A is a A°-factroid of A. It follows, then, that $\sqrt{0}$ is an A°-factroid.

Example 4.6 generalizes as follows.

PROPOSITION 4.7. *Suppose that A is commutative. Any intersection $\bigcap \mathfrak{q}_i$ of primary ideals $\mathfrak{q}_i$ of A is an $(A \smallsetminus \bigcup \sqrt{\mathfrak{q}_i})$-factroid of A. In particular, any intersection $\bigcap \mathfrak{p}_i$ of prime ideals $\mathfrak{p}_i$ of A is an $(A \smallsetminus \bigcup \mathfrak{p}_i)$-factroid of A.*

PROOF. Let $I = \bigcap \mathfrak{q}_i$. If $w \in A \smallsetminus \bigcup \sqrt{\mathfrak{q}_i}$ and $r \in A$ with $wr \in I$, then, for all i, one has $w \notin \sqrt{\mathfrak{q}_i}$, whence $r \in \mathfrak{q}_i$, so that $r \in I$. □

The following result essentially follows from Proposition 3.10 and [**Kis63**, Lemmas 3.1 and 8.1].

PROPOSITION 4.8 (cf. [**Kis63**, Lemmas 3.1 and 8.1]). *Suppose that A is commutative, and let I be an ideal of A.*

(1) *I is proper if and only if $W(I) \subseteq A \setminus I$, with equality if and only if I is prime.*
(2) *A prime ideal $\mathfrak{p}$ of A contains I if and only if*
$$W(\mathfrak{p}) \subseteq \{a \in A \mid (\sqrt{I} :_A a) \subseteq \mathfrak{p}\}$$
with equality if and only if $\mathfrak{p}$ is minimal over I.
(3) *One has*
$$W(\sqrt{I}) = \{a \in A \mid (\sqrt{I} :_A a) = \sqrt{I}\} = A \setminus \bigcup \mathrm{Min}(I),$$
where $\mathrm{Min}(I)$ is the set of all prime ideals of A that are minimal over I.
(4) *One has*
$$W(\sqrt{0}) = \{a \in A \mid (\sqrt{0} :_A a) = \sqrt{0}\} = A^\circ.$$
Equivalently, A° is the largest subset T of A such that $\sqrt{0}$ is a T-factroid of A.

PROOF. By Proposition 3.10, $W(I)$ is the set of all nonzerodivisors of A mod I. It follows that $W(I) \subseteq A \setminus I$ if and only if I is proper. But I is prime if and only if I is proper and every element of $A \setminus I$ is a nonzerodivisor mod I, which in turn holds if and only if $A \setminus I \subseteq W(I) \subseteq A \setminus I$. This proves statement (1). Statement (2) is follows from [**Kis63**, Lemmas 3.1 and 8.1] applied to the ring A/I. Finally, from statements (2) and (1) and the fact that $\sqrt{I} = \bigcap_{\mathfrak{p} \in \mathrm{Min}(I)} \mathfrak{p}$ it follows that
$$W(\sqrt{I}) = \bigcap_{\mathfrak{p} \in \mathrm{Min}(I)} W(\mathfrak{p}) = \bigcap_{\mathfrak{p} \in \mathrm{Min}(I)} (A \setminus \mathfrak{p}) = A \setminus \bigcup \mathrm{Min}(I),$$
thus proving statement (3), from which (4) immediately follows. Alternatively, statement (3) follows by lifting the equality $\mathrm{reg}(A/\sqrt{I}) = (A/\sqrt{I})^\circ$ to A (which holds since $A/\sqrt{I}$ is a reduced ring) and then applying the identity $W_A(\sqrt{I}) = \mathrm{reg}_{A/\sqrt{I}}(A)$ from Proposition 3.10. □

PROPOSITION 4.9. *A nontrivial commutative ring A is an integral domain if and only if A is the only nonzero ideal of A that is also a factroid of A.*

PROOF. Suppose that A is an integral domain. Since the ring A is nontrivial, one has $A \neq \{0\}$. Moreover, if I is both a nonzero ideal and a factroid of A, then $r \in I$ for some nonzero (hence regular) r, which implies $ar \in I$, and therefore $a \in I$, for any $a \in A$, whence $I = A$. Conversely, suppose that A is the only nonzero ideal of A that is also a factroid of A. Since the ring A is nontrivial, A has some minimal prime ideal P. Then P is a proper ideal of A that is also a factroid of A, and therefore $P = (0)$. Thus (0) is prime, whence A is an integral domain. □

PROPOSITION 4.10. *One has $[1]_A^{W \cap A^\times} = \mathbb{Z}\langle (W \cap A^\times)^{-1} \rangle$. Moreover, if $1 \in W$ and F is a W-factroid of A, then the following are equivalent:*

(a) $W \cap F \neq \varnothing$.
(b) $1 \in F$.
(c) $[1]_A^{W \cap A^\times} \subseteq F$.
(d) $(W \cap A^\times)^{-1} \subseteq F$.

PROOF. By Proposition 3.6, $[1]_A^{W\cap A^\times} = \mathbb{Z}\langle (W\cap A^\times)^{-1}\rangle$. Let F be a W-factroid of A containing an element w of W. Since $w1 \in F$, one has $1 \in F$. Conversely, if $1 \in F$, then F contains an element of W. Finally, $1 \in F$ if and only if $[1]_A^{W\cap A^\times} \subseteq F$, since F is a $(W \cap A^\times)$-factroid of A. □

COROLLARY 4.11. *One has* $[1]_A^{A^\times} = \mathbb{Z}\langle A^\times\rangle$. *Moreover, the following are equivalent for a factroid* F *of* A*:*

(a) F *contains a left-regular element of* A.
(b) $1 \in F$.
(c) $A^\times \subseteq F$.
(d) $\mathbb{Z}\langle A^\times\rangle \subseteq F$.
(e) *Any sum of finitely many units of* A *lies in* F.

From the above, we obtain the following. Note that the rings in question are called *S-rings* in [**Rap74**].

THEOREM 4.12. *Let* A *be a ring for which every element is a sum of units (e.g., a local ring or any quotient or localization of* $\mathbb{Z}$ *at a multiplicative subset). Then* A *is the only factroid of* A *that contains a left-regular element. Thus, if* A *is moreover an integral domain, then the only factroids of* A *are* $\{0\}$ *and* A.

Our next result implies that, for any ring A of algebraic numbers, the only factroids of A are $\{0\}$ and A.

PROPOSITION 4.13. *Let* D *be an integral domain with quotient field* K*, let* L/K *be any algebraic field extension, and let* A *be any subring of* L *containing* D*. Then for any* $\mathrm{reg}(A)$*-factroid* F *of any* A*-submodule* M *of* L *such that* F *contains* D*,* F *also contains* A*. In particular,* A *is the only factroid of* A *containing* D.

PROOF. Let M be an A-submodule of L, let F be any $\mathrm{reg}(A)$-factroid of M such that $D \subseteq F$, and let $0 \neq \alpha \in A$. Since α is algebraic over K, we can let $0 \neq h \in K[x]$ be the unique monic minimal polynomial of α. Choose any nonzero $c \in D$ such that $g := ch \in D[x]$. Since g is irreducible in $K[x]$ and $\alpha \neq 0$, the polynomial x does not divide g, so $0 \neq g(0) \in D \subseteq F$. But $g = -xf + g(0)$ for some polynomial $f \in D[x]$, so evaluating at α we get $\alpha f(\alpha) = g(0) \in F$. Since $g(0) \neq 0$, we have $f(\alpha) \in A \setminus \{0\} = \mathrm{reg}(A)$, so $\alpha \in F$. Since $\alpha \in A \setminus \{0\}$ was arbitrary, we have $A \subseteq F$. The final statement of the proposition follows from the special case $M = A$. □

COROLLARY 4.14. *If* A *is any subring of the algebraic closure of* $\mathbb{Q}$*, then the only factroids of* A *are* $\{0\}$ *and* A.

PROOF. If F is any nonzero factroid of A, then there is some $0 \neq u \in F$, so that $u \cdot 1 = u \in F$, whence $1 \in F$, so $\mathbb{Z} \subseteq F$ since $\mathbb{Z}$ is the subgroup of A generated by 1. The result then follows from Proposition 4.13. □

The proof of the following proposition is postponed until Section 8, as it is an immediate corollary of Theorem 8.3, which generalizes it to the Euclidean domains.

PROPOSITION 4.15. *Let* $A = k[x]$*, where* k *is a field. The factroids of* A *are precisely the sets*

$$C_n := \{f \in A \mid \deg f \leq n\}$$

for all $n \in \mathbb{N} \cup \{\pm\infty\}$*. Moreover,* A *is not principal, but each* C_n *for* $n < \infty$ *is principal, generated by any element of* A *of degree* n.

EXAMPLE 4.16. In light of Examples 4.1 and 4.2, it is natural to ask whether we can restrict our attention to saturated multiplicative sets. That is, is it always true that the W-factroids of A and $\widetilde{W}$-factroids of A coincide, where $\widetilde{W}$ is the saturation of W? The answer is no. For a counterexample, let $A = k[x]$, let W be the set of nonzero polynomials of even degree, and let $F = \operatorname{span}_k\{1, x^2\}$. Note that $\widetilde{W} = A^\circ$, since the square of any polynomial has even degree. Hence by Proposition 4.15, F is not a $\widetilde{W}$-factroid of A. However, we claim that F is a W-factroid of A.

To see this, let $w \in W$ and $f \in A$ such that $wf \in F$. Write $w = ax^2 + b$, with $a, b \in k$ and not both zero. If $a = 0$, then $w = b \in k^\times$ and since F is a k-vector space and $bf \in F$, we have $f = b^{-1}bf \in F$. Thus, we may assume $a \neq 0$, so that $\deg w = 2$. But since $wf \in F$, we have $2 + \deg f = \deg w + \deg f = \deg(wf) \leq 2$. Thus, $\deg f = 0$ so $f \in k \subseteq F$.

Note that $W(F)$ cannot be saturated, since, if it were, then since $W \subseteq W(F)$ we would have $\widetilde{W} \subseteq W(F)$ and then F would be a $\widetilde{W}$-factroid of A. So this example shows that $W(F)$ need not be saturated.

In fact, $W(F)$ is precisely the set of all nonzero polynomials in $k[x]$ of degree not equal to 1. To see the containment $\supseteq$, observe that any nonzero polynomial of degree 0 or 2 is in $F \subseteq W(F)$ while, if $\deg f \geq 3$ and $fg \in F$, then $g = 0 \in F$ (since degrees are additive on nonzero polynomials), so that $f \in W(F)$. For the opposite inclusion, let f be a polynomial of degree 1. Write $f = ax + b$, where $0 \neq a \in k$ and $b \in k$. Then $(ax - b)f = a^2x^2 - b^2 \in F$, but $ax - b \notin F$, so $ax + b \notin W(F)$.

Note also that $A(F) = (F :_A F) = k$, since multiplying any positive-degree polynomial by $x^2 \in A$ would force it out of F.

EXAMPLE 4.17. Let $A = k[x_1, \ldots, x_n]$, where k is a field and $n \geq 2$. Let f be an element of A such $f + c$ is irreducible for all $c \in k$ (e.g., $f = x_1^2 + x_2^3$). Then the principal factroid of A generated by f is the k-span of $\{1, f\}$.

In order to see this, it suffices to show that $U := \langle 1, f\rangle$ is a factroid of A, since it is obvious that any factroid containing f must contain U . So let $g \in A^\circ$ and $h \in (U :_A g)$. Then $gh \in U$. That is, there exist $a, b \in k$ with $gh = a + bf$. If $b = 0$, then $h \in k \subseteq U$, since any factor of a constant must be a constant. If $b \neq 0$, then $\frac{g}{b} \cdot h = f + \frac{a}{b}$, which is irreducible by assumption, so either $h \in k$ (in which case $h \in U$) or $\frac{g}{b} \in k$. In the latter case, it follows that $g \in k^\times$, so that $h = \frac{b}{g} \cdot f + \frac{a}{g} \in U$. Thus, $(U :_A g) \subseteq U$, so that since $g \in A^\circ$ was arbitrary, U is a factroid of A. Therefore, $U = [f]_A$.

Next, we determine the factroids of a direct product $A \times B$ of integral domains A and B in the case where $A \times B$ has more than one unit.

LEMMA 4.18. *Let A and B be rings, let $V \subseteq A$ and $W \subseteq B$ be multiplicative submonoids, let M be an A-module and N a B-module. Let F be a V-factroid of M and G a W-factroid of N. Then $F \times G$ is a $(V \times W)$-factroid of $M \times N$.*

PROOF. Let $(v, w) \in V \times W$ and $(m, n) \in M \times N$ with $(v, w)(m, n) = (vm, wn) \in F \times G$. Then $vm \in F$, so $m \in F$ since F is a V-factroid of M, and $wn \in G$, so $n \in G$ since G is a W-factroid of N. Thus, $(m, n) \in F \times G$. □

THEOREM 4.19. *Let A and B be integral domains. Assume that the ring $A \times B$ has more than one unit. Then the factroids of $A \times B$ are precisely those subgroups of the form $F \times G$, where F is a factroid of A and G a factroid of B.*

PROOF. First, recall that $\operatorname{reg}(A \times B) = \operatorname{reg}(A) \times \operatorname{reg}(B) = A^\circ \times B^\circ$. If F is a factroid of A and G of B, it then follows from Lemma 4.18 that $F \times G$ is a factroid of $A \times B$.

Conversely, let H be a factroid of $A \times B$. Let $\pi_1 : A \times B \to A$ and $\pi_2 : A \times B \to B$ be the projections. Set $F := \pi_1(H)$ and $G := \pi_2(H)$. We first show that F and G are factroids of their ambient rings. To see this, let $x \in A$ and $a \in A^\circ$ with $ax \in F$. Then there is some $y \in G$ with $(ax, y) \in H$. That is, $(a, 1)(x, y) \in H$, so since $(a, 1) \in (A \times B)^\circ$ and H is a factroid, we have $(x, y) \in H$, whence $x \in F$. Thus F is a factroid of A, and by symmetry G is a factroid of B.

Obviously $H \subseteq F \times G$. Now let $f \in F$. Then there is some $y \in G$ with $(f, y) \in H$. Since $(A \times B)^\times = A^\times \times B^\times$ is not a singleton, either $A^\times$ or $B^\times$ is not a singleton. Without loss of generality assume the former, and let $1 \neq \alpha \in A^\times$. Then $(\alpha^{-1}, 1)(\alpha f, y) = (f, y) \in H$, so since $(\alpha^{-1}, 1) \in \operatorname{reg}(A \times B)$, we have $(\alpha f, y) \in H$. Subtracting, we have $(1-\alpha, 1)(f, 0) = ((1-\alpha)f, 0) = (f - \alpha f, y - y) = (f, y) - (\alpha f, y) \in H$, so since $(1 - \alpha, 1) \in A^\circ \times B^\circ = (A \times B)^\circ$, we have $(f, 0) \in H$. We have shown that $F \times 0 \subseteq H$.

Now let $g \in G$. Then there is some $x \in F$ with $(x, g) \in H$. But since $(x, 0) \in H$, it follows that $(0, g) = (x, g) - (x, 0) \in H$.

Finally, take any $(f, g) \in F \times G$. Since $(f, 0), (0, g) \in H$, we have $(f, g) = (f, 0) + (0, g) \in H$. Thus, $H = F \times G$. □

EXAMPLE 4.20. The condition on the unit groups in Theorem 4.19 is necessary. For instance, let $A = B = \mathbb{F}_2$. Then $H = \mathbb{F}_2 \cdot (1, 1) = \{(0, 0), (1, 1)\}$ is a factroid of $\mathbb{F}_2 \times \mathbb{F}_2$ but is not the product of a pair of factroids.

REMARK 4.21. An approach such as that in Lemma 4.18 and Theorem 4.19 will not characterize the W-factroids of a product of rings for arbitrary W. This is because not all multiplicative sets in $A \times B$ are Cartesian products.

PROBLEM 4.22. Characterize the W-factroids of the rings $\mathbb{Z} \times \mathbb{Z}$ (resp., $\mathbb{Q} \times \mathbb{Q}$) for its various multiplicative subsets W.

PROBLEM 4.23. Characterize the factroids of the domain $k[x, y]$, where k is a field.

PROBLEM 4.24. Let K be a number field. Characterize the W-factroids of K (resp., $\mathcal{O}_K$) for its various multiplicative subsets W.

PROBLEM 4.25. Characterize the W-factroids of the division ring $K = \mathbb{Q}(i, j)$ of all quaternions for its various multiplicative subsets W.

5. Generating W-factroids of a module

In this section, we describe various ways to produce W-factroids of a module.

The set of all W-factroids of M is a partially ordered set under the relation $\subseteq$. By Proposition 2.10, the poset of all W-factroids of M has arbitrary infima and is therefore a complete poset. The supremum of a set of W-factroids is equal to the W-factroid generated by their union. By our next proposition, sups of directed subsets are just unions.

LEMMA 5.1. *Let $T \subseteq A$. An arbitrary union of T-saturated subsets of M is a T-saturated subset of M. More generally, the T-saturation of the union of an arbitrary collection of T-subsets of M is the union of their T-saturations in M.*

PROOF. First, we claim that $(\bigcup_{\alpha\in\Lambda} S_\alpha :_M t) = \bigcup_{\alpha\in\Lambda}(S_\alpha :_M t)$ for any $t \in A$ and any collection $\{S_\alpha \mid \alpha \in \Lambda\}$ of subsets S_α of M. To prove this identity, note that the containment $\supseteq$ is clear. Let $x \in (S :_M t)$, where $S = \bigcup_{\alpha\in\Lambda} S_\alpha$. Then $tx \in S$, whence $tx \in S_\beta$ for some $\beta \in \Lambda$, so that $x \in (S_\beta :_M t) \subseteq \bigcup_{\alpha\in\Lambda}(S_\alpha :_M t)$. This proves the identity. Finally, one has

$$\begin{aligned}\operatorname{Sat}_M^T(S) &= \bigcup_{w\in\langle T\rangle} (S :_M w)\\ &= \bigcup_{w\in\langle T\rangle}\bigcup_{\alpha\in\Lambda} (S_\alpha :_M w)\\ &= \bigcup_{\alpha\in\Lambda}\bigcup_{w\in\langle T\rangle} (S_\alpha :_M w)\\ &= \bigcup_{\alpha\in\Lambda} \operatorname{Sat}_M^T(S_\alpha).\end{aligned}$$

This completes the proof. □

PROPOSITION 5.2. *Let $T \subseteq A$. Any directed union of T-factroids of M is a T-factroid of M. More precisely, let $(\Lambda, \leq)$ be a directed set, and let $\{F_\alpha \mid \alpha \in \Lambda\}$ be a a collection of T-factroids of M such that $F_\alpha \subseteq F_\beta$ whenever $\alpha \leq \beta$. Then the union $\bigcup_{\alpha\in\Lambda} F_\alpha$ of the F_α is a T-factroid of M and is therefore the smallest T-factroid of M containing F_α for all α.*

PROOF. This follows from the lemma and the fact corresponding fact for $\mathbb{Z}$-submodules (i.e., for additive subgroups) of M. □

EXAMPLE 5.3. The sum of two principal factroids of a commutative ring need not be a factroid. Let $A = k[x,y]$, where k is a field. Let $F = \operatorname{span}_k\{1, x\}$ and $G = \operatorname{span}_k\{1, y^2 - x\}$. Then since $x + c$ and $y^2 - x + c$ are irreducible for all $c \in k$, both F and G are principal factroids of A, by Example 4.17. Since y is a factor of $y^2 = x + (y^2 - x)$, any factroid of A that contains F and G must contain y. But $F + G = \operatorname{span}_k\{1, x, y^2\}$ does not contain y.

Hereafter, per Proposition 2.9, we focus our attention on W-factroids of M, where W is a multiplicative subset of A.

One has two simple operations of "residuation," as described in Lemmas 5.4 and 5.6.

LEMMA 5.4. *For any W-factroid F of M and any subset $T \subseteq Z(A)$, $(F :_M T)$ is a W-factroid of M.*

PROOF. To see that $(F :_M T)$ is a subgroup of M, first note it contains 0. Next let $x, y \in (F :_M T)$. Then for any $t \in T$, we have $tx, ty \in F$. Since F is a subgroup of M, $t(x - y) = tx - ty \in F$, so since $t \in T$ was arbitrary, $x - y \in (F :_M T)$.

Now let $x \in M$ and $w \in W$ such that $wx \in (F :_M T)$. Then for any $t \in T$, we have $wtx = twx \in F$. Since $w \in W$ and F is a W-factroid, $tx \in F$. Since $t \in T$ was arbitrary, $x \in (F :_M T)$. □

REMARK 5.5. Lemma 5.4 says, equivalently, that $W((F :_M T)) \supseteq W(F)$ for any multiplicative set W, any W-factroid of M, and any subset $T \subseteq Z(A)$. In fact the containment can be proper. Indeed, let $A = M = \mathbb{Z}$, $F = 6\mathbb{Z}$, and $T = \{2\}$, so that $(F :_M T) = 3\mathbb{Z}$. Then $2 \in W((F :_M T)) \smallsetminus W(F)$.

LEMMA 5.6. *For any W-factroid F of M and any subset $S \subseteq M$, $(F :_A S)$ is a W-factroid of A.*

PROOF. For the subgroup property, first note that it contains 0. Next let $a, b \in (F :_A S)$. Then for any $x \in S$, we have $ax, bx \in F$, so that since F is a group, $(a-b)x = ax - bx \in F$. Since $x \in S$ was arbitrary, $a - b \in (F :_A S)$

Now let $a \in A$ and $w \in W$ such that $wa \in (F :_A S)$. Then for any $x \in S$, we have $wax \in F$. Thus, $ax \in (F :_M w) \subseteq F$ since F is a W-factroid. Since $x \in S$ was arbitrary, $a \in (F :_A S)$. Therefore $(F :_A S)$ is a W-factroid of A. □

In the next few results, we show that generation of W-factroids interacts well with multiplication of subsets of A and subsets of M, especially when the subset of A is in the center of A. This can be recast (though we shall not do so here) in the language of *modules over a quantale*, a notion important in lattice theory (see e.g. [**Ros94**]).

PROPOSITION 5.7. *Let $S \subseteq M$ and $T \subseteq A$. One has*

$$[T]_A^W S \subseteq [TS]_M^W.$$

Moreover, if $T \subseteq Z(A)$, then

$$[T]_A^W [S]_M^W \subseteq [TS]_M^W.$$

PROOF. Clearly one has $T \subseteq ([TS]_M^W :_A S)$, where the colon over A is a W-factroid of A by Lemma 5.6. It follows that $[T]_A^W \subseteq ([TS]_M^W :_A S)$, which is equivalent to the first statement. Suppose that $T \subseteq Z(A)$. Since $S \subseteq ([TS]_M^W :_M T)$ and the colon over M is a W-factroid of M by Lemma 5.4, one has $[S]_M^W \subseteq ([TS]_M^W :_A T)$, and therefore $T[S]_M^W \subseteq [TS]_M^W$. It follows, then, that

$$[T]_A^W [S]_M^W \subseteq [T[S]_M^W]_M^W \subseteq [[TS]_M^M]_M^W = [TS]_M^W.$$

This proves the second statement. □

COROLLARY 5.8. *Let $x \in M$ and $T \subseteq A$. Then $[T]_A^W x \subseteq [Tx]_M^W$.*

COROLLARY 5.9. *Let $S \subseteq M$ and $a \in Z(A)$. Then*

$$a[S]_M^W \subseteq [a]_A^W [S]_M^W \subseteq [aS]_M^W.$$

Equalities hold if a is a unit of A.

PROOF. The first claim is clear, and the second follows by applying the first to $aS \subseteq M$ and $a^{-1} \in Z(A)$. □

EXAMPLE 5.10. The condition that $a \in Z(A)$ is necessary in Corollary 5.9. To see this, consider the ring $A = k\langle x, y\rangle$ of polynomials in noncommuting variables x, y over a field k. Set $S := k[x]y$, and $W := \{1, x, x^2, x^3, \ldots\}$ = the multiplicative submonoid of A generated by x. If F is a W-factroid containing S, then it must contain $k[x] \oplus k[x]y$. But the latter set is already a W-factroid of A. Hence, $[S]_A^W = k[x] \oplus k[x]y$. On the other hand, $yS = yk[x]y$ is already a W-factroid of A, since it is an additive subgroup such that no element of it has a left W-factor other than 1. Thus, $[yS]_A^W = yS = yk[x]y$. Hence, $yx \in y[S]_A^W \setminus [yS]_A^W$.

COROLLARY 5.11. *Let V be a multiplicative submonoid of $Z(A)$. Then $[V]_A^W$ is a $\mathbb{Z}\langle V\rangle$-subalgebra of A and is the smallest subring of A containing V that is a W-factroid of A. Let F be any W-factroid of M. Then the following conditions are equivalent.*

(a) $VF \subseteq F$.

(b) *F is a V-submodule of M.*

(c) *F is a $\mathbb{Z}\langle V\rangle$-submodule of M.*
(d) *F is $[V]_A^W$-submodule of M.*
(e) *$V \subseteq A(F)$.*
(f) *$[V]_A^W \subseteq A(F)$.*
(g) *$\mathbb{Z}\langle V\rangle \subseteq A(F)$.*
(h) *$A(F)$ is a $[V]_A^W$-subalgebra of A.*
(i) *$A(F)$ is a $\mathbb{Z}\langle V\rangle$-subalgebra of A.*

PROOF. By Proposition 5.7, one has $1 \in [V]_A^W$ and $[V]_A^W[V]_A^W \subseteq [VV]_A^W \subseteq [V]_A^W$, and therefore $[V]_A^W$ is a subring of A. Moreover, since $V[V]_A^W \subseteq [V]_A^W$, the ring $[V]_A^W$ is also a V-submodule, hence a $\mathbb{Z}\langle V\rangle$-subalgebra, of A. The equivalence of the nine conditions is easy to check. □

COROLLARY 5.12. *The smallest W-factroid $[1]_A^W$ of A containing 1 is the smallest subring of A that is a W-factroid of A. Let F be any W-factroid of M. Then F is a $[1]_A^W$-submodule of M, $[1]_A^W$ is a subring of $A(F)$, and $A(F)$ is a $[1]_A^W$-subalgebra of A.*

Next, we provide a more explicit construction of the T-factroid $[S]_M^T$ of M generated by a subset S of M.

DEFINITION 5.13. Let $T \subseteq A$ and $S \subseteq M$. We inductively define additive subgroups $F_i^T(S)$ of M for all positive integers i, as follows. First, let $F_1^T(S)$ denote the additive subgroup of M generated by the T-saturation $\mathrm{Sat}_M^T(S)$ of S in M. For $i \geq 2$, we inductively define $F_i^T(S) := F_1^T(F_{i-1}^T(S))$. In other words, $F_i^T(S)$ is the i-fold application of the operation of F_1^T to S. Since $S \subseteq F_1^T(S)$, one has $S \subseteq F_1^T(S) \subseteq F_2^T(S) \subseteq \cdots$. Moreover, it is clear that S is a T-factroid of M if and only if $F_1^T(S) \subseteq S$, which is in turn equivalent to the inclusions $F_i^T(S) \subseteq S$ holding for all $i \geq 1$.

PROPOSITION 5.14. *Let $T \subseteq A$ and $S \subseteq M$. Then $[S]_M^T = \bigcup_{i\geq 1} F_i^T(S)$, where $F_i^T(S)$ is as in Definition 5.13.*

PROOF. Set $F(S) := \bigcup_{i\geq 1} F_i^T(S)$. We need to show that $F(S)$ is a T-factroid of M, and that $F(S)$ is contained in any T-factroid that contains S.

For the latter, let G be a T-factroid of M that contains S. Then $F_1^T(S) \subseteq F_1^T(G) \subseteq G$, so that $F_i^T(S) \subseteq G$ for all i, whence $F(S) \subseteq G$.

For the former, first note that $F(S)$ is a directed union of additive subgroups (or $\{1\}$-factroids) of M and is therefore an additive subgroup of M. Moreover, by Proposition 5.2, $\mathrm{Sat}_M^T(F(S)) = \bigcup_{i\geq 1} \mathrm{Sat}_M^T(F_i^T(S)) \subseteq \bigcup_{i\geq 1} F_{i+1}^T(S) = F(S)$, so that $F(S)$ is T-saturated. Therefore, $F(S)$ is a T-factroid of M.

Here is an alternative and more explicit proof. Let $W := \langle T\rangle$, and let $x, y \in F(S)$. Then since the union is nested, there is some i with $x, y \in F_i^T(S)$. Thus, $x - y \in F_i^T(S) \subseteq F(S)$. Next, let $z \in M$ and $w \in W$ such that $wz \in F(S)$. Then there is some i with $wz \in F_i^T(S)$, so that $z \in F_1^T(F_i^T(S)) = F_{i+1}^T(S) \subseteq F(S)$, completing the proof that $F(S)$ is a W-factroid. Hence $[S]_M^T \subseteq F(S)$. □

The corollary below provides an alternative proof of Corollary 5.9.

COROLLARY 5.15. *Let $T \subseteq A$ and $S \subseteq M$. Let $F_i^T(-)$ be as in Definition 5.13.*
(1) *For any $i \geq 0$, one has $F_i^T(T)S \subseteq F_i^T(TS)$. Hence, $[T]_A^T S \subseteq [TS]_M^W$.*

(2) *Suppose* $T \subseteq Z(A)$. *Then for any* $i \geq 0$, $TF_i^T(S) \subseteq F_i^T(TS)$. *Hence* $T[S]_M^T \subseteq [TS]_M^T$.

PROOF. In both cases, the second statement follows from the first, by Proposition 5.14. So it suffices to prove the first statement of each of (1), (2).

(1): Let $r \in F_1^W(T)$ and $s \in S$. Then there exist $a_j \in A$, $w_j \in W$ with $r = \sum_{j=1}^n a_j$ and $\pm w_j a_j \in T$. But then $rs = \sum_{j=1}^n a_j s$, and $\pm w_j(a_j s) \in TS$. Hence, $rs \in F_1^W(TS)$, and hence $F_1^W(T)S \subseteq F_1^W(TS)$. Replacing T by $F_i^W(T)$ inductively, one obtains the first statement.

(2): Since $T \subseteq Z(A)$, it is clear that, for any $b \in A$ and $t \in T$, we have $t(S :_M b) \subseteq (TS :_M b)$. So let $z \in F_1^W(S)$. Then there exist $z_j \in M$ and $w_j \in W$ with $z = \sum z_j$ and $z_j \in \pm(S :_M w_j)$. Hence, for any $t \in T$, we have $tz_j \in \pm(TS :_M w_j)$, and $tz = \sum_j tz_j$, so $tz \in F_1^W(TS)$. Replacing S by $F_i^W(S)$ inductively, one obtains the first statement. □

EXAMPLE 5.16. There are many examples where $[S]_M^W \supsetneq F_1^W(S)$. For instance, let $A = M = k[x]$, $W = A \setminus \{0\}$, and $S = \{f\}$, where k is a field and f is a monic irreducible polynomial in $k[x]$. The monic factors of f are 1 and f, so that $F_1^W(f) = \operatorname{span}_k\{1, f\}$. But by Proposition 4.15 one has $[f]^W = \operatorname{span}_k\{1, x, x^2, \ldots, x^n\}$, where $n = \deg f$. Thus, one has $F_1^W(f) = [f]_A^W$ if and only if $\deg f \leq 1$.

6. T-regular W-factroids of a module

In this section, we consider another general construction of factroids, which generally gives a larger set than $[-]_M^W$. Here, T will be an arbitrary subset of A, and W a multiplicative subset of A.

DEFINITION 6.1. Let T be a subset of A. A W-factroid F of M is T-*regular* if $([hF]_M^W :_M h) = F$ for all $h \in T$, or, equivalently, if $x \in F$ for any $x \in M$ such that $hx \in [hF]_M^W$ for some $h \in T$.

REMARK 6.2. If $T' \subseteq T$, then every T-regular W-factroid is T'-regular. If $T \subseteq Z(A)^\times = A^\times \cap Z(A)$, then any W-factroid is T-regular, by Corollary 5.9.

Observe also any W-factroid F of M is $(W \cap A(F))$-regular. Indeed, let $h \in W \cap A(F)$ and $x \in M$ with $hx \in [hF]_M^W$. Since $h \in W$ and $[hF]_M^W$ is a W-factroid, one has $x \in [hF]_M^W$. Then, since $h \in A(F) = (F :_A F)$, one has $hF \subseteq F$ and therefore $x \in [F]_M^W = F$.

Since an additive subgroup F of M is a W-submodule of M if and only if $W \cap A(F) = W$, a W-factroid of M that is also a W-submodule of M is automatically W-regular.

REMARK 6.3. Let F be a W-factroid of M. Let $T_M^W(F)$ be the set of all $h \in A$ such that $([hF]_M^W :_M h) = F$. Then F is T-regular if and only if $T \subseteq T_M^W(F)$; that is, $T_M^W(F)$ is the largest subset T of A such that the W-factroid F is T-regular. Note that

$$T_M^W(F) = \bigcup_{a \in Z(A)} (T_M^W(F) :_A a) \supseteq Z(A)^\times.$$

In other words, $T_A^W(F)$ is $Z(A)$-saturated; equivalently, F is T-regular if and only if F is $\operatorname{Sat}_M^{Z(A)}(T)$-regular. Moreover, by the previous remark one has

$$(W \cap A^\times)^{-1} \subseteq W \cap A(F) \subseteq T_M^W(F).$$

In other words, every W-factroid is $(W \cap A(F))$-regular, and therefore also $(W \cap A^\times)^{-1}$-regular.

PROBLEM 6.4. Must one have $A^\times \subseteq T_M^W(F)$? Equivalently, must any W-factroid be $A^\times$-regular?

PROBLEM 6.5. Must $T_M^W(F)$ be multiplicatively closed? Equivalently, is a W-factroid T-regular if and only if it is $\langle T\rangle$-regular, where $\langle T\rangle$ is the submonoid of A generated by T?

LEMMA 6.6. *Let $T \subseteq A$. Any intersection of T-regular W-factroids of an A-module M is T-regular.*

PROOF. Let $\{F_\alpha\}_{\alpha\in\Lambda}$ be a set of T-regular W-factroids in M, and let $F = \bigcap_\alpha F_\alpha$. Let $h \in T$ and $x \in M$ such that $hx \in [hF]_M^W$. Since $[-]_M^W$ is order-preserving and $hF \subseteq hF_\alpha$ for all α, it follows that $hx \in [hF_\alpha]_M^W$, so that $x \in ([hF_\alpha]_M^W :_M h) = F_\alpha$ Thus, $x \in \bigcap_\alpha F_\alpha = F$. □

Consequently, we can make the following definition.

DEFINITION 6.7. Let $T \subseteq A$. For any subset S of M, the *T-regular W-factroid of M generated by S* is the smallest T-regular W-factroid

$$[S]_M^{W,T} := \bigcap\{F \mid F \text{ is a } T\text{-regular } W\text{-factroid of } M \text{ and } S \subseteq F\}.$$

of M containing S. When M is understood, we abbreviate $[S]_M^{W,T}$ to $[S]^{W,T}$. When $S = \{x\}$ for some $x \in M$, we write $[x]_M^{W,T}$ or $[x]^{W,T}$ in place of $[\{x\}]_M^{W,T}$.

REMARK 6.8. Let $S \subseteq M$. If $T \subseteq A^\times$, then $[S]_M^{W,T} = [S]_M^W$. Moreover, if $T' \subseteq T \subseteq A$, then $[S]_M^{W,T'} \subseteq [S]_M^{W,T}$. Consequently, one has $[S]_M^{W,T} \supseteq [S]_M^W$ for all $T \subseteq A$.

REMARK 6.9. If A is commutative, then $[\varnothing]_M^{W,W} = [\varnothing]_M^W = \ker(M \to W^{-1}M)$ is the smallest (W-regular) W-factroid of M.

In Definition 5.13 and Proposition 5.14, we showed a way to construct $[S]_M^W$ from *below* – i.e., as a union of certain subsets of M rather than an intersection. The following construction of $G_M^{W,T}(S)$ is an attempt at doing a similar thing for T-regular W-factroids. However, as we shall see, this is only fully successful when T is a multiplicative submonoid of the center of A. Thus, in the commutative case the story is satisfying, but in the noncommutative case, questions remain.

DEFINITION 6.10. Let $T \subseteq A$ and $S \subseteq M$. Set

$$G_M^{W,T}(S) = \bigcup_{h\in T}([hS]_M^W :_M h) = \{x \in M : hx \in [hS]_M^W \text{ for some } h \in T\}.$$

When $T = W$, we write $G_M^W(S)$ for $G_M^{W,T}(S)$. When M is understood, we write $G^{W,T}(S)$ for $G_M^{W,T}(S)$ and $G^W(S)$ for $G_M^W(S)$. We also set the obvious conventions when S is a singleton.

LEMMA 6.11. *Let $T \subseteq A$ and $S \subseteq M$. Then $G_M^{W,T}(S) \subseteq [S]_M^{W,T}$.*

PROOF. Let G be any T-regular W-factroid containing S. Let $x \in G_M^{W,T}(S)$; then for some $h \in T$, $hx \in [hS]_M^W \subseteq [hG]_M^W$, so that $x \in ([hG]_M^W :_M h) = G$ since G is T-regular. The lemma follows. □

LEMMA 6.12. *Let $T \subseteq A$, $S \subseteq M$, and $c \in Z(A)$. Then $cG^{W,T}(S) \subseteq G^{W,T}(cS)$. Equality holds if c is a unit of A.*

PROOF. Let $x \in G^{W,T}(S)$. Then for some $h \in T$, we have $hx \in [hS]_M^W$. Then $hcx = chx \in c[hS]_M^W \subseteq [chS]_M^W = [hcS]_M^W$ by Corollary 5.9. Thus, $cx \in ([hcS]_M^W :_M h) \subseteq G^{W,T}(cS)$. Applying the given inclusion to $c^{-1} \in Z(A)$ and $cS \subseteq M$, one sees that the reverse inclusion holds if c is a unit of A. □

PROPOSITION 6.13. *Let W be a multiplicative subset of A, let V be a multiplicative submonoid of $Z(A)$, and let S be a subset of M. Then $G_M^{W,V}(S)$ is the smallest V-regular W-factroid of M containing S. Equivalently, $[S]_M^{W,V} = G_M^{W,V}(S)$.*

PROOF. First we prove $G = G^{W,V}(S)$ is a W-factroid of M.

Let $z, z' \in G$. Then there exist $h, h' \in V$ with $hz \in [hS]_M^W$ and $h'z' \in [h'S]_M^W$. It then follows from Corollary 5.9 that $hh'z, hh'z' \in [hh'S]_M^W$. Thus, $hh'(z - z') = hh'z - hh'z' \in [hh'S]_M^W$, where $hh' \in V$, whence $z - z' \in G$. Since G is nonempty, it follows that it is an additive subgroup of M.

Now let $x \in M$ and $r \in W$ such that $rx \in G$. Then there is some $h \in V$ with $rhx = hrx \in [hS]_M^W$. Thus, $hx \in ([hS]_M^W :_M r) \subseteq [hS]_M^W$ (since $[hS]_M^W$ is a W-factroid), whence $x \in G$. Thus, G is a W-factroid of M.

Next, to see that G is V-regular, let $h \in V$ and $z \in ([hG]_M^W :_M h)$. Then $hz \in [hG]_M^W$. But $hG = hG^{W,V}(S) \subseteq G^{W,V}(hS)$ by Lemma 6.12, so since $[-]_M^W$ is order-preserving, we have $hz \in [G^{W,V}(hS)]_M^W = G^{W,V}(hS)$, with the last equality because $G^{W,V}(hS)$ is a W-factroid. Thus for some $h' \in V$, we have $h'hz \in [h'hS]_M^W$, whence $z \in ([h'hS]_M^W :_M h'h) \subseteq G$ since $h'h \in V$.

Finally, since G is a V-regular W-factroid of M containing S, and since $[S]_M^{W,V} \supseteq G$ by Lemma 6.11, equality must hold. □

EXAMPLE 6.14. Equality need not hold in Proposition 6.13 when $V \not\subseteq Z(A)$. Indeed, $G_M^{W,V}(S)$ need not even be additively closed. For an example of this, let A and M be as in Example 2.16, let $W = \{1\}$ (so that $[-]_M^W$ is just additive subgroup generation), let $S = \{0\}$, and let $V = \langle x, y \rangle$. Then $(1,0) \in (0 :_M x) = ([0]_M^W :_M x)$, and $(0,1) \in (0 :_M y) = ([0]_M^W :_M y)$, so both are in $G_M^{W,V}(0)$, but for any $g \in V$, $(\bar{g}, \bar{g}) = g \cdot (1,1) \notin [g0]_M^W = 0$, so $(1,1) = (1,0) + (0,1) \notin G_M^{W,V}(0)$.

PROBLEM 6.15. Find a formula for $[S]_M^{W,T}$ analogous to the formula $[S]_M^W = \bigcup_{n=1}^{\infty} F_i^W(S)$ that works for arbitrary subsets T of A.

EXAMPLE 6.16. All of the factroids C_n of $A = k[x]$, as classified in Proposition 4.15, are $\mathrm{reg}(A)$-regular, since the degree function is additive. Equivalently, one has $[S]_A^{\mathrm{reg}(A)} = G_A^{\mathrm{reg}(A)}(S)$ for all subsets S of A.

REMARK 6.17. It is reasonable to ask: is it always true that $G_A^W(S) = [S]_A^W$? The answer is no, even when A is a domain, $W = \mathrm{reg}(A)$, and S is a singleton. See Example 7.13 for a specific example.

7. W-Egyptian fractions

In this section, we provide some applications of these factroid constructions to W-Egyptian fractions and W-reciprocal complements, defined as follows.

DEFINITION 7.1. Let A be a commutative ring and W a multiplicative set. Then the set of all sums $\{\frac{1}{w_1} + \cdots + \frac{1}{w_n} \in W^{-1}A \mid n \in \mathbb{N},\ w_i \in W\}$ is the nonunital (i.e., not necessarily unital) subring $R_W(A)$ of A generated by the subset $W^{-1} = \{\frac{1}{w} \mid w \in W\}$ of $W^{-1}A$, which we call the *W-reciprocal complement* of A. If $1 \in W$, then $R_W(A)$ is a unital subring of A.

An element of $W^{-1}A$ is *W-Egyptian* in the sense of [**Eps23**] if it lies in $R_W(A)$. An element a of A is *W-Egyptian* if $a/1$ is W-Egyptian. A ring A is said to be *W-Egyptian* if every element of A (or, equivalently, of $W^{-1}A$) is W-Egyptian. This condition holds if and only if $R_W(A) = W^{-1}A$, which in turn is equivalent to the condition that the image of the localization map $A \to W^{-1}A$ is contained in $R_W(A)$.

More generally, let M be an A-module and $S \subseteq M$ a subset. we let $R_W^S(M)$ be all sums in $W^{-1}M$ of elements of the form $\frac{s}{w}$, where $s \in S$ and $w \in W$. Note that if S' is the image of S in the localization map $M \to W^{-1}M$, then $R_W^S(M)$ is the $R_W(A)$-submodule of $W^{-1}M$ generated by S'.

REMARK 7.2. Let A, M, S, W be as above. As per the notation in Corollary 3.8, let $\frac{S}{1} := \{s/1 \in W^{-1}M \mid s \in S\}$, i.e., the image of the set S in the localization map $M \to W^{-1}M$. Then $R_W^S(M) = R_W(A)\frac{S}{1}$. That is, it is the $R_W(A)$-submodule of $W^{-1}M$ generated by $\frac{S}{1}$.

Note also that if $M = A$ and $S = \{1\}$, then $R_W^S(M) = R_W(A)$.

EXAMPLE 7.3. Let A be an integral domain with fraction field K. The subring $R(A) := R_{\mathrm{reg}(A)}(A)$ of K is the *reciprocal complement of A* (in the sense of [**Eps25**, Definition 2.1]), an element of K is said to *A-Egyptian* (or *Egyptian* if the context is clear) if it is reg(A)-Egyptian, and the domain A is said to be *Egyptian* (in the sense of [**GLO24**]) if A is reg(A)-Egyptian.

LEMMA 7.4. *Let A be a commutative ring, W a multiplicative submonoid of A, M an A-module, and $S \subseteq M$. Let $x \in M$ and $b \in W$ with $x \in [bS]_M^W$. Then $x/b \in R_W^S(M)$.*

PROOF. We show inductively on i that if $x \in F_i^W(bS)$, then $x/b \in R_W^S(M)$. We may assume $x \neq 0$.

Case $i = 0$: Then $x = bs$ for some $s \in S$, so $x/b = s/1 \in R_W^S(M)$.

Case $i > 0$, given true for smaller i: Then there are nonzero $x_j \in M$, $c_j \in W$ such that $x = \sum_{j=1}^n \pm x_j$ and $c_j x_j \in F_{i-1}^W(bS)$. By inductive hypothesis, we have $\frac{c_j x_j}{b} \in R_W^S(M)$. Write $\frac{c_j x_j}{b} = \sum_{h=1}^{t_j} \frac{s_{hj}}{u_{hj}}$, where $t_j \in \mathbb{N}$, $s_{hj} \in S$, and $u_{hj} \in W$. Then

$$\frac{x}{b} = \sum_{j=1}^{n} \pm\frac{x_j}{b} = \sum_{j=1}^{n} \frac{\pm 1}{c_j}\frac{c_j x_j}{b} = \sum_{j=1}^{n} \frac{\pm 1}{c_j}\sum_{h=1}^{t_j}\frac{s_{hj}}{u_{hj}} = \sum_{j=1}^{n}\sum_{h=1}^{t_j}\frac{\pm s_{hj}}{c_j u_{hj}} \in R_W^S(M). \qquad \square$$

We then have the following result characterizing elements of $R_W^S(M)$.

THEOREM 7.5. *Let A be a commutative ring, let W be a multiplicative submonoid of A, let M be an A-module, let $S \subseteq M$ a subset, and let $\alpha \in W^{-1}M$. The following are equivalent:*

(a) $\alpha \in R_W^S(M)$.
(b) *There exist* $x \in M$ *and* $b \in W$ *such that* $\alpha = x/b$ *and* $x \in [bS]_M^W$.
(c) *There exist* $x \in M$ *and* $b \in W$ *such that* $\alpha = x/b$ *and* $x \in G_M^W(bS)$.
(d) *For any* $x \in M$ *and* $b \in W$ *such that* $\alpha = x/b$, *we have* $x \in G_M^W(bS)$.
(e) *For any* $x \in M$ *and* $b \in W$ *such that* $\alpha = x/b$, *there is some* $h \in W$ *with* $hx \in F_1^W(hbS)$.

Consequently, in the special case where W *consists of* M*-regular elements of* A, *so that* M *can be identified with its image in* $L := W^{-1}M$, *we have*

$$R_W^S(M) = \bigcup_{b \in W} \left([bS]_M^W :_L b\right).$$

PROOF. (e) $\Rightarrow$ (d): This holds because $F_1^W(hbS) \subseteq [hbS]_M^W$.

(d) $\Rightarrow$ (c): Clear.

(c) $\Rightarrow$ (b): Since $x \in G_M^W(bS)$, there is some $h \in W$ with $hx \in [hbS]_M^W$, but $hx/hb = x/b = \alpha$, so (b) follows.

(b) $\Rightarrow$ (a): This follows from Lemma 7.4.

(a) $\Rightarrow$ (e): Let $x \in M$ and $b \in W$ such that $\alpha = x/b$. There exist $r_1, \ldots, r_m \in W$ and $s_1, \ldots, s_m \in S$ such that

$$\frac{x}{b} = \sum_{i=1}^{m} \frac{s_i}{r_i} = \frac{\sum_{i=1}^{m} \widehat{r_i} s_i}{r},$$

where $\widehat{r_i} = \prod_{j \neq i} r_j$ and $r = \prod_{i=1}^{m} r_i = r_j \widehat{r_j}$ for all j. Then there is some $w \in W$ such that $wrx = wb(\sum_{i=1}^{m} \widehat{r_i} s_i) = \sum_{i=1}^{m} (w\widehat{r_i}) bs_i$. Since $w\widehat{r_i} r_i = wr$ for all i, we have $w\widehat{r_i} \in F_1^W(wr)$, so that since $bs_i \in bS$ for each i, we have $wrx \in F_1^W(wr)bS \subseteq F_1^W(wrbS)$ by Corollary 5.15. □

Applied to the case $M = R$, $S = \{1\}$, we obtain the following:

COROLLARY 7.6. *Let* A *be a commutative ring, let* W *be a multiplicative submonoid of* A, *and let* $\alpha \in W^{-1}A$. *The following are equivalent:*

(a) $\alpha \in R_W(A)$.
(b) *There exist* $a \in A$ *and* $b \in W$ *such that* $\alpha = a/b$ *and* $a \in [b]^W$.
(c) *There exist* $a \in A$ *and* $b \in W$ *such that* $\alpha = a/b$ *and* $a \in G^W(b)$.
(d) *For any* $a \in A$ *and* $b \in W$ *such that* $\alpha = a/b$, *we have* $a \in G^W(b)$.
(e) *For any* $a \in A$ *and* $b \in W$ *such that* $\alpha = a/b$, *there is some* $h \in W$ *with* $ha \in F_1^W(hb)$.

Consequently, in the special case where W *consists of regular elements of* A *(e.g. if* A *is a domain and* $0 \notin W$*), so that* A *can be identified with its image in* $Q = W^{-1}A$, *we have*

$$R_W(A) = \bigcup_{b \in W} \left([b]^W :_Q b\right).$$

Next, we obtain a description of $G_M^W(S)$ in terms of modules of reciprocals:

COROLLARY 7.7. *Let* A *be a commutative ring,* $W \subseteq A$ *a multiplicative submonoid,* M *an* A*-module and* $S \subseteq M$ *a subset. Then* $G_M^W(S) = \bigcup_{h \in W}([hS]_M^W :_A h)$ *is the set of elements* $x \in M$ *such that* $x/1 \in R_W^S(M)$. *In other words, if* $\ell : M \to W^{-1}M$ *is the localization map, then*

$$G_M^W(S) = \ell^{-1}(R_W^S(M)).$$

COROLLARY 7.8. *Let A be a commutative ring, let W be a multiplicative submonoid of A, and let $b \in W$. Then $G^W(b) = \bigcup_{h\in W}([hb]^W :_A h)$ is the set of the numerators in A of the elements of $R_W(A)$ that can be written with denominator b. In other words, the smallest W-regular W-factroid of A containing b is given by*

$$G^W(b) = \{a \in A \mid a/b \in R_W(A)\}.$$

Equivalently, for any $a/b \in W^{-1}A$, where $a \in A$ and $b \in W$, one has $a/b \in R_W(A)$ if and only if $a \in G^W(b)$.

COROLLARY 7.9. *Let A be a commutative ring, and let W be a multiplicative submonoid of A. Then the set of all W-Egyptian elements of A is the smallest W-regular W-factroid $G^W(1) = \bigcup_{b\in W}\left([b]^W :_A b\right)$ of A containing 1. That is,*

$$G^W(1) = A \cap R_W(A).$$

Consequently, the following are equivalent:

(a) *A is W-Egyptian.*
(b) *$G^W(1) = A$.*
(c) *A is the only W-regular W-factroid of A containing 1.*

Let $\frac{W}{1}$ denote the image of W in $W^{-1}A$. Since $R_W(A)$ is the set of all $\frac{W}{1}$-Egyptian elements of $W^{-1}A$, and since every $\frac{W}{1}$-factroid of $W^{-1}A$ is $\frac{W}{1}$-regular (as the elements of $\frac{W}{1}$ are all units of $W^{-1}A$), the corollary above yields the following.

COROLLARY 7.10. *Let A be a commutative ring, let W be a multiplicative submonoid of A, let $Q = W^{-1}A$, let $\frac{W}{1}$ denote the image of W in Q, and let $\frac{1}{W} = \left(\frac{W}{1}\right)^{-1}$ denote the monoid of reciprocals in Q of the elements of $\frac{W}{1}$. One has*

$$R_W(A) = \mathbb{Z}\langle \tfrac{1}{W}\rangle = [1]_Q^{\frac{W}{1}} = G_Q^{\frac{W}{1}}(1).$$

Equivalently, $R_W(A)$ is the principal $\frac{W}{1}$-factroid of Q generated by 1 (which is $\frac{W}{1}$-regular). Consequently, the following are equivalent:

(a) *A is W-Egyptian.*
(b) *Q is the only $\frac{W}{1}$-factroid of Q containing 1.*
(c) *$Q = [1]_Q^{\frac{W}{1}}$.*

COROLLARY 7.11. *Let A be an integral domain with quotient field K, so that $\operatorname{reg}(A) = A^\circ = A \setminus \{0\}$. Then the following are equivalent:*

(a) *A is Egyptian.*
(b) *A is the only $\operatorname{reg}(A)$-regular $\operatorname{reg}(A)$-factroid of A containing 1.*
(c) *K is the only $\operatorname{reg}(A)$-factroid of K containing 1.*

More generally, the Egyptian elements of A (resp., K) are precisely the elements of the smallest $\operatorname{reg}(A)$-regular $\operatorname{reg}(A)$-factroid $G_A^{\operatorname{reg}(A)}(1)$ (resp., the smallest $\operatorname{reg}(A)$-factroid $[1]_K^{\operatorname{reg}(A)}$) of A (resp., K) containing 1.

REMARK 7.12. Corollaries 7.10 and 7.11 above also follow from Corollary 3.8 and Remark 6.2.

EXAMPLE 7.13 (Guerrieri). Let $A = \mathbb{F}_2[x,y]$, $W = \operatorname{reg}(A)$, and $f = (x+y^2)(y+x^2)$. Then $xy \in G^W(f) \setminus [f]_A^W$.

To see this, write $F = [f]_A^W$; we will use that F is a W-factroid. Write $g = x+y^2$ and $h = y+x^2$, so that $f = gh$. Then $g, h \in F$ since $f \in F$, so $(x+y)(1+x+y) =$

$x+y+(x+y)^2 = x+y+x^2+y^2 = f+g \in F$. Thus, $x+y \in F$, so $y(1+y) = y+y^2 = (x+y)+(x+y^2) = (x+y)+g \in F$; thus, $y \in F$. Similarly, $x(1+x) = (x+y)+h \in F$, so that $x \in F$. Then $x+g = y^2 \in F$ and $y+h = x^2 \in F$. For a set S let $\langle S \rangle$ be the $\mathbb{F}_2$-vector space spanned by S. We have $\langle 1,x,y,x^2,y^2,f \rangle \subseteq F$; to show equality, it suffices to show that $\langle 1,x,y,x^2,y^2,f \rangle$ is itself a W-factroid. For this, note that $\langle 1,x,y,x^2,y^2 \rangle$ has 32 elements; one can see by exhaustive search that for any element u of $\langle 1,x,y,x^2,y^2 \rangle$, either $u+f$ is irreducible, or $u+f$ is a product of elements of $\langle 1,x,y,x^2,y^2 \rangle$.

On the other hand, we have

$$\frac{1+xy}{f} = \frac{1}{xy} + \frac{1}{x(y+x^2)} + \frac{1}{y(x+y^2)} \in R_W(A).$$

Hence, by Theorem 7.5, $1+xy \in G^W(f)$, so that since $1 \in G^W(f)$, we have $xy \in G^W(f)$.

8. Euclidean domains

In this section, we study the factroids of any Euclidean domain.

First we review some preliminaries. A function $\sigma : A \to \mathbb{N} \cup \{-\infty\}$ on an integral domain A is a *Euclidean function on A* [**Hun80**, Definition 3.8] if, for all $a, b \in A$ with $b \neq 0$, we have the following:

- $\sigma(a) = -\infty$ if and only if $a = 0$,
- There exist $q, r \in A$ such that $a = bq + r$ and $\sigma(r) < \sigma(b)$, and
- $\sigma(a) \leq \sigma(ab)$.

If σ is a Euclidean function on A and $a \in A$, we say that $\sigma(a)$ is the *σ-degree of a*. Also (see, e.g., [**Eps25**, Proposition 2.6]), for any nonzero $a, b \in A$ one has $\sigma(ab) = \sigma(b)$ if and only if a is a unit, and therefore $\sigma(1) = \sigma(u) < \sigma(a)$ for all units u and all nonzero nonunits a.

A function on A that satisfies all but the third condition is also, by many authors (e.g. [**DF04**, §8.1]), said to be a "Euclidean function" on A. Due to the conflict in terminology (which is allowed to exist because both definitions give rise to the same class of domains [**Mot49**]) we will give a different name to Euclidean functions in this more general sense; we will call such a function a *generalized Euclidean function.* Among the generalized Euclidean functions, those that also satisfy the third condition are said to be *submultiplicative.* For any generalized Euclidean function σ, the corresponding function $\rho(a) = \min\{\sigma(ba) : b \in A \setminus \{0\}\}$ is a submultiplicative generalized Euclidean function on A, i.e. a Euclidean function in the sense of our definition of Eudlidean functions.

An integral domain A is said to be *Euclidean* if there is some generalized Euclidean, or, equivalently, some Euclidean function, on A.

DEFINITION 8.1. We say that a Euclidean function σ on A is *non-archimedean* if for all $a, b \in A$, $\sigma(a+b) \leq \max\{\sigma(a), \sigma(b)\}$.

EXAMPLE 8.2. The degree function deg is a non-archimedean Euclidean function on the Euclidean domain $k[x]$, for any field k. However, the Euclidean domain $\mathbb{Z}$ does not admit a non-archimedean Euclidean function.

Our first result, in a limited sense, characterizes the factroids of any Euclidean domain. However, it does so more explicitly for those Euclidean domains that admit a non-archimedean Euclidean function.

THEOREM 8.3. *Let A be a Euclidean domain with a generalized Euclidean function $\sigma : A \to \mathbb{N} \cup \{-\infty\}$. For all $n \in \mathbb{N} \cup \{\pm\infty\}$, let*

$$C_n := \{a \in A : \sigma(a) \leq n\}.$$

(1) *Let $F \neq A$ be a factroid of A. Then there is some $n \in \sigma(A)$ such that for any $x \in A$ with $\sigma(x) = n$, we have $F = C_n = [x]_A$.*
(2) *σ is a Euclidean function on A if and only if all of the sets C_n are* reg(A)-*saturated.*
(3) *σ is non-archimedean if and only if the sets C_n, for $n \in \sigma(A) \cup \{\infty\}$, are all additive subgroups of A.*

Hence, σ is

- *a Euclidean function if and only if the factroids of A are precisely the sets C_n that are additive subgroups of A, and*
- *a non-archimedean Euclidean function if and only if the factroids of A are all the sets C_n.*

PROOF. First, note that for any $n \in \mathbb{N} \cup \{\pm\infty\}$, we have $C_n = C_s$, where $s = \min\{t \in \sigma(A) \cup \{\pm\infty\} \mid t \geq n\}$.

To prove (1), it suffices to show that $C_n \subseteq F$ for any factroid F of A containing some element of A of σ-degree at least n, where $n \in \mathbb{N} \cup \{-\infty\}$. We prove this by induction on n, the case $n = -\infty$ being trivial. Let $n \geq 0$, and suppose that for all $\ell < n$ with $\ell \in \mathbb{N} \cup \{-\infty\}$, we have $C_\ell \subseteq F$ whenever F contains some element of σ-degree at least ℓ. Let $f \in F$ of σ-degree at least n. Let $0 \neq g \in F$ with $\sigma(g) \leq n$. Then there exist $q, r \in A$ with $f = qg + r$ and $\sigma(r) < \sigma(g)$. Since $f \neq r$ (as they have different σ-degrees), we have $q \neq 0$. Since $\sigma(r) < \sigma(g) = n \leq \sigma(f)$ and $f \in F$, it follows by inductive hypothesis that $C_{\sigma(r)} \subseteq F$, whence $r \in F$. Thus, $qg = f - r \in F$, so since $q \neq 0$ and F is a factroid, we have $g \in F$.

To prove (2), suppose that σ is a Euclidean function, and let $n \in \sigma(A) \cup \{\infty\}$. Let $a \in A$ and $b \in \text{reg}(A)$ with $ba \in C_n$. Then $\sigma(a) \leq \sigma(ba) \leq n$, so that $a \in C_n$. Thus the sets C_n are all reg(A)-saturated. Conversely, suppose that all of the sets C_n are reg(A)-saturated. Let $a, b \in A$ with $b \neq 0$. Let $n = \sigma(ab)$. Since C_n is reg(A)-saturated and $ab \in C_n$, one has $a \in C_n$, and therefore $\sigma(a) \leq n = \sigma(ab)$. Therefore σ is a Euclidean function on A.

To prove (3), suppose that σ is non-archimedean. For any n, let $a, b \in C_n$. Then $\sigma(a \pm b) \leq \max\{\sigma(a), \sigma(b)\} \leq n$, so $a \pm b \in C_n$. Thus, C_n is an additive subgroup of A. Conversely, suppose that each C_n is an additive subgroup of A. Let $a, b \in A$, and suppose without loss of generality that $\sigma(a) \leq n := \sigma(b)$. Then $a, b \in C_n$, whence $a + b \in C_n$, so that $\sigma(a + b) \leq n = \max(\sigma(a), \sigma(b))$. □

Proposition 4.15, whose proof we postponed, is an immediate corollary of Theorem 8.3 above. The following proposition is a partial converse of Proposition 4.15. We define 2^σ to be the function $a \mapsto 2^{\sigma(a)}$.

PROPOSITION 8.4. *For any Euclidean function σ on an integral domain A, the following conditions are equivalent:*

(a) *σ is non-archimedean and satifies the condition $\sigma(ab) = \sigma(a) + \sigma(b)$ for all $a, b \in A$.*
(b) *The function $2^\sigma : A \to \mathbb{N}$ is a non-archimedean norm on the domain A.*
(c) *The function 2^σ is a norm on the domain A, and the prime subring of A is contained in $\{a \in A : \sigma(a) \leq \sigma(1)\} = A^\times \cup \{0\}$.*

(d) *The function* 2^σ *is a norm on the domain* A, *and* A *contains a field.*
(e) *The factroids of* A *are the sets*

$$C_n := \{a \in A : \sigma(a) \le n\}$$

for all $n \in \sigma(A) \cup \{\infty\}$ *(which are distinct for all* $n \in \sigma(A)$*), and the level sets*

$$D_n := \{a \in A : \sigma(a) = n\}$$

satisfy

$$D_m D_n \subseteq D_{m+n}$$

for all $m, n \in \sigma(A)$.

Moreover, a domain A *admits a Euclidean function satisfying the equivalent conditions above if and only if* A *is a field or* $A \cong k[x]$ *for some field* k.

PROOF. The equivalence of conditions (a)–(c) follows from standard facts about norms on integral domains, and the equivalence of (c) and (d) follows from the fact that $\{a \in A \mid \sigma(a) \le \sigma(1)\} = A^\times \cup \{0\}$. The equivalence of conditions (a) and (e) follows from Theorem 8.3. Finally, suppose that σ satisfies all of the given conditions. Then the function $\delta = 2^\sigma$ is a Euclidean function on A in the sense of [**Jac85**, Section 2.15], and therefore, by Exercise 6 of that section, A is a field or $A \cong k[x]$ for some field k. The converse is clear. □

In fact, the following stronger result is possible and is due to user527492 on MathOverflow. See `https://mathoverflow.net/questions/485986/non-archimedean-euclidean-domains/486062#486062`.

PROPOSITION 8.5 (user527492 on MathOverflow). *An integral domain* A *admits a non-archimedean Euclidean function on* A *if and only if* $A = k$ *or* $A \cong k[x]$, *where* $k = A^\times \cup \{0\}$ *is a field.*

PROOF. Suppose that $A \ne k$, so that A has an nonzero nonunit t of minimal σ-degree. Let $\varphi: k[x] \to A$ be the unique k-algebra map sending $x \mapsto t$. We first show that φ is surjective. Note that k is in the image of φ by construction, so it suffices to show that $A \setminus k$ is also in the image of φ. Suppose, for the sake of contradiction, that the image of φ does not contain $A \setminus k$. Then there exists an element $y \in A \setminus k$ not in the image of φ of minimal σ-degree. Since σ is Euclidean, there exist $q, r \in A$ such that $y = qt + r$, where $\sigma(r) < \sigma(t)$. This implies that $r \in k$. If $q = 0$, then $y = r \in k$, which is a contradiction. Therefore, since σ is a non-archimedean Euclidean function,

$$\sigma(q) < \sigma(qt) = \sigma(y - r) \le \max(\sigma(y), \sigma(-r)) = \sigma(y).$$

By minimality, $q = \varphi(f)$ for some $f \in k[x]$, which implies $y = \varphi(fx + r)$, which is a contradiction.

We now show that φ is injective. Suppose, for the sake of contradiction, that $\varphi(f) = 0$, where $f = a_n x^n + \cdots + a_1 x + a_0 \ne 0$. Let k be least such that $a_k \ne 0$. Then $-a_k = \varphi(-a_k) = (\varphi(a_n)t^{n-k-1} + \cdots + \varphi(a_{k+1}))t$, implying that t is a unit, which is a contradiction. Hence φ is also injective, and therefore an isomorphism. □

PROBLEM 8.6. Does there exist a Euclidean domain that admits a non-archimedean generalized Euclidean function and yet is neither a field nor isomorphic to $k[x]$ for some field k?

9. Unit-additive rings and sublocalizable rings

In this section, in which all rings are assumed to be commutative, we show that the recently discovered notion of *unit-additivity* is intimately connected to the subring $[1]_A^{A^\circ}$ of a ring A. We also introduce a common generalization of the local rings an the unit-additive rings that we call the *sublocalizable rings*, and we show, analogously, their intimate connection to the subring $[1]_A^{A^\times} = \mathbb{Z}\langle A^\times\rangle$ of a ring A generated by the units of A.

DEFINITION 9.1 ([**ES25**]). A ring A is *unit-additive* if for any units $u, v \in A^\times$, $u + v$ is either a unit or nilpotent.

EXAMPLE 9.2. In [**ES25**], the second named author and Jay Shapiro provided numerous examples to show that unit-additivity is a very natural condition. For instance:

- If $A \subseteq B$ is an integral extension of rings and B is unit-additive, then so is A [**ES25**, Proposition 2.8].
- Any Boolean ring is unit-additive [**ES25**, Example 2.14].
- Unit-additivity characterizes a kind of simplicity in large classes of Euclidean domains. Let A be an Euclidean domain that is not a field and is finitely generated as a k-algebra, where k is a maximal subfield of A. If $k = \mathbb{F}_2$ or k is algebraically closed, then A is unit-additive if and only if $A \cong k[x]$ [**ES25**, Theorem 2.15].
- Unit-additivity can capture properties of monoids. Indeed, let A be a nonzero ring of characteristic zero, and let M be a cancellative commutative monoid. Then the monoid algebra $A[M]$ is unit-additive if and only if A is unit-additive and M is torsion-free with no nontrivial invertible elements (see [**ES25**, Theorem 3.6]).
- Unit-additivity is related to the fundamental theorem of algebra. In fact, let X be an irreducible variety over an algebraically closed field k, and let A be its coordinate ring. Then A is unit-additive if and only if every polynomial mapping $X \to \mathbb{A}_k^1$ has a root (see [**ES25**, Remark 4.4]).

EXAMPLE 9.3. Let A be a Euclidean domain with a Euclidean function σ. Then $C_{\sigma(1)} = A^\times \cup \{0\} = A^\times \cup \sqrt{0}$ is a multiplicative submonoid of A. Moreover, by Theorem 8.3(3) and [**ES25**, Proposition 2.1], the following are equivalent:

(a) A is unit-additive.
(b) $C_{\sigma(1)}$ is additively closed.
(c) $C_{\sigma(1)}$ is a factroid of A.
(d) $C_{\sigma(1)}$ is a subring of A.
(e) $C_{\sigma(1)}$ is a subfield of A.

For any ring A, let $J(A)$ denote the Jacobson radical of A.

LEMMA 9.4. *Let A be a ring. Then $A^\times \cup J(A) \subseteq \mathbb{Z}\langle A^\times\rangle = [1]_A^{A^\times}$. Consequently, the units of A are the same as those of its subring $R := [1]_A^{A^\times}$, and $J(A) \cap R \subseteq J(R)$.*

PROOF. The fact that $\mathbb{Z}\langle A^\times\rangle = [1]_A^{A^\times}$ follows from Corollary 4.11. Let $a \in J(A)$. Then $1 + a$ is a unit, so since $A^\times \subseteq R$, it follows that $1 + a \in R$. Since also $1 \in R$, we have $a = (1 + a) - 1 \in R$. Now, let $a \in J(A) \cap R$. Then $1 + ra \in A^\times = R^\times$ for all $r \in A$, hence for all $r \in R$, whence $a \in J(R)$. □

PROPOSITION 9.5. *Let A be a ring. Then $A^\times \cup \sqrt{(0)} \subseteq [1]_A^{A^\circ}$; equivalently, the units and nilpotent elements of A are the same as those of its subring $[1]_A^{A^\circ}$. Moreover, the inclusion is an equality if and only if A is unit-additive, if and only if $[1]_A^{A^\circ}$ is unit-additive, if and only if the inclusions*

$$A^\times \cup \sqrt{(0)} \subseteq A^\times \cup J(A) \subseteq [1]_A^{A^\times} \subseteq [1]_A^{\operatorname{reg}(A)} \subseteq [1]_A^{A^\circ}.$$

are equalities.

PROOF. The given inclusions hold because of Lemma 9.4. Therefore, if equalities hold, then $A^\times \cup \sqrt{(0)}$ is a subring of A, whence by [**ES25**, Proposition 2.1] A is unit-additive. Suppose, conversely, that A is unit-additive. Then $B = A^\times \cup \sqrt{(0)}$ is a subring of A by [**ES25**, Proposition 2.1], so we only need to show that B is an A°-saturated subset of A. So let $a \in A^\circ$ and $x \in A$ with $ax \in B$. Then either $ax \in A^\times$ or $ax \in \sqrt{(0)}$. If $ax \in A^\times$, then x is a unit, so $x \in A^\times \subseteq B$. Finally, suppose $ax \in \sqrt{(0)}$. Then for any minimal prime $\mathfrak{p}$ of A, we have $ax \in \mathfrak{p}$, but $a \notin \mathfrak{p}$, so $x \in \mathfrak{p}$. Since $\mathfrak{p}$ was arbitrary, $x \in \sqrt{(0)} \subseteq B$. □

EXAMPLE 9.6. Although, for any integral domain D, its subring $R := [1]_D^{\operatorname{reg}(D)}$ contains $[1]_D^{D^\times} = \mathbb{Z}\langle D^\times\rangle$, equality need not hold. For instance, let $D = \mathbb{Z}[x, 2/x] \subseteq \mathbb{Q}(x)$. Since the only units of D are 1 and -1, the subring of D generated by $D^\times$ is just $\mathbb{Z}$. But since $x(2/x) = 2 \in R$, one has $x, 2/x \in R$, and therefore $R = D$.

CONSTRUCTION 9.7 ([**ES25**, Definition 6.4]). Let A be an integral domain. For each $i \geq 0$, define multiplicatively closed sets V_i and W_i inductively as follows: $W_0 = \{1\}$, $V_0 = A^\times$. For each $i \geq 1$, we let W_i be the set of nonzero sums of elements of V_{i-1}, which by [**ES25**, Lemma 6.3] is a multiplicatively closed set. We then let V_i be the $\operatorname{reg}(A)$-saturation of W_i. Then set $U := \bigcup_i W_i = \bigcup_i V_i$. By [**ES25**, Proposition 6.2 and Theorem 8.1], U is a saturated multiplicatively closed set that is closed under nonzero sums. By [**ES25**, Proposition 8.4], $U^{-1}A$ is the unique smallest overring of A that is unit-additive, called the *unit-additive closure* of A and written A^{ua} (see [**ES25**, Definition 8.2]).

THEOREM 9.8. *Let A be an integral domain, with fraction field K. Let A^{ua} and U be as in Construction 9.7. Then $[1]_A^{\operatorname{reg}(A)} = [1]_A^{A^\circ} = \mathbb{Z}\langle U\rangle = U \cup \{0\}$, and therefore $A^{\mathrm{ua}} = ([1]_A^{\operatorname{reg}(A)} \smallsetminus \{0\})^{-1}A$. Moreover, one has $[1]_A^{\operatorname{reg}(A)} = A$ if and only if K is the only unit-additive overring of A.*

PROOF. First we verify that $\mathbb{Z}\langle U\rangle = U \cup \{0\}$. It is obvious that "$\supseteq$" holds. For the opposite containment, let $0 \neq x \in \mathbb{Z}\langle U\rangle$. Then $x = \sum_{j=1}^n u_j$ for some $u_j \in U$. But then, using the notation of Construction 9.7, there is some i such that $u_1, \ldots, u_n \in V_i$. Hence, $x \in W_{i+1} \cup \{0\} \subseteq U \cup \{0\}$.

To show that $[1]_A^{\operatorname{reg}(A)} \subseteq U \cup \{0\}$, by Corollary 5.12 it suffices to show that the ring $U \cup \{0\}$ is a $\operatorname{reg}(A)$-saturated subring of A. For this, let $a, x \in A$ with $ax \in U \cup \{0\}$ and $a \in \operatorname{reg}(A)$. If $ax = 0$, then since A is a domain and $a \neq 0$, we have $x = 0$. Otherwise, $ax \in U$, so there is some i with $ax \in W_i$. But then by definition $x \in V_i \subseteq U$. In either case $x \in U \cup \{0\}$, so this subring of A is $\operatorname{reg}(A)$-saturated.

For the opposite containment, we show by induction on i that $V_i \subseteq [1]_A^{\operatorname{reg}(A)}$ for all $i \geq 0$, the case $i = 0$ being trivial. So let $i \geq 0$ and assume $V_i \subseteq [1]_A^{\operatorname{reg}(A)}$. Then any sum of elements of V_i is in $[1]_A^{\operatorname{reg}(A)}$, so that $W_{i+1} \subseteq [1]_A^{\operatorname{reg}(A)}$. Now let

$x \in V_{i+1}$. Then there is some $0 \neq a \in A$ with $ax \in W_{i+1} \subseteq [1]_A^{\mathrm{reg}(A)}$. Then since this is a $\mathrm{reg}(A)$-factroid, it follows that $x \in [1]_A^{\mathrm{reg}(A)}$, completing the induction. Thus, $U \cup \{0\} = (\bigcup_{i \geq 0} V_i) \cup \{0\} \subseteq [1]_A^{\mathrm{reg}(A)}$.

For the final statement, suppose $[1]_A^{\mathrm{reg}(A)} = A$. Then every nonzero element of A is in U, so $K = (A \smallsetminus \{0\})^{-1}A = U^{-1}A = A^{\mathrm{ua}}$. Thus, K is the only overring of A that is unit-additive. Conversely suppose K is the only unit-additive overring of A that is unit-additive, so that $U^{-1}A = A^{\mathrm{ua}} = K$. Let $0 \neq a \in A$. Then $\frac{1}{a} \in K = U^{-1}A$, so there exist $b \in A$ and $u \in U$ with $\frac{1}{a} = \frac{b}{u}$ as elements of K. Thus, $ba = u \in U$. In particular, $ba \neq 0$ since $0 \notin U$, so $b \neq 0$. Since U is $\mathrm{reg}(A)$-saturated and $ba \in U$, we have $a \in U$. Thus, $A = U \cup \{0\} = [1]_A^{\mathrm{reg}(A)}$. □

The definition of unit-additive rings inspires the following generalization.

DEFINITION 9.9. A ring A is *sublocalizable* if it is nontrivial and $A^\times + A^\times \subseteq A^\times \cup J(A)$.

The following proposition generalizes [**ES25**, Proposition 2.1] and gives several equivalent conditions for a ring to be sublocalizable. In particular, the term is motivated by the equivalent condition (i).

PROPOSITION 9.10 (cf. [**ES25**, Proposition 2.1]). *Let A be a nontrivial ring. The following conditions are equivalent:*

(a) *A is sublocalizable.*
(b) $1 + A^\times \subseteq A^\times \cup J(A)$.
(c) *$A^\times \cup J(A)$ is a subring of A.*
(d) *There exists a proper ideal I of A such that $A^\times \cup I$ is a subring of A.*
(e) *$J(A)$ is the unique proper ideal I of A such that $A^\times \cup I$ is a subring of A.*
(f) *$B \smallsetminus A^\times$ is an ideal of A for some subring B of A containing $A^\times$ (in which case $B = A^\times \cup J(A)$ and $B \smallsetminus A^\times = J(A)$ are unique).*
(g) *$A^\times \cup J(A)$ is a local subring of A (with maximal ideal $J(A)$ and having the same units as A).*
(h) *A has a (unique) local subring with maximal ideal $J(A)$ and having the same units as A.*
(i) *A has a (unique) local subring whose maximal ideal is also an ideal of A and whose units are the same as those of A.*
(j) *A has a (unique) local subring B with $B^\times = A^\times$ and $J(B) = J(A)$.*
(k) *$A/J(A)$ is sublocalizable.*
(l) *$A/J(A)$ is unit-additive.*

PROOF. Clearly (a) implies (b). Suppose that (b) holds. Then

$$A^\times + A^\times = A^\times(1 + A^\times) \subseteq A^\times(A^\times \cup J(A)) = A^\times \cup J(A),$$

whence (a) holds. Therefore (a) and (b) are equivalent.

Note that, for any ideal I of A, one has $u + I \subseteq A^\times$ for some unit u of A if and only if $A^\times + I = A^\times$, if and only if $I \subseteq J(A)$. Suppose that (a) holds. Then one has

$$\begin{aligned}(A^\times \cup J(A)) + (A^\times \cup J(A)) &= (A^\times + A^\times) \cup (A^\times + J(A)) \cup (J(A) + J(A)) \\ &\subseteq (A^\times \cup J(A)) \cup A^\times \cup J(A) \\ &= A^\times \cup J(A)\end{aligned}$$

and

$$(A^\times \cup J(A))(A^\times \cup J(A)) = A^\times A^\times \cup A^\times J(A) \cup J(A)J(A) = A^\times \cup J(A).$$

It follows that conditions (a) and (c) are equivalent.

Suppose that $A^\times \cup I$ is a subring of I for some proper ideal I of A. We show that $I = J(A)$. Note first that $1 + I \subseteq A^\times \cup I$ and $1 + I$ is disjoint from I since I is proper. Therefore $1 + I \subseteq A^\times$, whence $I \subseteq J(A)$. To prove the reverse containment, note that $-1 + J(A) \subseteq A^\times$ and therefore $J(A) \subseteq 1 + A^\times \subseteq A^\times \cup I$, whence $J(A) \subseteq I$ since $J(A)$ is disjoint from $A^\times$. It follows, then, that $I = J(A)$. All told, this shows that conditions (a)–(j) are equivalent.

Finally, one has $J(A/J(A)) = (0)$ and therefore $\operatorname{nilrad}(A/J(A)) = (0)$, while also $(A/J(A))^\times = \{a+J(A) : a \in A^\times\}$ and $(A/\operatorname{nilrad}(A))^\times = \{a+\operatorname{nilrad}(A) : a \in A^\times\}$, from which it follows that (j)–(l) are equivalent. □

If a ring A is sublocalizable, then we say that the *sublocalization of* A is the unique local subring of A, namely $A^\times \cup J(A)$, whose maximal ideal is also an ideal of A and whose units are the same as those of A. Since any local ring is generated by its group of units, if A is sublocalizable, then its sublocalization B is equal to $\mathbb{Z}\langle B^\times\rangle = \mathbb{Z}\langle A^\times\rangle = [1]_A^{A^\times}$. More generally, a ring A having a local subring containing the units of A has a unique such subring, namely $\mathbb{Z}\langle A^\times\rangle$. Thus, we have the following.

COROLLARY 9.11. *Let A be a sublocalizable ring, with sublocalization B. Then B is the unique local subring of A that contains $A^\times$. Moreover, B is the smallest subring $\mathbb{Z}\langle A^\times\rangle = [1]_A^{A^\times}$ of A containing $A^\times$.*

EXAMPLE 9.12. A ring A having a (unique) local subring sharing the units of A need not be sublocalizable. For instance, let B be any reduced local ring other than a field, and let $A = B[x]$. By [**AM69**, Exercises 2 and 4 of Chapter 1], $A^\times = B^\times$ and $J(A) = (0)$. But $A^\times \cup J(A) = B^\times \cup (0)$ is not additively closed, since for any nonzero nonunit b of B, we have $-1, 1+b \in B^\times = A^\times$, so $b \in (A^\times + A^\times) \setminus (A^\times \cup J(A))$. Thus, A is not sublocalizable.

COROLLARY 9.13. *Let I be proper ideal of a ring A. The following conditions are equivalent:*

(a) *A is sublocalizable and $I = J(A)$.*
(b) *A is sublocalizable, with sublocalization $A^\times \cup I$.*
(c) *$A^\times \cup I$ is a subring of A.*
(d) *$A^\times \cup I$ is a local subring of A (with maximal ideal I and having the same units as A).*
(e) *$A^\times + A^\times \subseteq A^\times \cup I$ and $I \subseteq J(A)$.*
(f) *A/I is unit-additive and reduced, and $I \subseteq J(A)$.*

COROLLARY 9.14. *A ring A is unit-additive if and only if A is sublocalizable and $J(A) = \sqrt{(0)}$.*

REMARK 9.15. A ring A such that $J(A) = \sqrt{(0)}$ is said to be an *NJ ring* [**JRW19**].

REMARK 9.16. A ring A is a local ring if and only if $A \setminus A^\times$ is an ideal of A, in which case $A \setminus A^\times = J(A)$ is the unique maximal ideal of A. Thus, a local ring is equivalently a sublocalizable ring A with sublocalization A, or equivalently a ring

A such that $A \smallsetminus A^\times = J(A)$. Therefore, the sublocalizable rings can be thought of as a common generalization of the local rings and the unit-additive rings.

A ring is A both is local and unit-additive if and only if $A \smallsetminus A^\times = \sqrt{(0)}$, if and only if A has a unique prime ideal, if and only if A is local and zero dimensional. Consequently, a ring A is unit-additive if and only if it has a unique zero dimensional local subring whose (unique) prime ideal is an ideal of A and whose units are the same as those of A, if and only if it is sublocalizable with zero dimensional sublocalization. Thus, one could say that the sublocalizable rings are to the local rings as the unit-additive rings are to the zero dimensional local rings.

A ring A is *semiprimitive* if $J(A) = (0)$. Since the nilradical is always contained in the Jabobson radical, every semiprimitive ring is reduced. A ring A is semiprimitive and sublocalizable if and only if $A^\times + A^\times \subseteq A^\times \cup \{0\}$, if and only if A is reduced and unit-additive, if and only if A is sublocalizable with sublocalization a field, if and only if A has a subfield whose units are the same as those of A. Thus, among the semiprimitive rings, the sublocalizable and the unit-additive rings coincide.

Proposition 9.5 generalizes as follows.

COROLLARY 9.17. *Let A be a ring and I a proper ideal of A, and let T be any subset of A with $A^\times \subseteq T \subseteq W(I)$. Then the subring $[1]^T_A$ of A contains I as an ideal and has the same units as A. In particular, $A^\times \cup I \subseteq [1]^T_A$. Moreover, the following conditions are equivalent:*

(a) *A is sublocalizable and $I = J(A)$.*
(b) *The inclusion $A^\times \cup I \subseteq [1]^T_A$ is an equality.*
(c) *The ring $[1]^T_A$ is sublocalizable and $I = J([1]^T_A)$.*
(d) *The ring $[1]^T_A$ is local, with maximal ideal I.*
(e) *$[1]^T_A \smallsetminus I = A^\times$.*
(f) *$[1]^T_A \smallsetminus I = ([1]^T_A)^\times$.*
(g) *A has a (unique) local subring (namely $[1]^T_A$) having the same units as A and maximal ideal I.*
(h) *A is sublocalizable, with sublocalization $[1]^T_A$, which in turn has maximal ideal I.*

If the equivalent conditions above hold, then $I = \sqrt[A]{I} = J(A)$, $[1]^{A^\times}_A = [1]^T_A = [1]^{W(J(A))}_A$, and $W_A([1]^{A^\times}_A) = W_A(J(A)) = W(I)$ is the set of all elements of A that do not lie in any prime ideal of A that is minimal over $J(A)$.

PROOF. Since $A^\times \subseteq T$, one has $A^\times \subseteq [1]^{A^\times}_A \subseteq [1]^T_A$, and therefore $[1]^T_A$ has the same units as A. Moreover, one has $I \subseteq [1]^T_A I \subseteq [I]^T_A = I$ since I is a T-factroid of A, and therefore I is an ideal of the ring $[1]^T_A$.

Suppose that (a) holds, whence, by Corollary 9.13, $B := A^\times \cup I$ is a subring of A. Moreover, both $A^\times$ and I are T-saturated subsets of A, and therefore so is B. It follows, then, that B is a T-factroid of A containing 1 and contained in $[1]^T_A$, whence (b) holds. Conversely, if (b) holds, then (a) holds, again by Corollary 9.13. Moreover, the remaining equivalences follow, along with the equalities $[1]^{A^\times}_A = [1]^T_A = [1]^{W(J(A))}_A$ and $I = \sqrt{I} = J(A)$, by Proposition 9.10. Finally, the equality $W_A(B) = W_A(I)$, where $B = [1]^{A^\times}_A$, readily follows from the facts that $B = B^\times \cup I$ is a disjoint union, $B^\times = A^\times$, and $ax \in A^\times$ if and only if $a, x \in A^\times$, for any $a, x \in A^\times$. $\square$

EXAMPLE 9.18. The condition that the rings $\mathbb{Z}\langle A^\times\rangle = [1]_A^{A^\times} \subseteq [1]_A^{\mathrm{reg}(A)}$ and $[1]_A^{W(J(A))}$ are all equal and local does not guarantee that A is sublocalizable, even when A is an integral domain. To see this, let B be any local domain and $A = B[x]$, as in Example 9.12. We have $A^\times = B^\times$ and $J(A) = (0)$, so that $W(J(A)) = \mathrm{reg}(A)$ and $A^\times = B^\times$, and therefore $B = \mathbb{Z}\langle B^\times\rangle = \mathbb{Z}\langle A^\times\rangle$. Moreover, note that B is $\mathrm{reg}(A)$-saturated. This follows by consideration of degrees in the variable x, since if $f, g \in A$ with $0 \neq f$ and $fg \in B$, then $0 = \deg(fg) = \deg(f) + \deg(g)$, whence $\deg(g) = 0$ so that $g \in B$. Thus, we have $[1]_A^{W(J(A))} = [1]_A^{\mathrm{reg}(A)} = B$. But by Example 9.12, A is not sublocalizable.

The following corollary generalizes to sublocalizable rings known results on how the local and unit-additive properties behave with respect to integral extensions (see [**Mat86**, Lemma 2 of Section 9] and [**ES25**, Proposition 2.8]).

COROLLARY 9.19 (cf. [**ES25**, Proposition 2.8]). *Let $A \subseteq B$ be an integral extension of rings. Suppose that B is sublocalizable. Then A is sublocalizable, with unique local subring $[1]_A^{A^\times} = [1]_B^{B^\times} \cap A$ that has maximal ideal $J(A) = J(B) \cap A$ and units $A^\times = B^\times \cap A$.*

PROOF. The equalities $J(A) = J(B) \cap A$ and $A^\times = B^\times \cap A$ are well-known properties of integral extensions (see [**AM69**, Exercise 5 of Chapter 5]). Suppose that B is sublocalizable. Let $u, w \in A^\times$. Then either $u + w \in B^\times \cap A = A^\times$ or $u + w \in J(B) \cap A = J(A)$, so that A is sublocalizable. Consequently, one has $[1]_A^{A^\times} = A^\times \cup J(A) = (B^\times \cap A) \cup (J(B) \cap A) = (B^\times \cup J(B)) \cap A = [1]_B^{B^\times} \cap A$. This completes the proof. □

PROBLEM 9.20. Undertake a study of sublocalizable rings, e.g., generalize the results of [**ES25**] to the extent possible.

10. Factroids of submodules, quotients, and preservation along homomorphisms

A natural question to ask is how factroids of modules transfer or are preserved under ring and module homomorphisms. In general, preimages are more well-behaved than images.

PROPOSITION 10.1. *Let $\varphi : A \to B$ be a ring homomorphism and $T \subseteq B$ a subset.*

(1) *If T is a multiplicative subset (resp., submonoid) of B, then $\varphi^{-1}(T)$ is a multiplicative subset (resp., submonoid) of A.*
(2) *For any T-factroid F of B, $\varphi^{-1}(F)$ is a $\varphi^{-1}(T)$-factroid of A.*
(3) *For any additive subgroup F of B, one has $\varphi^{-1}(W(F)) \subseteq W(\varphi^{-1}(F))$.*
(4) *For any B-module N and any T-factroid G of N (as a B-module), G is a $\varphi^{-1}(T)$-factroid of N as an A-module of N by restriction of scalars along φ.*
(5) *For any B-module N and any additive subgroup G of N, one has $\varphi^{-1}(W_N(G)) \subseteq W_{\varphi N}(G)$, where ${}_\varphi N$ denotes the A-module N obtained by restriction of scalars along φ.*

PROOF. Statement (1) is clear.

Let F be a T-factroid of B. It is clear that $\varphi^{-1}(F)$ is an additive subgroup of A. Let $r, a \in A$ with $r \in \varphi^{-1}(T)$ and $ra \in \varphi^{-1}(F)$. Then $\varphi(r)\varphi(a) = \varphi(ra) \in F$, so since $\varphi(r) \in T$, one has $\varphi(a) \in F$, that is, $a \in \varphi^{-1}(F)$. This proves statement (2). Statement (3) then follows by applying statement (2) to the set $T = W(F)$.

To prove statement (4), let $r \in \varphi^{-1}(T)$ and $x \in N$ with $rx \in G$. By definition, $rx = \varphi(r)x$, so since $\varphi(r) \in T$, we have $x \in G$. This proves (4), and statement (5) then follows by applying (4) to the set $T = W_N(G)$. □

Below is the analogue of Proposition 10.1 for module homomorphisms.

PROPOSITION 10.2. *Let A be a ring and $h : M \to N$ an A-module homomorphism. Then, for any subset T of A and for any T-factroid G of N, $h^{-1}(G)$ is a T-factroid of M. Equivalently, for any additive subgroup G of N, one has $W_N(G) \subseteq W_M(h^{-1}(G))$.*

PROOF. Clearly $\varphi^{-1}(G)$ is an additive subgroup of M. Let $a \in T$ and $x \in M$ such that $ax \in h^{-1}(G)$. Then $ah(x) = h(ax) \in G$, so since G is a T-factroid, $h(x) \in G$, whence $x \in h^{-1}(G)$. Thus, $\varphi^{-1}(G)$ is a T-factroid of M. □

PROPOSITION 10.3. *Let $\varphi : A \to B$ be a ring homomorphism, $T \subseteq B$ a subset, M a B-module, L an A-submodule of the A-module M formed by restriction of scalars along φ, and $S \subseteq L$ a subset. Then $[S]_L^{\varphi^{-1}(T)} \subseteq [S]_M^T$.*

PROOF. Set $U := \varphi^{-1}(T)$. Let G be a T-factroid of M containing S, where M is thought of as a B-module. Then by Proposition 10.1, G is a U-factroid of M, where M is thought of as an A-module by restriction of scalars along φ. Let $j : L \hookrightarrow M$ be the (A-linear) inclusion map. Then by Proposition 10.2, $G \cap L = j^{-1}(G)$ is a U-factroid of L containing S. Hence, $[S]_L^U \subseteq G \cap L \subseteq G$. Since G was arbitrary, $[S]_L^U \subseteq [S]_M^T$. □

EXAMPLE 10.4. Although factroids are well-behaved under inverse images of ring homomorphisms (see Proposition 10.1(2)), they are not well-behaved under forward images. That is, if $\varphi : A \to B$ is a ring homomorphism, $W \subseteq A$ a multiplicative set, and F a W-factroid of A, $\varphi(F)$ is not necessarily a $\varphi(W)$-factroid of B.

For a surjective example, let $A = k[x,y]$, $B = k[x]$, and $\varphi : A \twoheadrightarrow B$ the k-algebra map that sends $x \mapsto x$ and $y \mapsto 0$. Let $W = k[x] \setminus \{0\} \subseteq A$. Let F be the vector space spanned by 1 and $x^2 + y^3$. Then by Example 4.17, F is a W-factroid. But $\varphi(F) = \operatorname{span}_k\{1, x^2\}$ is not a $\varphi(W)$-factroid of B, since $x^2 \in \varphi(F)$ and $x \in \varphi(W)$, but $x \notin \varphi(F)$.

For an injective example, let $A = k[x,y]$, $B = k[x,y,z]/(x^2 - yz)$, and $\varphi : A \to B$ the obvious injective ring homomorphism. Let $F = k[x] \subseteq A$ and $W = A \setminus \{0\} = \operatorname{reg}(A)$. Then F is a W-factroid of A, but $\varphi(F)$ isn't a $\varphi(W)$-factroid of B, since $yz = x^2 \in \varphi(F)$ and $y \in \varphi(W)$ but $z \notin \varphi(F)$.

The following proposition connects the T-factroids of M to those of the A-submodules of M.

PROPOSITION 10.5. *Let N be an A-submodule of M and F an additive subgroup of N. Then one has $W_N(F) \cap W_M(N) \subseteq W_M(F) \subseteq W_N(F)$. Equivalently: for any subset T of A, if F is a T-factroid of M, then F is T-factroid of N; and if F is a T-factroid of N and N is a T-factroid of M, then F is a T-factroid of M.*

PROOF. The proof is straightforward. □

The following proposition describes the T-factroids of any quotient of M.

PROPOSITION 10.6. *For any subset T of A and any A-submodule N of M, the association $F \mapsto F/N$ defines a bijection from the T-factroids of M containing N to the T-factroids of the A-module M/N. Consequently, one has $W_M(F) = W_{M/N}(F/N)$ for any additive subgroup F of M containing N.*

PROOF. The proof is straightforward. □

Since any T-factroid of M contains $[0]_M^T$, one has the following corollary.

COROLLARY 10.7. *For any subset T of A, the association $F \mapsto F/[0]_M^T$ defines a bijection from the T-factroids of M to the T-factroids of the A-module $M/[0]_M^T$. Moreover, the A-module $M/[0]_M^T$ is the universal quotient N of M such that every element of T is regular on N, or equivalently such that the zero submodule of N is a T-factroid of N.*

As a consequence of Corollary 10.7, given a subset T of A, there is no harm in assuming that the zero submodule of M is a T-factroid of M, or equivalently that every element of T is regular on M. Thus, there is no harm in assuming that the localization map $M \to \langle T\rangle^{-1}M$ is injective, i.e., that M is (canonically isomorphic to) an A-submodule of $\langle T\rangle^{-1}M$. This has some implications for Theorem 7.5 and its corollaries. For example, let A be any commutative ring, W a multiplicative set, and M an A-module. Set $\bar{M} := M/[0]_M^W$, $\bar{S} :=$ the image of the set S under the projection $M \twoheadrightarrow \bar{M}$, and $L := W^{-1}M = W^{-1}\bar{M}$. Then by combining Corollary 10.7 with the last part of Theorem 7.5, we have $R_W^S(M) = \bigcup_{b\in W}([b\bar{S}]_{\bar{M}}^W :_L b)$.

Finally, as a consequence of Propositions 10.5 and 10.6, one has the following partial analogue of Proposition 10.2 for images of T-factroids under a module homomorphism.

COROLLARY 10.8. *Let T be a subset of A, let $h : M \to N$ be any homomorphism of A-modules, and let F be an additive subgroup of M containing* $\ker h$. *Then one has $W_M(F) \cap W_N(\operatorname{im} h) \subseteq W_N(h(F)) \subseteq W_M(F) = W_{\operatorname{im} h}(h(F))$. Equivalently: if F is a T-factroid of M and* $\operatorname{im} h$ *is a T-factroid of N, then $h(F)$ is a T-factroid of N; and if $h(F)$ is a T-factroid of N, then F is a T-factroid of M.*

EXAMPLE 10.9. Recall the surjective map from Example 10.4, where $A = k[x,y]$, $F = \operatorname{span}_k\{x^2 + y^3, 1\}$, $B = k[x]$, $T := k[x] \setminus \{0\} \subseteq A$, and $\varphi : A \to B$ is the k-algebra homomorphism with $x \mapsto x$ and $y \mapsto 0$. Then setting $M := A$, and $N := B$ thought of as an A-module by restriction of scalars, this provides an example of a surjective A-linear map $h = \varphi : M \to N$ and a W-factroid F of M such that $h(F)$ is not a T-factroid of N. Of course, F does not contain $\ker h$.

11. Finiteness in graded settings

In this section, we show that for rings similar to polynomial rings over a field, factroids tend to be bounded in degree, and hence are finite-dimensional as vector spaces over the base field.

DEFINITION 11.1. Recall that for an $\mathbb{N}$-graded or $\mathbb{Z}$-graded abelian group $(G,+)$, the *degree* of a nonzero (but not necessarily homogeneous) element $g \in G$ can be defined as follows. We have a unique decomposition $g = \sum_{i=c}^{d} g_i$, where $c \le d$ in $\mathbb{N}$ (resp. $\mathbb{Z}$), each g_i is in the homogeneous degree i graded component of G, $g_c \neq 0$, and $g_d \neq 0$. Then $\deg g := d$.

LEMMA 11.2 (Guerrieri). *Let A be an $\mathbb{N}$-graded ring, $W \subseteq A$ a multiplicative set that does not contain 0, and M a torsion-free $\mathbb{Z}$-graded A-module. Let $S \subseteq M$ and $d \in \mathbb{Z}$ such that for all $s \in S$, $\deg s \leq d$. Then for all $u \in [S]_M^W$, $\deg u \leq d$.*

PROOF. By Definition 5.13 and induction, it suffices to show that $\deg u \leq d$ for all $u \in F_1^W(S)$. For any $u \in F_1^W(S)$, there exist $u_i \in M$ and $r_i \in W$ such that $u = \sum_{i=1}^n u_i$ and $r_i u_i \in S$ for all i. Since r_i has nonnegative degree and degrees are additive (by torsion-freeness of M), for each i we have $\deg(u_i) \leq \deg(r_i) + \deg(u_i) = \deg(r_i u_i) \leq d$. Thus, $\deg(u) \leq \max_{1 \leq i \leq n} \deg(u_i) \leq d$. □

THEOREM 11.3. *Let B be a Noetherian $\mathbb{N}$-graded domain over a field $B_0 = k$, and let A be a k-subalgebra of B. Let $W \subseteq B$ be a multiplicative set and let $V = W \cap A$. Let M be a finitely generated torsion-free $\mathbb{N}$-graded B-module, and let L be a nonzero A-submodule of M. Then any finitely generated V-factroid F of L is finite-dimensional as a k-vector space, with every element of no bigger degree than any element of a factroid generating set of F.*

PROOF. Let $S = \{s_1, \dots, s_m\}$ be a generating set for the V-factroid F, so that $F = [s_1, \dots, s_m]_L^V$. For each $1 \leq i \leq m$, let $d_i = \deg(s_i)$, and let $d = \max\{d_i \mid 1 \leq i \leq m\}$. It follows from Lemma 11.2 that every element of $[S]_M^W$ has degree $\leq d$. Since M is a Noetherian graded B-module, every $M_i = \{x \in M \mid x \text{ is homogeneous of degree } i\}$ is a finitely generated k-vector space. Since $[S]_M^W \subseteq \bigoplus_{i=0}^d M_i$, it follows that $[S]_M^W$ is a finite-dimensional k-vector space. By Proposition 10.3, we have $F = [S]_L^V \subseteq [S]_M^W$, so that F is also finite-dimensional over k, with all elements of degree $\leq d$. □

The above then applies when B is a finitely generated k-algebra domain, and A is a k-subalgebra of B (we might even have $A = B$), where we set $M = B$ and $L = A$. Hence, by finding linearly independent elements of F over k within the degree bound, one will eventually arrive at a finite vector-space basis of F over k.

Acknowledgments

First we would like to thank the anonymous referee, whose careful reading of and astute suggestions regarding the paper improved it substantially. Next, we thank Lorenzo Guerrieri for Example 7.13, Lemma 11.2, and the case of Lemma 7.4 where $M = A$ and $S = \{1\}$. Moreover, he had the initial insight that led to the discovery of factroids. Finally, we are grateful to the moderators of mathoverflow for creating the forum that paired us as coauthors. Without that forum, this paper would not exist.

References

[AM69] M. F. Atiyah and I. G. Macdonald, *Introduction to commutative algebra*, Addison-Wesley Publishing Co., Reading, Mass.-London-Don Mills, Ont., 1969. MR242802

[BE22] T. F. Bloom and C. Elsholtz, *Egyptian fractions*, Nieuw Arch. Wiskd. (5) **23** (2022), no. 4, 237–245. MR4544007

[BZ51] R. A. Beaumont and H. S. Zuckerman, *A characterization of the subgroups of the additive rationals*, Pacific J. Math. **1** (1951), 169–177. MR44522

[DF04] D. S. Dummit and R. M. Foote, *Abstract algebra*, 3rd ed., John Wiley & Sons, Inc., Hoboken, NJ, 2004. MR2286236

[DG66] M. Dunton and R. E. Grimm, *Fibonacci on Egyptian fractions*, Fibonacci Quart. **4** (1966), 339–354. MR207503

[EGL24] N. Epstein, L. Guerrieri, and K. A. Loper, *The reciprocal complement of a polynomial ring in several variables over a field*, Pacific J. Math. **338** (2025), no. 2, 267–293, DOI 10.2140/pjm.2025.338.267. MR4950937

[Eps23] N. Epstein, *Rings that are Egyptian with respect to a multiplicative set*, Comm. Algebra **52** (2024), no. 3, 1277–1282, DOI 10.1080/00927872.2023.2260872. MR4709590

[Eps24a] N. Epstein, *Egyptian integral domains*, Ric. Mat. **73** (2024), no. 5, 2901–2910, DOI 10.1007/s11587-023-00813-w. MR4815274

[Eps24b] N. Epstein, *Rational functions as sums of reciprocals of polynomials*, Amer. Math. Monthly **131** (2024), no. 9, 794–801, DOI 10.1080/00029890.2024.2379749. MR4815297

[Eps25] N. Epstein, *The unit fractions from a Euclidean domain generate a DVR*, Ric. Mat. **74** (2025), no. 4, 2307–2313, DOI 10.1007/s11587-024-00922-0. MR4961474

[ES14] N. Epstein and J. Shapiro, *Strong Krull primes and flat modules*, J. Pure Appl. Algebra **218** (2014), no. 9, 1712–1729, DOI 10.1016/j.jpaa.2014.01.009. MR3188867

[ES25] N. Epstein and J. Shapiro, *Rings where a non-nilpotent sum of units is a unit*, J. Algebra **672** (2025), 120–144, DOI 10.1016/j.jalgebra.2025.02.019. MR4878790

[GLO24] L. Guerrieri, K. A. Loper, and G. Oman, *From ancient Egyptian fractions to modern algebra*, J. Algebra Appl. (2024), 14 pages, DOI 10.1142/S0219498826500787.

[Gue25] L. Guerrieri, *The reciprocal complements of classes of integral domains*, J. Algebra **682** (2025), 188–214, DOI 10.1016/j.jalgebra.2025.05.021. MR4924204

[HH90] M. Hochster and C. Huneke, *Tight closure, invariant theory, and the Briançon-Skoda theorem*, J. Amer. Math. Soc. **3** (1990), no. 1, 31–116, DOI 10.2307/1990984. MR1017784

[Hun80] T. W. Hungerford, *Algebra*, Graduate Texts in Mathematics, vol. 73, Springer-Verlag, New York-Berlin, 1980. Reprint of the 1974 original. MR600654

[Jac85] N. Jacobson, *Basic algebra. I*, 2nd ed., W. H. Freeman and Company, New York, 1985. MR780184

[JRW19] M. Jiang, Y. Wang, and Y. Ren, *Extensions and topological conditions of NJ rings*, Turkish J. Math. **43** (2019), no. 1, 44–62, DOI 10.3906/mat-1805-103. MR3909275

[Kis63] J. Kist, *Minimal prime ideals in commutative semigroups*, Proc. London Math. Soc. (3) **13** (1963), 31–50, DOI 10.1112/plms/s3-13.1.31. MR143837

[Mat86] H. Matsumura, *Commutative ring theory*, Cambridge Studies in Advanced Mathematics, vol. 8, Cambridge University Press, Cambridge, 1986. Translated from the Japanese by M. Reid. MR879273

[Mot49] Th. Motzkin, *The Euclidean algorithm*, Bull. Amer. Math. Soc. **55** (1949), 1142–1146, DOI 10.1090/S0002-9904-1949-09344-8. MR32592

[Rap74] R. Raphael, *Rings which are generated by their units*, J. Algebra **28** (1974), 199–205, DOI 10.1016/0021-8693(74)90032-5. MR342554

[Rei14] D. Reimer, *Count like an Egyptian*, Princeton University Press, Princeton, NJ, 2014. A hands-on introduction to ancient mathematics, DOI 10.1515/9781400851416. MR3186542

[Ros94] K. I. Rosenthal, *Modules over a quantale and models for the operator* ! *in linear logic* (English, with French summary), Cahiers Topologie Géom. Différentielle Catég. **35** (1994), no. 4, 329–333. MR1307265

[vAvL63] P. J. van Albada and J. H. van Lint, *Reciprocal bases for the integers*, Amer. Math. Monthly **70** (1963), 170–174, DOI 10.2307/2312886. MR147448

Department of Mathematics, California State University, Channel Islands, Camarillo, California 93012

Department of Mathematical Sciences, George Mason University, Fairfax, Virginia 22030

Email address: `nepstei2@gmu.edu`

Contemporary Mathematics
Volume **836**, 2026
https://doi.org/10.1090/conm/836/16711

The n-total graph of a commutative ring

Djamila AitElhadi and Ayman Badawi

ABSTRACT. Let R be a commutative ring with $1 \neq 0$, $Z(R)$ be the set of all zero-divisors of R, and $n \geq 1$. This paper introduces the n-total graph of a commutative ring R. The n-total graph of a commutative ring R, denoted by $n - T(R)$, is an undirected simple graph with vertex set R, such that two vertices x, y in R are connected by an edge if $x^n + y^n$ in $Z(R)$. Note that if $n = 1$, then the 1-total graph of R is the total graph of R in the sense of Anderson-Badawi's paper on the total graph of a commutative ring. In this paper, we study some graph properties and theoretical ring structure.

1. Introduction

Graphs represent relations defined by taking ordered pairs from a set of vertices. Such ordered pairs are called edges. Graphs are very beneficial in physical, biological, and social applications. They are also used in computer scientific contexts like networks and data structure representations. In addition, Graphs are utilized in other mathematical fields, such as knot theory, where each type of knot can be associated with a graph.

Graphs over rings were first seriously researched during the late nineties of the last century. While the discussion was mainly about graphs over rings, there has been research on graphs over groups, semi-rings, modules over rings, etc. For a recent reference on graphs from rings, see the book [**AABC**] by Anderson, Asir, Badawi, and Chelvam. In the paper [**BE**] published in 1988 titled 'Colouring of Commutative rings' by Istvan Beck, the idea of the colouring of commutative rings was presented. This idea established a connection between graph theory and commutative ring theory that is evident in many studies such as [**AN**] and [**SB**]. The paper establishes the graph of R by defining the relation x is adjacent to y if and only if $xy = 0$ in a commutative R. The study is mainly concerned with characterizing and discussing finitely colourable rings and presenting certain conditions for which a ring R is finitely colourable. Anderson and Livingston introduce the zero divisor graph $\Gamma(R)$ of a commutative ring R in [**AL**] by defining the vertices of the graph as $V = Z(R) - \{0\}$ and defining the edges by the relation $x \in V$ is adjacent to $y \in V$ if and only if $xy = 0$. The zero divisor graph was investigated in several studies, such as [**AAS**], [**ALS**], and [**M**]. Another body of work that is well-studied in the field was the work of David F. Anderson and Ayman Badawi in [**AB**] on the

2020 *Mathematics Subject Classification.* Primary 13A15; Secondary 13B99, 05C99.

Key words and phrases. Total graph, zerodivisor graph, generalized total graph, annihilator graph, zerodivisor, connected, diameter, girth.

©2026 American Mathematical Society

total graph of a commutative ring R defined by taking the vertices $V = \{x \in R\}$ and defining the edges by x is adjacent to y if and only if $x + y \in Z(R)$.

Let R be a commutative ring with $1 \neq 0$ and let $n \geq 1$ be an integer. In this paper, we define the n-total graph of R by taking the vertices $V = R$ and defining the edges by $x \in V$ is adjacent to $y \in V$ if and only if $x^n + y^n \in Z(R)$. We denote the n-total graph of R by $n - T(R)$. The literature on graphs from rings is rich. We cannot state them all; for example see [**AAM**]–[**SB**]. A complete list of references up to 2021 that might interest the reader is [**AABC**].

We recall some definitions that are needed in this paper. Let G be a (simple) graph. We say that G is *connected* if there is a path between any two distinct vertices of G. At the other extreme, we say that G is *totally disconnected* if no two vertices of G are adjacent (i.e., no vertices of G are connected by one edge). For vertices x and y of G, the *distance* between x and y, denoted by $d(x, y)$, is defined to be the length of the shortest path from x to y ($d(x, x) = 0$ and if there is no path between x and y, then $d(x, y) = \infty$). The *diameter* of G is $\text{diam}(G) = \sup\{d(x, y) \mid x \text{ and } y \text{ are vertices of } G\}$. The *girth* of G, denoted by $gr(G)$, is the length of a shortest cycle in G ($gr(G) = \infty$ if G contains no cycles). We denote the complete graph on n vertices by K_n and the complete bipartite graph on m and n vertices by $K_{m,n}$ (we allow m and n to be infinite cardinals). We say that a (induced) subgraph G_1 is a *component* of a graph G if G_1 is connected and no vertex of G_1 is adjacent (in G) to any vertex not in G_1. If G is the union of m components, we say G is a decomposition of m components, and we write $G = \bigoplus_{i=1}^{m} G_i$ and if $G_1 = \cdots = G_m$, then we write $G = \bigoplus_{i=1}^{m} G_1$.

Throughout this paper, all rings are commutative with $1 \neq 0$. Let R be a commutative ring. Then $Z(R)$ denotes its set of all zero-divisor, $Nil(R)$ denotes its ideal of nilpotent elements, $Reg(R)$ denotes its set of non-zero-divisors (i.e., $Reg(R) = R \setminus Z(R)$), and $U(R)$ denotes its group of units. For $A \subseteq R$, let $A^* = A - \{0\}$. We say that R is *reduced* if $Nil(R) = \{0\}$. As usual, $\mathbb{Z}, \mathbb{Z}_n$, and F_q will denote the integers, integers modulo n, and the finite field with q elements, respectively.

2. Results and analysis

Let $n \geq 2$. First, we will show that the n-total graph of a ring differs from the total graph of a ring, as in the Anderson-Badawi paper. We have the following examples.

EXAMPLE 2.1. The total graph of $\mathbb{Z}_{49}$ is a decomposition of four components, $K_7 \oplus K_{7,7} \oplus K_{7,7} \oplus K_{7,7}$. Meanwhile, the 3-total graph of $\mathbb{Z}_{49}$ is a decomposition of two components, $K_7 \oplus K_{21,21}$.

EXAMPLE 2.2. The total graph of $\mathbb{Z}_7$ has 4 components, while the $3 - T(\mathbb{Z}_7)$ has 2 components, and the $2 - T(\mathbb{Z}_7)$ has 7 components

EXAMPLE 2.3. The total graph of $\mathbb{Z}_{13}$ has 7 components, the 6-total graph of $\mathbb{Z}_{13}$ has 2 components, and the 2 total graph of $\mathbb{Z}_{13}$ has 4 components

2.1. The case where $Z(R)$ is an ideal.

Let R be a commutative ring such that $Z(R)$ is an ideal of R. Then $Z(R)$ is a prime ideal of R; therefore, $R/Z(R)$ is an integral domain. Moreover, when R is a finite commutative ring, $R/Z(R)$ becomes a field as $Z(R) = Nil(R)$ is the only maximal ideal of the ring. Trivially, the induced subgraph $n-T(Z(R))$ is a complete component of the $n-T(R)$. Thus, the main focus of this section will be on $n - T(Reg(R))$. We will generalize the results seen in [**AB**] for finite commutative rings where $Z(R)$ is an ideal of R. The following two lemmas are known; we omit their proofs.

LEMMA 2.4. *Let F_q be a finite field. Then, $U(F_q) = F_q^*$ is a cyclic group. Furthermore, for each positive integer $m \geq 1$, if the equation $x^m = a$ has a solution in F_q^* for some $a \in F_q^*$, then there are precisely $gcd(m, |F_q^*|)$ distinct solutions in F_q^*.*

LEMMA 2.5. *Let F_q be a finite field, $a \in F_q^*$, and $m \geq 1$ be an integer. Then $x^m = a$ has a solution in F_q if and only if $a \in S_m = \{b^m \mid b \in F_q^*\}$. Furthermore, S_m is a cyclic subgroup of F_q^* of order $\frac{|F_q^*|}{gcd(m,|F_q^*|)}$.*

The following theorem generalizes [**AB**, Theorem 2.2 (1)] for finite commutative rings.

THEOREM 2.6. *Let R be a finite commutative ring such that $Z(R)$ is an ideal of R and $2 \in Z(R)$. Let $n \geq 1$ be an integer, $|Z(R)| = \alpha$, $|R/Z(R)| = \beta$, $\gamma = \alpha \cdot gcd(n, \beta - 1)$, and $d = \frac{\beta-1}{gcd(n,\beta-1)}$. Then the $n - T(R) = K_\alpha \oplus \bigoplus_{i=1}^{d} K_\gamma$. Note that $|R| = \alpha\beta$.*

PROOF. Let $n \geq 1$ be an integer. Assume that $2 \in Z(R)$, and let $x \in Reg(R)$ (i.e., $x \in U(R)$). Since $2 \in Z(R)$, $(x + z_1)^n + (x + z_2)^n = 2x^n + z_3 \in Z(R)$ for some $z_3 \in Z(R)$ for all $z_1, z_2 \in Z(R)$. Hence, each coset $x + Z(R)$ elements form a K_α subgraph of the $n - T(R)$. Since $F = |R/Z(R)|$ is a finite field, for each $a \in S_n(F) = \{b^n \mid b \in F^*\}$, the set $C_n(a) = \{w \in F^* \mid w^n = a\}$ has exactly $gcd(n, \beta - 1)$ cosets of $Z(R)$ by Lemma 2.4 and Lemma 2.5. Let $D_a = \cup_{v \in C_n(a)} v$. Then D_a is a subset of R, $|D_a| = \gamma$, and every two vertices in D_a are adjacent since $2 \in Z(R)$. Hence, D_a is a K_γ component of the $n - T(R)$. Since $|S_n(F)| = d$ by Lemma 2.5 and $Z(R)$ forms a K_α component of the $n - T(R)$, we conclude that the $n - T(R) = K_\alpha \oplus \bigoplus_{i=1}^{d} K_\gamma$. □

EXAMPLE 2.7. Let $p = 2$, $m = 2^i$, $c = 2^{i-1}$, and let $n \geq 1$ be an integer. The $n - T(\mathbb{Z}_m) = K_c \oplus K_c$ where the $n - T(Reg(\mathbb{Z}_m)) = K_c$.

EXAMPLE 2.8. Let $R = \mathbb{Z}_8$. We have $2 \in Z(R)$, $|Z(R)| = 4$ and $|R/Z(R)| = 2$. The 2-total graph of $\mathbb{Z}_8$ is $K_4 \oplus K_4$. Figure 1 shows the n-total graph of $\mathbb{Z}_8$ for every integer $n \geq 1$.

The following theorem generalizes [**AB**, Theorem 2.2 (2)] for finite commutative rings.

THEOREM 2.9. *Let $n \geq 1$ be an integer and R be a finite commutative ring such that $Z(R)$ is an ideal of R and $2 \notin Z(R)$. Let $|Z(R)| = \alpha$, $|R/Z(R)| = \beta$, $\gamma = \alpha \cdot gcd(n, \beta - 1)$, and $d = \frac{\beta-1}{gcd(n,\beta-1)}$. Then*

(1) *The $n - T(Reg(R))$ is totally disconnected if and only if d is odd.*
(2) *If d is even, then the $n - T(Reg(R)) = \bigoplus_{i=1}^{d/2} K_{\gamma,\gamma}$, and hence the $n - T(R) = K_\alpha \oplus \bigoplus_{i=1}^{d/2} K_{\gamma,\gamma}$.*

PROOF. Let $n \geq 1$ be an integer. Note that $F = R/Z(R)$ is a finite field and $a^{\beta-1} = 1 + Z(R)$ for every $a \in F^*$.

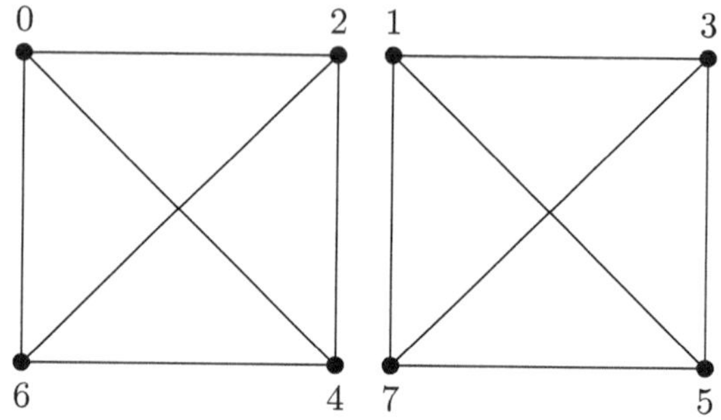

FIGURE 1. $n - T(\mathbb{Z}_8)$ for every integer $n \geq 1$

(1) Suppose the $n - T(Reg(R))$ is totally disconnected. Assume that d is even. Then $S_n(F) = \{b^n \mid b \in F^*\}$ is the unique cyclic subgroup of F^* of order d by Lemma 2.5. Since d is even, $-1 + Z(R) \in S_n(F)$. Since $2 \notin Z(R)$, $1 + Z(R) \neq -1 + Z(R)$. Hence there are $x = a + Z(R), y = b + Z(R) \in F^*$ for some $a, b \in Reg(R)$ such that $x^n + y^n = (a^n + b^n) + Z(R) = Z(R)$ in F. Thus $a^n + b^n \in Z(R)$; a contradiction since the $n - T(R)$ is totally disconnected.

For the converse, suppose d is odd. Note that $dn = k \cdot (\beta - 1)$ for some integer $k \geq 1$. Assume that the $n - T(Reg(R))$ is not totally disconnected. Then, there exist $x, y \in Reg(R)$ such that $x^n + y^n \in Z(R)$. Thus $x^n + Z(R) = -y^n + Z(R)$ in F. Hence $(x^n)^d + Z(R) = (-y^n)^d + Z(R)$ in F. Thus $x^{nd} + Z(R) = (-1^d \cdot y^{nd}) + Z(R)$ in F. Since $nd = k(\beta - 1)$ and d is odd, we have $1 + Z(R) = -1 + Z(R)$ in F. Hence $2 \in Z(R)$, a contradiction. Thus the $n - T(R)$ is totally disconnected.

(2) Let $a \in Reg(R)$. Since $2 \notin Z(R)$, $a^n + a^n = 2a^n \notin Z(R)$. Therefore, no two vertices in $a + Z(R)$ are connected in the $n - T(Reg(R))$. Since $S_n(F) = \{b^n \mid b \in F^*\}$ is the unique cyclic subgroup of F^* of order d by Lemma 2.5 and d is even, we conclude that $-a \in S_n(F)$ for every $a \in S_n(F)$. Since $F = |R/Z(R)|$ is a finite field, for each $a \in S_n(F) = \{b^n \mid b \in F^*\}$, the set $C_n(a) = \{w \in F^* \mid w^n = a\}$ has exactly $gcd(n, \beta - 1)$ cosets of $Z(R)$ by Lemma 2.4 and Lemma 2.5. Let $A = \cup_{v \in C_n(a)} v$ and $B = \cup_{w \in C_n(-a)} w$. Then $A \cup B$ is a subset of $Reg(R)$, and $|A| = |B| = \gamma$. Note that every two vertices of A are not adjacent, and every two vertices of B are not adjacent. However, every vertex in A is adjacent to every vertex in B. Thus $A \cup B$ elements form a $K_{\gamma,\gamma}$ component of the $n - T(Reg(R))$. Thus the $n - T(Reg(R)) = \bigoplus_{i=1}^{d/2} K_{\gamma,\gamma}$. Since $Z(R)$ elements form a K_α component of the $n - T(R)$. We conclude that the $n - T(R) = K_\alpha \oplus \bigoplus_{i=1}^{d/2} K_{\gamma,\gamma}$. □

EXAMPLE 2.10. Let $n \geq 1$ be an integer, p be an odd prime, $m = p^i$, $R = \mathbb{Z}_m$, $\alpha = p^{i-1} = |Z(\mathbb{Z}_m)|$, $\beta = |R/Z(R)|$, $\gamma = gcd(n, \beta - 1)\alpha$, and $d = (\beta - 1)/gcd(n, \beta - 1) = (p - 1)/gcd(n, p - 1)$.

(1) The $n - T(Reg(\mathbb{Z}_m))$ is totally disconnected if and only if d is odd.
(2) If d is even, then the $n - T(Reg(R)) = \bigoplus_{i=1}^{d/2} K_{\gamma,\gamma}$, and hence the $n - T(R) = K_\alpha \oplus \bigoplus_{i=1}^{d/2} K_{\gamma,\gamma}$.
(3) Let $R = \mathbb{Z}_{169}$ and $n = 3$. Let α, β, γ, and d as in Theorem 2.9. Then $\alpha = \beta = 13$, $\gamma = 39$, and $d = 4$. Then the $3 - T(R) = K_{39,39} \oplus K_{39,39} \oplus K_{13}$.

EXAMPLE 2.11. Let $R = \mathbb{Z}_9$ and $n = 2$. We have $\alpha = |Z(R)| = 3$, $\beta = |R/Z(R)| = 3$, $\gamma = gcd(n, \beta - 1)\alpha = 3gcd(2,2) = 6$, and $d = (\beta - 1)/gcd(n, \beta - 1) = 2/2 = 1$. Then the $2 - T(Reg(R))$ is a totally disconnected graph with 6 vertices. However, The total graph of $Reg(R) = 1 - T(Reg(R)) = K_{3,3}$. Thus the $2 - T(R) = K_{3,3} \oplus K_3$

COROLLARY 2.12. *Let $n \geq 1$ be an integer and R be a finite commutative ring such that $2 \notin Z(R)$. Let $|Z(R)| = \alpha$, $|R/Z(R)| = \beta$, and $\gamma = \alpha \cdot gcd(n, \beta - 1) = (\beta - 1)/2$. Then the $n - T(Reg(R))$ is connected if and only if $gcd(n, \beta - 1) = (\beta - 1)/2$. Furthermore, if the $n - T(Reg(R))$ is connected, then the $n - T(R) = K_\alpha \oplus K_{\gamma,\gamma}$, where the $n - T(Reg(R)) = K_{\gamma,\gamma}$ is connected.*

PROOF. Let $d = (\beta - 1)/gcd(n, \beta - 1)$. Suppose that the $gcd(n, \beta - 1) = (\beta - 1)/2$. Then $d = 2$. Hence by Theorem 2.9, the $n - T(Reg(R)) = K_{\gamma,\gamma}$ is connected. Conversely, suppose the $n - T(Reg(R))$ is connected. Since the $n - T(Reg(R)) = \bigoplus_{i=1}^{d/2} K_{\gamma,\gamma}$ by Theorem 2.9, we conclude that $d = 2$. Thus the $gcd(n, \beta - 1) = (\beta - 1)/2$. □

EXAMPLE 2.13. Let $n \geq 1$ be an integer, p be an odd prime, $m = p^i$, $R = \mathbb{Z}_m$, $\alpha = p^{i-1} = |Z(\mathbb{Z}_m)|$, $\beta = |R/Z(R)| = p$, $\gamma = gcd(n, \beta - 1)\alpha = gcd(n, p - 1)\alpha$, and $d = (\beta - 1)/gcd(n, \beta - 1) = (p - 1)/gcd(n, p - 1)$. Then the $n - T(Reg(R))$ is connected if and only if $gcd(n, p-1) = (p-1)/2$. Furthermore, if the $n - T(Reg(R))$ is connected, then the $n - T(R) = K_\alpha \oplus K_{\gamma,\gamma}$, where the $n - T(Reg(R)) = K_{\gamma,\gamma}$ is connected.

EXAMPLE 2.14. The $2 - T(Reg(\mathbb{Z}_{25}))$ is connected since $gcd(2, 5 - 1) = 2$. This is unlike $1 - T(Reg(\mathbb{Z}_{25}))$, which is not connected and has two complete bipartite components.

The following theorem generalizes [**AB**, Theorem 2.5] for finite commutative rings.

THEOREM 2.15. *Let $n \geq 1$ be an integer and R be a finite commutative ring such that $Z(R)$ is an ideal of R. Then*

(1) $\text{diam}(n - T(Reg(R))) = 0$, 1, 2, *or* ∞.
(2) $gr(n - T(Reg(R))) = 3$, 4, *or* ∞.

PROOF.

(1) Suppose the $n - T(Reg(R))$ is connected. Then the $n - T(Reg(R))$ is a singleton, a complete graph, or a complete bipartite graph by Theorem 2.6 and Theorem 2.9. Therefore, $\text{diam}(n - T(R)) \leq 2$.
(2) Suppose the $n - T(Reg(R))$ has a cycle. Then the $n - T(Reg(R))$ is a complete graph or a disjoint union of complete bipartite graphs by Theorem 2.6 and Theorem 2.9. Hence, if a component of the $n - T(Reg(R))$ has more than two vertices, then $gr(n - T(Reg(R)) = 3$ or 4. If each component of the $n - T(Reg(R))$ has two vertices or fewer, then $gr(n - T(Reg(R)) = \infty$. □

EXAMPLE 2.16. Let $n \geq 1$ be an integer. Then

(1) $\text{diam}(2 - T(Reg(\mathbb{Z}_2))) = 0$ and $gr(n - T(Reg(\mathbb{Z}_2))) = \infty$ as $Reg(\mathbb{Z}_2) = \{1\}$ is a singleton.
(2) $\text{diam}(2 - T(Reg(\mathbb{Z}_8))) = 1$, $gr(n - T(Reg(\mathbb{Z}_8))) = 3$, and the $gr(2 - T(Reg(\mathbb{Z}_{25}))) = 4$.

FIGURE 2. $3 - T(Reg(\mathbb{Z}_6))$

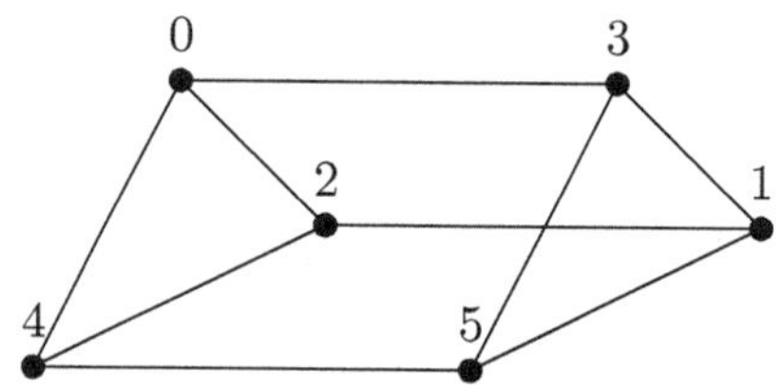

FIGURE 3. $3 - T(\mathbb{Z}_6)$

2.2. The case where $Z(R)$ is not an ideal.

THEOREM 2.17. *Suppose R is a commutative ring such that $Z(R)$ is not an ideal of R. Let $n \geq 1$ be an integer. Then the $n - T(R)$ is connected if and only if there exists a path from 0 to 1 in the $n - T(R)$.*

PROOF. If the $n - T(R)$ is connected, then there is a path from 0 to 1. Conversely, suppose there exists a path from 0 to 1 in the $n - T(R)$, say $0 - w_1 - w_2 - \cdots - w_q = 1$. Then for any $w_i - w_j$, $w_i^n + w_j^n \in Z(R)$. Let $v \in R^*$. Then $(vw_i)^n + (vw_j)^n = v^n w_i^n + v^n w_j^n = v^n(w_i^n + w_j^n) \in Z(R)$. Hence, $0 - vw_1 - \cdots - vw_q = v$ is a path from 0 to v for all v in R, and therefore, there is a path between any two vertices in R. Therefore, the $n - T(R)$ is connected. □

The next result generalizes [**AB**, Theorem 3.1, (3)] for odd integers.

THEOREM 2.18. *Let R be a commutative ring such that $Z(R)$ is not an ideal of R. Let $n \geq 1$ be an odd integer. If the $n - T(Reg(R))$ is connected, then the $n - T(R)$ is connected.*

PROOF. Suppose the $n - T(Reg(R))$ is connected. Since $Z(R)$ is not an ideal of R, there are $w, z \in Z(R)$ such that $w - z = r \in Reg(R)$. Therefore, we have $(w - z)^n = r^n = \sum_{k=1}^{n} \binom{n}{k}(-1)^k w^k z^{n-k} = \sum_{k=1}^{n-1} \binom{n}{k}(-1)^k w^k z^{n-k} - z^n = wq - z^n$ for some $q \in R$. Therefore, we have $r^n + z^n = wq \in Z(R)$. Since $0, z \in Z(R)$, every vertex of $Z(R)$ is connected to z. Since z is connected to $r \in Reg(R)$ and the $n - T(Reg(R))$ is connected, we conclude the $n - T(R)$ is connected. □

EXAMPLE 2.19. The $3 - T(\mathbb{Z}_6)$ is connected, as shown in Figure 3. Note that the $3 - T(Reg(\mathbb{Z}_6))$ is connected, as shown in Figure 2.

The next result generalizes [**AB**, Theorem 3.1, (3)] for even integers with an extra condition.

THEOREM 2.20. *Let $n \geq 2$ be an even integer, R be a commutative ring such that $Z(R)$ is not an ideal of R and suppose there is a $u \in R$ such that $u^n = -1$. If the $n - T(Reg(R))$ is connected, then the $n - T(R)$ is connected.*

PROOF. Suppose the $n - T(Reg(R))$ is connected. Since $Z(R)$ is not an ideal of R, there are $w, v \in Z(R)$ such that $w + v = r \in Reg(R)$. Let $z = u^{-1}v$. Hence, $w + uz = r \in Reg(R)$. Therefore, we have $(w + uz)^n = r^n = \sum_{k=1}^{n} \binom{n}{k} u^k w^k z^{n-k} =$

$\sum_{k=1}^{n-1} \binom{n}{k} u^k w^k z^{n-k} - z^n = wq - z^n$ for some $q \in R$. Therefore, we have $r^n + z^n = wq \in Z(R)$. Since $0, z \in Z(R)$, every vertex of $Z(R)$ is connected to z. Since z is connected to $r \in Reg(R)$ and the $n - T(Reg(R))$ is connected, we conclude the $n - T(R)$ is connected. □

EXAMPLE 2.21. The condition $u^n = -1$ is sufficient but unnecessary for this to hold. Let $R = \mathbb{Z}_{14}$. The $2 - T(Reg(R)))$ is connected (and thus the $2 - T(R)$ is connected). However, we do not have any element $u \in R$ such that $u^2 = -1$.

The next result generalizes [**AB**, Theorem 3.3] for odd integers.

THEOREM 2.22. *Let R be a commutative ring such that $Z(R)$ is not an ideal of R, and $n \geq 1$ be an odd integer. The $n - T(R)$ is connected if and only if $R = (z_1, z_2, \ldots, z_m)$ where $z_i \in Z(R)$ for each $1 \leq i \leq m$.*

PROOF. Suppose the $n - T(R)$ is connected. Then there is a path $0 - w_1 - w_2 - \cdots - w_q - 1$ in $n - T(R)$. This implies that $w_1^n, w_1^n + w_2^n, \ldots, w_{q-1}^n + w_q^n, w_q^n + 1 \in Z(R)$. Hence, $1 \in (w_1, w_1^n + w_2^n, dots, w_{q-1}^n + w_q^n, w_q^n + 1) \subseteq (Z(R))$. Hence, $R = (Z(R))$

Conversely, suppose that $(Z(R)) = R$. Then, we must show a path from 0 to x in $n - T(R)$ for any nonzero x in R. By hypothesis, we have $x = v_1 + \cdots + v_q$ for some $v_1, \ldots, v_q \in Z(R)$. Let $w_0 = 0$ and $w_j = (-1)^{q+j}(v_1 + \cdots + v_j)$ for each integer $1 \leq j \leq q$. Since n is odd, we have $(-1)^{n(q+j)} = (-1)^{q+j}$ and hence we have $w_j^n = (-1)^{q+j}(v_1 + \cdots + v_j)^n = w_j^n + w_{j+1}^n = v_{j+1}y \in Z(R)$ for some $y \in R$. Thus, $0 - w_1 - w_2 - \cdots - w_q = x$ is a path from 0 to x in $n - T(R)$. Let $0 \neq a, b \in R$. By the previous argument, there are paths from a to 0 and from 0 to b. Therefore, there is a path from a to b in $n - T(R)$; hence, $n - T(R)$ is connected. □

REMARK 2.23. In view of the proof of Theorem 2.25, if the $n - T(R)$ is connected for some integer $n \geq 1$, then $R = (z_1, z_2, \ldots, z_m)$, where $z_i \in Z(R)$ for each $1 \leq i \leq m$.

The next result generalizes [**AB**, Theorem 3.3] for even integers with an extra condition.

THEOREM 2.24. *Suppose R is a commutative ring such that $Z(R)$ is not an ideal of R. Let n be an even integer, and suppose there is $u \in R$ such that $u^n = -1$. Then the $n - T(R)$ is connected if and only if $R = (z_1, z_2, \ldots, z_m)$, where $z_i \in Z(R)$ for each $1 \leq i \leq m$.*

PROOF. Suppose the $n - T(R)$ is connected. By Remark 2.23, $R = (z_1, z_2, \ldots, z_m)$, where $z_i \in Z(R)$ for each $1 \leq i \leq m$. Suppose that $(Z(R)) = R$. It suffices to show a path from 0 to 1 in the $n - T(R)$. By hypothesis, we have $1 = v_1 + \cdots + v_q$ for some $v_1, \ldots, v_q \in Z(R)$. Let $w_0 = 0$ and $w_j = u^{q+j}(v_1 + \cdots + v_j)$ for each integer $1 \leq j \leq q$. Since $u^n = -1$, we have $(u)^{n(q+j)} = (-1)^{q+j}$ and hence we have $w_j^n = (-1)^{q+j}(v_1 + \cdots + v_j)^n = v_j^n + v_{j+1}^n = v_{j+1}y \in Z(R)$ for some $y \in R$. Thus, $0 - w_1 - w_2 - \cdots - w_q = 1$ is a path from 0 to 1 in the $n - T(R)$. Hence, the $n - T(R)$ is connected. □

The next result generalizes [**AB**, Theorem 3.4] for odd integers.

THEOREM 2.25. *Let R be a commutative ring such that $Z(R)$ is not an ideal of R, and n be an odd integer. Suppose that $R = (Z(R))$ (i.e. the $n - T(R)$ is connected). Then* $\operatorname{diam}(n - T(R)) = m$, *where m is the minimum number of zero* *divisors* z_1, ..., z_m *such that* $(z_1, \ldots, z_m) = R$. *Moreover,* $\operatorname{diam}(n - T(R)) = d(0, 1)$.

PROOF. First, we show that any path from 0 to 1 in R has a length of at least m. Suppose that $0 - b_1 - \cdots - b_{k-1} - 1$ is a path from 0 to 1 in $n - T(R)$ of length k. Then $b_1^n, b_1^n + b_2^n, \ldots, b_{k-1}^n + 1 \in Z(R)$ and hence $1 \in (b_1^n, b_1^n + b_2^n, \ldots, b_{k-1}^n + 1) \subseteq (Z(R))$. Thus, $k \geq m$. Let x and y be distinct elements in R. We will show that there is a path in R with a length less than or equal to m. Let $1 = z_1 + z_2 + \cdots + z_m$ for z_1, ..., $z_m \in Z(R)$. If m is even, define $z = x + y$. If m is odd, define $z = x - y$. In any case, let $d_0 = x$. For each $1 \leq k \leq m$ let $d_k = -(x + z(z_1 + \cdots + z_k))$ if k is odd, and $d_k = x + z(z_1 + \cdots + z_k)$ if k is even. Then $d_k^n + d_{k+1}^n = z_{k+1}q \in Z(R)$ for some $q \in R$. Also, $d_m = y$. Therefore, $x - d_1 - \cdots - d_m - 1 - y$ is a path of length at most m. In particular, a shortest path between 0 and 1 would have length m. Therefore, $\operatorname{diam}(n - T(R)) = m$. □

The next result generalizes [**AB**, Theorem 3.4] for even integers with an extra condition.

THEOREM 2.26. *Let R be a commutative ring such that $Z(R)$ is not an ideal of R. and n be an even integer. Suppose that $R = (Z(R))$ and there exists $u \in R$ such that $u^n = -1$. Then* $\operatorname{diam}(n - T(R)) = m$ *where m is the minimum number of zero divisors z_1, ..., z_m such that $(z_1, \ldots, z_m) = R$. Moreover,* $\operatorname{diam}(n - T(R)) = d(0, 1)$.

PROOF. First, we show that any path from 0 to 1 in R has a length of at least m. Suppose that $0 - b_1 - \cdots - b_{k-1} - 1$ is a path from 0 to 1 in $n - T(R)$ of length k. Then $b_1^n, b_1^n + b_2^n, \ldots, b_{k-1}^n + 1 \in Z(R)$ and hence $1 \in (b_1^n, b_1^n + b_2^n, \ldots, b_{k-1}^n + 1) \subseteq (Z(R))$. Thus, $k \geq m$. Let x and y be distinct elements in R. We will show that there is a path in R with a length less than or equal to m. Let $1 = z_1 + z_2 + \cdots + z_m$ for z_1, ..., $z_m \in Z(R)$. If m is even, define $z = x + y$. If m is odd, define $z = x - y$. In any case, let $d_0 = x$. For each $1 \leq k \leq m$ let $d_k = u(x + z(z_1 + \cdots + z_k))$ if k is odd and $d_k = x + z(z_1 + \cdots + z_k)$ if k is even. Then $d_k^n + d_{k+1}^n = z_{k+1}q \in Z(R)$ for some $q \in R$. Also, $d_m = y$. Therefore, $x - d_1 - \cdots - d_m - 1 - y$ is a path of length at most m. In particular, a shortest path between 0 and 1 would have length m. Therefore, $\operatorname{diam}(n - T(R)) = m$. □

The following results generalizes [**AB**, Corollary 3.5].

COROLLARY 2.27. *Let R be a commutative ring such that $Z(R)$ is not an ideal of R. if* $\operatorname{diam}(n - T(R)) = m$, *then* $\operatorname{diam}(n - T(Reg(R))) \geq m - 2$

PROOF. Since $\operatorname{diam}(n - T(R)) = d(0, 1) = m$, let $0 - b_1 - \cdots - b_{m-1} - 1$ be a path from 0 to 1 in R. Clearly, $b_1 \in Z(R)$. Suppose that for some $2 \leq i \leq m - 1$ we have b_i in $Z(R)$. Therefore, we can have the path $0 - b_i - \cdots - 1$ in $n - T(R)$ of length less than m. This is a contradiction. Thus, $b_i \in Reg(R)$ for each $2 \leq i \leq m - 1$ and hence $b_2 - \cdots - 1$ is a shortest path between b_2 and 1 in the $n - T(Reg(R))$ of length $m - 2$. Therefore $\operatorname{diam}(n - T(Reg(R))) \geq m - 2$. □

EXAMPLE 2.28. Let $R = \mathbb{Z}_{21}$. Every $x \in Reg(R)$, x is not connected to any other element in R. Hence, The $2 - T(R)$ is not connected.

EXAMPLE 2.29. It is not necessary to have $x \in R$ such that $x^n = -1$ for $n - T(R)$ to be connected when n is even. Let $R = \mathbb{Z}_6$. There is no $x \in R$ such

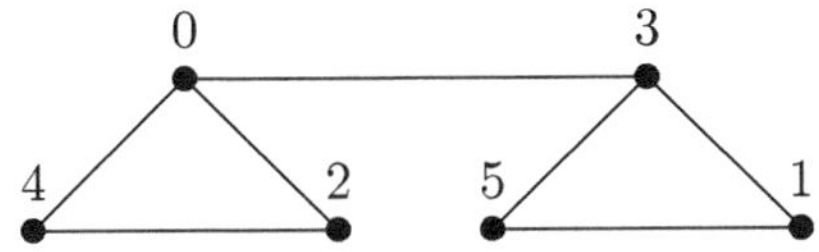

FIGURE 4. $2-T(\mathbb{Z}_6)$

that $x^2=-1$. However, the $2-T(R)$ is connected as shown in Figure 4. Note that $d(0,1)=2$, but $\operatorname{diam}(2-T(R))=3\neq d(0,2)$.

THEOREM 2.30. *Let $R=R_1\times\cdot\times R_k$ for some $k\geq 2$ where R_i is a commutative ring with $1\neq 0$ for each $1\leq i\leq k$. If $n\geq 1$ is an odd integer, then the $n-T(R)$ is connected. Moreover,* $\operatorname{diam}(n-T(R))=2$.

PROOF. Suppose $x,y\in R$. Then $x=(x_1,\ldots,x_k)$ and $y=(y_1,\ldots,y_k)$. Since n is odd, $x_i^n+(-x_i)^n=0$ (Similarly $y_j^n+(-y_j)^n=0$). Let $r=(r_1,\ldots,-x_i,\ldots,-y_j,\ldots,r_k)$. Then clearly, x^n+r^n is a zero divisor, and r^n+y^n is a zero divisor. Hence, $x-r-y$ is always a path in the $n-T(R)$. Therefore, the $n-T(R)$ is connected and $\operatorname{diam}(n-T(R))=2$. □

THEOREM 2.31. *Let $n\geq 2$ be an even integer and $R=R_1\times\cdots\times R_k$ for some $k\geq 2$ where R_i is a commutative ring with $1\neq 0$ for each $1\leq i\leq k$. Then the $n-T(R)$ is connected if and only if there is an x in R_j for some $1\leq j\leq k$ such that $x^n+1\in Z(R_j)$. Furthermore, if there exists a $u\in R$ such that $u^n=-1\in R$ (i.e., $u^n+1=0\in Z(R)$), then the $n-T(R)$ is connected and* $\operatorname{diam}(n-T(R))=2$.

PROOF. Suppose the $n-T(R)$ is connected. Then there exists a path from $(0,\ldots,0)$ to $(1,\ldots,1)$ in the $n-T(R)$, say $(0,\ldots,0)-\cdots-(x_1,\ldots,x_k)-(1,\ldots,1)$. Hence $x_i^n+1\in Z(R_i)$ for some $1\leq i\leq k$.

Conversely, suppose there is an x in R_j for some $1\leq j\leq k$ such that $x^n+1\in Z(R_j)$. Without the loss of generality, we may assume that $j=1$. Thus, $(0,\ldots,0)-(x,0,\ldots,0)-(1,\ldots,1)$ is a path of length 2 in the $n-T(R)$. Hence, the $n-T(R)$ is connected by Theorem 2.17.

Now, suppose that there exists a $u\in R$ such that $u^n=-1\in R$. Since $R=((1,0,\cdots,0),(0,1,\cdots,1))$, the $n-T(R)$ is connected and $\operatorname{diam}(n-T(R))=d((0,\ldots,0),(1,\ldots,1))=2$ by Theorem 2.26. □

COROLLARY 2.32. *Let $n\geq 2$ be an even integer and $R=R_1\times\cdots\times R_k$ for some $k\geq 2$ where R_i is an integral domain for each $1\leq i\leq k$. Then the $n-T(R)$ is connected if and only if there is an x in R_j for some $1\leq j\leq k$ such that $x^n=-1\in R_j$ (i.e. $x^n+1=0\in Z(R_j)$). Furthermore, if the $n-T(R)$ is connected and there exists a $u\in U(R)$ such that $u^n=-1\in R$ (i.e., $u^n+1=0\in Z(R)$), then* $\operatorname{diam}(n-T(R))=2$.

PROOF. This follows directly from Theorem 2.31 and the fact that R_i has no zero divisors except for 0 for each $1\leq i\leq k$. □

See Example 2.29, $R=Z_6$ is ring-isomorphic to $A=Z_2\times Z_3$. Note that $d((0,0),(1,1))=2$ in the $2-T(A)$, but $\operatorname{diam}(2-T(A))=3$.

EXAMPLE 2.33. $R=\mathbb{Z}_3\times\mathbb{Z}_3$. There is no $x\in R$ such that $x^2=-1$, and the $2-T(R)$ is not connected. See Figure 5. However, the $2-T(\mathbb{Z}_{14})$ is connected, and there is no $x\in\mathbb{Z}_{14}$ such that $x^2=-1$.

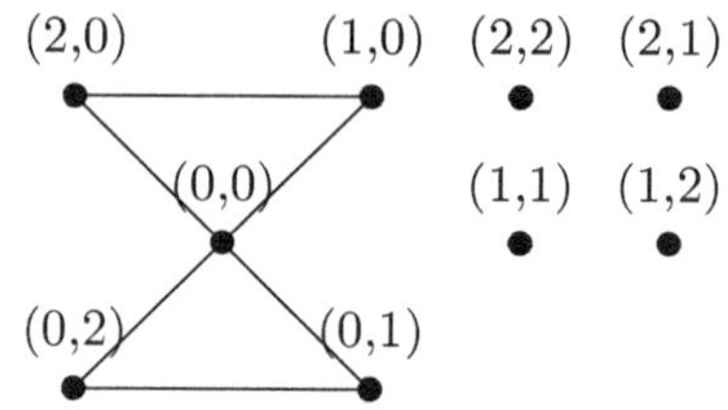

FIGURE 5. $2 - T(\mathbb{Z}_3 \times \mathbb{Z}_3)$

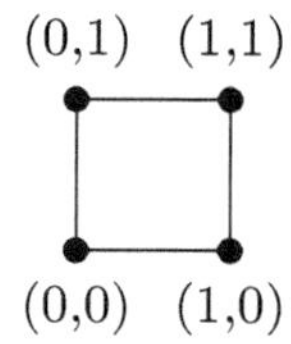

FIGURE 6. $2 - T(\mathbb{Z}_2 \times \mathbb{Z}_2)$

In view of Example 2.33, we have the following result.

COROLLARY 2.34. *Let n be an even integer and $R = R_1 \times \cdot \times R_k$ for some $k \geq 2$ where R_i is an integral domain for each $1 \leq i \leq k$. Then the $n - T(R)$ is connected if and only if there is an x in R_j for some $1 \leq j \leq k$ such that $x^n = -1$ in R_j (i.e., $x^n + 1 = 0 \in Z(R)$).*

PROOF. This follows directly from Theorem 2.31 and the fact that R_i has no zero divisors except for 0 for each $1 \leq i \leq k$. □

EXAMPLE 2.35. The $2 - T(\mathbb{Z}_5 \times \mathbb{Z}_3)$ is connected, since in $\mathbb{Z}_5$ we have $2^2 = 4 = -1 \in \mathbb{Z}_5$.

EXAMPLE 2.36. The $2 - T(\mathbb{Z}_2 \times \mathbb{Z}_2)$ is connected, since in $\mathbb{Z}_2$ we have $1^2 = 1 = -1 \in \mathbb{Z}_2$. This is shown in Figure 6.

In the following result, we generalize [**AB**, Example 3.8] but our proof differs entirely from that in [**AB**]

THEOREM 2.37. *For any integer $m \geq 2$, there exists a ring R such that for every integer $n \geq 1$, the $n - T(R)$ is connected and* $\mathrm{diam}(n - T(R)) = m$.

PROOF. Let $n \geq 1$, $A = \mathbb{Z}_2[X_1, X_2, \ldots, X_{m-1}, W_1, W_2, \ldots, W_m]$, $I = (W_1X_1, \ldots, W_{m-1}X_{m-1}, W_m(X_1 + \cdots + X_{m-1} + 1), \{W_iW_j | 1 \leq i < j \leq m\})$, and $R = A/I$. Denote $x_i = X_i + I \in R$ and $w_i = W_i + I \in R$. Then $(-1)^n = 1$. We can construct the following path from 0 to 1 in the $n - T(R)$

$$0 - x_1 - (x_1 + x_2) - \cdots - (x_1 + x_2 + \cdots + x_{m-1}) - 1 \tag{2.1}$$

Note that this is the shortest path from 0 to 1 of length m in the $n - T(R)$ since any sum of the x_i's is not a zero divisor in R, and thus we can not make a shorter path. Therefore, by Theorem 2.25 (In the case of odd n) and Theorem 2.26 (In the case of even n), $\mathrm{diam}(n - T(R)) = m$. □

Acknowledgment

The authors thank the referee for helpful comments.

References

[AAM] A. Abdollahi, S. Akbari, and H. R. Maimani, *Non-commuting graph of a group*, J. Algebra **298** (2006), no. 2, 468–492, DOI 10.1016/j.jalgebra.2006.02.015. MR2217623

[AB] M. Abdulla and A. Badawi, *On the dot product graph of a commutative ring II*, Int. Electron. J. Algebra **28** (2020), 61–74, DOI 10.24330/ieja.768135. MR4129767

[AKMM] S. Akbari, D. Kiani, F. Mohammadi, and S. Moradi, *The total graph and regular graph of a commutative ring*, J. Pure Appl. Algebra **213** (2009), no. 12, 2224–2228, DOI 10.1016/j.jpaa.2009.03.013. MR2553598

[AN] D. D. Anderson and M. Naseer, *Beck's coloring of a commutative ring*, J. Algebra **159** (1993), no. 2, 500–514, DOI 10.1006/jabr.1993.1171. MR1231228

[AABC] D. F. Anderson, T. Asir, A. Badawi, and T. Tamizh Chelvam, *Graphs from rings*, Springer, Cham, [2021] ©2021, DOI 10.1007/978-3-030-88410-9. MR4385024

[AAS] D. F. Anderson, M. C. Axtell, and J. A. Stickles Jr., *Zero-divisor graphs in commutative rings*, Commutative algebra—Noetherian and non-Noetherian perspectives, Springer, New York, 2011, pp. 23–45, DOI 10.1007/978-1-4419-6990-3_2. MR2762487

[AB] D. F. Anderson and A. Badawi, *The total graph of a commutative ring*, J. Algebra **320** (2008), no. 7, 2706–2719, DOI 10.1016/j.jalgebra.2008.06.028. MR2441996

[AB(a)] D. F. Anderson and A. Badawi, *On the total graph of a commutative ring without the zero element*, J. Algebra Appl. **11** (2012), no. 4, 1250074, 18, DOI 10.1142/S0219498812500740. MR2959423

[AB(b)] D. F. Anderson and A. Badawi, *The generalized total graph of a commutative ring*, J. Algebra Appl. **12** (2013), no. 5, 1250212, 18, DOI 10.1142/S021949881250212X. MR3055569

[ALS] D. F. Anderson, R. Levy, and J. Shapiro, *Zero-divisor graphs, von Neumann regular rings, and Boolean algebras*, J. Pure Appl. Algebra **180** (2003), no. 3, 221–241, DOI 10.1016/S0022-4049(02)00250-5. MR1966657

[AL] D. F. Anderson and P. S. Livingston, *The zero-divisor graph of a commutative ring*, J. Algebra **217** (1999), no. 2, 434–447, DOI 10.1006/jabr.1998.7840. MR1700509

[B(a)] A. Badawi, *On the annihilator graph of a commutative ring*, Comm. Algebra **42** (2014), no. 1, 108–121, DOI 10.1080/00927872.2012.707262. MR3169557

[B(b)] A. Badawi, *On the total graph of a ring and its related graphs: A survey*, Commutative algebra, Springer, New York, 2014, pp. 39–54. MR3330211

[B(c)] A. Badawi, *On the dot product graph of a commutative ring*, Comm. Algebra **43** (2015), no. 1, 43–50, DOI 10.1080/00927872.2014.897188. MR3240402

[BE] A. Badawi and Y. El-Ashi, *Graph of linear transformations over* $\mathbb{R}$, Algebra and related topics with applications, Springer Proc. Math. Stat., vol. 392, Springer, Singapore, [2022] ©2022, pp. 401–412, DOI 10.1007/978-981-19-3898-6_31. MR4540737

[BR] A. Badawi and R. Rissner, *Ramsey numbers of partial order graphs (comparability graphs) and implications in ring theory*, Open Math. **18** (2020), no. 1, 1645–1657, DOI 10.1515/math-2020-0085. MR4200340

[BE] I. Beck, *Coloring of commutative rings*, J. Algebra **116** (1988), no. 1, 208–226, DOI 10.1016/0021-8693(88)90202-5. MR944156

[D] A. Das, *Non-zero component union graph of a finite-dimensional vector space*, Linear Multilinear Algebra **65** (2017), no. 6, 1276–1287, DOI 10.1080/03081087.2016.1234577. MR3615543

[M] S. B. Mulay, *Cycles and symmetries of zero-divisors*, Comm. Algebra **30** (2002), no. 7, 3533–3558, DOI 10.1081/AGB-120004502. MR1915011

[SB] P. K. Sharma and S. M. Bhatwadekar, *A note on graphical representation of rings*, J. Algebra **176** (1995), no. 1, 124–127, DOI 10.1006/jabr.1995.1236. MR1345297

Department of Mathematics and Statistics, American University of Sharjah, P.O. Box 26666, Sharjah, United Arab Emirates
Email address: g00098182@alumni.aus.edu

Department of Mathematics and Statistics, American University of Sharjah, P.O. Box 26666, Sharjah, United Arab Emirates
Email address: abadawi@aus.edu

Contemporary Mathematics
Volume **836**, 2026
https://doi.org/10.1090/conm/836/16712

Radii of convergence of algebraic power series

Joseph Swanson

ABSTRACT. This expository paper explores algebraic power series. We briefly discuss their connection to combinatorics and then look at them through an algebraic lens. In particular, we look at deriving solutions to bivariate equations, characteristics of algebraic power series, and investigate the potential radii of convergence for an algebraic power series.

1. Introduction

Algebraic power series have long been a fundamental object of study in combinatorics. A power series is algebraic if it satisfies a polynomial equations with coefficients in a given field. These series naturally arise in enumerative combinatorics, where generating functions encode counting sequences for combinatorial structures [**41, 45**]. For example, the enumeration of plane binary trees with n internal nodes involves the sequence: 1, 1, 2, 5, 14 . . . , the Catalan numbers [**27**]. The generating function of the Catalan numbers is

$$f(x) = 1 + x + 2x^2 + 5x^3 + 14x^4 + \cdots = \sum_{n=0}^{\infty} \frac{1}{n+1}\binom{2n}{n} x^n = \frac{1-\sqrt{1-4x}}{2x}.$$

This power series satisfies the relation

$$F(x, y) = xy^2 - y + 1 = 0.$$

Additionally, the Fibonacci generating function satisfies

$$F(x, y) = (1 - x - x^2)y - x.$$

These two examples partially illustrate how algebraic power series emerge naturally from combinatorial recursion relations [**18, 42**]. Algebraic power series also arise in the study of *directed lattice paths*. A directed lattice path is a sequence of steps on $\mathbb{Z} \times \mathbb{Z}$, where each step belongs to a fixed finite set of permitted moves and progresses rightward. For instance, a *Motzkin path* is a directed lattice path that starts at the origin, stays on or above the x-axis, and consists of up-steps $(1, 1)$, down-steps $(1, -1)$, and level steps $(1, 0)$. For example,

2020 *Mathematics Subject Classification.* Primary 13B22, Secondary 13B25.

©2026 American Mathematical Society

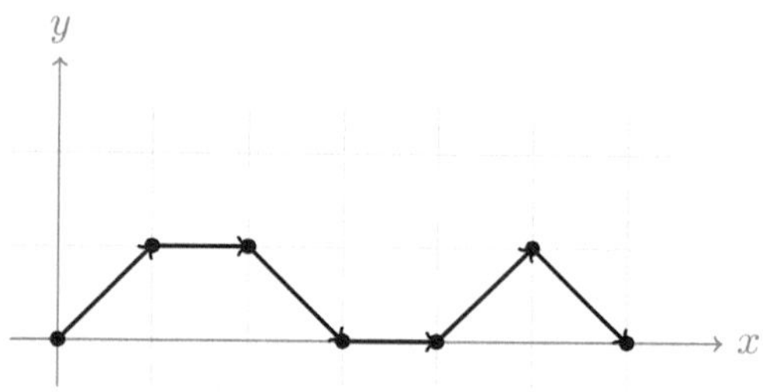

is a Motzkin path of length 6. The generating function of the number of Motzkin paths of lenth n is given by the power series expansion of

$$g(x) = \frac{1 - x - \sqrt{1 - 2x - 3x^2}}{2x^2}$$

which satisfies the relation

$$G(x, y) = x^4y^2 + (x^3 - x^2)y + x^2 = 0.$$

One may consider other paths that do not require the steps to stay above or end on the x-axis. Many more directed lattice paths exhibit generating functions that are algebraic. The curious reader is directed to [**5**, **42**] for further reading.

Once you have an algebraic power series, you can study its asymptotic behavior. A key feature of this behavior is the *critical exponent* α, which appears in the asymptotic form of the coefficients:

$$f_n \sim C \cdot A^n n^\alpha$$

(here $a_n \sim b_n$ equates to $\lim\limits_{n\to\infty} \frac{a_n}{b_n} = 1$) where C and A are constants and A^{-1} is typically the radius of convergence. As an illustration, consider the generating function of the Catalan numbers

$$f(x) = \sum_{n=0}^{\infty} \frac{1}{n+1}\binom{2n}{n} x^n.$$

Using Stirlings formula which states that

$$n! \sim \sqrt{2\pi n}\left(\frac{n}{e}\right)^n,$$

one can show that the coefficients of $f(x)$ satisfy the asymptotic behavior

$$f_n \sim \frac{4^n}{\sqrt{\pi n^3}}$$

Its critical exponent $\alpha = -\frac{3}{2}$ is guaranteed by the *Drmota–Lalley–Woods theorem*, which describes the universal behavior of generating functions defined by strongly connected systems [**3**, **4**, **17**].

In a different but related way, a power series is said to be D-finite(differentiably finite) if it satisfies a linear differential equation with polynomial coefficients. The study of D-finite power series provides further insights into the nature of generating functions and their closure properties under various operations [**31**, **40**]. More generally we have the following definition.

DEFINITION 1.1. Let R be a domain. A power series $\alpha(x) \in R[[x]]$ is said to be an *algebraic power series* if there exists $F(x, y) \in R[x][y]$, where

$$F(x, y) = f_n(x)y^n + f_{n-1}(x)y^{n-1} + \cdots + f_1(x)y + f_0(x) \quad (1.1)$$

such that $F(x, \alpha(x)) = 0$, where $F(x, y) \neq 0$. The set of all algebraic power series form a ring denoted $R_{\text{Alg}}[[x]]$. [**42**]

It is known that the radius of convergence of an algebraic power series over $\mathbb{Q}$ is the root of a monic polynomial with rational coefficients [**1**], [**23**], but it is still unknown which algebraic numbers appear as a radius of convergence. Banderier and Drmota considered this question from the viewpoint of a positive algebraic system of equations with rational coefficients [**4**] and conjectured that all algebraic numbers appear as a radius of convergence. We investigate this further, show which radii of convergence are known so far, and highlight some new results regarding them.

Since we are concerned with power series solutions to equations of the form (1.1), it is important to determine the conditions under which a given bivariate polynomial admits a power series solution. In general, given an equation of the form (1.1), it is difficult to tell whether a power series solution exists. There are some ways to alleviate this problem, and one way is to use the Implicit Function Theorem [**27**].

THEOREM 1.2 (Implicit Function Theorem). *Suppose $F(x, y) \in \mathbb{Q}[x][y]$ and consider the curve given by $F(x, y) = 0$. Assume that $F(0, y_0) = 0$ where $y_0 \in \mathbb{Q}$ and $F_y(0, y_0) \neq 0$. Then there exists a unique $\alpha(x) \in \mathbb{Q}[[x]]$ such that $F(x, \alpha(x)) = 0$.*

We quickly note that algebraic power series are closed under differentiation [**42**].

THEOREM 1.3. *Algebraic power series are closed under differentiation. That is, if $\alpha(x) \in R_{\text{Alg}}[[x]]$, then $\frac{d}{dx}(\alpha(x)) \in R_{\text{Alg}}[[x]]$.*

SKETCH OF PROOF. Assume that $\alpha(x) \in R_{\text{Alg}}[[x]]$. Then there exists $F(x, y) \in R[x][y]$ such that $F(x, \alpha(x)) = 0$. It is easy to show that by implicit differentiation,

$$\frac{dF}{dx} = \frac{\partial F}{\partial x} + \frac{dy}{dx}\frac{\partial F}{\partial y} = 0 \quad \Rightarrow \quad \frac{dy}{dx} = \frac{-\frac{\partial F}{\partial x}}{\frac{\partial F}{\partial y}},$$

hence,

$$\left.\frac{dy}{dx}\right|_{y=\alpha(x)} = \left.\left(\frac{-\frac{\partial F}{\partial x}}{\frac{\partial F}{\partial y}}\right)\right|_{y=\alpha(x)}. \tag{1.2}$$

Since $\alpha'(x)$ is a rational function of algebraic power series, it is an algebraic power series. □

REMARK 1.4. We point out that algebraic power series are not necessarily closed under integration. The simplest example is the geometric series

$$f(x) = \frac{1}{1-x}.$$

We have that,

$$\int f(x)\, dx = \int \frac{1}{1-x} dx = -\ln(|1-x|) + C.$$

As the natural logarithm is not algebraic, $\int f(x) dx \notin R_{\text{Alg}}[[x]]$. Notice that $h(x) = -\ln(1-x)$ does satisfy the differential equation

$$(1-x)y' - 1 = 0.$$

Any function that satisfies a linear differential equation in y with coefficients in $R[x]$ is said to be D-finite. D-finite series are closed under integration. For more on the topic of D-finite series, we refer the reader to [**30**, **31**, **40**].

It is worth noting that some algebraic power series are closed under integration. For example, let $g(x) = \sqrt{1+x}$. For simplicity, we will assume the constant of integration is always zero. Integrating $g(x)$ gives

$$G^{(1)}(x) := \int g(x)dx = \int \sqrt{1+x}\, dx = \frac{2}{3}(1+x)^{3/2}.$$

$G^{(1)}(x)$ satisfies

$$9y^2 - 4(1+x)^3 = 0.$$

Let $G^{(n)}(x)$ denote the nth integral of of $g(x)$. Then,

$$G^{(n)}(x) = \frac{2^n}{\prod_{i=1}^{n}(2i+1)}(x+1)^{(2n+1)/2}$$

satisfies

$$\prod_{i=1}^{n}(2i+1)^2y^2 - 2^{2n}(x+1)^{2n+1} = 0.$$

Given a bivariate equation, $F(x, y) = 0$, one can use the Newton-Puiseux algorithm to get all the series solutions.

2. Newton Puiseux algorithm

Let R be an integral domain and K its algebraic closure. We are interested in roots that satisfy an equation of the form

$$F(x, y) = f_n(x)y^n + f_{n-1}(x)y^{n-1} + \cdots + f_1(x)y + f_0(x) = 0. \tag{2.1}$$

We have seen that if $F_y(0, y_0) \neq 0$, then the implicit function theorem guarantees a power series solution. However, the ring of power series $R[[x]]$ is not enough to contain all of our roots. Even the ring allowing division by x, $R[[x]][\frac{1}{x}]$, does not contain all of the roots. We need to allow fractional exponents on our series. The set of fractional Laurent series are called *Puisuex series* and are denoted

$$R\langle\langle x\rangle\rangle = \left\{\sum_{i=m}^{\infty} r_i x^{i/n} | r_i \in R, m \in \mathbb{Z}, n \in \mathbb{N} \text{ fixed}\right\}.$$

If we let K be the algebraic closure of R, then $K\langle\langle x\rangle\rangle$ is algebraically closed. Thus, the n roots of equation (2.1) all lie in $K\langle\langle x\rangle\rangle$. This result is due to Victor Puiseux [**27**].

THEOREM 2.1. ***(Puiseux's Theorem)*** *The field $K\langle\langle x\rangle\rangle$ is algebraically closed.*

Given an equation in the form of (1.1), the solutions can be obtained term by term using the Newton-Puiseux algorithm, first used by Isaac Newton and then later by Victor Puiseux. We provide a loose sketch of the algorithm, and we refer the reader to [**10**] for a more in-depth view. We will use an example to establish working definitions needed to apply the algorithm.

EXAMPLE 2.2. We consider the set

$$S = \{(1,0), (3,0), (0,4), (0,1), (1,1), (4,3), (2,2)\}$$

and its graph

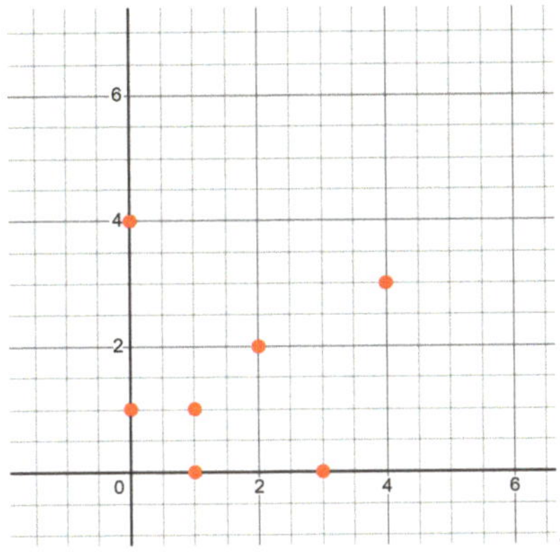

The *boundary of the convex hull* is the boundary of the smallest convex polygon containing all of S, which is given below in Figure (a). The *boundary of the lower hull* is the boundary of the lower section of the convex hull from the top left point to the lower right point of the convex hull, which is given below in Figure (b). An *edge* is a line of the convex hull that goes through at least two points, an example is given below in Figure (c).

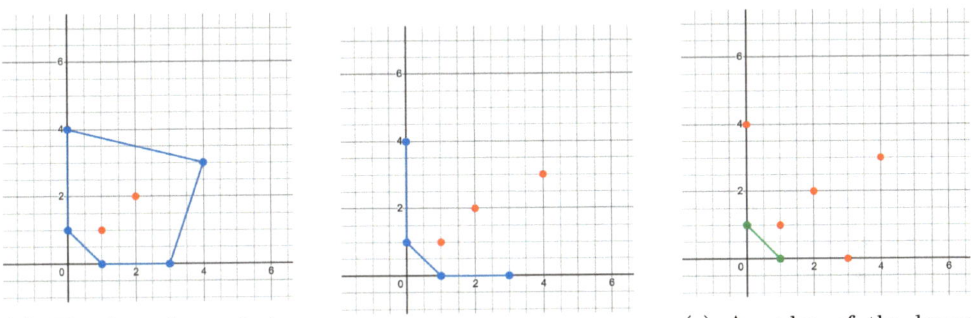

(a) The boundary of the convex hull

(b) The lower convex hull

(c) An edge of the lower hull

We will work over $\mathbb{C}$. The Newton-Puisuex Algorithm applied to

$$F(x,y) = \sum_{\substack{\alpha=0,\ldots,m \\ \beta=0,\ldots,n}} A_{\alpha,\beta} x^{\alpha} y^{\beta},$$

consists of the following steps:

(1) Compute and plot $\Delta(F) := \{(\alpha, \beta) \mid A_{\alpha,\beta} \neq 0\}$.
(2) Compute the boundary of the convex hull.
(3) Pick an edge, E, of the lower hull with a nonzero slope.
(4) For every point E passes through, set the sum of the corresponding monomials equal to zero and solve for y.
(5) Use $y = x^{a_1}$ found in step four and repeat the algorithm for $F(x, y + x^{a_1})$.

EXAMPLE 2.3. We will compute the first two terms of the power series expansion centered at $x = 0$ of the curve

$$y^2 - x^2 = (y^2 + x^2)^2$$

which satisfies $f(0) = 1$. Since we know that the constant term of the power series expansion is one, we start the algorithm at $1 + y$. We have that

$$F(x, 1 + y) = -3x^2 - x^4 - 2y - 4x^2y - 5y^2 - 2x^2y^2 - 4y^3 - y^4 = 0.$$

(1) In the first step, we compute and graph

$$\Delta(F) = \{(2, 0), (4, 0), (0, 1), (2, 1), (0, 2), (2, 2), (0, 3), (0, 4)\}.$$

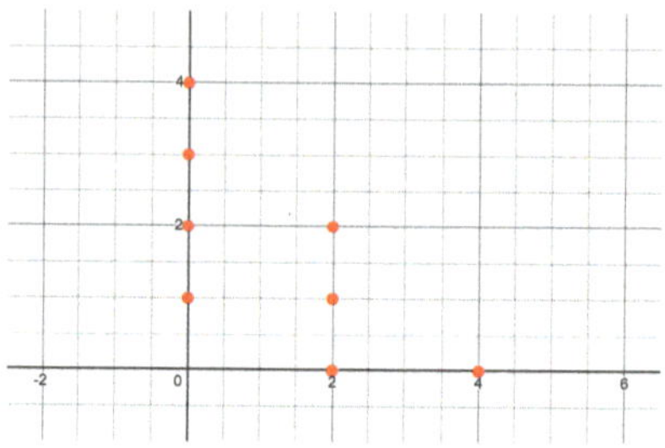

(2) Next we compute the boundary of the convex hull.

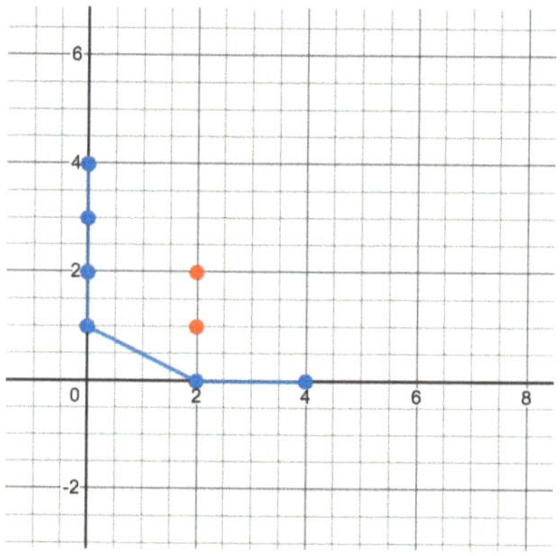

(3) Next we pick an edge of the boundary of the lower hull with a nonzero slope. We are forced to choose the edge that passes through $(0, 1)$ and $(2, 0)$.
(4) The point $(0, 1)$ corresponds to $-2y$ and the point $(2, 0)$ corresponds to $-3x^2$. Setting the sum of the corresponding monomials gives

$$-2y - 3x^2 = 0 \quad \Rightarrow \quad y = -\frac{3}{2}x^2.$$

We repeat the process with $F(x, 1 - \frac{3}{2}x^2 + y) = -x^2 + (1 - \frac{3}{2}x^2 + y)^2 - (x^2 + (1 - \frac{3}{2}x^2 + y)^2)^2$:

(1) We compute and graph

$$\Delta(F) = \{(4, 0), (6, 0), (8, 0), (0, 1), (2, 1), (4, 1), (6, 1), (0, 2), (2, 2), (4, 2), (0, 3), (2, 3), (0, 4)\}.$$

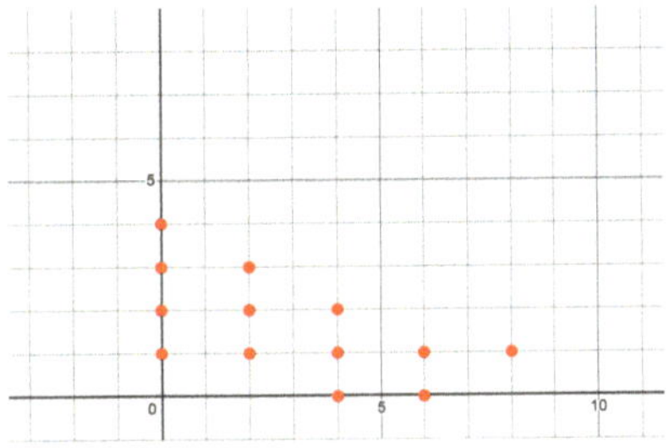

(2) Next we compute the boundary of the convex hull.

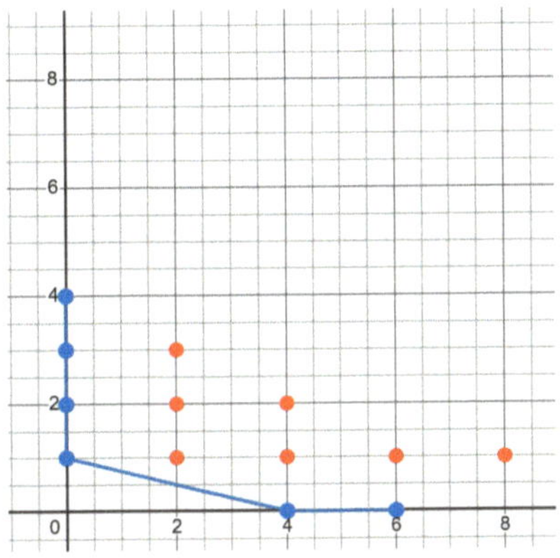

(3) We pick an edge of the boundary of the lower hull with a nonzero slope. We are forced to take the edge which passes through $(0,1)$ and $(4,0)$.
(4) The point $(0,1)$ corresponds to the term $-2y$ and the point $(4,0)$ corresponds to the term $\frac{25}{4}x^4$. Setting the sum of these monomials to zero yields

$$-2y - \frac{25}{4}x^4 \quad \Rightarrow \quad y = -\frac{25}{8}x^4.$$

Thus far we have determined that $y = 1 - 3/2x^2 - 25/8x^4$ are the first three terms of the power series solution. Graphically we have:

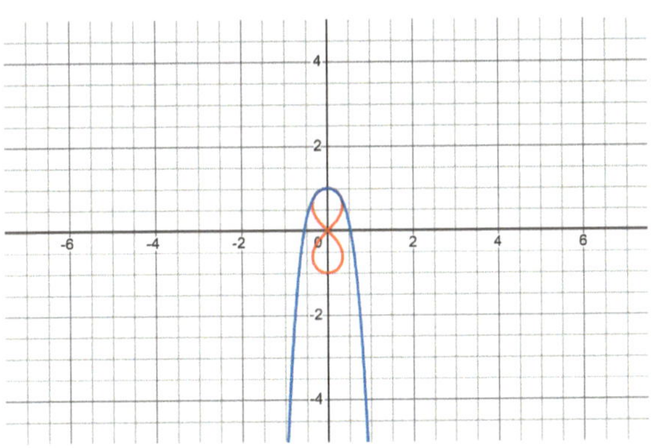

This algorithm also works for fractional power series. We will demonstrate this with our second example.

EXAMPLE 2.4. We consider the curve

$$F(x,y) = 2x\left(x^2+y^2\right) + \left(x^2+y^2\right)^2 - y^2.$$

(1) We compute $\Delta(F) = \{(3,0),(4,0),(0,2),(1,2),(2,2),(0,4)\}$ and provide its graph.

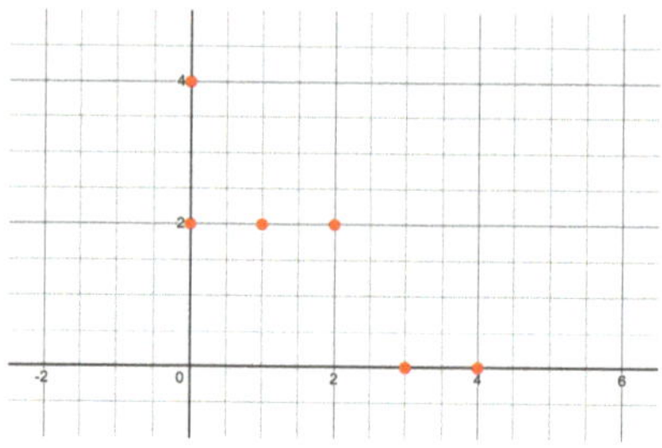

(2) Next we compute the boundary of the convex hull.

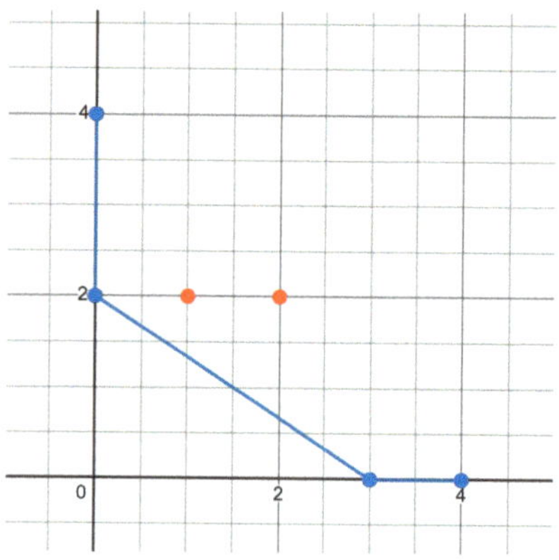

(3) We are forced to choose the edge which passes through $(0,2)$ and $(3,0)$.
(4) The point $(0,2)$ corresponds to $-y^2$ and the point $(3,0)$ corresponds to $2x^3$. Setting the sum of these monomials to zero yields

$$-y^2 + 2x^3 = 0 \quad \Rightarrow \quad y = \pm\sqrt{2}x^{3/2}.$$

Now we repeat the algorithm with $F(x, y+\sqrt{2}x^{3/2})$:

(1) $\Delta(F)= \{(4,0),(5,0),(6,0),(3/2,1),(5/2,1),(7/2,1),(9/2,1),(0,2),(1,2),(2,2),(3,2),(3/2,3),(0,4)\}$

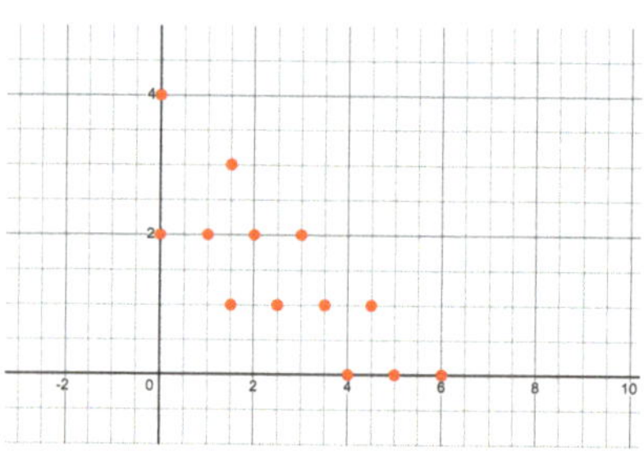

(2) Next we compute the boundary of the convex hull.

(3) Picking the edge which passes through $(0,2), (\frac{3}{2}, 1)$ leads us back to the negative Puiseux solution. Instead, we choose the second edge which passes through $(\frac{3}{2}, 1), (4,0)$.

(4) The point $(\frac{3}{2}, 1)$ corresponds to $2\sqrt{2}x^{3/2}y$ and the point $(4,0)$ corresponds to $-5x^4$. Setting the sum of these monomials to zero yields

$$-5x^4 + 2\sqrt{2}x^{3/2}y = 0 \quad \Rightarrow \quad y = \frac{5x^{5/2}}{2\sqrt{2}}.$$

So we have that $y = \sqrt{2}x^{3/2} + \frac{5x^{5/2}}{2\sqrt{2}}$ are the first two terms of the power series solution. Graphically, we have:

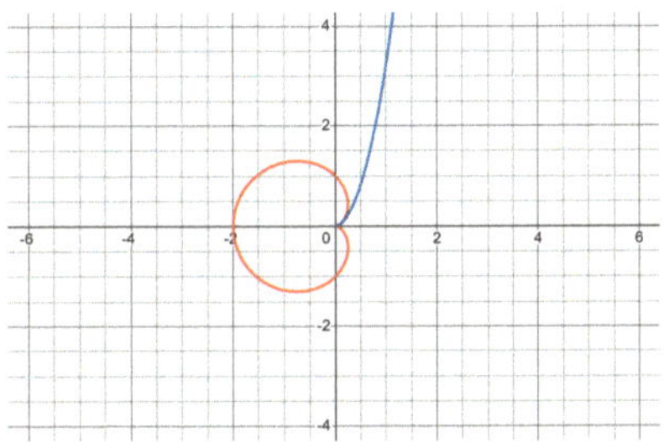

It has been shown in [**10**] that every solution to $F(x, y) = 0$ can be found by using this algorithm. Now that we can determine the series solutions, we turn our focus to exploring the possible radii of convergence for an algebraic power series.

3. Radius of convergence

It can be demonstrated that an algebraic power series must have a positive radius of convergence [**23**]. Furthermore, we know that polynomials are algebraic power series with an infinite radius of convergence, and, in fact, they are the only ones with this property. To show that an infinite radius of convergence implies a series is a polynomial, we present the following lemma and proposition [**29**].

LEMMA 3.1 (Cauchy's inequality). *Let*

$$f(z)=\sum_{i=0}^{\infty} a_i(z-a)^i$$

have radius of convergence $R>0$. *Then*

$$\left|\frac{f^{(n)}(a)}{n!}\right| \leq \frac{M(r)}{r^n}$$

for all $0<r<R$, *where*

$$M(r)=\sup\{|f(z)| \ : \ |z-a|=R\}.$$

PROPOSITION 3.2. *If* $f(z)$ *is entire and* $|f(z)|\leq M|z|^m$ *for an* $M>0$, $m\in\mathbb{N}_0$ *and for all* $z\in\mathbb{C}$ *such that* $|z|\geq R\in\mathbb{R}$, *then* $f(z)$ *is a polynomial of degree less than or equal to* m.

REMARK 3.3. We point out that Liouville's theorem is a direct corollary of Proposition 3.2 when $m=0$. Recall that Liouville's theorem provides a proof of the *Fundamental Theorem of Algebra*, which states that every non-constant polynomial with complex coefficients has at least one complex root. The statment of the theorem comes from [**29**].

THEOREM (Liouville). *Let* $f:\mathbb{C}\to\mathbb{C}$ *be an entire function that is bounded, i.e., there exists a constant* $M>0$ *such that* $|f(z)|\leq M$ *for all* $z\in\mathbb{C}$. *Then* f *is constant.*

We are now able to show that an algebraic power series has an infinite radius of convergence if and only if it is a polynomial.

THEOREM 3.4. *Let* R *be a subfield of the complex numbers and* $\alpha\in R_{\text{Alg}}[[x]]$. *Then* $\alpha(x)\in R[x]$ *if and only if radius of convergence of* $(\alpha)=\infty$.

PROOF. The forward direction is trivial. Conversely, suppose that radius of convergence of $(\alpha(x))=\infty$. By definition there exists an irreducible polynomial

$$F(x,y)=f_n(x)y^n+f_{n-1}(x)y^{n-1}+\cdots+f_1(x)y+f_0(x)\in R[x][y]$$

such that $F(x,\alpha)=0$. Let $m=\max_{0\leq i\leq n}\{\deg(f_i(x))\}$. Then there exists $M_1>0$ such that

$$\sum_{i=0}^{n}|f_i(x)|\leq |x|^{m+1} \text{ for } |x|\geq M_1.$$

Let

$$c=\begin{cases}|f_n(0)| & \text{if } f_n(0)\neq 0\\ 1 & \text{otherwise.}\end{cases}$$

There exists $M_2>0$ such that $c\leq |f_n(x)|$ for all $|x|\geq M$. Let $M=\max\{M_1,M_2\}$.
We have that

$$\begin{aligned} f_n(x)\alpha(x)^n &= -\left(\sum_{i=0}^{n-1} f_i(x)\alpha(x)^i\right) \\ &\Rightarrow |f_n(x)||\alpha(x)|^n=\left|\sum_{i=0}^{n-1} f_i(x)\alpha(x)^i\right|\leq \sum_{i=0}^{n-1}|f_i(x)||\alpha(x)|^i. \end{aligned}$$

If $|\alpha(x)| \leq 1$, then

$$|f_n(x)||\alpha(x)|^n \leq |f_n(x)||\alpha(x)| \leq |f_n(x)| \leq \sum_{i=0}^{n} |f_i(x)|.$$

If $|\alpha(x)| > 1$, then

$$|f_n(x)||\alpha(x)|^n \leq \sum_{i=0}^{n-1} |f_i(x)||\alpha(x)|^i \leq |\alpha(x)|^{n-1} \sum_{i=0}^{n-1} |f_i(x)|.$$

Therefore,

$$|f_n(x)||\alpha(x)| \leq \sum_{i=0}^{n-1} |f_i(x)| \leq \sum_{i=0}^{n} |f_i(x)|$$

holds regardless of the modulus of $\alpha(x)$. Thus, when $|x| \geq M$, it is the case that

$$c|\alpha(x)| \leq |f_n(x)||\alpha(x)| \leq \sum_{i=0}^{n} |f_i(x)|,$$

which implies that

$$|\alpha(x)| \leq \frac{1}{c} \sum_{i=0}^{n} |f_i(x)| \leq \frac{1}{c} |x|^{m+1}.$$

We note that $\alpha(x)$ is entire because its radius of convergence is infinite. Therefore, by Proposition 3.2, we have that $\alpha(x)$ is a polynomial, i.e., $\alpha(x) \in R[x]$. □

COROLLARY 3.5. *Any power series with an infinite radius of convergence that is not a polynomial is transcendental. For example,* e^x, $\sin(x)$, *and* $\cos(x)$ *are transcendental functions.*

With a necessary and sufficient condition for an algebraic power series to have an infinite radius of convergence, we will consider algebraic power series that are not polynomials. We first show that the radius of convergence for an algebraic power series is an algebraic number. In order to show this, we recall the resultant of two polynomials and the discriminate of a polynomial. We refer the reader to [**12**] and [**14**] for more details. Let R denote an integral domain.

DEFINITION 3.6. The resultant of $f(x) = \sum_{i=0}^{n} a_i x^i$ and $g(x) = \sum_{i=0}^{m} b_i x^i$ in $R[x]$ is the determinant of the Sylvester matrix:

$$\begin{pmatrix}
a_n & & & & b_m & & & \\
a_{n-1} & a_n & & & b_{m-1} & b_m & & \\
a_{n-2} & a_{n-1} & \ddots & & b_{m-2} & b_{m-1} & \ddots & \\
\vdots & a_2 & \ddots & a_n & \vdots & b_{m-2} & \ddots & b_m \\
a_0 & \vdots & \ddots & a_{n-1} & b_0 & \vdots & \ddots & b_{m-1} \\
 & a_0 & & a_{n-2} & & b_0 & & b_{m-2} \\
 & & \ddots & \vdots & & & \ddots & \vdots \\
 & & & a_0 & & & & b_0
\end{pmatrix}$$

We denote the result of f and g by $\operatorname{Res}(f, g)$.

The resultant has many useful properties [**14**]. One property we highlight is that $\operatorname{Res}(f, g) = 0$ if and only if f and g have a common root in $\mathbb{C}$.

EXAMPLE 3.7. The resultant of $f(x) = 1 + 2x + x^2$ and $g(x) = 3 + 3x$ is

$$\operatorname{Res}(f,g) = \begin{vmatrix} 1 & 3 & 0 \\ 2 & 3 & 3 \\ 1 & 0 & 3 \end{vmatrix} = 1 \cdot \begin{vmatrix} 3 & 3 \\ 0 & 3 \end{vmatrix} - 3 \cdot \begin{vmatrix} 2 & 3 \\ 1 & 3 \end{vmatrix} = 9 - 3(6-3) = 9 - 9 = 0$$

This shows that f and g have a common root. Upon inspection, one sees that $x = -1$ is a shared root of f and g.

When working with polynomials in two variables, we specify which indeterminate is viewed as the variable by adding a subscript and treating the other variable as a coefficient.

EXAMPLE 3.8. Let $F(x,y) = x^2y^2 + 1$ and $G(x,y) = (x+1)y$. Then

$$\begin{aligned} \operatorname{Res}_y(F,G) &= \begin{vmatrix} x^2 & x+1 & 0 \\ 0 & 0 & x+1 \\ 1 & 0 & 0 \end{vmatrix} \\ &= x^2 \cdot \begin{vmatrix} 0 & x+1 \\ 0 & 0 \end{vmatrix} - (x+1) \cdot \begin{vmatrix} 0 & x+1 \\ 1 & 0 \end{vmatrix} \\ &= 0 + (x+1)^2 \\ &= (x+1)^2 \end{aligned}$$

and

$$\operatorname{Res}_x(F,G) = \begin{vmatrix} y^2 & y & 0 \\ 0 & y & y \\ 1 & 0 & y \end{vmatrix} = y^2 \cdot \begin{vmatrix} y & y \\ 0 & y \end{vmatrix} - y \cdot \begin{vmatrix} 0 & y \\ 1 & y \end{vmatrix} = y^4 + y^2.$$

With the resultant defined we can now define the discriminant [**25**].

DEFINITION 3.9. Given a polynomial $f(x) = a_0 + a_1x + \cdots + a_nx^n$, the discriminant of f is

$$\operatorname{Disc}(f) = \frac{(-1)^{n(n-1)/2}}{a_n} \operatorname{Res}(f, f').$$

Because the discriminant is defined via the resultant, it inherits the property that $\operatorname{Disc}(f) = 0$ if and only if f and f' have a common root.

EXAMPLE 3.10. The derivative of $f(x) = ax^2 + bx + c$ is $f'(x) = 2ax + b$, and thus,

$$\begin{aligned} \operatorname{Disc}(f) &= \frac{(-1)^3}{a} \begin{vmatrix} a & 2a & 0 \\ b & b & 2a \\ c & 0 & b \end{vmatrix} \\ &= \frac{-1}{a}\left(a \cdot \begin{vmatrix} b & 2a \\ 0 & b \end{vmatrix} - 2a \cdot \begin{vmatrix} b & 2a \\ c & b \end{vmatrix}\right) \\ &= \frac{-1}{a}(ab^2 - (2ab^2 - 4a^2c)) \\ &= b^2 - 4ac. \end{aligned}$$

Again, when working with polynomials in two variables, we specify which indeterminate we differentiate with respect to by adding a subscript.

EXAMPLE 3.11. Let $F(x,y) = y^2 - x^2 + x^3$. Then $F_x(x,y) = -2x + 3x^2$ and $F_y(x,y) = 2y$. We have

$$\mathrm{Disc}_y(F) = \frac{(-1)^3}{1}\begin{vmatrix} 1 & 2 & 0 \\ 0 & 0 & 2 \\ -x^2+x^3 & 0 & 0 \end{vmatrix} = -1(-2(0-2(-x^2+x^3))) = -4(-x^2+x^3)$$

$$\mathrm{Disc}_x(F) = \frac{(-1)^{10}}{1}\begin{vmatrix} 1 & 0 & 3 & 0 & 0 \\ -1 & 1 & -2 & 3 & 0 \\ 0 & -1 & 0 & -2 & 3 \\ y^2 & 0 & 0 & 0 & -2 \\ 0 & y^2 & 0 & 0 & 0 \end{vmatrix} = -y^2(-4+27y^2).$$

REMARK 3.12. Given a bivariate polynomial $F(x,y) \in R[x][y]$ with $\deg_y(F) = n$ and $\deg_x(F) = m$, $\deg(\mathrm{Disc}_y(F)) \leq 2m(n-1)$.

We now return to showing that the radius of convergence for an algebraic power series is an algebraic number. Every algebraic power series is a root of a bivariate polynomial of the form

$$F(x,y) = f_n(x)y^n + f_{n-1}(x)y^{n-1} + \cdots + f_1(x)y + f_0(x).$$

The singularities of $F(x,y)$ are of three types [**1**], [**23**]:

(1) roots of $f_n(x)$
(2) $x_0 \in \mathbb{C}$ such that $F(x_0,y)$ has a root of multiplicity greater than one.
(3) points at infinity.

We only focus on the first two scenarios since we have dealt with an infinite radius of convergence.

The singularities from the first two cases mark potential radii of convergence for algebraic power series [**23**]. The first are all the roots of $f_n(x)$. If $F(x,y)$ has a repeated root, then $\mathrm{Disc}_y(F) = 0$, where $\mathrm{Disc}_y(F)$ is the discriminant of F with respect to y. Since the discriminant is a polynomial, there are only finitely many x-values that satisfy this condition. Given that singularities determine the possible radii of convergence and arise as roots of either $f_n(x)$ or $\mathrm{Disc}_y(F)$, it follows that the radius of convergence is an algebraic number.

It is natural to ask if every algebraic number appears as the radius of convergence of an algebraic power series when $R = \mathbb{Q}$. The question was asked under a different context by Banderier and Drmota in [**4**]. In that paper, the authors conjecture this to be true when working with a positive algebraic system of equations with rational coefficients, but this remains an open problem. In order to establish which algebraic numbers appear as radii of convergence of algebraic power series, we first establish some definitions and observations. We begin by reviewing the known results obtained thus far. To start, we look at properties of rational power series and the Hadamard product.

DEFINITION 3.13. A series $f(x)$ is a rational power series if

$$f(x) = \frac{g(x)}{h(x)}$$

where $g(x), h(x) \in \mathbb{Q}[x]$, $h(x) \neq 0$, and $h(0) \neq 0$.

Rational power series play an important role in determining which algebraic numbers appear as a radii of convergence for algebraic power series. We introduce an operation for power series called the Hadamard product [**40**].

DEFINITION 3.14. Let $f(x) = \sum_{i=0}^{\infty} a_i x^i$ and $g(x) = \sum_{i=0}^{\infty} b_i x^i$ be power series. The *Hadamard product* of f and g, denoted $f * g$, is

$$f * g = \sum_{i=0}^{\infty} a_i b_i x^i.$$

The radius of convergence is multiplicative in the following sense.

THEOREM 3.15. *If $f, g \in R[[x]]$ with nonzero finite radii of convergence R_1 and R_2 respectively, then radius of convergence of $(f * g) = R_1 R_2$* [**4**].

Given two rational power series, then their Hadamard product is also a rational power series [**40**]. The same question can be asked of algebraic power series. Unfortunately, the Hadamard product of two algebraic power series is not necessarily an algebraic power series. The Hadamard product of an algebraic power series and a rational power series, however, is an algebraic power series [**40**].

LEMMA 3.16. *If $\alpha(x) \in R_{\text{Alg}}[[x]]$, then $\alpha(\gamma x) \in R_{\text{Alg}}[[x]]$ for $\gamma \in R$.*

SKETCH OF PROOF. Suppose $\alpha(x) \in R_{\text{Alg}}[[x]]$. By definition, there exists $F(x, y) \in R[x][y]$ such that $F(x, \alpha(x)) = 0$. One can easily verify that $F(\gamma x, \alpha(\gamma x)) = 0$. Clearly $\alpha(\gamma x) \in R[[x]]$ and thus $\alpha(\gamma x) \in R_{\text{Alg}}[[x]]$. □

REMARK 3.17. If the radius of convergence for $a(x)$ is $0 < R < \infty$, then the radius of convergence of $a(\gamma x)$ is $\frac{R}{|\gamma|}$.

We define an important set regarding the radius of convergence of an algebraic power series.

$$\mathcal{C} := \left\{ R \mid r(x) = \frac{f(x)}{g(x)} \in \mathbb{Q}(x),\ g(0) \neq 0, \right.$$
$$\left. \text{and radius of convergence of } r(x) = R < \infty \right\}.$$

We know that every element of $\mathcal{C}$ appears as a radius of convergence for an algebraic power series because a rational power series $r(x) = \frac{f(x)}{g(x)}$ such that $g(0) \neq 0$ satisfies

$$g(x)y - f(x) = 0$$

and hence is algebraic. However, we only pick up the algebraic numbers that are the smallest modulus in comparison to the other roots of the minimal polynomial. That is, if $r \in \mathbb{C}$ and f_r is its minimal polynomial, and $|r| \leq |r_i|$ for every r_i that is a root of f_r, then $|r|$ is the radius of convergence for an algebraic power series.

EXAMPLE 3.18. Consider $y = x^2 - 2x - 49$. The roots of y are $x = 1 + 5\sqrt{2}, 1 - 5\sqrt{2}$. Therefore, $|1 - 5\sqrt{2}|$ is the radius of convergence of

$$f(x) = \frac{1}{x^2 - 2x - 49}$$

since $|1 - 5\sqrt{2}| < |1 + 5\sqrt{2}| = 1 + 5\sqrt{2}$. It is not straightforward, and requires more work to find an algebraic power series that has a radius of convergence $1 + 5\sqrt{2}$. We will resolve this later in the paper.

We aim to provide evidence that the modulus of every algebraic number except zero appears as the radius of convergence of an algebraic power series. While this remains an open question, we conjecture it to be true. We denote the algebraic numbers by $\mathbb{A}$.

CONJECTURE 3.19. *If $r \in \mathbb{A}\backslash\{0\}$, then there exists an algebraic power series $\alpha(x)$ such that radius of convergence of $(\alpha(x)) = |r|$.*

To support this claim, we observe that if an algebraic power series has a radius of convergence r, then there exists algebraic power series with radii of convergence $r^{1/n}$ and $|c|r$ where $c \in \mathcal{C}$.

THEOREM 3.20. *Given an algebraic power series with radius of convergence R, there also exists an algebraic power series of radius of convergence of the following:*

(1) *$R^{1/n}$ for $n \in \mathbb{N}$.*
(2) *cR where $c \in \mathcal{C}$.*

PROOF.

(1) Let $\alpha(x) \in R_{\text{Alg}}[[x]]$ with radius of convergence of $(\alpha(x)) = R < \infty$. There exists $F(x, y) \in \mathbb{Q}[x][y]$ with $F(x, \alpha(x)) = 0$. Then one can easily verify that $F(x^n, \alpha(x^n)) = 0$ so $\alpha(x^n) \in R_{\text{Alg}}[[x]]$ with radius of convergence of $(\alpha(x^n)) = R^{1/n}$.

(2) Let $\alpha(x) \in R_{\text{Alg}}[[x]]$ with radius of convergence of $(\alpha(x)) = R < \infty$. There exists $F(x, y) \in \mathbb{Q}[x][y]$ with $F(x, \alpha(x)) = 0$. There exists a rational power series, $r(x)$, with radius of convergence of $(r(x)) = c$ where $c \in \mathcal{C}$. The Hadamard product of $\alpha(x)$ and $r(x)$ is algebraic with a radius of convergence cR. □

This theorem, applied to the numbers in $\mathcal{C}$ does not provide any new radii of convergence. To see this let the absolute Galois closure of $\mathbb{Q}$ be denoted by $G := \text{Gal}(\mathbb{A}/\mathbb{Q})$. The set $\mathcal{C}$ is multiplicatively closed [**43**].

PROOF. Let $a, b \in \mathcal{C}$. Then $|a| \leq |g(a)|$ and $|b| \leq |g(b)|$ for all $g \in G$. Therefore, if $c \in R_{ab}$, then $c = g(ab)$ for some $g \in G$. Thus, we have that

$$|c| = |g(ab)| = |g(a)g(b)| = |g(a)||g(b)| \geq |a||b| = |ab|$$

and thus, $ab \in \mathcal{C}$. □

The theorem becomes useful when we get an element outside of $\mathcal{C}$. We will only consider quadratic numbers of the form $a + b\sqrt{n}$ where n is square free and $a, b \in \mathbb{Q}$. Referring back to example 3.18, we will show an algebraic power series with radius of convergence $1 + 5\sqrt{2}$.

EXAMPLE 3.21. We need $1+5\sqrt{2}$ to be a root of $\text{Disc}_y(F)$. If $1+5\sqrt{2}$ is a root, then $1 - 5\sqrt{2}$ is also a root. Since the radius of convergence of a power series is its closest singularity and $|1 - 5\sqrt{2}| < |1 + 5\sqrt{2}|$, this poses a problem. One solution is to separate these roots among the algebraic power series of $F(x, y)$. First, we will make our constant term $c = -1$(for reasons that will become apparent later on)

$$f(x) = \frac{1}{49}x^2 - \frac{2}{49}x - 1$$

has roots $x_1 = 1 - 5\sqrt{2}$ and $x_2 = 1 + 5\sqrt{2}$. We can separate these roots by considering

$$f_1 = \frac{1}{7}x + \sqrt{\frac{2}{49}x + 1}$$
$$f_2 = \frac{1}{7}x - \sqrt{\frac{2}{49}x + 1}$$

It is elementary to verify that

$$f_1(x_1) = 0 \quad f_1(x_2) \neq 0$$
$$f_2(x_2) = 0 \quad f_2(x_1) \neq 0$$

Now we want the roots of f_1 and f_2 to appear as singularities, so we consider

$$y_1 = \sqrt[3]{\frac{1}{7}x + \sqrt{\frac{2}{49}x + 1}}$$
$$y_2 = \sqrt[3]{\frac{1}{7}x - \sqrt{\frac{2}{49}x + 1}}.$$

Therefore,

$$F(x, y) = y^6 - \frac{2}{7}xy^3 + \frac{1}{49}x^2 - \frac{2}{49}x - 1$$

has the roots we want. We quickly observe that letting $c = -1$ is needed so that y_1 and y_2 admit power series expansions in $\mathbb{Q}$. Before we say we are done, you will notice that when we separated the roots of $f(x)$ using f_1 and f_2, we inadvertently introduced another singularity under the square root. The singularity occurs at $x = -\frac{49}{2}$. Since $1 + 5\sqrt{2} < |-\frac{49}{2}|$ there are no issues. The only singularities of y_2 are $x = -\frac{49}{2}$ and $x = 1 + 5\sqrt{2}$, therefore radius of convergence of $(y_2) = 1 + 5\sqrt{2}$.

We note that $1 + 5\sqrt{2} \notin \mathcal{C}$. With this additional radius of convergence we also get $r = (1 + 5\sqrt{2})^{1/n}$ where $n \in \mathbb{N}$ and $c(1 + 5\sqrt{2})$ where $c \in \mathcal{C}$ by theorem 3.20. While we gain access to a larger set of radii of convergence, we still do not obtain them all. Notably, it is easy to verify that $(1 + 5\sqrt{2})c \neq 1 + \sqrt{2}$ for any $c \in \mathcal{C}$.

In order to find other quadratic numbers, the idea of separating roots can be used, but we need to ensure the root we are looking for is the smallest root of all the singularities. Given

$$y_1 = \sqrt[3]{ax + \sqrt{bx + 1}}$$
$$y_2 = \sqrt[3]{ax - \sqrt{bx + 1}}$$

We need the additional requirement that the modulus of both roots of $a^2x^2 - bx - 1$ are smaller than the modulus of the root of $bx + 1$, i.e.,

$$\frac{|b| + \sqrt{b^2 + 4a^2}}{2a^2} < \frac{1}{|b|} \Rightarrow |a| > \sqrt{2}|b|.$$

As long as this requirement is satisfied we are guaranteed a radius of convergence not in $\mathcal{C}$. Additionally, with the method of separating roots, obtaining a radius of convergence expressed as $a + b\sqrt{n}$ for every $n \in \mathbb{N}$ is straightforward, however, it is still unknown whether each number in the form $a + b\sqrt{n}$ can be reached.

In conclusion, we have broadened the scope of possible radii of convergence to include quadratic numbers that were previously inaccessible. While the method of separating roots is effective for quadratic numbers, extending it to higher-degree algebraic numbers presents challenges.

References

[1] L. V. Ahlfors, *Complex analysis. An introduction to the theory of analytic functions of one complex variable*, McGraw-Hill Book Co., Inc., New York-Toronto-London, 1953. MR54016

[2] G. E. Andrews, P. Paule, A. Riese, and V. Strehl, *MacMahon's partition analysis. V. Bijections, recursions, and magic squares*, Algebraic combinatorics and applications (Gößweinstein, 1999), Springer, Berlin, 2001, pp. 1–39. MR1851939

[3] C. Banderier and M. Drmota, *Coefficients of algebraic functions: Formulae and asymptotics*, Discrete Mathematics & Theoretical Computer Science (2013), no. Proceedings.

[4] C. Banderier and M. Drmota, *Formulae and asymptotics for coefficients of algebraic functions*, Combin. Probab. Comput. **24** (2015), no. 1, 1–53, DOI 10.1017/S0963548314000728. MR3318039

[5] C. Banderier and P. Flajolet, *Basic analytic combinatorics of directed lattice paths*, Theoret. Comput. Sci. **281** (2002), no. 1-2, 37–80, DOI 10.1016/S0304-3975(02)00007-5. Selected papers in honour of Maurice Nivat. MR1909568

[6] A. Bertoni, M. Goldwurm, and M. Santini, *Random generation for finitely ambiguous context-free languages*, Theor. Inform. Appl. **35** (2001), no. 6, 499–512 (2002), DOI 10.1051/ita:2001128. A tribute to Aldo de Luca. MR1922291

[7] O. Bodini, A. Darrasse, and M. Soria, *Distances in random Apollonian network structures* (English, with English and French summaries), 20th Annual International Conference on Formal Power Series and Algebraic Combinatorics (FPSAC 2008), Discrete Math. Theor. Comput. Sci. Proc., AJ, Assoc. Discrete Math. Theor. Comput. Sci., Nancy, 2008, pp. 307–318. MR2721463

[8] O. Bodini and Y. Ponty, *Multi-dimensional Boltzmann sampling of languages*, 21st International Meeting on Probabilistic, Combinatorial, and Asymptotic Methods in the Analysis of Algorithms (AofA'10), Discrete Math. Theor. Comput. Sci. Proc., AM, Assoc. Discrete Math. Theor. Comput. Sci., Nancy, 2010, pp. 49–63. MR2735331

[9] A. Bostan, X. Caruso, and J. Roques, *Algebraic solutions of linear differential equations: An arithmetic approach*, Bull. Amer. Math. Soc. (N.S.) **61** (2024), no. 4, 609–658, DOI 10.1090/bull/1835. MR4803603

[10] E. Casas-Alvero, *Singularities of plane curves*, London Mathematical Society Lecture Note Series, vol. 276, Cambridge University Press, Cambridge, 2000, DOI 10.1017/CBO9780511569326. MR1782072

[11] E. Casas-Alvero, *Algebraic curves, the Brill and Noether way*, Universitext, Springer, Cham, [2019] ©2019, DOI 10.1007/978-3-030-29016-0. MR3971541

[12] D. A. Cox, J. Little, and D. O'Shea, *Ideals, varieties, and algorithms: An introduction to computational algebraic geometry and commutative algebra*, 4th ed., Undergraduate Texts in Mathematics, Springer, Cham, 2015, DOI 10.1007/978-3-319-16721-3. MR3330490

[13] D. A. Cox, *Galois theory*, 2nd ed., Pure and Applied Mathematics (Hoboken), John Wiley & Sons, Inc., Hoboken, NJ, 2012, DOI 10.1002/9781118218457. MR2919975

[14] D. A. Cox, J. Little, and D. O'Shea, *Using algebraic geometry*, 2nd ed., Graduate Texts in Mathematics, vol. 185, Springer, New York, 2005. MR2122859

[15] Desmos, *Graph of* $y = x^2 - 3x + 2$, Desmos Graphing Calculator, 2025, Accessed: 2025-03-01.

[16] D. S. Dummit and R. M. Foote, *Abstract algebra*, 3rd ed., John Wiley & Sons, Inc., Hoboken, NJ, 2004. MR2286236

[17] P. Flajolet and A. Odlyzko, *Singularity analysis of generating functions*, SIAM J. Discrete Math. **3** (1990), no. 2, 216–240, DOI 10.1137/0403019. MR1039294

[18] P. Flajolet and R. Sedgewick, *Analytic combinatorics*, Cambridge University Press, Cambridge, 2009, DOI 10.1017/CBO9780511801655. MR2483235

[19] J. B. Fraleigh, *A first course in abstract algebra*, Addison-Wesley Publishing Co., Reading, Mass.-London-Don Mills, Ont., 1967. MR225619

[20] R. L. Graham, D. E. Knuth, and O. Patashnik, *Concrete mathematics*, 2nd ed., Addison-Wesley Publishing Company, Reading, MA, 1994. A foundation for computer science. MR1397498

[21] L. C. Grove, *Algebra*, Dover Publications, Inc., Mineola, NY, 2004. Reprint of the 1983 original, with an errata list on pp. xv–xvi. MR2107088

[22] R. Hartshorne, *Algebraic geometry*, Graduate Texts in Mathematics, No. 52, Springer-Verlag, New York-Heidelberg, 1977. MR463157

[23] E. Hille, *Analytic function theory. Vol. II*, Introductions to Higher Mathematics, Ginn and Company, Boston, Mass.-New York-Toronto, 1962. MR0201608

[24] T. W. Hungerford, *Algebra*, Graduate Texts in Mathematics, vol. 73, Springer-Verlag, New York-Berlin, 1980. Reprint of the 1974 original. MR600654

[25] T. W. Hungerford, *Algebra*, vol. 73, Springer Science & Business Media, 2012.

[26] I. Kaplansky, *Commutative rings*, Conference on Commutative Algebra (Univ. Kansas, Lawrence, Kan., 1972), Lecture Notes in Math., Vol. 311, Springer, Berlin-New York, 1973, pp. 153–166. MR360550

[27] M. Kauers, *The concrete tetrahedron*, Proceedings of the 36th international symposium on Symbolic and algebraic computation, 2011, pp. 7–8. MR2768529

[28] S. G. Krantz and H. R. Parks, *The implicit function theorem: History, theory, and applications*, Birkhäuser Boston, Inc., Boston, MA, 2002, DOI 10.1007/978-1-4612-0059-8. MR1894435

[29] S. Lang, *Complex analysis*, vol. 103, Springer Science & Business Media, 2013.

[30] L. Lipshitz, *The diagonal of a D-finite power series is D-finite*, J. Algebra **113** (1988), no. 2, 373–378, DOI 10.1016/0021-8693(88)90166-4. MR929767

[31] L. Lipshitz, *D-finite power series*, J. Algebra **122** (1989), no. 2, 353–373, DOI 10.1016/0021-8693(89)90222-6. MR999079

[32] P. A. MacMahon, *Combinatory analysis*, Chelsea Publishing Co., New York, 1960. Two volumes (bound as one). MR141605

[33] MacTutor, *George birch jerrard.*

[34] D. Marques and C. G. Moreira, *A note on a complete solution of a problem posed by K. Mahler*, Bull. Aust. Math. Soc. **98** (2018), no. 1, 60–63, DOI 10.1017/S0004972718000187. MR3819972

[35] Arnold Ostebee, *Topics in number theory: In honor of b. gordon and s chowla*, The American Mathematical Monthly **106** (1999), no. 9, 876 (English), Copyright - Copyright Mathematical Association Of America Nov 1999; Last updated - 2017-11-08; CODEN - AMMYAE.

[36] I. Rivin, *Galois groups of generic polynomials*, 2015.

[37] R. E. Rodríguez, I. Kra, and J. P. Gilman, *Complex analysis: In the spirit of Lipman Bers*, 2nd ed., Graduate Texts in Mathematics, vol. 245, Springer, New York, 2013, DOI 10.1007/978-1-4419-7323-8. MR2986247

[38] M. Schönert et al., *GAP – Groups, Algorithms, and Programming – version 3 release 4 patchlevel 4*, 1997.

[39] S. Stahl, *Introductory modern algebra: A historical approach*, Wiley, 2013.

[40] R. P. Stanley, *Differentiably finite power series*, European J. Combin. **1** (1980), no. 2, 175–188, DOI 10.1016/S0195-6698(80)80051-5. MR587530

[41] R. P. Stanley, *Enumerative combinatorics. Vol. 1*, Cambridge Studies in Advanced Mathematics, vol. 49, Cambridge University Press, Cambridge, 1997. With a foreword by Gian-Carlo Rota; Corrected reprint of the 1986 original, DOI 10.1017/CBO9780511805967. MR1442260

[42] R. P. Stanley, *Enumerative combinatorics. Vol. 2*, Cambridge Studies in Advanced Mathematics, vol. 62, Cambridge University Press, Cambridge, 1999. With a foreword by Gian-Carlo Rota and appendix 1 by Sergey Fomin, DOI 10.1017/CBO9780511609589. MR1676282

[43] J. Swanson, *Algebraic and Integral Closure of a Polynomial Ring in Its Power Series Ring*, ProQuest LLC, Ann Arbor, MI, 2023. Thesis (Ph.D.)–Clemson University. MR4724624

[44] E. J. Wilczynski, *On the Form of the Power Series for an Algebraic Function*, Amer. Math. Monthly **26** (1919), no. 1, 9–12, DOI 10.2307/2974040. MR1519165

[45] H. S. Wilf, *generatingfunctionology*, Academic Press, Inc., Boston, MA, 1990. MR1034250

[46] D. Zeilberger, *A holonomic systems approach to special functions identities*, J. Comput. Appl. Math. **32** (1990), no. 3, 321–368, DOI 10.1016/0377-0427(90)90042-X. MR1090884

[47] D. Zill and P. Shanahan, *A first course in complex analysis with applications*, Jones & Bartlett Learning, 2009.

Email address: `joseph.swanson@stvincent.edu`

Contemporary Mathematics
Volume **836**, 2026
https://doi.org/10.1090/conm/836/16713

A length function of $\mathbb{Z}[X_1, \ldots, X_m]$-modules and Mahler measure

Davide Castelnovo, Dikran Dikranjan, Anna Giordano Bruno, Dario Spirito, and Simone Virili

Dedicated to the memory of Peter Vámos

ABSTRACT. The notion of algebraic entropy has a long history dating back to a very brief outline of the basic ideas in a paper by R. L. Adler, A. G. Konheim and M. H. McAndrew in 1965. The notion was subsequently developed by a long series of authors working in various contexts. For an abelian group A and an endomorphism $\phi\colon A \to A$, the algebraic entropy $h_{\mathrm{alg}}(\phi)$ is defined by $h_{\mathrm{alg}}(\phi) = \sup_{F \in \mathcal{F}(A)} \lim_{n\to\infty} \frac{1}{n} \log |F + \phi(F) + \ldots + \phi^{n-1}(F)|$, where $\mathcal{F}(A)$ is the family of all finite non-empty subsets of A. The algebraic entropy of an m-tuple of commuting endomorphisms of an abelian group is defined similarly.

Under the celebrated Kaplansky paradigm, abelian groups provided with an m-tuple of pairwise commuting endomorphisms can be identified with $\mathbb{Z}[X_1, \ldots, X_m]$-modules, so the algebraic entropy becomes a length function $\mathtt{h}$ of $\mathbb{Z}[X_1, \ldots, X_m]$-modules. We show that this length function $\mathtt{h}$ can be completely described via the Mahler measure of polynomials $f \in \mathbb{Z}[X_1, \ldots, X_m]$. As an application we obtain various uniqueness theorems for $\mathtt{h}$.

To ease the readers non-familiar with the dynamical aspects of the paper, we first provide a treatment of the entropy of only single endomorphisms, leaving the general case of entropy of a finite set of pairwise commuting endomorphisms for the final part.

1. Introduction

In their celebrated paper [**1**], Adler, Konheim and McAndrew introduced the notion of *topological entropy* $h_{\mathrm{top}}(f)$ of a continuous selfmap $f\colon X \to X$ of a compact space X (see §3.3 for the definition in the case when X is a compact group and f is a continuous endomorphism of X). At the end of the paper they introduced also the algebraic entropy $h_{\mathrm{alg}}(\phi)$ of an endomorphism $\phi\colon A \to A$ of an abelian group A. This definition of Adler, Konheim and McAndrew is suitable only for

2020 *Mathematics Subject Classification.* Primary 11R06, 13C60, 16D10, 16P40, 16S50, 18E40, 20K30, 37A35, 37B40.

Key words and phrases. Mahler measure, length function, algebraic entropy, topological entropy, Noetherian ring, polynomial ring, Yuzvinski Theorem, Bernoulli shift.

The third named author thanks the "National Group for Algebraic and Geometric Structures, and Their Applications" (GNSAGA - INdAM).

The fifth named author was partially supported by MINECO (grant PID2023-147110NB-I00) and by the Generalitat de Catalunya as part of the research group "Laboratori d'Interaccions entre Geometria, Àlgebra i Topologia (LIGAT)" (grant 2021SGR01015).

©2026 American Mathematical Society

torsion abelian groups as it is based on the images of the finite subgroups of an abelian group A. Later, M. D. Weiss wrote a brief article [**40**] establishing some basic properties of h_{alg} and a remarkable connection between h_{alg} and h_{top}; indeed, when $\phi\colon A \to A$ is an endomorphism of an abelian group A which is torsion, the algebraic entropy of ϕ coincides with the topological entropy of its Pontryagin dual $\widehat{\phi}$ (which makes sense since $\widehat{A}$ is compact), i.e., $h_{\mathrm{alg}}(\phi) = h_{\mathrm{top}}(\widehat{\phi})$. This kind of result was named "Bridge Theorem" by L. Salce.

Later Peters [**29**] considered a notion of algebraic entropy based on finite subsets instead of finite subgroups, but his definition was tailored for automorphisms $\phi\colon A \to A$ of abelian groups A (inverse images of finite subsets F of A were used in place of images, compare with (3.1) and (3.2)). Peters [**29**] proved a Bridge Theorem for automorphisms of countable not necessarily torsion abelian groups.

The study of the algebraic entropy restarted only thirty years later in [**16**], where the torsion case was thoroughly investigated. As the definition of the algebraic entropy for automorphisms from [**29**] could not be used for endomorphisms, it was appropriately modified to cover also arbitrary endomorphisms in [**13**]. Moreover, a Bridge Theorem was proved in this setting in [**11**] (see Theorem 3.12), extending both Weiss' and Peters' ones.

In 1967 Yuzvinski [**41**] discovered the remarkable connection of the Mahler measure with the topological (measure) entropy. The counterpart of this connection (named Algebraic Yuzvinski Formula), regarding the connection between the algebraic entropy and the Mahler measure, was proved in [**18**] (see Theorem 3.7 below). More detail on the algebraic entropy can be found in [**11, 12, 14, 18**] and elsewhere.

Another notion closely related to algebraic entropy is that of length function, introduced by Northcott and Reufel [**27**] and deeply studied by Vámos [**38**] (see §2 for the definition of length function). A fundamental tool for the study of length functions and more specifically for this paper is a decomposition for the length functions of R-modules, where R is a commutative Noetherian ring, proved by Vámos in [**38**] (see Theorem A.18): there is a standard way to write a given length function as a linear combination of a suitable family of "basic length functions", each associated with a prime ideal of R. Theorem 2.13 below is a variation of Vámos' result, that produces sharper (i.e., less redundant) decompositions in general, that we apply in the proofs of the main results of this paper.

The connection between the algebraic entropy of endomorphisms of abelian groups and length functions of $\mathrm{Mod}_{\mathbb{Z}[X]}$ was pointed out and studied in [**31**]. Following a well-known approach of Kaplansky [**21**], an endomorphism of an abelian group A can be viewed in a standard way as an R-module structure on A, where R is the polynomial ring $\mathbb{Z}[X]$. As pointed out by Peter Vámos, after this identification, the entropy function h_{alg} defined on endomorphisms of abelian groups turns out to be a length function of $\mathrm{Mod}_{\mathbb{Z}[X]}$, that we denote by $\mathtt{h}$ in the sequel.

The definition of the algebraic entropy h_{alg} was extended to actions of a cancellative amenable semigroup S on abelian groups in [**9**] (see §4.1). Also in this more general case there is an equivalence between abelian groups provided with an S-action and $\mathbb{Z}[S]$-modules. The non-trivial fact that the algebraic entropy of S-actions on abelian groups is a length function $\mathtt{h}$ on the category of $\mathbb{Z}[S]$-modules follows from recent results from [**15**] (for $S = \mathbb{N}^m$ see also [**10**]).

In this vein, with $m\in\mathbb{N}_+$ and $R=\mathbb{Z}[X_1,\dots,X_m]$, we discuss the relation between this length function $\mathtt{h}$ of R-modules corresponding to h_{alg}, and the (logarithmic) Mahler measure $\mathsf{m}(f)$ of polynomials $f(X_1,\dots,X_m)\in R$ (see (4.2) for the definition). More precisely, since R is Noetherian, it suffices to compute $\mathtt{h}(R/\mathfrak{p})$ for $\mathfrak{p}\in\operatorname{Spec}(R)$ (see Theorem 2.13). Here, and in the sequel, the m-tuple of pairwise commuting endomorphisms of $R/\mathfrak{p}$ is simply the m-tuple of multiplications by $X_i+\mathfrak{p}$ in $R/\mathfrak{p}$, $i=1,\dots,m$. We refer to the related $\mathbb{N}^m$-action on $R/\mathfrak{p}$, as the *standard* $\mathbb{N}^m$-action on $R/\mathfrak{p}$.

If A is a commutative ring and $\mathfrak{p}$ a prime ideal of A, the *height* $ht(\mathfrak{p})$ of $\mathfrak{p}$ is the supremum of the length of the chains $\mathfrak{p}_1\subset\mathfrak{p}_2\subset\cdots\subset\mathfrak{p}$ of prime ideals of A terminating at $\mathfrak{p}$, or equivalently the Krull dimension of the localization $A_\mathfrak{p}$ [**42**, Chapter 4, §14]. When $R=\mathbb{Z}[X_1,\dots,X_m]$, the prime ideals $\mathfrak{p}$ of R with $ht(\mathfrak{p})=1$ are principal, and thus the following theorem contains all the necessary information for computing h_{alg}, by Theorem 2.13.

THEOREM 1. *Let $R=\mathbb{Z}[X_1,\dots,X_m]$. Then:*

(a) $\mathtt{h}(R/(f))=\mathsf{m}(f)$ *for every non-zero* $f\in R$*;*
(b) $\mathtt{h}(R)=\infty$ *and* $\mathtt{h}(R/\mathfrak{p})=0$ *when $\mathfrak{p}$ is a prime ideal of R with $ht(\mathfrak{p})>1$.*

The proof of this theorem is given in §4.4.

The next corollary follows Theorem 1(a) and the definition of Mahler measure, as far as item (a) is concerned, and Boyd's Theorem 4.9 for item (b) (for the definition of extended cyclotomic polynomial see Definition 4.7).

COROLLARY A. *Let $R=\mathbb{Z}[X_1,\dots,X_m]$ and $f\in R$.*

(a) *If $f\neq 0$, then* $\mathtt{h}(R/(f))<\infty$.
(b) $\mathtt{h}(R/(f))$ *vanishes precisely when f is an extended cyclotomic polynomial.*

Another important consequence of Theorem 1 is the following uniqueness result for the algebraic entropy for m-tuples of pairwise commuting endomorphisms of abelian groups (see Theorem 3.1 for the definitions of the three properties involved in the corollary).

For $m\in\mathbb{N}_+$ and a prime p, identifying the cyclic group $\mathbb{Z}(p):=\mathbb{Z}/p\mathbb{Z}$ with the finite field $\mathbb{F}_p$ and abelian groups of exponent p with $\mathbb{F}_p$-vector spaces (so the abelian group $\bigoplus_{\mathbb{N}^m}\mathbb{Z}(p)$ with the underlying abelian group of the polynomial ring $\mathbb{F}_p[X_1,\dots,X_m]$), define the *multiple right Bernoulli shift* $\sigma_{p,m}\colon\bigoplus_{\mathbb{N}^m}\mathbb{Z}(p)\to\bigoplus_{\mathbb{N}^m}\mathbb{Z}(p)$ simply as the m-tuple of multiplications by X_i in $\mathbb{F}_p[X_1,\dots,X_m]$, $i=1,\dots,m$. Obviously, $\sigma_{p,1}\colon\bigoplus_{\mathbb{N}}\mathbb{Z}(p)\to\bigoplus_{\mathbb{N}}\mathbb{Z}(p)$, is the celebrated *right Bernoulli shift*, defined by $(x_0,x_1,\dots)\mapsto(0,x_0,x_1,\dots)$). We omit m, and simply write σ_p, when no confusion is possible.

COROLLARY B (Uniqueness Theorem 1). *Let $R=\mathbb{Z}[X_1,\dots,X_m]$. Every function h' defined on all m-tuples of pairwise commuting endomorphisms of abelian groups satisfying the Invariance property, the Continuity Axiom, the Addition Theorem and $h'(R/\mathfrak{p})=\mathsf{m}(f)$ for all principal prime ideals $\mathfrak{p}=(f)\neq 0$ of R, must coincide with the algebraic entropy* $\mathtt{h}$.

This result presents several novelties. First of all it proves uniqueness of h_{alg} for a finite number of commuting endomorphisms, to be compared with the known result for a single endomorphism in Theorem 3.9. Secondly, there is a new approach even in the case $m=1$, i.e., $R=\mathbb{Z}[X]$. Namely, in Theorem 3.9 one finds, beyond the Invariance property, the Continuity Axiom, the Addition Theorem, *two normalization conditions*, one given by the Algebraic Yuzvinski Formula (Theorem 3.7) and the other one by the value $h_{\mathrm{alg}}(\sigma_p) = \log p$ on the multiple right Bernoulli shift σ_p. In Corollary B these two normalizations are both covered by the *single condition* $h'(R/\mathfrak{p}) = \mathsf{m}(f)$ for all principal prime ideals $\mathfrak{p} = (f)$ of R, since the constant polynomial p, where p is a prime, has $\mathsf{m}(p) = \log p$.

The following is another uniqueness result for the algebraic entropy of m-tuples of pairwise commuting endomorphisms of abelian groups of prime exponent. The final assertion of this theorem covers [**3**, Theorem 4.4] (so, [**3**, Theorem 4.1] and [**3**, Corollary 4.5] as well).

THEOREM 2 (Uniqueness Theorem 2). *Let p be a prime and $R_p = \mathbb{F}_p[X_1, \dots, X_m]$. Every function h' defined on all m-tuples of pairwise commuting endomorphisms of abelian groups of exponent p (i.e., $\mathbb{F}_p$-vector spaces) satisfying the Invariance property, the Continuity Axiom, the Addition Theorem and $h'(R_p) = \log p$ coincides with rk_{R_p}. In particular, the algebraic entropy $\mathtt{h}$ coincides with rk_{R_p}.*

The next, third, uniqueness theorem cannot be obtained as an immediate consequence of Corollary B (it will be deduced from Theorem 2 in §4).

THEOREM 3 (Uniqueness Theorem 3). *Every function h' defined on all m-tuples of pairwise commuting endomorphisms of torsion abelian groups satisfying the Invariance property, the Continuity Axiom, the Addition Theorem and $h'(\sigma_p) = \log p$ for all primes p, must coincide with the algebraic entropy $\mathtt{h}$.*

The algebraic entropy h_{alg} of single endomorphisms satisfies the so called Logarithmic Law, namely, for every endomorphism $\phi\colon A \to A$, $h_{\mathrm{alg}}(\phi^n) = n h_{\mathrm{alg}}(\phi)$ for every $n \in \mathbb{N}$. Then Theorem 3 improves on the Uniqueness Theorem from [**16**], where h' was assumed to satisfy additionally the Logarithmic Law, so $h'(\phi^n) = nh'(\phi)$.

The proof of all the main results stated in the introduction will be given in §4.4.

This paper grew over the talk given by the second named author at the conference "Advances in Commutative Ring Theory" held in Catania in 2024. Therefore it is adapted to an audience oriented to commutative algebra, less familiar with dynamical systems. This is why, after §2 dedicated to length functions, §3 treats only the entropy of *single* endomorphisms, providing the flavour of this phenomenon in a relatively simple situation. Then §4 is focused on the general case of entropy of an arbitrary finite set of pairwise commuting endomorphisms.

Synopsis. In §2 we recall the notion of length function and its basic properties. Moreover, we state a useful decomposition theorem for length functions (Theorem 2.13, proved in the Appendix) inspired by (and in some cases sharper than) the celebrated Vámos' decomposition theorem.

In §3 we recall the basic properties of the entropy function h_{alg} for single endomorphisms of abelian groups; in particular §3.3 is dedicated to the Uniqueness Theorem (Theorem 3.9) of h_{alg} and its connection to topological entropy. Then, in §3.4 an alternative proof of Theorem 3.9 is given by using the properties of length functions.

§4 culminates with the proofs of the main results stated in the introduction. To this end we recall in §4.1 the general definition of the algebraic entropy h_{alg} for actions of cancellative commutative semigroups on abelian groups, that we apply to the case of $\mathbb{N}^m$-actions, that is, the case of m commuting endomorphisms of an abelian group. In §4.2 we recall the Bridge Theorem for actions of commutative cancellative monoids and Ore localizations. To the notion of Mahler measure for polynomials of m variables is dedicated §4.3. Finally, in §4.4 we give the proof of Theorems 1, 2 and 3, as well as that of Corollary B.

In the Appendix we provide more detail on length functions, culminating in a general Decomposition Theorem A.14 from which we obtain the proof of Theorem 2.13 and Vámos' decomposition theorem (Theorem A.18), as well as a comparison between these results.

2. Length functions

We fix throughout this section a commutative Noetherian ring R, and we denote by Mod_R (resp., mod_R) the category of (finitely generated) R-modules. Furthermore, given a module $M \in \mathrm{Mod}_R$, we denote by $\mathcal{L}(M)$ its lattice of submodules, and we let $\mathcal{L}_f(M) := \mathcal{L}(M) \cap \mathrm{mod}_R$ be the sublattice of finitely generated submodules.

DEFINITION 2.1. For $\mathcal{C} \in \{\mathrm{Mod}_R, \mathrm{mod}_R\}$, an *additive function* of $\mathcal{C}$ is a map $L\colon \mathcal{C} \to \mathbb{R}_{\geq 0} \cup \{\infty\}$ that satisfies the following conditions:

(1) $L(0) = 0$;
(2) $L(M) = L(N) + L(\overline{M})$, whenever $0 \to N \to M \to \overline{M} \to 0$ is exact in $\mathcal{C}$.

We refer to condition (2) as the *additivity* of L.

Observe that, if there is $M \in \mathcal{C}$ such that $L(M) < \infty$, one deduces that $L(0) = L(M) - L(M) = 0$ just applying condition (2) to the short exact sequence $0 \to M \xrightarrow{\mathrm{id}_M} M \to 0 \to 0$. Hence, condition (1) becomes redundant in this case.

In the following lemma we collect some consequences of the definition of additive function. The proof is straightforward and so it is left to the reader.

LEMMA 2.2. *Let L be an additive function of $\mathcal{C} \in \{\mathrm{Mod}_R, \mathrm{mod}_R\}$. Then:*

(a) *$L(M) = L(M')$, whenever $M \cong M'$ in $\mathcal{C}$;*
(b) *$L(M) \geq L(N)$ and $L(M) \geq L(M/N)$, for all $N \leq M \in \mathcal{C}$;*
(c) *$L(M) = \sum_{i=1}^{n} L(M_i/M_{i-1})$, if $0 = M_0 \leq M_1 \leq \cdots \leq M_n = M \in \mathcal{C}$.*

DEFINITION 2.3. A map $L\colon \mathrm{Mod}_R \to \mathbb{R}_{\geq 0} \cup \{\infty\}$ is *upper continuous* if

$$L(M) = \sup\{L(F) : F \in \mathcal{L}_f(M)\}, \qquad \text{for every } M \in \mathrm{Mod}_R.$$

If L is both additive and upper continuous, then L is called a *length function.*

The following lemma is just a reformulation of the definition of length function. We leave the proof to the reader.

LEMMA 2.4. *For an additive function L of Mod_R the following are equivalent:*

(a) *L is a length function;*
(b) *for every module M and every directed system $\{M_i : i \in I\}$ of submodules such that $\sum_{i\in I} M_i = M$, we have $L(M) = \sup\{L(M_i) : i \in I\}$.*

As their name suggests, abstract length functions were modeled by Northcott and Reufel [**27**] on the so-called composition length of modules. We introduce this prototypical function in §2.1, while the reader is addressed to §A.1 for other concrete examples.

2.1. Prime series and composition series of modules.

DEFINITION 2.5. A *series* of $M \in \mathrm{Mod}_R$ is just a finite chain of submodules:

$$\sigma\colon \qquad 0 = M_0 \leq M_1 \leq \cdots \leq M_n = M, \quad \text{with } n \in \mathbb{N}. \tag{2.1}$$

The modules M_i/M_{i-1}, for $i = 1, \ldots, n$, are called the *factors* of σ, and its *length* $\ell(\sigma) \in \mathbb{N}$ is the number of strict inequalities (or, equivalently, of non-trivial factors).

Let $L\colon \mathrm{Mod}_R \to \mathbb{R}_{\geq 0} \cup \{\infty\}$ be a length function and observe that, by upper continuity, L is completely determined by its restriction to mod_R. Moreover, each $M \in \mathrm{mod}_R$ admits a series with cyclic factors, so that $L(M)$ is the sum of the values of L on these cyclics (see Lemma 2.2(c)). Hence, if $L'\colon \mathrm{Mod}_R \to \mathbb{R}_{\geq 0} \cup \{\infty\}$ is a second length function of Mod_R, then

$$(L = L') \quad \Leftrightarrow \quad \big(L(R/I) = L'(R/I), \text{ for all } I \in \mathcal{L}(R)\big). \tag{2.2}$$

In particular, the collection of all length functions of Mod_R is a *set* (as opposed to a proper class).

For general rings, one cannot improve much on (2.2) (see Remark A.16), but for R commutative and Noetherian it is usually redundant to compare the values of two length on all cyclics. To describe a sharper criterion we need some notation:

- $\mathrm{Spec}(R) := \{\text{prime ideals of } R\}$ denotes the *spectrum* of R, that we view as a poset, partially ordered by inclusion;
- if $\mathfrak{p} \leq \mathfrak{q} \in \mathrm{Spec}(R)$, $\mathfrak{q}$ is a *specialization* of $\mathfrak{p}$, and $\mathfrak{p}$ a *generalization* of $\mathfrak{q}$;
- $\mathrm{Max}(R) \subseteq \mathrm{Spec}(R)$ is the *maximal spectrum* of R.
- $\mathrm{Spec}_1(R) := \{\mathfrak{p} \in \mathrm{Spec}(R) : ht(\mathfrak{p}) = 1\}$.

DEFINITION 2.6. Given $M \in \mathrm{Mod}_R$, its series σ in (2.1) is said to be:

- a *composition series* if M_i/M_{i-1} is a simple module, for all $i = 1, \ldots, n$;
- a *prime series* if $M_i/M_{i-1} \cong R/\mathfrak{p}_i$, with $\mathfrak{p}_i \in \mathrm{Spec}(R)$, for all $i = 1, \ldots, n$.

EXAMPLE 2.7. A given $M \in \mathrm{Mod}_R$ has a composition series if and only if it is both Artinian and Noetherian. In this case, by the Jordan-Holder Theorem, all composition series of M have the same length.

The *composition length* $\ell\colon \mathrm{Mod}_R \to \mathbb{R}_{\geq 0} \cup \{\infty\}$ is defined by $\ell(M) := \ell(\sigma) \in \mathbb{N}$, if M has a composition series σ, and $\ell(M) := \infty$, otherwise. Clearly, $\ell(M) = 0$ if and only if $M = 0$, $\ell(M) = 1$ precisely if M is simple, while $\ell(M) < \infty$ if and only if M is both Noetherian and Artinian. It is a standard result in module theory that ℓ is a length function (see, e.g., [**17**]).

While composition series have factors isomorphic to $R/\mathfrak{p}$, with $\mathfrak{p} \in \mathrm{Max}(R)$, for a prime series we allow all $\mathfrak{p}$ in $\mathrm{Spec}(R)$.

LEMMA 2.8. *Let R be a Noetherian ring, then any $M \in \mathrm{mod}_R$ has a prime series. In particular, if L and L' are two length functions of* Mod_R*, then:*

$$(L = L') \quad \Leftrightarrow \quad \big(L(R/\mathfrak{p}) = L'(R/\mathfrak{p}), \textit{ for all } \mathfrak{p} \in \mathrm{Spec}(R)\big).$$

PROOF. For the existence of prime series see [**37**, Lemma 00L0], while for the second statement proceed as for (2.2) but knowing that, for each $M \in \mathrm{mod}_R$, one can find a series with cyclic factors which is actually a prime series. □

2.2. A decomposition theorem for length functions. The goal of this subsection is to discuss a standard decomposition for the length functions of Mod_R. In other words, we want to find a standard way to write a given length function as a linear combination of a suitable family of "basic length functions", each associated with a prime ideal of R. One such result was proved by Vámos in [**38**] (see Theorem A.18); Theorem 2.13 below is a variation on Vámos' result, that produces sharper (i.e., less redundant) decompositions in general. We refer to §A.4 of the Appendix for an in-depth comparison of the two results.

First of all we need to build, for each $\mathfrak{p} \in \mathrm{Spec}(R)$, the corresponding "basic length function" $\ell_{\mathfrak{p}}$ mentioned in the above discussion, which will be obtained from the composition length of Example 2.7 via the following pullback construction:

LEMMA 2.9. *Let S be a ring, $F\colon \mathrm{Mod}_R \to \mathrm{Mod}_S$ a functor, and suppose that F is exact (it sends short exact sequences to short exact sequences) and that it commutes with direct limits. Then, any length function L of Mod_S induces the following length function of Mod_R, called the* pullback of L along F*:*

$$L_F\colon \mathrm{Mod}_R \to \mathbb{R}_{\geq 0} \cup \{\infty\}, \quad \textit{such that} \quad L_F(M) := L(F(M)), \textit{ for all } M \in \mathrm{Mod}_R.$$

Observe that if $F\colon \mathrm{Mod}_R \to \mathrm{Mod}_S$ is a left adjoint, then it is right-exact and it commutes with direct limits, i.e., whenever a left adjoint is left-exact it satisfies the conditions of Lemma 2.9. For example, if M is an R-S-bimodule, then $- \otimes_R M\colon \mathrm{Mod}_R \to \mathrm{Mod}_S$ is a left adjoint, which is left-exact if and only if M is flat as an R-module. In particular, for each $\mathfrak{p} \in \mathrm{Spec}(R)$, the localization $R \to R_{\mathfrak{p}}$ makes $R_{\mathfrak{p}}$ into a flat R-module, and so the functor $- \otimes_R R_{\mathfrak{p}}\colon \mathrm{Mod}_R \to \mathrm{Mod}_{R_{\mathfrak{p}}}$ satisfies the conditions of Lemma 2.9.

DEFINITION 2.10. For each $\mathfrak{p} \in \mathrm{Spec}(R)$, let $\ell_{\mathfrak{p}} := \ell \circ (- \otimes_R R_{\mathfrak{p}})$ be the pullback of the composition length $\ell\colon \mathrm{Mod}_{R_{\mathfrak{p}}} \to \mathbb{R}_{\geq 0} \cup \{\infty\}$ along $- \otimes_R R_{\mathfrak{p}}\colon \mathrm{Mod}_R \to \mathrm{Mod}_{R_{\mathfrak{p}}}$.

In the following definition we associate to a length function L of Mod_R several subsets of $\mathrm{Spec}(R)$, each reflecting some aspect of L. In particular, they determine if a given "basic length function" $\ell_{\mathfrak{p}}$ appears in each decomposition of L (both for Vámos' original decomposition (A.6), and for the ones of Theorem 2.13).

DEFINITION 2.11. Given a length function L of Mod_R, define:

$$\begin{aligned}
\mathrm{Supp}(L) &:= \{\mathfrak{p} \in \mathrm{Spec}(R) : L(R/\mathfrak{p}) > 0\},\\
\mathrm{Ass}(L) &:= \max \mathrm{Supp}(L) := \{\text{maximal elements in } \mathrm{Supp}(L)\},\\
\mathcal{Z}(L) &:= \{\mathfrak{p} \in \mathrm{Spec}(R) : L(R/\mathfrak{p}) = 0\} = \mathrm{Spec}(R) \setminus \mathrm{Supp}(L),\\
\mathrm{Fin}(L) &:= \{\mathfrak{p} \in \mathrm{Spec}(R) : L(R/\mathfrak{p}) < \infty\},\\
\mathrm{Inf}(L) &:= \{\mathfrak{p} \in \mathrm{Spec}(R) : L(R/\mathfrak{p}) = \infty\} = \mathrm{Spec}(R) \setminus \mathrm{Fin}(L).
\end{aligned}$$

Observe that both sets $\mathcal{Z}(L) \subseteq \mathrm{Fin}(L)$ are closed under specialization and, therefore, their complements $\mathrm{Supp}(L) \supseteq \mathrm{Inf}(L)$ in $\mathrm{Spec}(R)$ are closed under generalization. Since the generalization closed subsets of $\mathrm{Spec}(R)$ are precisely the generalization closures of their maximal elements, it follows that $\mathrm{Ass}(L)$ completely determines $\mathrm{Supp}(L)$ as follows: $\mathrm{Supp}(L) = \{\mathfrak{p} \in \mathrm{Spec}(R) : \mathfrak{p} \subseteq \mathfrak{q} \text{ for some } \mathfrak{q} \in \mathrm{Ass}(L)\}$. In particular, $\mathrm{Supp}(L) = \emptyset$ if and only if $\mathrm{Ass}(L) = \emptyset$; in the Noetherian case, by Lemma 2.8, this only occurs for the zero length function $L = 0$ (see §A.1).

The following result, which is [**38**, Lemma 3], uncovers the following less immediate relations between the above subsets of the spectrum: $\mathfrak{q} \in \mathcal{Z}(L)$ whenever it is a strict specialization of some $\mathfrak{p} \in \mathrm{Fin}(L)$, while $\mathfrak{p} \in \mathrm{Inf}(L)$ whenever it is a strict generalization of some $\mathfrak{q} \in \mathrm{Supp}(L)$.

LEMMA 2.12. *Let L be a length function of* Mod_R, *and take* $\mathfrak{p} \subsetneq \mathfrak{q} \in \mathrm{Spec}(R)$. *If* $L(R/\mathfrak{p}) < \infty$, *then* $L(R/\mathfrak{q}) = 0$. *Consequently, if* $L(R/\mathfrak{q}) \neq 0$, *then* $L(R/\mathfrak{p}) = \infty$.

PROOF. The short exact sequence $0 \to \mathfrak{q}/\mathfrak{p} \to R/\mathfrak{p} \to R/\mathfrak{q} \to 0$, yields $L(R/\mathfrak{p}) = L(R/\mathfrak{q}) + L(\mathfrak{q}/\mathfrak{p}) \geq L(\mathfrak{q}/\mathfrak{p})$. Moreover, as $R/\mathfrak{p}$ is a domain, its non-trivial ideal $\mathfrak{q}/\mathfrak{p}$ contains a copy of $R/\mathfrak{p}$, showing that $L(\mathfrak{q}/\mathfrak{p}) = L(R/\mathfrak{p}) < \infty$. Hence, combining these observations, $L(R/\mathfrak{q}) = L(R/\mathfrak{p}) - L(\mathfrak{q}/\mathfrak{p}) = L(R/\mathfrak{p}) - L(R/\mathfrak{p}) = 0$.

The second assertion immediately follows from the first one. □

We can finally state our main decomposition result. We refer to §A.3 of the Appendix for the proof of Theorem 2.13.

THEOREM 2.13. *Let R be a commutative Noetherian ring and let L be a length function of* Mod_R. *Then*

$$L = \sum_{\mathfrak{p} \in \mathrm{Supp}(L)} L(R/\mathfrak{p}) \cdot \ell_{\mathfrak{p}} = \sum_{\mathfrak{p} \in \mathrm{Ass}(L)} L(R/\mathfrak{p}) \cdot \ell_{\mathfrak{p}}, \tag{2.3}$$

with the convention $\infty \cdot 0 = 0$.

2.3. h-like length functions on Noetherian domains. Throughout this subsection, let D be a (commutative) Noetherian domain with field of fractions Q. Our goal is to study a particular class of length functions of Mod_D, which we call h-*like*, of which the algebraic entropy on $D := \mathbb{Z}[X_1, \ldots, X_m]$ is a prominent example.

DEFINITION 2.14. A length function L of Mod_D is called h-*like* if $\mathrm{Inf}(L) \subseteq \{0\}$ or, equivalently, if $\mathrm{Fin}(L) \supseteq \mathrm{Spec}(D) \setminus \{0\}$.

Recall that the class of *torsion* D-modules can be defined as follows:

$$\mathrm{Tor}_D := \{M \in \mathrm{Mod}_D : M \otimes_D Q = 0\}.$$

EXAMPLE 2.15. The torsion-free rank $\mathrm{rk}_D = \dim_Q \circ (- \otimes_D Q)$ (see Example A.2) is h-like, since $\mathrm{Inf}(\mathrm{rk}_D) = \emptyset$ and $\mathrm{Supp}(\mathrm{rk}_D) = \{0\}$. Moreover, if L is a length function of Mod_D such that $\mathrm{Supp}(L) = \{0\}$, then either $L(D) = \infty$ and so $L = \chi_{\mathrm{Tor}_D}$ (see Examples A.1), or $0 < L(D) < \infty$ and so $L = L(D) \cdot \mathrm{rk}_D$ by Example A.2.

EXAMPLE 2.16. Suppose that L is an h-like length function that is singular, that is, L has range contained in $\{0, \infty\}$. We have two cases:

- either $L(D) = 0$, and so $L = 0$ is the zero length function;
- or $L(D) = \infty$, and so $L = \infty \cdot \mathrm{rk}_D = \chi_{\mathrm{Tor}_D}$ (see Examples A.1 and A.2).

Observe that $L = \chi_{\mathrm{Tor}_D}$ if and only if $\mathrm{Supp}(L) = \mathrm{Inf}(L) = \{0\}$. In fact, $M \in \mathrm{Tor}_D$ if and only if for every $x \in M$, $\mathrm{Ann}(x) \neq 0$; now, $\mathrm{Supp}(L) = \{0\}$ implies $L(D/I) = 0$ for every ideal $I \neq 0$ of D, so $M \in \mathrm{Tor}_D$ if and only if $L(Dx) = L(D/\mathrm{Ann}(x)) = 0$ (as $Dx \cong D/\mathrm{Ann}(x)$) for every $x \in M$, that is, $L(M) = 0$ by Lemma 2.4.

Given $F \in \mathrm{mod}_D$, take $0 = F_0 \leq F_1 \leq \cdots \leq F_n = F$ a prime series (see Lemma 2.8), with $F_i/F_{i-1} \cong D/\mathfrak{p}_i$ and $\mathfrak{p}_i \in \mathrm{Spec}(D)$, for all $i = 1, \ldots, n$; then:

$$(F \in \mathrm{mod}_D \cap \mathrm{Tor}_D) \quad \Leftrightarrow \quad (\mathfrak{p}_i \neq 0, \text{ for all } i = 1, \ldots, n). \tag{2.4}$$

PROPOSITION 2.17. *Let L be a length function of* Mod_D. *Then, the following conditions are equivalent:*

(a) *L is* $\mathtt{h}$*-like, that is, $L(D/\mathfrak{p}) < \infty$, for all $\mathfrak{p} \in \mathrm{Spec}(D) \setminus \{0\}$;*
(b) *$\mathrm{Spec}_1(D) \subseteq \mathrm{Fin}(L)$, that is, $L(D/\mathfrak{p}) < \infty$, for all $\mathfrak{p} \in \mathrm{Spec}_1(D)$;*
(c) *$L(F) < \infty$, for all $F \in \mathrm{Tor}_D \cap \mathrm{mod}_D$.*

In particular, if L is $\mathtt{h}$*-like, then $L(D/\mathfrak{p}) = 0$ for any $\mathfrak{p} \in \mathrm{Spec}(D)$ with $\mathrm{ht}(\mathfrak{p}) > 1$, i.e., $\mathrm{Supp}(L) \subseteq \mathrm{Spec}_1(D) \cup \{0\}$.*

PROOF. (a)$\Leftrightarrow$(b) Observe that both $\mathrm{Fin}(L)$ and $\mathrm{Spec}(D) \setminus \{0\}$ are closed under specialization. Furthermore, $\mathrm{Spec}_1(D)$ is the subset of minimal elements in $\mathrm{Spec}(D) \setminus \{0\}$, which coincides with the closure of $\mathrm{Spec}_1(D)$ under specialization. Hence, $\mathrm{Spec}(D) \setminus \{0\} \subseteq \mathrm{Fin}(L)$ if and only if $\mathrm{Spec}_1(D) \subseteq \mathrm{Fin}(L)$.

(c)$\Rightarrow$(a) Use that, for $\mathfrak{p} \in \mathrm{Spec}(D)$, clearly $D/\mathfrak{p} \in \mathrm{Tor}_D$ if and only if $\mathfrak{p} \neq 0$.

(a)$\Rightarrow$(c) Any $F \in \mathrm{Tor}_D \cap \mathrm{mod}_D$ has a prime series as discussed in (2.4). Thus, $L(F) = \sum_{i=1}^n L(F_i/F_{i-1}) = \sum_{i=1}^n L(D/\mathfrak{p}_i)$, with $\mathfrak{p}_i \neq 0$, for all $i = 1, \ldots, n$. Moreover, $L(D/\mathfrak{p}_i) < \infty$, for all $i = 1, \ldots, n$ by condition (a), and so $L(F) < \infty$.

Finally, if the height of $\mathfrak{p} \in \mathrm{Spec}(D)$ is strictly bigger than 1, then there is some $\mathfrak{q} \in \mathrm{Supp}_1(D)$ such that $\mathfrak{q} \subsetneq \mathfrak{p}$. Hence, if L is $\mathtt{h}$-like, and so $L(D/\mathfrak{q}) < \infty$, then $L(D/\mathfrak{p}) = 0$ by Lemma 2.12. □

COROLLARY 2.18. *Let L and L' be length functions of* Mod_D *such that:*

(a) *L is* $\mathtt{h}$*-like;*
(b) *$L'(D/\mathfrak{p}) = L(D/\mathfrak{p})$ for all $\mathfrak{p} \in \mathrm{Spec}_1(D)$.*

Then, L' is also $\mathtt{h}$*-like. Furthermore, $L = L'$ if and only if $L(D) = L'(D)$. In particular, $L = L'$ whenever $\mathrm{Supp}(L) \cap \mathrm{Spec}_1(D) \neq \emptyset$.*

PROOF. By Proposition 2.17 and condition (a), $\mathrm{Spec}_1(D) \subseteq \mathrm{Fin}(L)$, which implies $\mathrm{Spec}_1(D) \subseteq \mathrm{Fin}(L')$ by condition (b), and so also that L' is $\mathtt{h}$-like. By Proposition 2.17, the supports of L and L' are both contained in $\mathrm{Spec}_1(D) \cup \{0\}$. In particular, $L = L'$ if and only if $L(D/\mathfrak{p}) = L'(D/\mathfrak{p})$, for all $\mathfrak{p} \in \mathrm{Spec}_1(D) \cup \{0\}$. By condition (b), this becomes equivalent to $L(D) = L'(D)$. Finally, if $\mathrm{Supp}(L) \cap \mathrm{Spec}_1(D) \neq \emptyset$, there is $\mathfrak{p} \in \mathrm{Spec}_1(D)$ such that $0 \neq L(D/\mathfrak{p}) = L'(D/\mathfrak{p}) < \infty$. By Lemma 2.12, we can conclude that $L(D) = \infty = L'(D)$. □

3. Entropy of a single endomorphism

Let A be an abelian group and let $\mathcal{F}(A)$ be the set of its non-empty finite subsets. Given an endomorphism $\phi\colon A \to A$ and $F \in \mathcal{F}(A)$, for every $n \in \mathbb{N}_+$ let

$$T_n(\phi, F) := F + \phi(F) + \ldots + \phi^{n-1}(F).$$

Fekete's subadditivity lemma [**7**] guarantees the existence of the limit

$$H(\phi, F) := \lim_{n\to\infty} \frac{\log |T_n(\phi, F)|}{n}. \tag{3.1}$$

Since obviously $|T_n(\phi, F)| \leq |F|^n$, the value $H(\phi, F)$ is always finite, bounded by $\log |F|$. Moreover, the quantity $H(\phi, F)$ measures the growth rate of the function $n \mapsto |T_n(\phi, F)|$, characterizing the dynamical behaviour of the endomorphism ϕ over the subset F. The *algebraic entropy* of ϕ is then defined as

$$h_{\mathrm{alg}}(\phi) := \sup\{H(\phi, F) : F \in \mathcal{F}(A)\}. \tag{3.2}$$

3.1. Properties of the algebraic entropy of an endomorphism of an abelian group. In the next theorem we gather the triad of most relevant basic properties of the entropy function h_{alg}. Recall that, given an endomorphism $\phi\colon A \to A$, a subgroup N of G is *ϕ-invariant* if $\phi(N) \subseteq N$.

THEOREM 3.1. *Let A be an abelian group and $\phi\colon A \to A$ an endomorphism.*

(1) [Invariance] $h_{\mathrm{alg}}(\psi^{-1}\phi\psi) = h_{\mathrm{alg}}(\phi)$ *for every abelian group G and every isomorphism $\psi\colon G \to A$.*

(2) [Continuity Axiom] *If $(H_i)_{i\in I}$ is a directed system of ϕ-invariant subgroups of A with $\bigcup_{i\in I} H_i = A$, then*

$$h_{\mathrm{alg}}(\phi) = \lim_{i\in I} h_{\mathrm{alg}}(\phi\restriction_{H_i}) = \sup_{i\in I} h_{\mathrm{alg}}(\phi\restriction_{H_i}).$$

(3) [Addition Theorem] *If N is a ϕ-invariant subgroup of A, then $h_{\mathrm{alg}}(\phi) = h_{\mathrm{alg}}(\phi\restriction_N) + h_{\mathrm{alg}}(\bar\phi_{A/N})$, where $\bar\phi_{A/N}\colon A/N \to A/N$ is the endomorphism of A/N induced by ϕ.*

While (1) and (2) are almost immediate, the proof of (3) required a lot of time and effort. It was proved first for torsion abelian groups in [**16**], while the proof of the general case turned out to be much harder and appeared much later only in [**13**]. The Addition Theorem allows for a separate study of the algebraic entropy for torsion abelian groups and for torsion-free abelian groups since for an endomorphism $\phi\colon A \to A$ of an abelian group A,

$$h_{\mathrm{alg}}(\phi) = h_{\mathrm{alg}}(\phi\restriction_{t(A)}) + h_{\mathrm{alg}}(\bar\phi_{A/t(A)}),$$

for $t(A)$ the torsion part of A (that is fully invariant in A), and $A/t(A)$ torsion-free.

In this vein of gradual reduction, the next property reduces the computation of the algebraic entropy of endomorphisms of torsion-free abelian groups to the case of endomorphisms of vector spaces over $\mathbb{Q}$.

THEOREM 3.2. *If $D(A)$ is the divisible hull of a torsion-free abelian group A, then $h_{\mathrm{alg}}(\widetilde\phi) = h_{\mathrm{alg}}(\phi)$ for any endomorphism $\phi\colon A \to A$ of A and its (unique) extension $\widetilde\phi\colon D(A) \to D(A)$.*

EXAMPLE 3.3.

(a) If A is a finite abelian group, then $h_{\mathrm{alg}}(\phi) = 0$ for every endomorphism $\phi\colon A \to A$ since $|T_n(\phi, F)| \leq |A|$ for every $F \in \mathcal{F}(A)$ and $n \in \mathbb{N}$.

(b) $h_{\mathrm{alg}}(\mathrm{id}_A) = 0$ for every abelian group A. This follows from the fact that the function $n \mapsto |T_n(\phi, F)|$ has polynomial growth, hence $\lim_{n\to\infty} \frac{1}{n} \log |T_n(\phi, F)| = 0$.

(c) Let p be a prime and $A = \bigoplus_{\mathbb{N}} \mathbb{Z}(p)$ be the countable vector space over the finite field $\mathbb{Z}/p\mathbb{Z}$. The *right Bernoulli shift* $\sigma_p\colon A \to A$ has $h_{\mathrm{alg}}(\sigma_p) = \log p$, where $\sigma_p\colon (x_0, x_1, \ldots) \mapsto (0, x_0, x_1, \ldots)$. Indeed, for any finite subset $F \subseteq A$ there is m in $\mathbb{N}_+$ such that $F \leq F_m := \bigoplus_{i=0}^{m-1} \mathbb{Z}(p) \leq A$. Since $T_n(\sigma_p, F_m) = F_{m+n-1}$ for every $n \in \mathbb{N}_+$ and $\log |F_m| = m \log p$, one gets $h_{\mathrm{alg}}(\sigma_p) = \lim_{n\to\infty} \frac{(m+n-1)}{n} \log p = \log p$.

3.2. The Mahler measure and the Algebraic Yuzvinski Formula. The next property of h_{alg} needs a new ingredient, namely the Mahler measure, that we recall here. The Mahler measure was defined independently by Lehmer [**24**] and Mahler [**26**] in two different equivalent forms.

DEFINITION 3.4. For a polynomial $f(X) = a_0X^n + a_1X^{n-1} + \dots + a_n \in \mathbb{C}[X]$ and its complex roots $\lambda_1, \dots, \lambda_n$, the *Mahler measure* of f is

$$\mathsf{M}(f) := |a_0| \cdot \prod_{|\lambda_i|\geq 1} |\lambda_i|.$$

The *logarithmic Mahler measure* is $\mathsf{m}(f) := \log \mathsf{M}(f) = \log|a_0| + \sum_{|\lambda_i|\geq 1} \log|\lambda_i|$.

REMARK 3.5. We follow the convention that the product of an empty family of real numbers is equal to 1. In particular, if $f(X) = a_0X^n + a_1X^{n-1} + \dots + a_n \in \mathbb{C}[X]$ is a polynomial of degree 0 or of degree $n > 0$ and with complex roots $\lambda_1, \dots, \lambda_n$ such that $|\lambda_i| < 1$ for all $i = 1, \dots, n$, we have $\mathsf{M}(f) = |a_0|$ and $\mathsf{m}(f) = \log|a_0|$.
In the sequel $f(X) \in \mathbb{Z}[X]$, so this occurs only when $f(X) = a_0X^n$.

Let $\Phi_n(X)$ be the nth cyclotomic polynomial, i.e., the irreducible monic polynomial having as roots all the primitive n^{th} roots of unity. Obviously, $\mathsf{m}(\Phi_n) = 0$, since Φ_n is monic. We shall see below that this property can be partially inverted (see Theorem 3.6).

First of all, let us recall *Jensen's formula* [**20**] (see also [**2**]): if $\alpha \neq 0$ then

$$\int_0^1 \log|e^{2\pi it} - \alpha|dt = \max\{0, \log|\alpha|\}$$

Using this, one can easily deduce the following remarkable and useful equivalent form of the Mahler measure: for $f \in \mathbb{C}[X]$,

$$\mathsf{m}(f) = \int_0^1 \log|f(e^{2\pi it})|dt. \tag{3.3}$$

This, in turn, entails that the Mahler measure M is multiplicative (i.e., $\mathsf{M}(fg) = \mathsf{M}(f)\mathsf{M}(g)$ for $f, g \in \mathbb{C}[X]$), and that it satisfies the following theorem of Kronecker, which describes when the Mahler measure vanishes:

THEOREM 3.6 (Kronecker [**23**]). *If $f(X)$ is an irreducible monic integer polynomial then $\mathsf{m}(f) = 0$ if and only if $f(X) = X$ or $f(X)$ is a cyclotomic polynomial.*

We can extend the definition of Mahler measure to endomorphisms of $\mathbb{Q}^n$. Indeed, any endomorphism $\phi\colon \mathbb{Q}^n \to \mathbb{Q}^n$ can be represented by an $n \times n$ matrix over $\mathbb{Q}$, which has a primitive characteristic polynomial f_ϕ over $\mathbb{Z}$. We define the (logarithmic) Mahler measure of ϕ as $\mathsf{M}(\phi) := \mathsf{M}(f_\phi)$ (resp., $\mathsf{m}(\phi) := \mathsf{m}(f_\phi)$).

The following theorem, inspired by Yuzvinski's formula for the topological entropy [**41**], was proved in [**18**]. The case of automorphisms of $\mathbb{Z}^n$ (which easily follows from Theorem 3.7) was settled earlier in [**39**].

THEOREM 3.7 (Algebraic Yuzvinski Formula [**18**]). *$h_{\text{alg}}(\phi) = \mathsf{m}(\phi)$ for every endomorphism ϕ of $\mathbb{Q}^n$ and $n \in \mathbb{N}_+$.*

EXAMPLE 3.8. Applying the above theorem one can immediately see that for coprime integers $a, b > 0$ the endomorphism $\phi\colon \mathbb{Q} \to \mathbb{Q}$, $x \mapsto \frac{a}{b}x$, has $h_{\text{alg}}(\phi) = \max\{\log a, \log b\}$ (this can also be computed directly). Therefore, if $\phi\colon \mathbb{Z} \to \mathbb{Z}$ is the multiplication by a, then $h_{\text{alg}}(\phi) = \log a$. Indeed $D(\mathbb{Z}) = \mathbb{Q}$ and by Theorem 3.2, $h_{\text{alg}}(\phi) = h_{\text{alg}}(\widetilde{\phi})$ for the extension $\widetilde{\phi}\colon \mathbb{Q} \to \mathbb{Q}$ which is again the multiplication by a, so that $h_{\text{alg}}(\widetilde{\phi}) = \log a$.

3.3. Uniqueness of the algebraic entropy and its connection to the topological entropy. The properties given in the previous section completely determine the algebraic entropy as a function defined on endomorphisms of abelian groups:

THEOREM 3.9 ([**13**]). *Every function h' defined on all endomorphisms of abelian groups and satisfying:*

(a) *the Invariance property;*
(b) *the Continuity Axiom;*
(c) *the Addition Theorem;*
(d) $h'(\sigma_p) = \log p$ *for every prime* p *(Bernoulli normalization);*
(e) *the Algebraic Yuzvinski Formula;*

must coincide with the algebraic entropy h_{alg}.

Two proofs of this theorem can be found in [**13**]. A modification of the second one, using length functions, is given in §3.4.

In order to reveal the connection of h_{alg} to the topological entropy, first we have to introduce the notion of topological entropy $h_{\mathrm{top}}(\psi)$ for a continuous endomorphism $\psi\colon K \to K$ of a compact group K endowed with its Haar measure μ. The original definition in [**1**] was tailored for arbitrary self-maps of topological spaces.

Let $\mathcal{C}(K)$ be the family of all compact neighborhoods of 1 in K. For $n \in \mathbb{N}_+$, the nth *ψ-cotrajectory* of a fixed $U \in \mathcal{C}(K)$ is

$$C_n(\psi, U) := U \cap \psi^{-1}(U) \cap \cdots \cap \psi^{-n+1}(U).$$

Given a continuous endomorphism $\psi\colon K \to K$ and $U \in \mathcal{C}(K)$,

$$H_{\mathrm{top}}(\psi, U) := \limsup_{n\to\infty} -\frac{\log \mu(C_n(\psi, U))}{n},$$

and finally the *topological entropy* $h_{\mathrm{top}}(\psi)$ is

$$h_{\mathrm{top}}(\psi) := \sup\{H_{\mathrm{top}}(\psi, U) : U \in \mathcal{C}(K)\}.$$

REMARK 3.10. We recall that a *measure space* is a triple $(X, \mathfrak{B}, \mu)$, where X is a set, $\mathfrak{B}$ is a σ-algebra over X (the elements of $\mathfrak{B}$ are called *measurable sets*) and $\mu\colon \mathfrak{B} \to \mathbb{R}_{\geq 0} \cup \{\infty\}$ is a measure. Moreover $(X, \mathfrak{B}, \mu)$ is called *probability space* if $\mu(X) = 1$. When there is no possibility of confusion we denote $(X, \mathfrak{B}, \mu)$ simply by X. The selfmap $\psi\colon X \to X$ is a *measure preserving trasformation* if $\mu(\psi^{-1}(B)) = \mu(B)$ for every $B \in \mathfrak{B}$. For a measure space X and a measure preserving transformation $\psi\colon X \to X$ Kolmogorov [**22**] and Sinai [**34**] defined the *measure entropy* $h_{mes}(\psi)$.

Every compact group K equipped with its Haar measure is a probability space. Making use of the uniqueness of the Haar measure, Halmos [**19**] proved that if $\psi\colon K \to K$ is a continuous endomorphism of a compact group K, then ψ is measure preserving precisely when ψ is surjective. So, in this case one can compute both the topological entropy and the measure entropy of ψ. As an application of his Uniqueness Theorem for the topological entropy, Stoyanov [**36**] proved that the topological entropy and the measure entropy of ψ coincide.

In the sequel $\mathbb{T}$ is the compact group of complex numbers of modulus 1. For an abelian group A, considered as a discrete group, the *Pontryagin dual* $\widehat{A}$ is the group $\mathrm{Hom}(A, \mathbb{T})$ of all homomorphisms $A \to \mathbb{T}$ (called *characters* of A) endowed with the topology induced by the compact group $\mathbb{T}^A$ equipped with the compact product topology. Since $\mathrm{Hom}(A,\mathbb{T})$ is a closed subgroup of $\mathbb{T}^A$, the dual $\widehat{A}$ of A is compact. (Usually, the Pontryagin dual $\widehat{A}$ is defined for arbitrary locally compact abelian groups A, $\widehat{A}$ is still locally compact and $\widehat{\widehat{A}}\cong A$; this topological isomorphism, known as Pontryagin-van Kampen duality theorem, is natural. However, this level of generality it is not needed in this paper, so we prefer to define $\widehat{A}$ only for discrete abelian groups A.)

For an endomorphism $\phi\colon A\to A$, the dual endomorphism $\widehat{\phi}\colon \widehat{A}\to\widehat{A}$, defined by

$$\widehat{\phi}\colon \widehat{A}\to\widehat{A},\quad \chi\mapsto\chi\circ\phi \text{ for } \chi\in\widehat{A},$$

is continuous. Therefore, $h_{\mathrm{top}}(\widehat{\phi})$ is defined and it is natural to ask if there is any connection between $h_{\mathrm{top}}(\widehat{\phi})$ and $h_{\mathrm{alg}}(\phi)$. Some initial answers were provided in the seventies of the last century [**29, 40**]:

THEOREM 3.11. *If $\phi\colon A\to A$ is an endomorphism of an abelian group A, then $h_{\mathrm{alg}}(\phi)=h_{\mathrm{top}}(\widehat{\phi})$ in the following two cases:*

(a) (Weiss [**40**]) *when A is torsion;*

(b) (Peters [**29**]) *when A is countable and ϕ is an automorphism.*

Much later all three restraints (torsion, countable and automorphism) in the above theorem were simultaneously removed:

THEOREM 3.12 (Bridge Theorem [**11**]). *If $\phi\colon A\to A$ is an endomorphism of an abelian group A, then $h_{\mathrm{alg}}(\phi)=h_{\mathrm{top}}(\widehat{\phi})$.*

The proof of this theorem makes use of all five properties in Theorem 3.9. A much easier alternative proof was obtained in [**13**] using the Uniqueness Theorem (Theorem 3.9) and the Algebraic Yuzvinski Formula (Theorem 3.7).

3.4. An alternative proof of the Uniqueness Theorem using length functions. Following Kaplansky [**21**], there is an equivalence between abelian groups A provided with an endomorphism and $\mathbb{Z}[X]$-modules A, namely, an endomorphism of an abelian group A can be viewed in a standard way as a $\mathbb{Z}[X]$-module structure on A, putting $X\cdot a:=\phi(a)$ for every $a\in A$.

Thus, given a function h' defined on all endomorphisms of abelian groups that satisfies the Invariance property, the Addition Theorem (to give additivity) and the Continuity Axiom (to get the upper continuity as a length function), and given $A\in\mathrm{Mod}_{\mathbb{Z}[X]}$, we can set $h'_0(A)=h'(\phi)$ where $\phi\colon A\to A$ is the multiplication by X; then, h'_0 is a length function of $\mathbb{Z}[X]$. For $h'=h_{\mathrm{alg}}$, we prefer to briefly use $\mathtt{h}$ in place of $(h_{\mathrm{alg}})_0$, i.e., $\mathtt{h}(A)=h_{\mathrm{alg}}(\phi)$.

Applying Theorem 2.13 to $R=\mathbb{Z}[X]$, we obtain an alternative proof of the uniqueness of $\mathtt{h}$, namely, of Theorem 3.9. One can also prove the same result by applying Corollary 2.18 as in the proof of Corollary B. The proof that we present at the end of this subsection has the advantage of combining dynamical aspects with properties of length functions.

Before proceeding further, we need some results on length functions.

REMARK 3.13. Let R be a commutative ring and I an ideal of R. The subcategory $\mathcal{C}_I$ of Mod_R formed by the modules whose annihilator contains I is isomorphic to the category of R/I-modules; thus, the restriction of a length function L of Mod_R to $\mathcal{C}_I$ is just a length function of $\mathrm{Mod}_{R/I}$.

LEMMA 3.14. *Let R be a commutative Noetherian ring, I an ideal of R and let L be a length function of* Mod_R. *Let S be a multiplicatively closed subset of R and let $\lambda\colon R \to S^{-1}R$ be the localization map. If $I = \lambda^{-1}(S^{-1}I)$, then $L(R/I) = L(S^{-1}R/S^{-1}I)$.*

PROOF. As a first step, let us suppose that $I = 0$. In such a case we have

$$0 = \lambda^{-1}(S^{-1}0) = \ker(\lambda)$$

so that λ is injective and thus we can consider R as a subring of $S^{-1}R$. Now, S, endowed with the divisibility order is a directed set and $S^{-1}R$ is the direct limit of the family $\{s^{-1}R\}_{s\in S}$, so that, by Lemma 2.4, $L(S^{-1}R) = \sup\{L(s^{-1}R) : s \in R\}$. But, for every $s \in S$, $s^{-1}R$ is isomorphic to R and so $L(s^{-1}R) = L(R)$.

If $I \neq 0$, we can apply Remark 3.13 and the previous argument to the induced map $\bar{\lambda}\colon R/I \to S^{-1}R/S^{-1}I$ in order to conclude. □

COROLLARY 3.15. *Let R be a commutative ring, $\mathfrak{p} \in \mathrm{Spec}(R)$ and S a multiplicatively closed set with $S \cap \mathfrak{p} = \emptyset$. Then, $L(R/\mathfrak{p}) = L(S^{-1}R/S^{-1}\mathfrak{p}) = L(R_\mathfrak{p}/\mathfrak{p}R_\mathfrak{p})$.*

PROOF. The fact that $L(R/\mathfrak{p}) = L(S^{-1}R/S^{-1}\mathfrak{p})$ follows from Lemma 3.14. The fact that $L(R/\mathfrak{p}) = L(R_\mathfrak{p}/\mathfrak{p}R_\mathfrak{p})$ follows taking $S = R \setminus \mathfrak{p}$. □

COROLLARY 3.16. *Let $R = \mathbb{Z}[X_1, \dots, X_m]$.*

(a) *If I is an ideal of R with $I \cap \mathbb{Z} = (0)$, then $L(R/I) = L(R'/IR')$, where $R' = \mathbb{Q}[X_1, \dots, X_m]$.*
(b) *If $\mathfrak{p}$ is a prime ideal of R with $a := X_1 \dots X_m \notin \mathfrak{p}$ and R_a is the ring R localized at a (i.e., $R_a = S^{-1}R$ with $S = \{1, a, a^2, \dots\}$), then $L(R/\mathfrak{p}) = L(R_a/\mathfrak{p}R_a)$.*

PROOF. (a) If $I \cap \mathbb{Z} = 0$, then I is disjoint from the multiplicatively closed set $\mathbb{Z} \setminus \{0\}$. It follows that $IR' \cap R = I$, and the claim follows from Lemma 3.14.

(b) As $\mathfrak{p}R_a \cap R = \mathfrak{p}$ (since $\mathfrak{p}$ is disjoint from the multiplicatively closed set S), Lemma 3.14 applies again. □

AN ALTERNATIVE PROOF OF THEOREM 3.9. Let h' be a function satisfying the properties of Theorem 3.9. By the discussion at the beginning of this subsection, the first three properties of the theorem allow us to consider h' as a length function of $\mathbb{Z}[X]$ (we are conflating h' with what we denoted above by h'_0); we claim that the two scaling properties (Bernoulli normalization and the Algebraic Yuzvinski Formula) uniquely determine h'. To do so, we show that they uniquely determine $h'(\mathbb{Z}[X]/\mathfrak{p})$ for every prime ideal $\mathfrak{p}$ of $\mathbb{Z}[X]$.

Suppose first that $\mathfrak{p} = p\mathbb{Z}[X]$ for some prime number p. Then, $R/\mathfrak{p} \cong \mathbb{Z}(p)^{(\mathbb{N})}$ and the induced endomorphism is the Bernoulli shift σ_p. Hence, $h'(\mathbb{Z}[X]/\mathfrak{p}) = \log p$ by Bernoulli normalization.

If $\mathfrak{p}$ is a maximal ideal of $\mathbb{Z}[X]$, then $\mathfrak{p}$ has height 2 and contains a prime number p, and thus the prime ideal $p\mathbb{Z}[X]$; by Lemma 2.12, $h'(\mathbb{Z}[X]/\mathfrak{p}) = 0$ since $h'(\mathbb{Z}[X]/p\mathbb{Z}[X])$ is finite and nonzero. The same Lemma 2.12 implies that if $\mathfrak{p} = 0$ then $h'(\mathbb{Z}[X]/\mathfrak{p}) = \infty$.

It remains to consider the case when $\mathfrak{p}$ is a prime ideal of height 1 and $\mathfrak{p} \neq p\mathbb{Z}[X]$ for every prime number p. Since $\mathbb{Z}[X]$ is a unique factorization domain, $\mathfrak{p} = (f)$ for some irreducible polynomial $f \in \mathbb{Z}[X]$ of degree $n > 0$. Since $\mathfrak{p} \cap \mathbb{Z} = 0$, we can apply Corollary 3.16 to obtain $h'(\mathbb{Z}[X]/\mathfrak{p}) = h'(\mathbb{Q}[X]/(f))$. Then, $\mathbb{Q}[X]/(f) \simeq \mathbb{Q}[\theta]$, where θ is any root of f, and the endomorphism ϕ induced by the multiplication by X is just the multiplication by θ. As $\mathbb{Q}[\theta]\cong\mathbb{Q}^n$ as abelian groups, one obtains a linear map $\phi\colon \mathbb{Q}^n\to\mathbb{Q}^n$ with minimal polynomial f. Therefore, the Algebraic Yuzvinski Formula gives $h'(\mathbb{Q}[\theta])=\mathsf{m}(f)$. Thus $h'(\mathbb{Z}[X]/\mathfrak{p})=\mathsf{m}(f)$.

Therefore, $h'=\mathtt{h}$ by Theorem 2.13. □

The above argument gives a proof of Theorem 1 for $m=1$. Moreover, it shows that h' satisfying the assumptions of Theorem 3.9 is an $\mathtt{h}$-like length function of $\mathrm{Mod}_{\mathbb{Z}[X]}$; in particular, $\mathtt{h}$ is itself an $\mathtt{h}$-like length function of $\mathrm{Mod}_{\mathbb{Z}[X]}$.

4. Higher-dimensional entropy, Ore localizations and Mahler measure

This final section is focused on *higher dimensional entropy*, i.e., the entropy of an m-tuple of commuting endomorphisms or, equivalently, that of an action of $\mathbb{N}^m$. The main results of this section are the promised proofs of Theorems 1, 2, 3 and Corollary B.

4.1. The algebraic entropy of an m-tuple of commuting endomorphisms of an abelian group. Ceccherini-Silberstein, Coornaert and Krieger [**7**] defined the topological entropy h_{top} of actions of cancellative amenable semigroups S on compact spaces. Following the same line, the second and third named authors and Fornasiero [**9**] defined the algebraic entropy h_{alg} of actions of cancellative amenable semigroups S on abelian groups, extending the case of $\mathbb{N}$-actions (single endomorphisms).

The notion of amenable semigroup relies on that of Følner nets $\{F_i\}_{i\in I}$ of finite subsets of S (see [**8**] for the general definition). Since we are interested in the case when $S=\mathbb{N}^m$, below we use the Følner sequence $\{[0,n]^m\}_{n\in\mathbb{N}}$ of $\mathbb{N}^m$. Moreover, we recall the definition of algebraic entropy for cancellative commutative semigroups S (these are amenable and so admit Følner nets $\{F_i\}_{i\in I}$) and actions $S\overset{\alpha}{\curvearrowright}A$ with A an abelian group.

For a non-empty subset X of A and for every $F\in\mathcal{F}(S)$, the *α-trajectory of X with respect to F* is

$$T_F(\alpha,X):=\sum_{s\in F}\alpha(s)(X).$$

For $X\in\mathcal{F}(A)$, $T_F(\alpha,X)$ is finite as well for every $F\in\mathcal{F}(S)$, and the *algebraic entropy of α with respect to X* is

$$H(\alpha,X):=\lim_{i\in I}\frac{\log|T_{F_i}(\alpha,X)|}{|F_i|},\tag{4.1}$$

where $\{F_i\}_{i\in I}$ is a Følner net of S. The *algebraic entropy of α* is

$$h_{\mathrm{alg}}(\alpha):=\sup\{H(\alpha,X):X\in\mathcal{F}(A)\}.$$

The limit (4.1) defining $H(\alpha,X)$ exists and its value does not depend on the choice of the Følner net $\{F_i\}_{i\in I}$ in view of a well known fact due to Ornstein and Weiss [**28**]. Moreover, for every $X\in\mathcal{F}(A)$, $H(\alpha,X)\le\log|X|$ is bounded.

We apply the above definition to $S=\mathbb{N}^m$ with Følner sequence $\{F_n\}_{n\in\mathbb{N}}$ given by $F_n=[0,n]^m$, for $n\in\mathbb{N}$:

DEFINITION 4.1. Let $m \in \mathbb{N}_+$, A be an abelian group and $\phi_1, \ldots, \phi_m$ commuting endomorphisms of A. The *algebraic entropy* of this m-tuple ϕ is

$$h_{\mathrm{alg}}(\phi) := \sup_{F \in \mathcal{F}(A)} \lim_{n\to\infty} \frac{1}{(n+1)^m} \log \left| \sum_{k_1=0}^{n} \cdots \sum_{k_m=0}^{n} (\phi^{k_1} \circ \ldots \circ \phi^{k_m})(F) \right|.$$

According to the already mentioned Kaplansky paradigm, an $\mathbb{N}^m$-action on an abelian group A can be viewed in a standard way as an R-module structure on A, where $R = \mathbb{Z}[X_1, \ldots, X_m]$; in other words, every m-tuple $\phi_1, \ldots, \phi_m$ of pairwise commuting endomorphisms of A defines a structure of an R-module on A. Viceversa, every R-module A has an m-tuple of pairwise commuting endomorphisms given by the multiplication by X_i, $i = 1, \ldots, m$.

As in the case $m = 1$, one can prove that h_{alg} has the same properties (the Addition Theorem proved in this case in [**10**]) which ensure that it induces a length function of Mod_R, that we denote by $\mathtt{h}$ in the sequel. Since R is Noetherian, one may limit the computation of $\mathtt{h}$ to the case of cyclic R-modules $R/\mathfrak{p}$, with $\mathfrak{p} \in \mathrm{Spec}(R)$ by Theorem 2.13.

EXAMPLE 4.2. For $R = \mathbb{Z}[X_1, \ldots, X_m]$ and $\mathfrak{p} = (X_m)$ we show that $\mathtt{h}(R/\mathfrak{p}) = 0$.

With $k, n \in \mathbb{N}_+$, $\alpha = (a_1, \ldots, a_{m-1}) \in [0,n]^{m-1}$, and $\beta\colon [0,n]^{m-1} \to [-k,k]$ (where we briefly write b_α in place of $\beta(\alpha)$ for every $\alpha \in [0,n]^{m-1}$), let

$$f_\beta(X_1, \ldots, X_{m-1}) = \sum_{\alpha \in [0,k]^{m-1}} b_\alpha X_1^{a_1} \ldots X_{m-1}^{a_{m-1}} \quad \text{and}$$

$$Y_k = \{ f_\beta(X_1, \ldots, X_{m-1}) + \mathfrak{p} : \beta \in [-k,k]^{[0,k]^{m-1}} \} \subseteq R/\mathfrak{p},$$

so that $T_{F_n}(\sigma, Y_k) = \sum_{\alpha \in [0,n]^{m-1}} X_1^{a_1} \ldots X_{m-1}^{a_{m-1}} Y_k \subseteq Y_{k+n}$, where $F_n = [0,n]^m$. Indeed, $X_n \cdot R/\mathfrak{p} = 0$, hence only summands of the form $z \cdot Y_k$, with $z = X_1^{a_1} \ldots X_{m-1}^{a_{m-1}}$ remain in $T_{F_n}(\sigma, Y_k)$ and their sum is obviously contained in Y_{k+n}. So,

$$|F_n| = (n+1)^m \text{ and } \log |T_{F_n}(\sigma, Y_n)| \leq \log |Y_{n+k}| \leq (n+k+1)^{m-1} \log(2n+2k+1).$$

Therefore,

$$H(\sigma, Y_k) = \lim_{n\to\infty} \frac{\log |T_{F_n}(\sigma, Y_k)|}{|F_n|} \leq \lim_{n\to\infty} \frac{(n+k+1)^{m-1} \log(2n+2k+1)}{(n+1)^m} = 0.$$

Since this holds true for every $k \in \mathbb{N}_+$ and since every finite subset of $R/\mathfrak{p}$ is contained in some Y_k, we deduce that $\mathtt{h}(R/\mathfrak{p}) = h_{\mathrm{alg}}(\sigma) = 0$.

4.2. The Bridge Theorem for actions of commutative cancellative monoids S and Ore localizations. Let S be a cancellative commutative semigroup and A an abelian group, and consider the action $S \overset{\alpha}{\curvearrowright} A$. Then α induces an action $S \overset{\widehat{\alpha}}{\curvearrowright} \widehat{A}$, defined by $\widehat{\alpha}(s) = \widehat{\alpha(s)}\colon \widehat{A} \to \widehat{A}$ for every $s \in S$. Recently, a Bridge Theorem was established in the setting of amenable monoid actions:

THEOREM 4.3 (Bridge Theorem [**15**]). *If $S \overset{\alpha}{\curvearrowright} A$ is an action of a cancellative commutative monoid S on an abelian group A and $\widehat{\alpha}$ is the dual action of S on the compact Pontryagin dual of A, then $h_{\mathrm{top}}(\widehat{\alpha}) = h_{\mathrm{alg}}(\alpha)$.*

And also an Addition Theorem:

THEOREM 4.4 (Addition Theorem [**15**]). *If $S \overset{\alpha}{\curvearrowright} A$ is an action of a cancellative right amenable monoid S on an abelian group A and N is a α-invariant subgroup* *of A (i.e., $\alpha(s)(N) \subseteq N$ for every $s \in S$), then $h_{\mathrm{alg}}(\alpha) = h_{\mathrm{alg}}(\alpha_{\restriction N}) + h_{\mathrm{alg}}(\overline{\alpha})$ where $S \overset{\overline{\alpha}}{\curvearrowright} A/N$ is the action induced by α on A/N.*

What follows holds in the general case of cancellative amenable monoids (see [**15**]), but we present it here in the more familiar context of commutative ones, as we apply it to $\mathbb{N}^m$.

Commutative cancellative monoids S have a group of fractions G. Every action $G \overset{\alpha}{\curvearrowright} A$ produces, via restriction, an action $S \overset{\alpha_{\restriction S}}{\curvearrowright} A$ with $h_{\mathrm{alg}}(\alpha_{\restriction S}) = h_{\mathrm{alg}}(\alpha)$. Moreover, as proved in [**15**], every action $S \overset{\alpha}{\curvearrowright} A$, can be extended to an action $G \overset{\alpha^*}{\curvearrowright} A^*$ (named *Ore localization of* α), where $A^* = A \otimes_{\mathbb{Z}[S]} \mathbb{Z}[G]$.

THEOREM 4.5 ([**15**]). *Let S be a cancellative commutative monoid and G its group of fractions. Let $S \overset{\alpha}{\curvearrowright} A$ be an action of S on an abelian group A. Then $h_{\mathrm{alg}}(\alpha^*) = h_{\mathrm{alg}}(\alpha)$ in the above notation.*

This localization procedure will be applied here only for $S = \mathbb{N}^m$, with group of fractions $G = \mathbb{Z}^m$. Now a G-action on an abelian group A is an assignment of an m-tuple of commuting automorphisms of A, so this produces a specific structure of an R^Q-module on A, where $R^Q = \mathbb{Z}[X_1^{\pm 1}, \ldots, X_m^{\pm 1}]$ is the ring of Laurent polynomials. On the other hand, an R^Q-module M has its standard action $\mathbb{Z}^m \overset{\varsigma}{\curvearrowright} M$ obtained by multiplication on M by each of the m variables. After this identification, the entropy function h_{alg} defined on $\mathbb{Z}^m$-actions becomes a length function of Mod_{R^Q}.

Furthermore, for $M \in \mathrm{Mod}_{\mathbb{Z}[X_1,\ldots,X_m]}$, $\mathtt{h}(M) = \mathtt{h}(M \otimes R^Q)$. It is worth paying some attention to the case when $M \otimes R^Q$ vanishes, e.g., when $M = R^Q/\mathfrak{p}$ and $\mathfrak{p} = (X_i)$ for some $i = 1, \ldots, m$. But as shown in Example 4.2, in this case also $\mathtt{h}(M)$ vanishes, so the equality $\mathtt{h}(M) = \mathtt{h}(M \otimes R^Q)$ remains true also in this case.

4.3. Higher-dimensional Mahler measure and Boyd's theorem. As we recalled above in (3.3), according to Jensen's formula, for $f \in \mathbb{C}[X]$

$$\mathsf{M}(f) = \exp\left(\int_0^1 \log|f(e^{2\pi i t})|dt\right).$$

This form of the Mahler measure allows for the definition of the higher-dimensional Mahler measure for $f(X_1, \ldots, X_m) \in \mathbb{C}[X_1, \ldots, X_m]$ by putting

$$\mathsf{M}_m(f) = \exp\left(\int_0^1 \cdots \int_0^1 \log|f(e^{2\pi i t_1}, \ldots, e^{2\pi i t_m})|dt_1 \ldots dt_m\right) \tag{4.2}$$

and $\mathsf{m}_m(f) = \log \mathsf{M}_m(f)$.

When $f(X_1, \ldots, X_m) = f(X_i)$ depends only on one of the variables, then $\mathsf{M}_m(f) = \mathsf{M}(f)$ and $\mathsf{m}_m(f) = \mathsf{m}(f)$. (In particular, $\mathsf{m}_m(X_i - p) = \mathsf{m}(X_i - p) = \log p$.) In the sequel, we omit the index m in M_m and m_m when no confusion is possible. On the other hand, when $f(X_1, \ldots, X_m) = X_1^{k_1} \ldots X_m^{k_m}$ is a monomial, then obviously $\mathsf{M}(f) = 1$, so $\mathsf{m}(f) = 0$.

EXAMPLE 4.6. The following examples (due to Smyth – see [**6**, Appendix 1]) show that $\mathsf{m}(f) > 0$ may occur "frequently".

(a) $\mathsf{m}(1 + X_1 + X_2) = \frac{3\sqrt{3}}{4\pi} \sum_{n=1}^{\infty} \left(\frac{n}{3}\right) \frac{1}{n^2}$, where $\left(\frac{n}{3}\right)$ is the Legendre symbol.

(b) $\mathsf{m}(1 + X_1 + X_2 + X_3) = \frac{7\zeta(3)}{4\pi^2}$, where ζ is the Riemann zeta function.

In both cases $\mathsf{m}(f) = \log \mathsf{M}(f) > 0$.

The next step is to clarify when, in the spirit of Kronecker's theorem in the one-dimensional case (Theorem 3.6), $\mathsf{m}(f) = 0$. The following notion, introduced by Boyd, is closely related to the vanishing of Mahler measure. Let us recall that $\Phi_n(X)$ in $\mathbb{Z}[X]$ is the nth cyclotomic polynomial, $n \in \mathbb{N}_+$.

DEFINITION 4.7. An *extended cyclotomic polynomial* in $R = \mathbb{Z}[X_1, \ldots, X_m]$ is a polynomial of the form

$$f(X_1, \ldots, X_m) = X_1^{k_1} \ldots, X_m^{k_m} \Phi_l(X_1^{s_1} \ldots X_m^{s_m}), \tag{4.3}$$

where $m \in \mathbb{N}_+$, $k_1, \ldots, k_m, s_1, \ldots, s_m \in \mathbb{Z}$ and $n \in \mathbb{N}_+$.

EXAMPLE 4.8. Easy examples of extended cyclotomic polynomials are:

(a) $X_1^m - X_2^k = X_2^k(X_2^m X_2^{-k} - 1) = X_2^k \Phi_1(X_1^m X_2^{-k})$; and

(b) $X_1^2 + X_1X_2 + X_2^2 = X_2^2((X_1X_2^{-1})^2 + X_1X_2^{-1} + 1) = X_2^2\Phi_6(X_1X_2^{-1})$.

THEOREM 4.9 (Boyd 1981). *Let* $f \in \mathbb{C}[X_1, \ldots, X_m]$. *Then* $\mathsf{m}(f) = 0$ *if and only if* f *is an extended cyclotomic polynomial.*

In particular, $\mathsf{m}(f) = 0$ for the polynomials from Example 4.8.

Following the notation of §4.2, $R^Q = \mathbb{Z}[X_1^{\pm 1}, \ldots, X_m^{\pm 1}]$ will be the ring of Laurent polynomials. For a cyclic module $M = R^Q/\mathfrak{f}$ with respect to a principal ideal $\mathfrak{f} = (f)$ of R^Q, Lind, Schmidt and Ward [**25**] proved the following fundamental fact.

THEOREM 4.10. *For* $f \in R^Q$, $\mathfrak{f} = (f)$ *and the standard action*

$$\mathbb{Z}^m \overset{\varsigma}{\curvearrowright} R^Q/\mathfrak{f}, \tag{4.4}$$

the dual action $\widehat{\varsigma}$ *of* $\mathbb{Z}^m$ *on* $\widehat{R^Q/\mathfrak{f}}$ *has* $h_{\text{top}}(\widehat{\varsigma}) = \mathsf{m}(f)$.

Combining this theorem with the Bridge Theorem 4.3, ensuring $h_{\text{alg}}(\varsigma) = h_{\text{top}}(\widehat{\varsigma})$, we deduce:

COROLLARY 4.11. *For* $0 \neq f \in R^Q$, $\mathfrak{f} = (f)$ *and the standard action* (4.4), *one has* $h_{\text{alg}}(\varsigma) = \mathsf{m}(f)$.

4.4. Proofs of Theorems 1, 2, 3 and Corollary B.

REMARK 4.12. We are interested in establishing the equality $\mathtt{h}(R/\mathfrak{p}) = \mathsf{m}(f)$, where $R = \mathbb{Z}[X_1, \ldots, X_m]$ and $\mathfrak{p} = (f) \in \operatorname{Spec}(R)$, and Corollary 3.16 will be one of the essential steps in doing so (see Theorem 1 and its proof below). But Corollary 3.16 establishes in this setting that $\mathtt{h}(R/\mathfrak{p}) = \mathtt{h}(R_a/\mathfrak{p}R_a)$, where R_a is the ring R localized at a (i.e., $R_a = S^{-1}R$ with $S = \{1, a, a^2, \ldots\}$), in case $\mathfrak{p}$ is a prime ideal of R not containing $a := X_1 \ldots X_m$.

We need to know that $\mathtt{h}(R/\mathfrak{p}) = \mathsf{m}(f)$ also for those prime ideals $\mathfrak{p} = (f)$ with $X_1 \ldots X_m \in \mathfrak{p}$, i.e., those of the form $\mathfrak{p} = (X_i)$ for some $i = 1, \ldots, m$. In this case the ideal $S^{-1}\mathfrak{p} = R_a$ "blows-up" in R_a, so $\mathtt{h}(R_a/\mathfrak{p}R_a) = 0$, and one cannot apply Corollary 3.16 to conclude that $\mathtt{h}(R/\mathfrak{p}) = 0$ as well. Nevertheless, the equality $\mathtt{h}(R/\mathfrak{p}) = 0$ is granted by Example 4.2, as $\mathsf{m}(X_i) = 0$ for every $i = 1, \ldots, m$.

Therefore, the equality $\mathtt{h}(R/\mathfrak{p}) = \mathsf{m}(f)$ remains true for all non-zero principal prime ideals $\mathfrak{p} = (f)$ of R.

PROOF OF THEOREM 1.

(a) Consider $\mathfrak{p} \in \mathrm{Spec}_1(R)$ and $a := X_1 \ldots X_m \notin \mathfrak{p}$. Then the ring R localized at a is precisely R^Q. By Corollary 3.16(b), $\mathtt{h}(R/\mathfrak{p}) = \mathtt{h}(R^Q/\mathfrak{p}R^Q)$, and by Corollary 4.11, $\mathtt{h}(R^Q/\mathfrak{p}R^Q) = h_{\mathrm{alg}}(\beta) = \mathsf{m}(f)$, where $\mathfrak{p} = (f)$.

Our assertion was proved for all primes $\mathfrak{p} = (f)$ which do not contain a. In case $a \in \mathfrak{p}$, there exists $i = 1, \ldots, m$ such that $X_i \in \mathfrak{p}$, since $\mathfrak{p}$ is prime. Then $f \mid X_i$ and we deduce that $\mathfrak{p} = (X_i)$ since X_i is irreducible. Now Example 4.2 gives $h_{\mathrm{alg}}(R/\mathfrak{p}) = 0$. As obviously $\mathsf{m}(f) = 0$, this establishes the desired equality $\mathtt{h}(R/\mathfrak{p}) = \mathsf{m}(f)$ also in this case.

The above argument used essentially the fact that f is irreducible (equivalently, $\mathfrak{p} = (f)$ is prime). To finish the proof of item (a) of the theorem in the general case, consider the standard action $\mathbb{N}^m \overset{\sigma}{\curvearrowright} R/Rf$. Since $R/fR \otimes R^Q \cong R^Q/fR^Q$, we can identify the standard action $\mathbb{Z}^m \overset{\varsigma}{\curvearrowright} R^Q/fR^Q$ with the action $\mathbb{Z}^m \overset{\sigma^*}{\curvearrowright} R^Q/fR^Q$, so $h_{\mathrm{alg}}(\sigma) = h_{\mathrm{alg}}(\sigma^*) = h_{\mathrm{alg}}(\varsigma) = \mathsf{m}(f)$, in view of Theorem 4.5 and Corollary 4.11. In particular, $\mathtt{h}(R/\mathfrak{p}) = \mathsf{m}(f)$. This ends up the proof of item (a).

(b) The following claim ensures the implication (a)$\Rightarrow$(b):

CLAIM 4.13. If $\mathfrak{p}$ is a prime ideal of $R = \mathbb{Z}[X_1, \ldots, X_m]$ with $ht(\mathfrak{p}) > 1$, then $\mathtt{h}(R/\mathfrak{p}) = 0$.

PROOF OF THE CLAIM. Assume that $\mathfrak{p}$ is a prime ideal of R with $ht(\mathfrak{p}) > 1$ and pick a non-zero $f \in \mathfrak{p}$. Factorize $f = f_1 \ldots f_r$ into irreducible factors. Then, without loss of generality we can assume $\mathfrak{q} = f_1 R$ is a prime ideal of height 1 contained in $\mathfrak{p}$. Then, $\mathtt{h}(R/\mathfrak{q}) < \infty$ by (a); Lemma 2.12 now guarantees that $\mathtt{h}(R/\mathfrak{p}) = 0$. □

□

PROOF OF THEOREM 2. Let R_p be $\mathbb{F}_p[X_1, \ldots, X_m]$. By Example A.2, every length function L with $0 < L(R_p) < \infty$ satisfies $L = L(R_p) \cdot \mathrm{rk}_{R_p}$. As $\mathtt{h}(R_p) = \log p$ (this can be deduced directly from the definition of h_{alg}, see for example [**3**, Theorem 3.3]), every function h' as in the hypothesis (i.e., satisfying the Invariance property, the Continuity Axiom, the Addition Theorem and $h'(R) = \log p$), defined on m-tuples of pairwise commuting endomorphisms of abelian groups of exponent p, coincides with $\mathtt{h}$. □

PROOF OF THEOREM 3. Now, let R be $\mathbb{Z}[X_1, \ldots, X_m]$ but we consider only torsion R-modules. In view of the Addition Theorem and the Continuity Axiom, it suffices to prove by induction that $h'(A) = \mathtt{h}(A)$ for every $n \in \mathbb{N}_+$ and every R-module A whose underlying abelian group has exponent p^n. For $n = 1$ the assertion follows from Theorem 2. If $n > 1$ consider an R-module A of exponent p^n and the endomorphism $f \colon A \to A$ defined by $f(x) = p^{n-1}x$ for every $x \in A$. Then $\operatorname{Im} f = p^{n-1}A$, while

$$\ker(f) = A[p^{n-1}] := \{x \in A : p^{n-1}x\} = 0.$$

Now $\ker(f)$ and $\operatorname{Im}(f)$ have exponents p^{n-1} and p, respectively. Thus by our inductive hypothesis $h'(\ker(f)) = \mathtt{h}(\ker(f))$ and $h'(\operatorname{Im}(f)) = \mathtt{h}(\operatorname{Im}(f))$. Applying the Addition Theorem to the short exact sequence $0 \to \ker(f) \to A \to \operatorname{Im}(f) \to 0$ we deduce that $h'(A) = \mathtt{h}(A)$. □

Take again R to be $\mathbb{Z}[X_1, \ldots, X_m]$. As always, for an ideal I of R, the quotient module R/I, considered as a dynamical system, is equipped with the m-tuple of endomorphisms given by the shifts $r + I \mapsto X_i r + I$ in R/I, $i = 1, 2, \ldots, m$.

Proof of Corollary B. We have to prove that every function h' as in the hypothesis (i.e., satisfying the Invariance property, the Continuity Axiom, the Addition Theorem and $h'(R/\mathfrak{p}) = \mathsf{m}(f)$ for every principal prime ideal $\mathfrak{p} = (f)$ of R) coincides with $\mathtt{h}$.

Since all non-zero principal prime ideals $\mathfrak{p}$ of R have $ht(\mathfrak{p}) = 1$ (i.e., $\mathfrak{p} \in \mathrm{Spec}_1(R)$), we deduce from Corollary A that $\mathtt{h}$ is an $\mathtt{h}$-like length function. Since the first part of our hypothesis implies that h' is a length function, while the second part and Theorem 1 imply $h'(R/\mathfrak{p}) = \mathtt{h}(R/\mathfrak{p})$ for all $\mathfrak{p} \in \mathrm{Spec}_1(R)$, we conclude with Corollary 2.18 that h' is an $\mathtt{h}$-like length function as well. According to Corollary 2.18, to prove that $h' = \mathtt{h}$ it suffices to check that $\mathtt{h}(R) = \infty = h'(R)$ and $h'(R/\mathfrak{q}) = \mathtt{h}(R/\mathfrak{q})$ for every $\mathfrak{q} \in \mathrm{Spec}_1(R)$. The second assertion is covered by our hypothesis, as already mentioned. It remains to verify that $\mathtt{h}(R) = \infty = h'(R)$. In view of Example A.2, $h'(R) = \infty$ if and only if $h' \neq C \cdot \mathrm{rk}_R$ for every $C > 0$, and similarly for $\mathtt{h}$. This is true, since for the prime ideal $(X_1 - p)R$

$$h'(R/(X_1 - p)R) = \mathsf{m}(X_1 - p) = \log p \neq 0 = C \cdot \mathrm{rk}(R/(X_1 - p)R)$$

for every $C > 0$. □

Appendix A. Length functions and the Decomposition Theorem 2.13

Fix, throughout this appendix, a commutative Noetherian ring R.

A.1. Examples of length functions. Following [**38**], we say that a length function L of Mod_R is *singular* if it takes values in $\{0, \infty\}$. For example consider the following two singular length functions:

- the *zero length function* $\mathsf{0}\colon \mathrm{Mod}_R \to \{0\}$, such that $\mathsf{0}(M) := 0$, for all $M \in \mathrm{Mod}_R$;
- the *improper length function* $\chi_\infty\colon \mathrm{Mod}_R \to \{0, \infty\}$, such that $\chi_\infty(0) := 0$, and $\chi_\infty(M) := \infty$, for all $0 \neq M \in \mathrm{Mod}_R$.

Observe that, if L is a length function of Mod_R, then $L = \mathsf{0}$ if and only if $L(R) = 0$, while $L = \chi_\infty$ if and only if $L(M) = \infty$ for any simple module $M \in \mathrm{Mod}_R$.

More generally, observe that a function $L\colon \mathrm{Mod}_R \to \{0, \infty\}$ is a (singular) length function if and only if the following is a hereditary torsion class:

$$\mathrm{Ker}(L) := \{M \in \mathrm{Mod}_R : L(M) = 0\} \subseteq \mathrm{Mod}_R\,.$$

Indeed, the additivity of L amounts to the requirement that $\mathrm{Ker}(L)$ is a Serre class (i.e., closed under submodules, quotients, and extensions), while the upper continuity of L is equivalent to ask that $\mathrm{Ker}(L)$ is closed under direct unions. In fact, the assignment $L \mapsto \mathrm{Ker}(L)$ defines a bijection from the set of singular length functions of Mod_R, to the family of hereditary torsion classes in Mod_R. The inverse of this map is defined as in the following example.

Example A.1. Whenever $\mathcal{T} \subseteq \mathrm{Mod}_R$ is a hereditary torsion class, we can define the following singular length function of $\mathrm{Mod}(R)$:

$$\chi_{\mathcal{T}}\colon \mathrm{Mod}_R \to \{0, \infty\}, \quad \text{such that} \quad \chi_{\mathcal{T}}(M) := 0 \quad \text{if and only if } M \in \mathcal{T};$$

while $\chi_{\mathcal{T}}(M) := \infty$, otherwise. As in the above discussion, the additivity of $\chi_{\mathcal{T}}$ holds since $\mathcal{T}$ is a Serre class, while its closure under direct limits implies the upper continuity of $\chi_{\mathcal{T}}$. In particular: $\chi_{\mathrm{Mod}_R} = 0$, while $\chi_{\{0\}} = \chi_\infty$.

Next we present some fundamental (non-singular) examples of length functions.

EXAMPLE A.2. Let D be a Noetherian domain with field of fractions Q; so that $-\otimes_D Q\colon \mathrm{Mod}_D \to \mathrm{Mod}_Q$ is exact and commutes with direct limits. By Lemma 2.9, the pullback along $-\otimes_D Q$ of the obvious length function $\dim_Q\colon \mathrm{Mod}_Q \to \mathbb{N}\cup\{\infty\}$ is a length function of Mod_D, called the *torsion free rank*

$$\mathrm{rk}_D\colon \mathrm{Mod}_D \to \mathbb{R}_{\geq 0}\cup\{\infty\} \qquad \text{such that} \qquad \mathrm{rk}_D(M) := \dim_Q(M\otimes_D Q).$$

Clearly, $\mathrm{rk}_D(D) = 1$. On the other hand, if L is a length function of Mod_D with $0 < L(D) < \infty$, then $L = L(D)\cdot \mathrm{rk}_D$ by [**27**, Theorem 2].

EXAMPLE A.3. Let $R = \mathbb{Z}$, so that Mod_R is the category of Abelian groups. The logarithm of the cardinality $\log|-|\colon \mathrm{Mod}_R \to \mathbb{R}_{\geq 0}\cup\{\infty\}$, where $\log|A| = \infty$ whenever A is not finite, is a length function.

Examples A.2 and A.3 are important because they are the first two concrete length functions used in defining algebraic entropies in [**16**] and [**32**], respectively.

We now describe a natural way to modify and combine some known length functions to obtain new ones.

DEFINITION A.4. Let $\mathcal{C} \in \{\mathrm{Mod}_R, \mathrm{mod}_R\}$ and let M be a generic module in $\mathcal{C}$.

(a) Given an additive function L of $\mathcal{C}$ and $\lambda \in \mathbb{R}_{\geq 0}\cup\{\infty\}$, we define:

$$\lambda L\colon \mathcal{C} \to \mathbb{R}_{\geq 0}\cup\{\infty\} \qquad \text{such that} \qquad \lambda L(M) = \lambda\cdot(L(M)),$$

with the convention that $\infty\cdot 0 = 0 = 0\cdot\infty$.

(b) If $(L_i)_{i\in I}$ is a family of additive functions of $\mathcal{C}$, we define:

$$\sum_{i\in I} L_i\colon \mathcal{C}\to\mathbb{R}_{\geq 0}\cup\{\infty\}, \text{ such that } \left(\sum_{i\in I} L_i\right)(M) := \sum_{i\in I} L_i(M),$$

defining $\sum_I \lambda_i := \sup_{F\in\mathcal{F}(I)} \sum_{i\in F}\lambda_i$, for any $(\lambda_i)_I \subseteq \mathbb{R}_{\geq 0}\cup\{\infty\}$.

The following lemma is an easy consequence of the definitions.

LEMMA A.5. *In the notation of Definition* A.4, λL *and* $\sum_{i\in I} L_i$ *are additive functions. Furthermore, if* $\mathcal{C} = \mathrm{Mod}_R$ *then* λL *and* $\sum_{i\in I} L_i$ *are length functions, provided* L *and* L_i *(for all* $i\in I$*) are length functions, respectively.*

In part (a) of Definition A.4 we used the convention $\infty\cdot 0 = 0 = 0\cdot\infty$. On the other hand, although we will always assume that $\infty\cdot 0 = 0$, Vámos assumes that $0\cdot\infty = \infty$ for his decomposition theorem in [**38**]. In the following lemma we explore some oddities of this alternative convention.

LEMMA A.6. *Let* L *be an additive function of* $\mathcal{C} \in \{\mathrm{mod}_R, \mathrm{Mod}_R\}$ *and suppose that* $0\cdot\infty = \infty$. *The following statements hold true:*

(a) *if* $\mathcal{C} = \mathrm{mod}_R$, *then both* $0\cdot L$ *and* $\infty\cdot L$ *are singular additive functions. Furthermore,* $0\cdot L(F) = 0$ *if* $L(F) < \infty$, *and* $0\cdot L(F) = \infty$ *otherwise;*

(b) *if* $\mathcal{C} = \mathrm{Mod}_R$ *and* L *is a length function, the function* $0\cdot L$ *is not a length function, whenever there exists* $F \in \mathrm{mod}_R$ *such that* $0 < L(F) < \infty$.

PROOF.

(a) is clear since the class $\{F \in \mathrm{mod}_R : L(F) < \infty\}$ is Serre.

(b) Let $M := F^{(\mathbb{N})}$ and observe that, the inclusion of any $K \in \mathcal{L}_f(M)$, factors through a finite coproduct of copies of F, and so $L(K) < \infty$. Hence, $0 \cdot L(K) = 0$ for any $K \in \mathcal{L}_f(M)$ but $0 \cdot L(M) = 0 \cdot \infty = \infty$, so $0 \cdot L$ is not upper continuous. □

A.2. Length functions of Mod_R **vs additive functions of** mod_R**.** Given a length function L of Mod_R, denote its restriction to mod_R by:

$$L_{\restriction \mathrm{mod}_R}\colon \mathrm{mod}_R \to \mathbb{R}_{\geq 0} \cup \{\infty\}, \tag{A.1}$$

which is clearly additive. Conversely, any additive function $L'\colon \mathrm{mod}_R \to \mathbb{R}_{\geq 0} \cup \{\infty\}$ can be extended to a function of Mod_R, as follows:

$$\widehat{L}'\colon \mathrm{Mod}_R \to \mathbb{R}_{\geq 0} \cup \{\infty\} \quad \text{such that} \quad \widehat{L}'(M) := \sup\{L(F) : F \in \mathcal{L}_f(M)\}, \tag{A.2}$$

for every $M \in \mathrm{Mod}_R$. The following lemma proves that $\widehat{L}'$ is a length function.

LEMMA A.7. *In the above notation, $\widehat{L}'$ is a length function of* Mod_R.

PROOF. Clearly, $\widehat{L}'$ is upper continuous and $\widehat{L}'(0) = 0$. To check additivity let $N \leq M \in \mathrm{Mod}_R$ and choose $F \leq \mathcal{L}_f(M)$, so $F \cap N \in \mathcal{L}_f(N)$ and $(F+N)/N \in \mathcal{L}_f(M/N)$. The additivity of L' on mod_R gives $L'(F) = L'(F \cap N) + L'((F+N)/N)$ which proves that $\widehat{L}'(M) \leq \widehat{L}'(N) + \widehat{L}'(M/N)$ by the arbitrariness of F.

On the other hand, let $F_1 \in \mathcal{L}_f(N)$ and $F_2^* \in \mathcal{L}_f(M/N)$. Choose a module $F_2 \in \mathcal{L}_f(M)$ such that $(F_2 + N)/N = F_2^*$ and set $F = F_1 + F_2$. Then

$$L'(F) = L'(F \cap N) + L'((F+N)/N) \geq L'(F_1) + L'(F_2^*),$$

since $F_2^* = (F+N)/N$ (being F_1 contained in N) and $F_1 \subseteq F \cap N$. Hence, $\widehat{L}'(M) \geq \widehat{L}'(N) + \widehat{L}'(M/N)$, as desired. □

In the next theorem we prove that any length function of Mod_R is completely determined by the values it takes on mod_R.

THEOREM A.8. *In the above notation, the assignments $L \mapsto L_{\restriction \mathrm{mod}_R}$ and $L' \mapsto \widehat{L}'$ in* (A.1) *and* (A.2)*, respectively, define inverse bijections between length functions of* Mod_R *and additive functions of* mod_R. *Furthermore, these bijections are "linear", i.e., they commute with the operations introduced in Definition* A.4.

PROOF. In the notation of the statement, $L_{\restriction \mathrm{mod}_R}$ is clearly additive and so the assignment $(-)_{\restriction \mathrm{mod}_R}$ is well-defined. Furthermore, Lemma A.7 proves that $\widehat{L}'$ is a length function and so also the map $(\widehat{-})$ is well-defined. It is now easy to see that these two maps are inverse to each other, and this concludes the proof. □

REMARK A.9. Given $\mathfrak{p} \in \mathrm{Spec}(R)$, let $\ell'_\mathfrak{p} := (\ell_\mathfrak{p})_{\restriction \mathrm{mod}_R}$, and fix the convention $0 \cdot \infty = \infty$. By Lemma A.6, $0 \cdot \ell'_\mathfrak{p}$ is an additive function of mod_R, and $0 \cdot \ell_\mathfrak{p} = 0 \cdot \widehat{\ell'_\mathfrak{p}}$ is not a length function of Mod_R. On the other hand, $\widehat{0 \cdot \ell'_\mathfrak{p}}$ is a length function, by Lemma A.7, so that $\widehat{0 \cdot \ell'_\mathfrak{p}} \neq 0 \cdot \ell_\mathfrak{p}$.

A.3. Decomposition of additive functions and uniqueness theorems. Let $L\colon \mathrm{mod}_R \to \mathbb{R}_{\geq 0} \cup \{\infty\}$ be an additive function, $\Sigma \subseteq \mathrm{Spec}(R)$, and let:

$$L_\Sigma := \sum_{\mathfrak{p} \in \Sigma} L(R/\mathfrak{p}) \ell_\mathfrak{p} \colon \mathrm{mod}_R \to \mathbb{R}_{\geq 0} \cup \{\infty\}. \tag{A.3}$$

As discussed through the end of §A.1, we always assume that $\infty \cdot 0 = 0$. On the other hand, there are two possible conventions one can adopt for the value of the indeterminate product $0 \cdot \infty$, each producing a different additive function.

DEFINITION A.10. In the above setting, we fix the following notations:

(a) L_Σ indicates the function obtained using the operations in Definition A.4, that is, adopting the conventions: $\infty \cdot 0 = 0$ and $0 \cdot \infty = 0$;
(b) L^V_Σ still indicates the linear combination $\sum_{\mathfrak{p}\in\Sigma} L(R/\mathfrak{p})\ell_\mathfrak{p}$, but now adopting Vámos' conventions: $\infty \cdot 0 = 0$ and $0 \cdot \infty = \infty$.

The following result gives some relations between functions of the form L_Σ, and those of the form $L^V_{\Sigma'}$, for suitable Σ, $\Sigma' \subseteq \mathrm{Spec}(R)$.

LEMMA A.11. *Let L be an additive function of* mod_R, *and* $\Sigma \subseteq \mathrm{Spec}(R)$. *Then,*

(1) *both L_Σ and L^V_Σ are additive functions of* mod_R*;*
(2) $L_\Sigma = L^V_\Sigma$, *whenever* $\Sigma \subseteq \mathrm{Supp}(L)$*;*
(3) $L_\Sigma = L_{\Sigma\cap\mathrm{Supp}(L)}$.

In particular, $L_\Sigma = L_{\Sigma\cap\mathrm{Supp}(L)} = L^V_{\Sigma\cap\mathrm{Supp}(L)}$.

PROOF.

(1) L_Σ is additive by Lemma A.5, while the additivity of L^V_Σ follows by a combination of Lemma A.6(a) and Lemma A.5.

(2) If $\Sigma \subseteq \mathrm{Supp}(L)$ then $L(R/\mathfrak{p}) \neq 0$ for all $\mathfrak{p} \in \Sigma$. In particular, for $M \in \mathrm{mod}_R$, the product $L(R/\mathfrak{p})\ell_\mathfrak{p}(M)$ is never equal to $0 \cdot \infty$.

(3) Given $\mathfrak{q} \in \Sigma \setminus \mathrm{Supp}(L)$, clearly $L(R/\mathfrak{q}) = 0$ and so $L(R/\mathfrak{q})\ell_\mathfrak{q}(M) = 0$ for $M \in \mathrm{mod}_R$, using that $0 \cdot \infty = 0$ whenever $\ell_\mathfrak{q}(M) = \infty$. Hence, we can erase form L_Σ all the summands of the form $L(R/\mathfrak{q})\ell_\mathfrak{q}$, for $\mathfrak{q} \in \Sigma \setminus \mathrm{Supp}(L)$ without affecting its values. Hence, $L_\Sigma = L_{\Sigma\cap\mathrm{Supp}(L)}$. □

As in [**38**, Lemma 4], now we compute the values $\ell_\mathfrak{p}(R/\mathfrak{q})$ (with $\mathfrak{p}, \mathfrak{q} \in \mathrm{Spec}(R)$). By Lemma 2.8, these values uniquely characterize $\ell_\mathfrak{p}$.

COROLLARY A.12. *Let* $\mathfrak{p}$ *and* $\mathfrak{q} \in \mathrm{Spec}(R)$. *Then*

$$\ell_\mathfrak{q}(R/\mathfrak{p}) = \begin{cases} 0 & \textit{if } \mathfrak{p} \not\subseteq \mathfrak{q}; \\ 1 & \textit{if } \mathfrak{p} = \mathfrak{q}; \\ \infty & \textit{if } \mathfrak{p} \subset \mathfrak{q}. \end{cases}$$

PROOF. If $\mathfrak{p} \not\subseteq \mathfrak{q}$ then $R/\mathfrak{p} \otimes_R R_\mathfrak{q} = 0$. On the other hand, if $\mathfrak{p} = \mathfrak{q}$ then $R/\mathfrak{p} \otimes_R R_\mathfrak{q} \cong R_\mathfrak{q}/\mathfrak{q}R_\mathfrak{q}$ is the unique (up to isomorphism) simple $R_\mathfrak{q}$-module. Finally, when $\mathfrak{p} \subsetneq \mathfrak{q}$, the thesis follows by Lemma 2.12. □

Given $\mathfrak{p} \in \mathrm{Spec}(R)$, we denote the closure of $\{\mathfrak{p}\}$ under specialization as follows:

$$V(\mathfrak{p}) := \{\mathfrak{q} \in \mathrm{Spec}(R) : \mathfrak{p} \leq \mathfrak{q}\} \subseteq \mathrm{Spec}(R).$$

LEMMA A.13. *Let L be an additive function of* mod_R *and* $\Sigma \subseteq \mathrm{Spec}(R)$. *Then,*

(a) $L^V_\Sigma(R/\mathfrak{p}) = L^V_{\Sigma\cap V(\mathfrak{p})}(R/\mathfrak{p})$, *for all* $\mathfrak{p} \in \mathrm{Spec}(R)$*;*
(b) $L(R/\mathfrak{p})\ell_\mathfrak{p}(R/\mathfrak{p}) = L(R/\mathfrak{p})$, *for all* $\mathfrak{p} \in \mathrm{Spec}(R)$*;*
(c) *if* $0 \cdot \infty = \infty$, *and* $\mathfrak{p} \lneq \mathfrak{q} \in \mathrm{Spec}(R)$, *then* $L(R/\mathfrak{q})\ell_\mathfrak{q}(R/\mathfrak{p}) = \infty$.

PROOF.

(a) For any $\mathfrak{p}' \in \Sigma \setminus V(\mathfrak{p})$, we have $\ell_{\mathfrak{p}'}(R/\mathfrak{p}) = 0$ by Corollary A.12, and so $L(R/\mathfrak{p}')\ell_{\mathfrak{p}'}(R/\mathfrak{p}) = 0$, which is clear if $L(R/\mathfrak{p}') < \infty$, and uses the convention $\infty \cdot 0 = 0$, otherwise. Thus, $L^V_\Sigma(R/\mathfrak{p}) = L^V_{\Sigma\setminus V(\mathfrak{p})}(R/\mathfrak{p}) + L^V_{\Sigma\cap V(\mathfrak{p})}(R/\mathfrak{p}) = L^V_{\Sigma\cap V(\mathfrak{p})}(R/\mathfrak{p})$ as, by the first part of the argument, $L^V_{\Sigma\setminus V(\mathfrak{p})}(R/\mathfrak{p}) = 0$.

(b) is trivial since $\ell_{\mathfrak{p}}(R/\mathfrak{p}) = 1$, by Corollary A.12.

(c) By Corollary A.12, $\ell_{\mathfrak{q}}(R/\mathfrak{p}) = \infty$, and so $L(R/\mathfrak{q})\ell_{\mathfrak{q}}(R/\mathfrak{p}) = \infty$, which is clear if $L(R/\mathfrak{q}) > 0$, and uses the convention $0 \cdot \infty = \infty$, otherwise. □

Given an additive function L of mod_R, in the following theorem we completely characterize the subsets Σ of the spectrum such that $L = L^V_\Sigma$.

THEOREM A.14. *Let L be an additive function of* mod_R, *and let* $\Sigma \subseteq \mathrm{Spec}(R)$. *Then, the following are equivalent:*

(a) $L = L^V_\Sigma$;

(b) *each prime* $\mathfrak{p} \in \mathrm{Spec}(R)$ *satisfies the following conditions:*

$$V(\mathfrak{p}) \cap \Sigma \begin{cases} \neq \emptyset & \text{if } \mathfrak{p} \in \mathrm{Inf}(L); \\ = \{\mathfrak{p}\} & \text{if } \mathfrak{p} \in \mathrm{Supp}(L) \cap \mathrm{Fin}(L); \\ \subseteq \{\mathfrak{p}\} & \text{if } \mathfrak{p} \in \mathcal{Z}(L). \end{cases} \tag{A.4}$$

PROOF. (a)⇒(b). Let $\mathfrak{p} \in \mathrm{Spec}(R)$ and suppose that $L^V_\Sigma(R/\mathfrak{p}) = L(R/\mathfrak{p})$. By Lemma A.13(a), we deduce that $L^V_{\Sigma\cap V(\mathfrak{p})}(R/\mathfrak{p}) = L(R/\mathfrak{p})$. Let us distinguish the following three cases:

- If $L(R/\mathfrak{p}) = \infty$, then $L^V_{\Sigma\cap V(\mathfrak{p})}(R/\mathfrak{p}) \neq 0$, and so $\Sigma \cap V(\mathfrak{p}) \neq \emptyset$;
- If $0 < L(R/\mathfrak{p}) < \infty$, then $L^V_{\Sigma\cap V(\mathfrak{p})}(R/\mathfrak{p}) = L(R/\mathfrak{p}) < \infty$ implies that $\Sigma \cap V(\mathfrak{p}) \subseteq \{\mathfrak{p}\}$ by Lemma A.13(c), while $0 < L(R/\mathfrak{p}) = L_{\Sigma\cap V(\mathfrak{p})}(R/\mathfrak{p})$ implies $\Sigma \cap V(\mathfrak{p}) \neq \emptyset$. Combining the two conditions gives $\Sigma \cap V(\mathfrak{p}) = \{\mathfrak{p}\}$. Conclude by Lemma A.13(b) that $L^V_\Sigma(R/\mathfrak{p}) = L(R/\mathfrak{p})$.
- If $L(R/\mathfrak{p}) = 0$, then $L_{\Sigma\cap V(\mathfrak{p})}(R/\mathfrak{p}) = 0$ and so, by Lemma A.13(c), $\mathfrak{p}$ does not have any strict specialization in Σ, i.e., $\Sigma \cap V(\mathfrak{p}) \subseteq \{\mathfrak{p}\}$.

(b)⇒(a). By Lemma 2.8, it is enough to verify that $L^V_\Sigma(R/\mathfrak{p}) = L(R/\mathfrak{p})$, for all $\mathfrak{p} \in \mathrm{Spec}(R)$. We distinguish the following three cases:

- If $L(R/\mathfrak{p}) = \infty$, then $V(\mathfrak{p}) \cap \Sigma \neq \emptyset$. So, either $\mathfrak{p} \in \Sigma$ or $\mathfrak{p} \subsetneq \mathfrak{q} \in \Sigma$: in the first case, $L^V_\Sigma(R/\mathfrak{p}) \geq L(R/\mathfrak{p})\ell_{\mathfrak{p}}(R/\mathfrak{p}) = \infty$ while, in the latter, $L^V_\Sigma(R/\mathfrak{p}) \geq L(R/\mathfrak{q})\ell_{\mathfrak{q}}(R/\mathfrak{p}) = \infty$, by Lemma A.13(c).
- If $0 < L(R/\mathfrak{p}) < \infty$, then $V(\mathfrak{p}) \cap \Sigma = \{\mathfrak{p}\}$ and, using Lemma A.13(b), $L^V_\Sigma(R/\mathfrak{p}) = L^V_{\{\mathfrak{p}\}}(R/\mathfrak{p}) = L(R/\mathfrak{p})$.
- If $L(R/\mathfrak{p}) = 0$, then $V(\mathfrak{p}) \cap \Sigma \subseteq \{\mathfrak{p}\}$. If $V(\mathfrak{p}) \cap \Sigma = \emptyset$, then $L^V_\Sigma(R/\mathfrak{p}) = 0$, while if $V(\mathfrak{p}) \cap \Sigma = \{\mathfrak{p}\}$, then $L^V_\Sigma(R/\mathfrak{p}) = L^V_{\{\mathfrak{p}\}}(R/\mathfrak{p}) = L(R/\mathfrak{p}) = 0$, by Lemma A.13(b). □

It is almost immediate to deduce Vámos' decomposition from the above theorem but we postpone this result to §A.4. On the other hand, using Lemma A.11 and Theorem A.14, we can now characterize those $\Sigma \subseteq \mathrm{Spec}(R)$ for which $L_\Sigma = L$.

COROLLARY A.15. *Let L be an additive function of* mod_R, *and let* $\Sigma \subseteq \mathrm{Spec}(R)$. *Then, the following are equivalent:*

(a) $L = L_\Sigma$;

(b) $\mathrm{Ass}(L) \subseteq \Sigma$.

PROOF. Using Lemma A.11(3) and Theorem A.14, it is not difficult to see that condition (a) is equivalent to:

(a′) $\Sigma \cap \mathrm{Supp}(L)$ *satisfies condition* (A.4) *in Theorem* A.14*(b).*

Hence, we just need to verify the equivalence "(b)⇔(a′)".

(b)⇒(a′). For simplicity, suppose that $\mathrm{Ass}(L) \subseteq \Sigma \subseteq \mathrm{Supp}(L)$, let $\mathfrak{p} \in \mathrm{Spec}(R)$, and consider the following three cases:

- if $\mathfrak{p} \in \mathcal{Z}(L)$, i.e., $\mathfrak{p} \notin \mathrm{Supp}(L)$ then $V(\mathfrak{p}) \cap \Sigma \subseteq V(\mathfrak{p}) \cap \mathrm{Supp}(L) = \emptyset$, since $\mathrm{Supp}(L)$ is generalization closed.
- if $\mathfrak{p} \in \mathrm{Supp}(L) \cap \mathrm{Fin}(L)$ then, by Lemma 2.12, $\mathfrak{p}$ is maximal in $\mathrm{Supp}(L)$, so $\mathfrak{p} \in \mathrm{Ass}(L) \subseteq \Sigma$. Furthermore, $\mathfrak{q} \in \mathcal{Z}(L)$, whenever $\mathfrak{p} \neq \mathfrak{q} \in V(\mathfrak{p})$, and so $\mathfrak{q} \notin \Sigma$, by the previous case. Summarizing: $V(\mathfrak{p}) \cap \Sigma = \{\mathfrak{p}\}$.
- if $\mathfrak{p} \in \mathrm{Inf}(L) \subseteq \mathrm{Supp}(L)$, let $\mathfrak{q} \in \mathrm{Ass}(L)$ be a maximal element of $\mathrm{Supp}(L)$ containing $\mathfrak{p}$, then: $\mathfrak{q} \in V(\mathfrak{p}) \cap \mathrm{Ass}(L) \subseteq V(\mathfrak{p}) \cap \Sigma \neq \emptyset$.

(a′)⇒(b). For simplicity, assume $\Sigma \subseteq \mathrm{Supp}(L)$. Now, if $\mathfrak{p} \in \mathrm{Ass}(L)$, then either $\mathfrak{p} \in \mathrm{Fin}(L) \cap \mathrm{Supp}(L)$ (Lemma 2.12), and so $\mathfrak{p} \in \{\mathfrak{p}\} = V(\mathfrak{p}) \cap \Sigma \subseteq \Sigma$, or $\mathfrak{p} \in \mathrm{Inf}(L)$ but $\mathfrak{q} \notin \mathrm{Supp}(L)$ whenever $\mathfrak{p} \neq \mathfrak{q} \in V(\mathfrak{p})$ (by the maximality of $\mathfrak{p}$ in $\mathrm{Supp}(L)$), and so $\Sigma \cap V(\mathfrak{p}) \neq \emptyset$ implies $\mathfrak{p} \in \Sigma$. □

PROOF OF THEOREM 2.13. Given a length function L of Mod_R, consider its restriction $L_{\restriction \mathrm{mod}_R}$, which is an additive function of mod_R. Observe that, by Corollary A.15, both $\mathrm{Ass}(L)$ and $\mathrm{Supp}(L)$ are subsets of $\mathrm{Spec}(R)$ that can be used for a decomposition of $L_{\restriction \mathrm{mod}_R}$, i.e., the following formulas hold for all $M \in \mathrm{mod}_R$:

$$L(M) = \sum_{\mathfrak{p} \in \mathrm{Supp}(L)} L(R/\mathfrak{p})\ell_{\mathfrak{p}}(M) = \sum_{\mathfrak{p} \in \mathrm{Ass}(L)} L(R/\mathfrak{p})\ell_{\mathfrak{p}}(M), \tag{A.5}$$

Using Theorem A.8, we extend the additive functions appearing in (A.5) from mod_R to Mod_R: as the extension process is "linear" (i.e., it respects linear combinations), we deduce that (A.5) remains true for all $M \in \mathrm{Mod}_R$, as desired. □

REMARK A.16. Theorem 2.13 may fail if R is not Noetherian. For example, let $(R, \mathfrak{m})$ be a 1-dimensional non-discrete valuation domain, so $\mathrm{Spec}(R) = \{0, \mathfrak{m}\}$, let Q be its field of fractions, and $v\colon Q \to \mathbb{R} \cup \{\infty\}$ a valuation relative to R. Consider the unique length function L_v of Mod_R such that $L_v(R/I) := \inf_{x \in I} v(x)$ (see [**27**, Theorem 4]): then $0 < L_v(R/I) < \infty$ for all $I \leq R$ such that $0 \neq I \lneq \mathfrak{m}$. On the other hand, $\ell_0(R/I) = 0$, and $\ell_{\mathfrak{m}}(R/I) = \infty$, for any such ideal and, therefore, L_v cannot be written as a linear combination of ℓ_0 and $\ell_{\mathfrak{m}}$.

A.4. A comparison with Vámos' original decomposition.

DEFINITION A.17 ([**38**]). Let L be an additive function of mod_R and define:

- $\mathcal{P}(L)$ as the set of minimal elements of $\mathrm{Fin}(L)$, and
- $\mathcal{Q}(L)$ as the subset of all maximal elements of $\mathrm{Inf}(L)$ that are not contained in any $\mathfrak{p} \in \mathcal{P}(L)$.

We can now give a complete proof of Vámos' decomposition theorem as a consequence of Theorem A.14.

THEOREM A.18. *Let R be a commutative Noetherian ring, and let L be an additive function of* mod_R*. Then, the set $\mathcal{P}(L) \cup \mathcal{Q}(L)$ satisfies condition* (A.4) *in*

Theorem A.14, *for all* $\mathfrak{p} \in \mathrm{Spec}(R)$. *In particular, for any* $M \in \mathrm{mod}_R$,

$$L(M) = \sum_{\mathfrak{p}\in\mathcal{P}(L)} L(R/\mathfrak{p}) \cdot \ell_{\mathfrak{p}}(M) + \sum_{\mathfrak{p}\in\mathcal{Q}(L)} \infty \cdot \ell_{\mathfrak{p}}(M), \tag{A.6}$$

with the following conventions: $0 \cdot \infty = \infty$ *and* $\infty \cdot 0 = 0$.

PROOF. Observe that the elements of $\mathcal{P}(L) \cup \mathcal{Q}(L)$ are not comparable, so $\mathcal{P}(L) \cup \mathcal{Q}(L)$ contains at most one element of $V(\mathfrak{p})$, for each $\mathfrak{p} \in \mathrm{Spec}(R)$. Furthermore, the construction of $\mathcal{P}(L)$ directly implies that $V(\mathfrak{p}) \cap (\mathcal{P}(L) \cup \mathcal{Q}(L)) \subseteq \{\mathfrak{p}\}$, for all $\mathfrak{p} \in \mathrm{Fin}(L)$ and so, in particular, for all $\mathfrak{p} \in \mathcal{Z}(L)$.

Similarly, $\{\mathfrak{p}\} = V(\mathfrak{p}) \cap \mathcal{P}(L) = V(\mathfrak{p}) \cap (\mathcal{P}(L) \cup \mathcal{Q}(L))$ if $\mathfrak{p} \in \mathrm{Supp}(L) \cap \mathrm{Fin}(L)$.

Finally, suppose that $L(R/\mathfrak{p}) = \infty$ and consider a maximal element $\mathfrak{q}$ of $\mathrm{Inf}(L)$ that contains $\mathfrak{p}$. By construction, either $\mathfrak{q} \in \mathcal{Q}(L)$, or there is $\mathfrak{q}' \in \mathcal{P}(L)$ such that $\mathfrak{p} \leq \mathfrak{q} \leq \mathfrak{q}'$. Hence, $V(\mathfrak{p}) \cap (\mathcal{P}(L) \cup \mathcal{Q}(L)) \neq \emptyset$ since either $\mathfrak{q} \in V(\mathfrak{p}) \cap \mathcal{Q}(L)$ or $\mathfrak{q}' \in V(\mathfrak{p}) \cap \mathcal{P}(L)$. □

Let us briefly discuss some of the properties of the above result and compare it with the decompositions obtained in Theorem 2.13. In some cases, the two decompositions coincide; take, for example, the composition length ℓ of mod_R: then $\mathrm{Ass}(\ell) = \mathcal{P}(\ell) \cup \mathcal{Q}(\ell) = \mathrm{Max}(R)$. On the other hand, there are also cases on the other extreme, for which $\mathrm{Ass}(L)$ and $\mathcal{P}(L) \cup \mathcal{Q}(L)$ are disjoint, as it happens in the following example.

EXAMPLE A.19. Let R be a commutative Noetherian ring of dimension at least 2, and consider the additive function $L := \sum_{\mathfrak{p}\in\mathrm{Spec}(R)\setminus\mathrm{Max}(R)} \infty \cdot \ell_{\mathfrak{p}}$ of mod_R. Observe that $L(R/\mathfrak{p}) = 0$ for all $\mathfrak{p} \in \mathrm{Max}(R)$, while $L(R/\mathfrak{p}) = \infty$ for all $\mathfrak{p} \in \mathrm{Spec}(R) \setminus \mathrm{Max}(R)$. Then, $\mathcal{P}(L) = \mathrm{Max}(R)$ and $\mathcal{Q}(L) = \emptyset$, but $\mathrm{Ass}(L)$ is the set of maximal elements of $\mathrm{Spec}(R) \setminus \mathrm{Max}(R)$. In particular, $(\mathcal{P}(L) \cup \mathcal{Q}(L)) \cap \mathrm{Ass}(L) = \emptyset$.

Another difference between the two decompositions is their redundancy. Indeed, it follows by Corollary A.15 that, in the decomposition of Theorem 2.13, there is no redundancy at all, in the sense that the equality $L = L_\Sigma$ holds for $\Sigma = \mathrm{Ass}(L)$ but it always fails for any smaller set. On the other hand, sometimes Vámos' decomposition may contain redundant summands that can be removed without changing the result, as it happens in the following example.

EXAMPLE A.20. Let $R = \mathbb{Z}[X]$. For every $n \in \mathbb{N}_+$, denote by $\mathfrak{p}_n$ the principal prime ideal (Φ_n) of R. By Kronecker's theorem (Theorem 3.6) and since $\mathtt{h}(R/\mathfrak{p}) = \mathsf{m}(f)$ when $\mathfrak{p} = (f) \in \mathrm{Spec}(R)$, we have $\mathcal{Z}(\mathtt{h}) = \{\mathfrak{p}_n : n \in \mathbb{N}_+\} \cup \mathrm{Max}(R)$, so

$$\mathrm{Supp}(\mathtt{h}) = \mathrm{Spec}(R) \setminus \mathcal{Z}(\mathtt{h}) = (\mathrm{Spec}_1(R) \cup \{0\}) \setminus \{\mathfrak{p}_n : n \in \mathbb{N}_+\}$$

and $\mathrm{Ass}(\mathtt{h}) = \{\mathfrak{p} \in \mathrm{Supp}(\mathtt{h}) : \mathfrak{p} \neq 0\}$, hence, $\mathrm{Supp}(\mathtt{h}) = \{0\} \cup \mathrm{Ass}(\mathtt{h})$. Moreover,

$$\mathcal{P}(\mathtt{h}) = \mathrm{Spec}_1(R) = \mathrm{Ass}(\mathtt{h}) \cup \{\mathfrak{p}_n : n \in \mathbb{N}_+\}.$$

So, in this context, Theorem A.18 gives

$$\mathtt{h} = \mathtt{h}^V_{\mathcal{P}(\mathtt{h})} = \sum_{\mathfrak{p}\in\mathcal{P}(\mathtt{h})} \mathtt{h}(R/\mathfrak{p}) \cdot \ell_{\mathfrak{p}} = \sum_{n\in\mathbb{N}_+} \mathtt{h}(R/\mathfrak{p}_n) \cdot \ell_{\mathfrak{p}_n} + \sum_{\mathfrak{p}\in\mathrm{Ass}(\mathtt{h})} \mathtt{h}(R/\mathfrak{p}) \cdot \ell_{\mathfrak{p}}. \tag{A.7}$$

Here the term $X = \sum_{n\in\mathbb{N}_+} \mathtt{h}(R/\mathfrak{p}_n) \cdot \ell_{\mathfrak{p}_n}$ is superfluous, as $\sum_{\mathfrak{p}\in\mathrm{Ass}(\mathtt{h})} \mathtt{h}(R/\mathfrak{p}) \cdot \ell_{\mathfrak{p}} = \mathtt{h}$ by Theorem 2.13. Nevertheless, this additional term X is non-zero. Indeed, for $\mathfrak{q} = 0$ one has $\ell_{\mathfrak{p}_n}(R/\mathfrak{q}) = \infty$, hence, $\mathtt{h}(R/\mathfrak{p}_n) \cdot \ell_{\mathfrak{p}_n}(R/\mathfrak{q}) = 0 \cdot \infty = \infty$ for every $n \in \mathbb{N}$ (due to the convention $0 \cdot \infty = \infty$ in Theorem A.18) and $X(R/\mathfrak{q}) = \infty$. On the other hand, for every $\mathfrak{p} \in \mathrm{Ass}(\mathtt{h})$, $\mathtt{h}(R/\mathfrak{p}) > 0$ and $\mathfrak{q} < \mathfrak{p}$, so $\ell_{\mathfrak{p}}(R) = \ell_{\mathfrak{p}}(R/\mathfrak{q}) = \infty$ and $\mathtt{h}(R/\mathfrak{p}_n) \cdot \ell_{\mathfrak{p}_n}(R/\mathfrak{q}) = \infty$ in the sum $\mathtt{h}_{\mathrm{Ass}(\mathtt{h})}(R/\mathfrak{p}) = \mathtt{h}(R/\mathfrak{q})$.

Acknowledgment

It is a pleasure to thank the referee for the very careful reading and the wealth of useful suggestions.

References

[1] R. L. Adler, A. G. Konheim, and M. H. McAndrew, *Topological entropy*, Trans. Amer. Math. Soc. **114** (1965), 309–319, DOI 10.2307/1994177. MR175106

[2] L. V. Ahlfors, *Complex analysis: An introduction to the theory of analytic functions of one complex variable*, 3rd ed., International Series in Pure and Applied Mathematics, McGraw-Hill Book Co., New York, 1978. MR510197

[3] A. Biś, D. Dikranjan, A. Giordano Bruno, and L. Stoyanov, *Algebraic entropies of commuting endomorphisms of torsion abelian groups*, Rend. Semin. Mat. Univ. Padova **144** (2020), 45–60, DOI 10.4171/rsmup/55. MR4186445

[4] R. Bowen, *Entropy for group endomorphisms and homogeneous spaces*, Trans. Amer. Math. Soc. **153** (1971), 401–414, DOI 10.2307/1995565. MR274707

[5] D. W. Boyd, *Kronecker's theorem and Lehmer's problem for polynomials in several variables*, J. Number Theory **13** (1981), no. 1, 116–121, DOI 10.1016/0022-314X(81)90033-0. MR602452

[6] D. W. Boyd, *Speculations concerning the range of Mahler's measure*, Canad. Math. Bull. **24** (1981), no. 4, 453–469, DOI 10.4153/CMB-1981-069-5. MR644535

[7] T. Ceccherini-Silberstein, M. Coornaert, and F. Krieger, *An analogue of Fekete's lemma for subadditive functions on cancellative amenable semigroups*, J. Anal. Math. **124** (2014), 59–81, DOI 10.1007/s11854-014-0027-4. MR3286049

[8] M. M. Day, *Amenable semigroups*, Illinois J. Math. **1** (1957), 509–544. MR92128

[9] D. Dikranjan, A. Fornasiero, and A. Giordano Bruno, *Algebraic entropy for amenable semigroup actions*, J. Algebra **556** (2020), 467–546, DOI 10.1016/j.jalgebra.2020.02.033. MR4088441

[10] D. Dikranjan, A. Fornasiero, A. Giordano Bruno, and F. Salizzoni, *The addition theorem for locally monotileable monoid actions*, J. Pure Appl. Algebra **227** (2023), no. 1, Paper No. 107113, 35, DOI 10.1016/j.jpaa.2022.107113. MR4459680

[11] D. Dikranjan and A. Giordano Bruno, *The connection between topological and algebraic entropy*, Topology Appl. **159** (2012), no. 13, 2980–2989, DOI 10.1016/j.topol.2012.05.009. MR2944770

[12] D. Dikranjan and A. Giordano Bruno, *The bridge theorem for totally disconnected LCA groups*, Topology Appl. **169** (2014), 21–32, DOI 10.1016/j.topol.2014.02.029. MR3199856

[13] D. Dikranjan and A. Giordano Bruno, *Entropy on abelian groups*, Adv. Math. **298** (2016), 612–653, DOI 10.1016/j.aim.2016.04.020. MR3505752

[14] D. Dikranjan and A. Giordano Bruno, *Entropy on normed semigroups (towards a unifying approach to entropy)*, Dissertationes Math. **542** (2019), 90, DOI 10.4064/dm791-2-2019. MR4016120

[15] D. Dikranjan, A. Giordano Bruno, and S. Virili, *Ore localization of amenable monoid actions and applications toward entropy—addition formulas and the bridge theorem*, Illinois J. Math. **67** (2023), no. 4, 715–787, DOI 10.1215/00192082-10950854. MR4678815

[16] D. Dikranjan, B. Goldsmith, L. Salce, and P. Zanardo, *Algebraic entropy for abelian groups*, Trans. Amer. Math. Soc. **361** (2009), no. 7, 3401–3434, DOI 10.1090/S0002-9947-09-04843-0. MR2491886

[17] A. Facchini, *Module theory*, Modern Birkhäuser Classics, Birkhäuser/Springer Basel AG, Basel, 1998. Endomorphism rings and direct sum decompositions in some classes of modules; [2012 reprint of the 1998 original] [MR1634015]. MR3025306

[18] A. Giordano Bruno and S. Virili, *Algebraic Yuzvinski formula*, J. Algebra **423** (2015), 114–147, DOI 10.1016/j.jalgebra.2014.10.007. MR3283712

[19] P. R. Halmos, *On automorphisms of compact groups*, Bull. Amer. Math. Soc. **49** (1943), 619–624, DOI 10.1090/S0002-9904-1943-07995-5. MR8647

[20] J. L. W. V. Jensen, *Sur un nouvel et important théorème de la théorie des fonctions* (French), Acta Math. **22** (1899), no. 1, 359–364, DOI 10.1007/BF02417878. MR1554908
[21] I. Kaplansky, *Infinite abelian groups*, University of Michigan Press, Ann Arbor, MI, 1954. MR65561
[22] A. N. Kolmogorov, *A new metric invariant of transient dynamical systems and automorphisms in Lebesgue spaces* (Russian), Dokl. Akad. Nauk SSSR (N.S.) **119** (1958), 861–864. MR103254
[23] L. Kronecker, *Zwei Sätze über Gleichungen mit ganzzahligen Coefficienten* (German), J. Reine Angew. Math. **53** (1857), 173–175, DOI 10.1515/crll.1857.53.173. MR1578994
[24] D. H. Lehmer, *Factorization of certain cyclotomic functions*, Ann. of Math. (2) **34** (1933), no. 3, 461–479, DOI 10.2307/1968172. MR1503118
[25] D. Lind, K. Schmidt, and T. Ward, *Mahler measure and entropy for commuting automorphisms of compact groups*, Invent. Math. **101** (1990), no. 3, 593–629, DOI 10.1007/BF01231517. MR1062797
[26] K. Mahler, *On some inequalities for polynomials in several variables*, J. London Math. Soc. **37** (1962), 341–344, DOI 10.1112/jlms/s1-37.1.341. MR138593
[27] D. G. Northcott and M. Reufel, *A generalization of the concept of length*, Quart. J. Math. Oxford Ser. (2) **16** (1965), 297–321, DOI 10.1093/qmath/16.4.297. MR195905
[28] D. S. Ornstein and B. Weiss, *Entropy and isomorphism theorems for actions of amenable groups*, J. Analyse Math. **48** (1987), 1–141, DOI 10.1007/BF02790325. MR910005
[29] J. Peters, *Entropy on discrete abelian groups*, Adv. in Math. **33** (1979), no. 1, 1–13, DOI 10.1016/S0001-8708(79)80007-9. MR540634
[30] L. Salce, *Some results on the algebraic entropy*, Groups and model theory, Contemp. Math., vol. 576, Amer. Math. Soc., Providence, RI, 2012, pp. 297–304, DOI 10.1090/conm/576/11336. MR2962892
[31] L. Salce, P. Vámos, and S. Virili, *Length functions, multiplicities and algebraic entropy*, Forum Math. **25** (2013), no. 2, 255–282, DOI 10.1515/form.2011.117. MR3031785
[32] L. Salce and P. Zanardo, *A general notion of algebraic entropy and the rank-entropy*, Forum Math. **21** (2009), no. 4, 579–599, DOI 10.1515/FORUM.2009.029. MR2541474
[33] K. Schmidt, *Dynamical systems of algebraic origin*, Progress in Mathematics, vol. 128, Birkhäuser Verlag, Basel, 1995. MR1345152
[34] Ja. Sinaĭ, *On the concept of entropy for a dynamic system* (Russian), Dokl. Akad. Nauk SSSR **124** (1959), 768–771. MR103256
[35] C. J. Smyth, *A Kronecker-type theorem for complex polynomials in several variables*, Canad. Math. Bull. **24** (1981), no. 4, 447–452, DOI 10.4153/CMB-1981-068-8. MR644534
[36] L. N. Stojanov, *Uniqueness of topological entropy for endomorphisms on compact groups* (English, with Italian summary), Boll. Un. Mat. Ital. B (7) **1** (1987), no. 3, 829–847. MR916296
[37] The Stacks project authors, *The Stacks project*, `https://stacks.math.columbia.edu`, (2025).
[38] P. Vámos, *Additive functions and duality over Noetherian rings*, Quart. J. Math. Oxford Ser. (2) **19** (1968), 43–55, DOI 10.1093/qmath/19.1.43. MR223434
[39] S. Virili, *Entropy for endomorphisms of LCA groups*, Topology Appl. **159** (2012), no. 9, 2546–2556, DOI 10.1016/j.topol.2011.02.017. MR2921844
[40] M. D. Weiss, *Algebraic and other entropies of group endomorphisms*, Math. Systems Theory **8** (1974/75), no. 3, 243–248, DOI 10.1007/BF01762672. MR385834
[41] S. A. Juzvinskiĭ, *Metric properties of the endomorphisms of compact groups* (Russian), Izv. Akad. Nauk SSSR Ser. Mat. **29** (1965), 1295–1328. MR194588
[42] O. Zariski and P. Samuel, *Commutative algebra. Vol. 1*, Graduate Texts in Mathematics, No. 28, Springer-Verlag, New York-Heidelberg-Berlin, 1975. With the cooperation of I. S. Cohen; Corrected reprinting of the 1958 edition. MR384768

Dipartimento di Informatica, Università di Pisa, Pisa, Italy
Email address: castelnovod@gmail.com

Dipartimento di Scienze Matematiche, Informatiche e Fisiche, Università di Udine, Udine, Italy
Email address: dikran.dikranjan@uniud.it

Dipartimento di Scienze Matematiche, Informatiche e Fisiche, Università di Udine, Udine, Italy
Email address: anna.giordanobruno@uniud.it

Dipartimento di Scienze Matematiche, Informatiche e Fisiche, Università di Udine, Udine, Italy
Email address: dario.spirito@uniud.it

Departament de Matemàtiques, Universitat Autònoma de Barcelona, Bellaterra (Barcelona), Spain
Email address: virili.simone@gmail.com, simone.virili@uab.cat

Contemporary Mathematics
Volume **836**, 2026
https://doi.org/10.1090/conm/836/16714

Weakly Arf property for quadratic quotients of the Rees algebra

Damiano Saccone

ABSTRACT. We provide a characterization of the weakly Arf property for a family of quadratic quotients of the Rees algebra and for double coverings of rings.

Introduction

The notion of weakly Arf property was recently introduced and developed by E. Celikbas, O. Celikbas, C. Ciupercă, N. Endo, S. Goto, R. Isobe, and N. Matsuoka in [**C**]. This property aims to generalize the concept of Arf rings, which were introduced by J. Lipman in 1971 (see [**L2**]), to study one-dimensional ring satisfying specific conditions related to the blow-up process of curve singularities. The origins of these ideas can be traced back to the work of C. Arf in 1949 (see [**A2**]), which provided an algebraic interpretation of the sequence of multiplicities of successive points of an algebraic branch.

An Arf ring is defined as a Noetherian semi-local in which the localization at every maximal ideal is a one-dimensional Cohen-Macaulay local ring. One possible formal definition is as follows:

DEFINITION. Let R be a Noetherian semi-local ring such that for every maximal ideal $\mathfrak{m}$ of R, the localization $R_{\mathfrak{m}}$ is a one-dimensional Cohen-Macaulay local ring. Then R is called an *Arf ring* if the following conditions hold:

(1) Every integrally closed regular ideal I in R has a principal reduction; that is, there exists $n \geq 0$ and $x \in I$ such that $I^{n+1} = xI^n$.
(2) If $x, y, z \in R$ with x a non-zerodivisor, and both $y/x, z/x$ are elements of $Q(R)$ that are integral over R, then their product $(yz)/x \in R$, where, $Q(R)$ represents the total ring of fractions of R.

Since condition (1) becomes superfluous when the ring is a one-dimensional Cohen-Macaulay local ring, the authors of [**C**] were interested in investigating the consequences of assuming only condition (2). In particular, they studied whether certain ring constructions (such as the idealization, amalgamated duplication, and some particular fiber products) yield weakly Arf rings (cf. [**C**, Theorem 5.2, Theorem 6.2, Proposition 6.11]).

2020 *Mathematics Subject Classification.* Primary 13A30, 13B22, 14H30; Secondary 14H20.
Key words and phrases. Weakly Arf rings, Rees algebra, covering of rings.

©2026 American Mathematical Society

Quadratic quotients of the Rees algebra have been introduced in [**BDS**], where their fundamental properties were observed and studied. These rings, defined as

$$R(I)_{a,b} = \frac{\mathcal{R}_+}{(f(t))R[t] \cap \mathcal{R}_+}$$

where $\mathcal{R}_+$ denotes the Rees algebra associated with the ring R and the ideal I, and $f(t) = t^2 + at + b \in R[t]$ is a monic quadratic polynomial, offer a generalization of other algebraic constructions such as Nagata's idealization and the amalgamated duplication. Therefore, in this paper we have studied under which hypotheses $R(I)_{a,b}$ satisfies the weakly Arf property. The paper is structured as follows:

In Section 1, we begin by recalling the preliminary concepts and definitions related to weakly Arf rings and quadratic quotients of the Rees algebra.

In Section 2, we focus on the study of the integral closure of quadratic quotients of the Rees algebra, in particular we show some cases where the integral closure $\overline{R(I)_{a,b}}$ coincides with $\overline{R} + \overline{R}t$.

In Section 3, we present the main results on the weakly Arf property. In particular, we prove in Theorem 3.7 that, under certain conditions, $R(I)_{a,b}$ is a weakly Arf ring if and only if R is a weakly Arf ring.

Finally, in Section 4, we extend the analysis to a similar construction, based on a fractional ideal I and the R-module $R \oplus I$ with multiplication defined by $(r, i)(s, j) = (rs + bij, rj + si)$, where $b \in R$ satisfies $bI^2 \subseteq R$. This construction was introduced in [**L1**] and is referred to as a "double covering of rings". We also investigate the conditions under which this ring satisfies the weakly Arf property, as detailed in Theorems 4.4 and 4.9.

1. Background

In this section, we introduce the notations and properties that will be used throughout the paper. In particular, we recall the definition of weakly Arf rings and quadratic quotients of the Rees algebra.

Let R be a commutative ring with a unity element, we denote the set of non-zerodivisors of R by $W(R)$ and its total ring of fractions by $Q(R)$. An element $u \in Q(R)$ is said to be *integral over* R if there is a monic polynomial $p(X) \in R[X]$ such that $p(u) = 0$. The set of all elements of $Q(R)$ that are integral over R is called the *integral closure of* R *in* $Q(R)$ and is denoted by $\overline{R}$. We say that R is *integrally closed* if $\overline{R} = R$. For two R-submodules M and N of $Q(R)$, we set $(M : N) = \{a \in Q(R) \mid aN \subseteq M\}$.

Let I be an ideal of R. An element $r \in R$ is said to be *integral over* I if r satisfies a relation

$$r^n + a_1 r^{n-1} + a_2 r^{n-2} + \cdots + a_{n-1} r + a_n = 0 \quad (n > 0)$$

with $a_i \in I^i$ for $i = 1, 2, \ldots, n$. The set of all elements of R that are integral over I is called the *integral closure of* I *in* R and it is denoted by $\overline{I}$. We say that I is *integrally closed* if $\overline{I} = I$. We say that I is a *regular ideal* if $I \cap W(R) \neq \emptyset$, i.e., if I contains a non-zerodivisor of R.

If I is a regular ideal of R, an element $x \in I$ is called a *principal reduction of* I if $xI^n = I^{n+1}$ for some integer $n > 0$. Principal reductions always exist when R is one-dimensional, local with an infinite residue field. For a one-dimensional ring, we can also define the *blow-up of* R *at* I as $R^I := \bigcup_{n \geq 0}(I^n : I^n)$. We say that I is *stable in* R if $R^I = I : I$.

When R is a semi-local ring, we can define the sequence of semi-local rings

$$R = R_0 \subseteq R_1 \subseteq R_2 \subseteq \cdots \subseteq \overline{R}$$

where, for each $i \geq 0$, R_{i+1} is the ring obtained by R blowing-up the Jacobson radical of R_i. A local ring of the form $(R_i)_{\mathfrak{m}}$, where $\mathfrak{m}$ is a maximal ideal in R_i, is said to be *infinitely near to* R.

The notion of Arf rings is related to the study of the curve singularities, as it is possible to give the following characterization ([**L2**, Theorem 2.2]):

THEOREM 1.1. *The following conditions on R are equivalent:*

(1) *R is an Arf ring.*
(2) *Every integrally closed regular ideal in R is stable.*
(3) *If S is any local ring infinitely near to R, then the embedding dimension of S is equal to the multiplicity of S.*

1.1. Weakly Arf rings. In [**C**], the authors develop the theory of weakly Arf rings by generalizing the notion of Arf rings. They provide the following definition:

DEFINITION 1.2. A ring R is said to be *weakly Arf* if for all $r, s, u \in R$ such that $u \in W(R)$ and $r/u, s/u \in \overline{R}$, it follows that $rs/u \in R$.

If R is a weakly Arf ring, we write "R is WA".

The difference between Arf and weakly Arf rings is determined by the existence of principal reductions of integrally closed ideals. The two notions are equivalent, for instance, in the case where R is a one-dimensional Cohen-Macaulay local ring with infinite residue field.

Clearly, every Arf ring is WA. However, the ring $k[[X,Y]]/(XY(X+Y))$, with $k = \mathbb{Z}/2\mathbb{Z}$, is an example of a one-dimensional local WA ring which is not an Arf ring ([**C**, Example 9.6]). Notice that, in this case, the residue field is finite.

As mentioned in the introduction, the authors have studied when idealization, amalgamated duplication, and some particular fiber products produce a WA ring. We will now present these results.

We recall that, given a ring R and an R-module M, the *idealization of M over R*, denoted by $R \ltimes M$, is the R-module $R \oplus M$ with multiplication given by $(r, m) \cdot (s, n) = (rs, rn + sm)$. When R is a Noetherian ring and M is a finitely generated torsion-free R-module, which can be seen as an R-submodule of $L = Q(R) \otimes_R M$, thus $Q(R \ltimes M) = Q(R) \ltimes L$ and $\overline{R \ltimes M} = \overline{R} \ltimes L$, we have the following theorem:

THEOREM 1.3 ([**C**, Theorem 5.2]). *The following conditions are equivalent:*

(1) *$R \ltimes M$ is WA.*
(2) *R is WA and $\overline{aR} \cdot M = aM$ for every $a \in W(R)$.*
(3) *R is WA and M is an $\overline{R}$-module.*

Finally, let R be a commutative ring and let I be an ideal of R. The *amalgamated duplication of R with respect to I* is the following subring of $R \times R$:

$$R \bowtie I := \{(r, r+i) \mid r \in R,\ i \in I\}$$

with product given by $(r, r+i) \cdot (s, s+j) = (rs, rs + rj + si + ij)$. When I is a regular ideal of R, we know that $Q(R \bowtie I) = Q(R) \bowtie Q(R)$ and $\overline{R \bowtie I} = \overline{R} \bowtie \overline{R}$ (see [**DF**, Corollary 3.3]). We present the following proposition related to these rings:

PROPOSITION 1.4 ([**C**, Proposition 6.11]). *The amalgamated duplication $R \bowtie I$ is WA if and only if so is R and I is also an ideal of $\overline{R}$.*

1.2. Quadratic quotients of the Rees Algebra. Let R be a commutative and unitary ring, let I be a proper ideal of R and let t be an indeterminate. We denote the *Rees algebra associated to R and I* by

$$\mathcal{R}_+ = \bigoplus_{n \geq 0} I^n t^n \subseteq R[t].$$

Given a monic quadratic polynomial $f(t) = t^2 + at + b \in R[t]$, the ring

$$R(I)_{a,b} := \frac{\mathcal{R}_+}{(f(t))R[t] \cap \mathcal{R}_+}$$

is called a *quadratic quotient of the Rees algebra.* These rings were investigated in [**BDS**], where it was noted, in particular, that $R(I)_{a,b} = R \oplus It$ as abelian groups, with the multiplication defined by

$$(r + it) \cdot (s + jt) = rs - bij + (rj + si - aij)t.$$

PROPOSITION 1.5 ([**BDS**, Proposition 1.3]). *The following ring extensions are integral:*

$$R \subseteq R(I)_{a,b} \subseteq \frac{R[t]}{(f(t))R[t]}.$$

In particular, the three rings have the same Krull dimension.

REMARK 1.6. As mentioned in the introduction, both idealization with respect to an ideal and duplication can be realized as a quadratic quotient of the Rees algebra (by taking polynomials t^2 or $t^2 - t$). More precisely, we have the following ring isomorphisms (see [**BDS**, Proposition 1.4]): $R(I)_{0,0} \cong R \ltimes I$ and $R(I)_{-1,0} \cong R \bowtie I$.

More generally, we also have these ring isomorphisms (see [**BDS**, Proposition 1.5]): $R(I)_{a,b} \cong R \ltimes I$ when $f(t) = (t - \xi)^2$ with $\xi \in R$, and $R(I)_{a,b} \cong R \bowtie I$ when $f(t) = (t - \xi)(t - \eta)$ with $(t - \xi)R[t]$ and $(t - \eta)R[t]$ comaximal ideals of $R[t]$.

2. Integral closure of the quadratic quotients of the Rees algebra

In order to understand the WA property for $R(I)_{a,b}$, it is necessary to determine $\overline{R(I)_{a,b}}$ explicitly. The next two propositions lay the groundwork for our analysis.

PROPOSITION 2.1 ([**BDS**, Proposition 1.7]). *Each element of $Q(R(I)_{a,b})$ is of the form $\frac{r+it}{u}$, where $u \in W(R)$.*

PROPOSITION 2.2 ([**BDS**, Corollary 1.8]). *If I is a regular ideal, then the rings $R(I)_{a,b}$ and $R[t]/(f(t))$ have the same total ring of fractions and integral closure.*

In particular, we have $Q(R(I)_{a,b}) = Q(R) + Q(R)t$.

Throughout the remainder of this paper, we shall always assume that I is a regular ideal and we will denote $Q(R(I)_{a,b})$ by Q.

It was shown in [**BDS**] that $\overline{R(I)_{a,b}}$ contains $\overline{R}[t]/(f(t))$. Moreover, the authors provide an example where this containment is strict. Consider the ring $R = \mathbb{Z}[X]/(X^2 + 4)$, which is not integrally closed. If denotes x the image of X in R, then the element $x/2 \in Q(R)$ is a root of the polynomial $T^2 + 1 \in R[T]$.

REMARK 2.3. Assume that R is an integral domain, then $Q(R) \subseteq Q(R)+Q(R)t$ defines a finite field extension, which may also be separable, for instance, when the polynomial $f(t)$ is irreducible. Moreover, if $\overline{R}$ is a Noetherian regular domain, and $f(t) \notin \mathfrak{m}$ for every maximal ideal $\mathfrak{m}$ of $\overline{R}[t]$, then by [**A1**, Corollary 1], we have $\overline{R[t]/(f(t))} = \overline{R} + \overline{R}t$. Therefore, in such cases, the integral closure of $R(I)_{a,b}$ is exactly $\overline{R} + \overline{R}t$.

Assuming only that R is a commutative ring. Before proceeding with this analysis, we recall the following result, which will be instrumental in our discussion.

LEMMA 2.4 ([**E**, Proposition 4.11]). *Let $R \subseteq S$ be a ring extension, and suppose that $g(X) \in R[X]$ is a monic polynomial such that $g(X) = h(X)l(X)$ in $S[X]$, with $h(X)$ and $l(X)$ monic. Then the coefficients of h and l are integral over R.*

Consequently, and independently of the polynomial $f(t)$, we have the following:

REMARK 2.5. Given $\alpha \in Q$, we can write $\alpha = r + st$ with $r, s \in Q(R)$. Then a direct computation shows that the following quadratic relation holds for α:

$$\alpha^2 - \alpha(2r - as) + r^2 + bs^2 - ars = 0. \tag{2.1}$$

Then α is a root of the polynomial $h(X) = X^2 + (2r - as)X + r^2 + bs^2 - ars$. On the other hand, since the extension $R \subseteq R(I)_{a,b}$ is integral, if α is integral over $R(I)_{a,b}$ it is also integral over R, hence it is a root of some monic polynomial $g(X)$ with coefficients in R. This implies that $h(X)$ must divide $g(X)$ over Q and, applying Lemma 2.4, we get $\alpha \in \overline{R(I)_{a,b}}$ if and only if $2r - as \in \overline{R}$ and $r^2 + bs^2 - ars \in \overline{R}$.

With the aim of better understanding how its integral closure is structured, we now examine certain special cases, specifically when the defining polynomial is either $f(t) = t^2 + b$ or $f(t) = t^2 + at$.

REMARK 2.6. Consider the case $f(t) = t^2 + b$. Assume that R contains a field of characteristic different from two. Then an element $\alpha \in \overline{R(I)_{0,b}}$ if and only if the coefficients of (2.1) are in $\overline{R}$. This means $2r \in \overline{R}$ and $r^2 + bs^2 \in \overline{R}$. Since R contains a field of characteristic different from two and $2r \in \overline{R}$, it follows that $r \in \overline{R}$. Consequently, from $r^2 + bs^2 \in \overline{R}$, we also have $bs^2 \in \overline{R}$.

We observe that if $\gamma \in Q(R)$ such that $\gamma^2 \in \overline{R}$, there is a monic polynomial $\varphi(X) \in R[X]$ such that $\varphi(\gamma^2) = 0$. Let us consider $\tilde{\varphi}(X) = \varphi(X^2) \in R[X]$, which is a monic polynomial and satisfies $\tilde{\varphi}(\gamma) = \varphi(\gamma^2) = 0$, thus $\gamma \in \overline{R}$.

In particular, since $b^2s^2 \in \overline{R}$, we get $bs \in \overline{R}$ and therefore $s \in (\overline{R} : b)$. Hence $\alpha \in \overline{R} + (\overline{R} : b)t$.

From this remark, we obtain the following result.

COROLLARY 2.7. *Let R be a commutative ring containing a field of characteristic different from two, I be a regular ideal of R, and $f(t) = t^2 + b$. Assume that $b(\overline{R} : b)^2 \subseteq \overline{R}$. Then, the integral closure of $R(I)_{0,b}$ is*

$$\overline{R(I)_{0,b}} = \overline{R} + (\overline{R} : b)t.$$

PROOF. From the above remark, we know that $\overline{R(I)_{0,b}} \subseteq \overline{R} + (\overline{R} : b)t$. Conversely, let $\alpha = r + st \in \overline{R} + (\overline{R} : b)t$. Since $s^2 \in (\overline{R} : b)^2$, we get $bs^2 \in \overline{R}$. Moreover, $r \in \overline{R}$, thus $2r, r^2 \in \overline{R}$. Now, if we consider $\alpha \in Q$ and use equation (2.1), we obtain that its coefficients are in $\overline{R}$, therefore α is integral over $R(I)_{0,b}$. Hence $\overline{R} + (\overline{R} : b)t \subseteq \overline{R(I)_{0,b}}$. □

REMARK 2.8. Now, consider the case $f(t) = t^2 + at$, we have $\alpha \in \overline{R(I)_{a,0}}$ if and only if the coefficients of (2.1) are in $\overline{R}$. This implies $2r - as \in \overline{R}$ and $r^2 - ars \in \overline{R}$. We notice that if $r \in \overline{R}$, we get $as \in \overline{R}$; therefore $s \in (\overline{R} : a)$.

Now suppose $r \notin \overline{R}$. Since

$$r^2 - (2r - as)r + (r^2 - ars) = 0,$$

we can consider the polynomial

$$p(X) = X^2 - (2r - as)X + (r^2 - ars) \in Q(R)[X],$$

which satisfies $p(\alpha) = p(r) = 0$. Moreover, if $\alpha \in \overline{R(I)_{a,0}}$, then there exists a monic polynomial $q(X) \in (R(I)_{a,0})[X]$ such that $q(\alpha) = 0$. If we take $q(X)$ of minimal degree, then $q(X) = h(X)p(X)$ for some $h(X) \in Q[X]$. Hence $q(r) = 0$, which implies $r \in \overline{R(I)_{a,0}}$. However, since the extension $R \subseteq R(I)_{a,0}$ is integral, it would follow that $r \in \overline{R}$, which is a contradiction. Hence, we must have $r \in \overline{R}$.

Thus, even in this case, we arrive at a conclusion analogous to the previous one.

COROLLARY 2.9. *Let R be a commutative ring, I be a regular ideal of R, and $f(t) = t^2 + at = t(t + a)$. Then, the integral closure of $R(I)_{a,0}$ is*

$$\overline{R(I)_{a,0}} = \overline{R} + (\overline{R} : a)t.$$

PROOF. The inclusion $\overline{R(I)_{a,0}} \subseteq \overline{R} + (\overline{R} : a)t$ follows from the remark above. Conversely, it is enough to observe that if $r, as \in \overline{R}$, then the coefficients $2r - as$ and $r^2 - ars$ of equation (2.1) are both in $\overline{R}$. Hence, $r + st \in \overline{R(I)_{a,0}}$. □

In general, if the polynomial $f(t) = t^2 + at + b$ is not of the form $t^2 + b$ or $t^2 + at$, we have the following result.

PROPOSITION 2.10. *Let R be a commutative ring, I be a regular ideal of R, and $f(t) = t^2 + at + b$. Let $L = (\overline{R} : a) \cap (\overline{R} : b) \subseteq Q(R)$, and assume that $bL^2 \subseteq \overline{R}$. Then $\overline{R} + Lt \subseteq \overline{R(I)_{a,b}}$.*

PROOF. Let $\alpha = r + st \in \overline{R} + Lt$; it follows that $r, as, bs, bs^2 \in \overline{R}$. Thus $2r - as \in \overline{R}$ and $r(r - as) + bs^2 = r^2 + bs^2 - ars \in \overline{R}$. Hence $\alpha \in \overline{R(I)_{a,b}}$, so $\overline{R} + Lt \subseteq \overline{R(I)_{a,b}}$. □

REMARK 2.11. For Nagata's idealization $R \ltimes I = R(I)_{0,0}$ and for the amalgamated duplication $R \bowtie I = R(I)_{-1,0}$ of R with respect to I, it is known that their integral closures are, respectively, $\overline{R \ltimes I} = \overline{R} + Q(R)t$ and $\overline{R \bowtie I} = \overline{R} + \overline{R}t$. On the other hand, we have $(\overline{R} : 0) = Q(R)$ and $(\overline{R} : -1) = \overline{R}$. In both cases, we have $\overline{R(I)_{a,b}} = \overline{R} + Lt$.

Therefore, it is reasonable to believe that $\overline{R(I)_{a,b}} = \overline{R} + Lt$ for any $a, b \in R$ with $bL^2 \subseteq \overline{R}$. However, in general, we were not able to prove the equality.

Now, we want to focus on understanding the integral closure of $R(I)_{a,b}$ when the polynomial $f(t)$ is reducible in $R[t]$. Therefore, we assume that there exist

$\xi, \eta \in R$ such that $f(t) = (t-\xi)(t-\eta)$. Then we have the following diagram:

$$\begin{array}{ccccc} Q & \xrightarrow{\sim} & \frac{Q(R)[t]}{(t-\xi)(t-\eta)} & \xrightarrow{\Psi} & Q(R) \times Q(R) \\ & & \uparrow & & \uparrow \\ & & \frac{R[t]}{(t-\xi)(t-\eta)} & \longrightarrow & R \times R \\ & & \uparrow & \overset{\Theta}{\nearrow} & \\ & & R(I)_{a,b} & & \end{array}$$

where Ψ is defined by $\Psi(\varphi(t)) = (\varphi(\xi), \varphi(\eta))$ for any $\varphi(t) \in Q(R)[t]$.

Assume that $\xi - \eta \in W(R)$. This implies that $\xi - \eta$ is invertible in $Q(R)$, and since $\xi - \eta = (t-\eta) - (t-\xi) \in Q(R)[t]$, it follows that the ideals $I_\xi = (t-\xi)Q(R)[t]$ and $I_\eta = (t-\eta)Q(R)[t]$ are comaximal in $Q(R)[t]$. Therefore, by the Chinese Remainder Theorem, Ψ is a ring isomorphism. Clearly, we can observe Θ as the restriction of Ψ to $R(I)_{a,b}$. In particular, Θ is injective. Hence, we need to compute $\mathrm{Im}(\Theta) = \Psi(R(I)_{a,b})$.

REMARK 2.12. Let $r + it \in R(I)_{a,b}$. Then $\Psi(r+it) = (r+i\xi, r+i\eta)$, so $\mathrm{Im}(\Theta) \subseteq K := \{(r+i\xi, r+i\eta) \mid r \in R,\ i \in I\}$.

On the other hand, if $(x,y) \in K$, there exists $r \in R$ and $i \in I$ such that $(x,y) = (r+i\xi, r+i\eta)$. If we assume $r+it \in R(I)_{a,b}$, then $\Psi(r+it) = (x,y)$, which implies $K \subseteq \mathrm{Im}(\Theta)$. Hence $\mathrm{Im}(\Theta) = K$.

Now, we aim to determine the integral closure of K in $Q(R) \times Q(R)$. This analysis will, in turn, provide insight into the integral closure of $R(I)_{a,b}$, since Ψ is injective. More precisely, we have $\overline{R(I)_{a,b}} = \Psi^{-1}(\overline{K})$. To proceed, let us consider the canonical projections $\pi_1 \colon Q(R) \times Q(R) \to Q(R)$ and $\pi_2 \colon Q(R) \times Q(R) \to Q(R)$ which project onto the first and second components, respectively.

LEMMA 2.13. *Under the above-mentioned conditions, the integral closure of* K *in* $Q(R) \times Q(R)$ *is* $\overline{K} = \overline{R} \times \overline{R}$.

PROOF. Let $\alpha = (r,s) \in Q(R) \times Q(R)$ be integral over K, and let $\varphi(t) \in K[t]$ be a monic polynomial such that $\varphi(\alpha) = 0$. Write

$$\varphi(t) = t^n + \sum_{h=0}^{n-1} a_h t^h \quad \text{with} \quad a_h \in K;$$

since every a_h belongs to K, we can write $a_h = (r_h + i_h\xi, r_h + i_h\eta)$ with $r_h \in R$ and $i_h \in I$. Therefore,

$$\begin{aligned} \varphi(t) &= t^n + \sum_{h=0}^{n-1} (r_h + i_h\xi, r_h + i_h\eta) t^h \\ &= \left(t^n + \sum_{h=0}^{n-1} (r_h + i_h\xi) t^h, t^n + \sum_{h=0}^{n-1} (r_h + i_h\eta) t^h \right). \end{aligned}$$

This shows that if (r,s) is integral over K, then r and s are integral over $\pi_1(K)$ and $\pi_2(K)$, respectively.

Conversely, let $\alpha = (r,s) \in \overline{\pi_1(K)} \times \overline{\pi_2(K)} \subseteq Q(R) \times Q(R)$. Then, there exist two monic polynomials $\varphi_1(t) \in \pi_1(K)[t]$ and $\varphi_2(t) \in \pi_2(K)[t]$ such that $\varphi_1(r) = 0$

and $\varphi_2(s) = 0$. We observe that, for every $(x, y) \in \pi_1(K) \times \pi_2(K) \subseteq R \times R$, there are $x', y' \in R$ such that $(x, x'), (y', y) \in K$. Therefore, if we write

$$\varphi_1(t) = t^n + a_{n-1}t^{n-1} + \cdots + a_1 t + a_0,$$

with $a'_h \in R$ such that $(a_h, a'_h) \in K$ for each $h = 0, 1, \ldots, n$, with $a_n = 1$. Hence, we may consider the polynomial

$$\varphi'_1(t) = (1, a'_n)t^n + \sum_{h=0}^{n-1}(a_h, a'_h)t^h \in K[t].$$

Moreover, since $\Theta(1) = (1, 1)$, we have $(1, 1) \in K$. Therefore, we may choose $a'_n = 1$, ensuring that $\varphi'_1(t)$ is a monic polynomial, in a similar way, we can consider the monic polynomial $\varphi'_2(t) \in K[t]$. Now, consider the polynomial $\varphi(t) = \varphi'_1(t)\varphi'_2(t) \in K[t]$, which is a monic polynomial satisfying $\varphi(\alpha) = 0$, which implies $\alpha \in \overline{K}$. This shows that $\overline{K} = \overline{\pi_1(K)} \times \overline{\pi_2(K)}$.

Since $\pi_1(K) = \pi_2(K) = R$, we have $\overline{\pi_1(K)} = \overline{\pi_2(K)} = \overline{R}$. Hence

$$\overline{K} = \overline{R} \times \overline{R}. \qquad \square$$

Now, we are ready to demonstrate the following result.

THEOREM 2.14. *Given a commutative ring R and a regular ideal I, suppose that the polynomial $f(t)$ can be factored as $f(t) = (t - \xi)(t - \eta)$ for some $\xi, \eta \in R$. If the difference $\xi - \eta$ is an element of $W(R)$ and possesses an inverse in R, then $\overline{R(I)_{a,b}} = \overline{R} + \overline{R}t$.*

PROOF. Let $(r, s) \in Q(R) \times Q(R)$. We observe that for any $w \in Q(R)$, we can write $r = (r - w\xi) + w\xi$ and $s = (s - w\eta) + w\eta$. In particular, if we take $w = (r - s)(\xi - \eta)^{-1}$, we get $r - w\xi = s - w\eta$. Indeed,

$$\begin{aligned} r - w\xi &= r - (r - s)(\xi - \eta)^{-1}\xi \\ &= (r(\xi - \eta) - (r - s)\xi)(\xi - \eta)^{-1} \\ &= (r\xi - r\eta - r\xi + s\xi)(\xi - \eta)^{-1} \\ &= (s\xi - r\eta)(\xi - \eta)^{-1}. \end{aligned}$$

And similarly:

$$\begin{aligned} s - w\eta &= s - (r - s)(\xi - \eta)^{-1}\eta \\ &= (s(\xi - \eta) - (r - s)\eta)(\xi - \eta)^{-1} \\ &= (s\xi - s\eta - r\eta + s\eta)(\xi - \eta)^{-1} \\ &= (s\xi - r\eta)(\xi - \eta)^{-1}. \end{aligned}$$

So, $r - w\xi = s - w\eta$. Therefore, $\Psi(r - w\xi + wt) = (r - w\xi + w\xi, s - w\eta + w\eta) = (r, s)$.

Let $r + st \in \overline{R(I)_{a,b}}$. Then $\Psi(r + st) = (r + s\xi, r + s\eta) \in \overline{R} \times \overline{R}$. Since $(\xi - \eta)^{-1} \in \overline{R}$ and $s(\xi - \eta) = (r + s\xi) - (r + s\eta) \in \overline{R}$, we must have $s \in \overline{R}$, which also implies $r \in \overline{R}$. Thus $r + st \in \overline{R} + \overline{R}t$, hence $\overline{R(I)_{a,b}} \subseteq \overline{R} + \overline{R}t$.

Conversely, let $r + st \in \overline{R} + \overline{R}t$. Then $\Psi(r + st) = (r + s\xi, r + s\eta) \in \overline{R} \times \overline{R}$. Therefore, $r + st \in \Psi^{-1}(\overline{R} \times \overline{R}) = \overline{R(I)_{a,b}}$, hence $\overline{R(I)_{a,b}} = \overline{R} + \overline{R}t$. $\square$

It would be interesting to understand whether weaker conditions on the polynomial $f(t)$ can be identified that characterize when $\overline{R(I)_{a,b}}$ coincides with $\overline{R} + \overline{R}t$. However, since this question is beyond the scope of the present paper, it will not be addressed here. Instead, we will work under the assumption that $\overline{R(I)_{a,b}} = \overline{R}+\overline{R}t$, which is more tractable for our study of the weakly Arf property.

3. Weakly Arf property

Before proceeding to the study the WA property, we observe certain facts about the relations between the set nonzero-divisors of R and $R(I)_{a,b}$.

REMARK 3.1. We notice that $W(R) \subseteq W(R(I)_{a,b})$. Indeed, let $r \in W(R)$. Suppose $r(s+jt) = rs + rjt = 0$ in $R(I)_{a,b}$, then $rs = rj = 0$ in R. Since $r \in W(R)$, it follows that $s = j = 0$. Hence $r \in W(R(I)_{a,b})$.

Now, suppose that R is a Noetherian, Cohen-Macaulay, local ring and that I is a maximal Cohen-Macaulay R-module; that is, I is a Cohen-Macaulay R-module with the same dimension as R, as an R-module (the dimension of I as an R-module is the Krull dimension of the ring $R/\operatorname{Ann}(I)$). Then, by [**BDS**, Proposition 2.7] $R(I)_{a,b}$ is a Cohen-Macaulay local ring, and the ideal (0) is pure in both R and $R(I)_{a,b}$. In the Cohen-Macaulay setting, it is well known that the set of zero-divisors is the union of all the minimal prime ideals of the zero ideal. For a ring R, we define $\operatorname{Min}(R)$ to be the set of the minimal prime ideals of the zero ideal.

REMARK 3.2. If the polynomial $f(t) = (t-\xi)(t-\eta)$ is reducible in $R[t]$, let $\mathfrak{p}$ be a prime ideal of R. By [**DS**, Lemma 1.1] and [**BDS**, Proposition 1.9], there are at most two prime ideals of $R(I)_{a,b}$ lying over $\mathfrak{p}$, and they are:

$$\mathfrak{q}_\xi := \{s+jt \mid s \in R,\ j \in I,\ s+j\xi \in \mathfrak{p}\}$$
$$\mathfrak{q}_\eta := \{s+jt \mid s \in R,\ j \in I,\ s+j\eta \in \mathfrak{p}\}.$$

LEMMA 3.3. *Assume that $R(I)_{a,b}$ is a Cohen-Macaulay local ring. Suppose further that $f(t) = (t-\xi)(t-\eta)$ for some $\xi, \eta \in R$. Then, an element $r+it \in R(I)_{a,b}$ is a zero-divisor if and only if at least one of $r+i\xi$ and $r+i\eta$ is a zero-divisor of R.*

In particular, $r+it \in W(R(I)_{a,b})$ implies $r+i\xi, r+i\eta \in W(R)$.

PROOF. Let $r+it \in R(I)_{a,b}$ be a zero-divisor. Then $r+it \in \mathfrak{q}$ for some prime ideal $\mathfrak{q} \in \operatorname{Min}(R(I)_{a,b})$. By [**BDS**, Remark 1.10], we have $\mathfrak{p} = \mathfrak{q} \cap R \in \operatorname{Min}(R)$. Since $f(t) = (t-\xi)(t-\eta)$, the above remark implies that $\mathfrak{q} = \mathfrak{q}_\xi$ or $\mathfrak{q} = \mathfrak{q}_\eta$. Therefore, at least one of $r+\xi i$ and $r+\eta i$ is a zero-divisor of R.

Conversely, let $r \in R$ and $i \in I$ be such that $r+i\xi$ (or $r+i\eta$) is a zero-divisor in R. Then $r+i\xi \in \mathfrak{p}$ for some $\mathfrak{p} \in \operatorname{Min}(R)$. Again by the above remark, we must have $\mathfrak{p} = \mathfrak{q}_\xi \cap R$ (or $\mathfrak{p} = \mathfrak{q}_\eta \cap R$). Therefore, $r+it \in \mathfrak{q}_\xi \cup \mathfrak{q}_\eta$ and $\mathfrak{q}_\xi, \mathfrak{q}_\eta \in \operatorname{Min}(R(I)_{a,b})$. Hence $r+it$ is a zero-divisor in $R(I)_{a,b}$. □

REMARK 3.4. If the polynomial $f(t)$ is irreducible and $\mathfrak{p}$ is a prime ideal of R, by [**DS**, Proposition 1.2], the only prime ideal of $R(I)_{a,b}$ lying over $\mathfrak{p}$ is the set

$$\mathfrak{q} = \{s+jt \mid s \in \mathfrak{p},\ j \in I \cap \mathfrak{p}\}.$$

Moreover, if $\mathfrak{p}$ is a minimal prime ideal of R and $\mathfrak{q}'$ is a prime ideal of $R(I)_{a,b}$ contained in $\mathfrak{q}$, then $\mathfrak{q}' \cap R \subseteq \mathfrak{q} \cap R = \mathfrak{p}$. Since $\mathfrak{p}$ is a minimal prime ideal, we must have $\mathfrak{q}' \cap R = \mathfrak{p}$. Hence, by incomparability, $\mathfrak{q}' = \mathfrak{q}$. That is, $\mathfrak{q}$ is a minimal prime ideal of $R(I)_{a,b}$.

LEMMA 3.5. *Assume that $R(I)_{a,b}$ is a Cohen-Macaulay local ring. Suppose further that the polynomial $f(t)$ is irreducible in $R[t]$. If an element $r+it \in R(I)_{a,b}$ is in $W(R(I)_{a,b})$, then $r \in W(R)$ or $r+i \in W(R)$.*

PROOF. We observe that, if $r + it \in R(I)_{a,b}$ is a zero-divisor, then $r + it \in \mathfrak{q}$ for some minimal prime ideal $\mathfrak{q}$ of $R(I)_{a,b}$. Let us consider the prime ideal of R, $\mathfrak{p} = \mathfrak{q} \cap R$. Then $\mathfrak{q}$ is the only prime ideal or $R(I)_{a,b}$ lying over $\mathfrak{p}$. Therefore, $r \in \mathfrak{p}$ and $i \in I \cap \mathfrak{p}$; in particular, r and i are zero-divisors of R.

Now, if $r + it \in W(R(I)_{a,b})$, then $r \in W(R)$ or $i \in W(R)$. If r and $r + i$ are zero-divisors, then $i = (r + i) - r$ is a zero-divisor. Hence $r + it$ is a zero-divisor, which is a contradiction. □

We can summarize the above discussion as follows:

COROLLARY 3.6. *Assume that $R(I)_{a,b}$ is a Cohen-Macaulay local ring. Then, if $r + it \in W(R(I)_{a,b})$, there is $j \in I$ such that $r + j \in W(R)$.*

Now, we are ready to prove our main result.

THEOREM 3.7. *Assume that $R(I)_{a,b}$ is a Cohen-Macaulay local ring. Suppose further that I is a regular ideal of R and also an ideal of $\overline{R}$. If $\overline{R(I)_{a,b}} = \overline{R} + \overline{R}t$, then $R(I)_{a,b}$ is a WA ring if and only if R is a WA ring.*

PROOF. We assume that $R(I)_{a,b}$ is a WA ring. Let $r, s, p \in R$ be such that $r \in W(R)$ and $s, p \in r\overline{R}$. Since $\overline{R} \subseteq \overline{R(I)_{a,b}}$, we have $s, p \in r(\overline{R(I)_{a,b}})$, with $r \in W(R(I)_{a,b})$. It follows that $sp/r = q + it$ for some $q \in R$ and $i \in I$, thus $sp = rq + rit \in R$ so that $ri = 0$ and $sp/r = q \in R$; hence R is a WA ring.

Conversely, we assume that R is a WA ring. Let $\alpha, \beta, \gamma \in R(I)_{a,b}$ such that $\alpha \in W(R(I)_{a,b})$ and $\beta/\alpha, \gamma/\alpha \in \overline{R} + \overline{R}t$; we write $\alpha = r + it$, $\beta/\alpha = x + yt$ and $\gamma/\alpha = s + pt$ with $r \in R$, $i \in I$ and $x, y, s, p \in \overline{R}$. Then

$$\beta = (r + it)(x + yt) = rx - biy + (ry + ix - aiy)t$$
$$\gamma = (r + it)(s + pt) = rs - bip + (rp + is - aip)t.$$

Since $I \subseteq R$ is an ideal of $\overline{R}$, we get $biy, bip, i(x - ay), i(s - ap) \in I$; it follows that $rx, rs \in R$ and $ry, rp \in I$. Let us compute $(\beta\gamma)/\alpha$:

$$\begin{aligned}\frac{\beta\gamma}{\alpha} &= (r + it)(x + yt)(s + pt)\\ &= (rx - biy + (ry + ix - aiy)t)(s + pt)\\ &= (rx - biy)s - b(ry + ix - aiy)p\\ &\quad + [(rx - biy)p + (ry + ix - aiy)(s - ap)]t.\end{aligned}$$

We have

$$biys + b(ry + ix - aiy)p \in R$$
$$(rx - biy)p + (ry + ix - aiy)(s - ap) \in I.$$

Hence, we only need to show that $rxs \in R$. Since $\alpha \in W(R(I)_{a,b})$, by Corollary 3.6, we can find $j \in I$ such that $u = r + j \in W(R)$. We write rxs as follows:

$$rxs = rxs + jxs - jxs = uxs - jxs.$$

Since $jxs \in I$, we need to prove that $uxs \in R$. Clearly $ux, us \in u\overline{R}$, therefore $ux/u \in \overline{R}$ and $us/u \in \overline{R}$. Moreover, $ux = (r + j)x = rx + jx \in R$ (similarly $us \in R$). Now, by the WA property of R we get

$$uxs = \frac{(ux)(us)}{u} \in R.$$

This shows that $\beta\gamma/\alpha \in R(I)_{a,b}$, hence $R(I)_{a,b}$ is a WA ring. □

4. Weakly Arf property for double coverings

Let R be an integral domain and let $b \in R$ be an element that is not a square in the field of fractions $Q(R)$. Let I be a fractional ideal of R such that $bI^2 \subseteq R$, where I^2 is the R-module generated by $\{i^2 \mid i \in I\}$. Recall that a fractional ideal of R is an R-submodule of $Q(R)$ such that there exists a non-zero element $a \in R$ such that $aI \subseteq R$.

We consider the ring

$$S = \frac{Q(R)[T]}{(T^2 - b)}.$$

Let t denote the image of T in S. We define the subring

$$R + It := \{r + it \mid r \in R,\ i \in I\} \subseteq S.$$

We observe that $R + It$, which is isomorphic to $R \oplus I$ as an R-module, is endowed with the multiplication

$$(r, i) \cdot (s, j) = (rs + ijb, rj + si).$$

The ring $R + It$, constructed in this way, is called a *double covering* of R with respect to b. These rings were introduced in [**L1**], where the following remarks and results are also proved. Note that, since b is not a square in $Q(R)$, the ring $R + It$ is also an integral domain.

For the moment, assume that R is integrally closed. Furthermore, assume that R contains a field k with $\mathrm{char}(k) \neq 2$. We define the following subset of $Q(R)$:

$$\tilde{I} := \{i \in Q(R) \mid bi^2 \in R\}.$$

REMARK 4.1. Note that $bij \in R$, for any $i, j \in \tilde{I}$. In fact, if $i, j \in \tilde{I}$, then $bi^2, bj^2 \in R$. Thus $(bi^2)(bj^2) = (bij)^2 \in R$, which means the element bij is integral over R. Since R is integrally closed by assumption, it follows that $bij \in R$.

REMARK 4.2. The set $\tilde{I}$ is a fractional ideal of R. Indeed, for any $i, j \in \tilde{I}$, we have $b(i-j)^2 = bi^2 - 2bij + bj^2 \in R$. By definition, $bi^2, bj^2 \in R$, and by the previous remark, $bij \in R$. Thus $b(i-j)^2 \in R$, which implies $i - j \in \tilde{I}$. Therefore, $\tilde{I}$ is an additive subgroup of $Q(R)$. Moreover, if $r \in R$ and $i \in \tilde{I}$, then $b(ri)^2 = r^2(bi^2) \in R$; hence $\tilde{I}$ is an R-module. Finally, note that for any $i \in \tilde{I}$, we have $(bi)^2 = b(bi^2) \in R$. This means bi is integral over R. Since R is integrally closed, $bi \in R$. This holds for all $i \in \tilde{I}$, so we have $b\tilde{I} \subseteq R$. As $b \neq 0$, $\tilde{I}$ is a fractional ideal of R.

The proof of the next proposition can be found in [**L1**, Proposition 4.17].

PROPOSITION 4.3. *Let R be an integrally closed domain containing a field k with* $\mathrm{char}(k) \neq 2$. *Then* $Q(R + It) = Q(R) + Q(R)t$ *and* $\overline{R + It} = R + \tilde{I}t$.

THEOREM 4.4. *Under the hypotheses of Proposition* 4.3, *the ring $R + It$ is a WA ring.*

PROOF. Let $r_h + i_h t \in R + It$ for $h = 1, 2, 3$ such that $\frac{r_h + i_h t}{r_1 + i_1 t} \in \overline{R + It}$ for $h = 2, 3$. By Proposition 4.3, we can write $\frac{r_h + i_h t}{r_1 + i_1 t} = \tilde{r}_h + \tilde{i}_h t$ for any $h = 2, 3$ with $\tilde{r}_h \in R$ and $\tilde{i}_h \in \tilde{I}$. We must show that the product $\frac{(r_2 + i_2 t)(r_3 + i_3 t)}{r_1 + i_1 t}$ is in $R + It$. We calculate

$$\frac{(r_2 + i_2 t)(r_3 + i_3 t)}{r_1 + i_1 t} = (r_1 + i_1 t)(\tilde{r}_2 + \tilde{i}_2 t)(\tilde{r}_3 + \tilde{i}_3 t) =$$

$$(r_1\tilde{r}_2\tilde{r}_3 + r_1\tilde{i}_2\tilde{i}_3 b + i_1\tilde{r}_2\tilde{i}_3 b + i_1\tilde{r}_3\tilde{i}_2 b) + (r_1\tilde{r}_2\tilde{i}_3 + r_1\tilde{i}_2\tilde{r}_3 + i_1\tilde{r}_2\tilde{r}_3 + i_1\tilde{i}_2\tilde{i}_3 b)t.$$

Clearly, $r_1\tilde{r}_2\tilde{r}_3 \in R$. By hypothesis $bI^2 \subseteq R$, so $I \subseteq \tilde{I}$; furthermore $\tilde{I}$ is an R-module such that $b\tilde{I}^2 \subseteq R$, thus $r_1\tilde{i}_2\tilde{i}_3 b + i_1\tilde{r}_2\tilde{i}_3 b + i_1\tilde{r}_3\tilde{i}_2 b \in R$. Now, $\tilde{r}_2\tilde{r}_3 i_1$ and $i_1\tilde{i}_2\tilde{i}_3 b$ are both in I, but $r_1\tilde{r}_2\tilde{i}_3 + r_1\tilde{i}_2\tilde{r}_3 \in \tilde{I}$. For $h = 2, 3$ we have $r_h + i_h t \in R + It$ and

$$\begin{aligned} r_h + i_h t &= (r_1 + i_1 t)(\tilde{r}_h + \tilde{i}_h t) \\ &= (r_1\tilde{r}_h + i_1\tilde{i}_h b) + (r_1\tilde{i}_h + \tilde{r}_h i_1)t \end{aligned}$$

therefore $r_1\tilde{i}_h + \tilde{r}_h i_1 \in I$. Since $\tilde{r}_h i_1 \in I$, we obtain

$$r_1\tilde{i}_h = (r_1\tilde{i}_h + \tilde{r}_h i_1) - \tilde{r}_h i_1 \in I.$$

Hence, $r_1\tilde{r}_2\tilde{i}_3 + r_1\tilde{i}_2\tilde{r}_3 + i_1\tilde{r}_2\tilde{r}_3 + i_1\tilde{i}_2\tilde{i}_3 b \in I$. This shows that $R + It$ is a WA ring. □

EXAMPLE 4.5. Let k be an algebraically closed field with $\operatorname{char}(k) \neq 2$. Let $R = k[X, Y]$, $I = R$, and $b = X^3$. Then the ring $R + It \cong k[X, Y, T]/(T^2 - X^3)$ is WA ring by Theorem 4.4. This agrees with the result in [**C**, Example 3.6 (1)]).

COROLLARY 4.6. *Let R be an integrally closed domain that contains a field k with* $\operatorname{char}(k) \neq 2$ *and let I be a proper ideal of R. Let $b \in R$ be an element that is not a square in $Q(R)$. Then $R(I)_{0,-b}$ is a WA ring.*

Now we study the case in which R is not necessarily integrally closed. Assume that R is an integral domain containing a field k with $\operatorname{char}(k) \neq 2$. Let $b \in R$ be an element that is not a square in $Q(R)$ and let I be a fractional ideal of R such that $bI^2 \subseteq R$. Let $R + It$ be as above and define the subset of $Q(R)$

$$\tilde{J} := \{j \in Q(R) \mid bj^2 \in \overline{R}\}.$$

REMARK 4.7. The set $\tilde{J}$ is a fractional ideal of $\overline{R}$. The arguments are similar to those for $\tilde{I}$: for any $i, j \in \tilde{J}$, we have $(bij)^2 = (bi^2)(bj^2) \in \overline{R}$, which implies $bij \in \overline{R}$. It follows that $b(i - j)^2 = bi^2 + bj^2 - 2bij \in \overline{R}$, hence $i - j \in \tilde{J}$. For any $r \in \overline{R}$, $ri \in \tilde{J}$. Thus, $\tilde{J}$ is an $\overline{R}$-module. Finally, $(bi)^2 = b(bi^2) \in \overline{R}$, so $bi \in \overline{R}$. Therefore, $b\tilde{J} \subseteq \overline{R}$, making $\tilde{J}$ a fractional ideal of $\overline{R}$. We also note that if R is integrally closed, then $\tilde{J} = \tilde{I}$.

In this case as well, in order to study the WA property, it is necessary to first determine the integral closure of $R + It$. As we proceed with this analysis, the Lemma 2.4 will prove crucial.

PROPOSITION 4.8. *Let R be an integral domain that contains a field k with* $\operatorname{char}(k) \neq 2$. *If R is not integrally close, then $\overline{R + It} = \overline{R} + \tilde{J}t$.*

PROOF. Every element $r + st \in Q(R) + Q(R)t$ is a root of the polynomial $p(X) = X^2 - 2rX + r^2 - bs^2 \in Q(R)[X]$, as shown by the identity:

$$(r + st)^2 - 2r(r + st) + r^2 - bs^2 = 0.$$

If an element $r + st$ belongs to $R + It$, its corresponding polynomial $p(X)$ has coefficients in R. This shows that every element of $R + It$ is integral over R, making the extension $R \subseteq R + It$ integral. Consequently, the integral closure of R in $Q(R) + Q(R)t$, denoted by $\check{R}$, coincides with $\overline{R + It}$.

We now show the inclusion $\overline{R} + \tilde{J}t \subseteq \overline{R + It}$. Let $r + st \in \overline{R} + \tilde{J}t$. By definition the coefficients of its associated polynomial $p(X)$ lie in $\overline{R}$. Therefore, $r + st$ is integral over $\overline{R}$. Since $\overline{R}$ is an integral extension of R, it follows by transitivity that $r + st$ is integral over R, which means $r + st \in \check{R} = \overline{R + It}$.

For the converse inclusion, let $r + st \in \overline{R + It}$. This means $r + st$ is integral over R, so it is a root of some monic polynomial $h(X) \in R[X]$. As $r + st$ is also a root of $p(X) \in Q(R)[X]$, $p(X)$ must divide $h(X)$ over $Q(R)$. By Lemma 2.4, the coefficients of $p(X)$ must be integral over R. Specifically, $2r \in \overline{R}$ and $r^2 - bs^2 \in \overline{R}$. Since R contains a field of characteristic different from 2, we can deduce that $r \in \overline{R}$, which in turn implies $bs^2 \in \overline{R}$. This shows that $r + st \in \overline{R} + \tilde{J}t$, completing the proof. □

We conclude this paper with the following result.

THEOREM 4.9. *Let R be an integral domain containing a field of characteristic different from two, let $b \in R$ be a non-square in $Q(R)$, and let I be a fractional ideal of R satisfying $b^2 I \subseteq R$. If $R + It$ is a WA ring, then so is R.*

PROOF. To prove that R is a WA ring, we must show that for any $x, y, z \in R$ with $x \neq 0$ and $y/x, z/x \in \overline{R}$, it follows that $yz/x \in R$.

The condition $y/x, z/x \in \overline{R}$ and the inclusion $\overline{R} \subseteq \overline{R + It}$ imply that $y/x, z/x \in \overline{R + It}$. Since $R + It$ is a WA ring by hypothesis, this implies $yz/x \in R + It$. An element in $R + It$ has the form $r + it$ for some $r \in R$ and $i \in I$. On the other hand, the element yz/x belongs to $Q(R)$. The equality $yz/x = r + it$ can only hold in the ring $Q(R) + Q(R)t$ if $i = 0$. Therefore, $yz/x \in R$, which concludes the proof. □

We conclude this section by observing that, in light of the results in [**C**], it is reasonable to conjecture that $R + It$ is a WA ring if and only if R is a WA ring and I is also an ideal of $\overline{R}$. However, we have not yet found a complete proof.

References

[A1] T. Albu, *On a paper of Uchida concerning simple finite extensions of Dedekind domains*, Osaka Math. J. **16** (1979), no. 1, 65–69. MR527017

[A2] C. Arf, *Une interprétation algébrique de la suite des ordres de multiplicité d'une branche algébrique* (French), Proc. London Math. Soc. (2) **50** (1948), 256–287, DOI 10.1112/plms/s2-50.4.256. MR31785

[BDS] V. Barucci, M. D'Anna, and F. Strazzanti, *A family of quotients of the Rees algebra*, Comm. Algebra **43** (2015), no. 1, 130–142, DOI 10.1080/00927872.2014.897549. MR3240409

[C] E. Celikbas, O. Celikbas, C. Ciupercă, N. Endo, S. Goto, R. Isobe, and N. Matsuoka, *On the ubiquity of Arf rings*, J. Commut. Algebra **15** (2023), no. 2, 177–231, DOI 10.1216/jca.2023.15.177. MR4611108

[DF] M. D'Anna and M. Fontana, *An amalgamated duplication of a ring along an ideal: The basic properties*, J. Algebra Appl. **6** (2007), no. 3, 443–459, DOI 10.1142/S0219498807002326. MR2337762

[DS] M. D'Anna and F. Strazzanti, *New algebraic properties of quadratic quotients of the Rees algebra*, J. Algebra Appl. **18** (2019), no. 3, 1950047, 14, DOI 10.1142/S0219498819500476. MR3924825

[E] D. Eisenbud, *Commutative algebra: With a view toward algebraic geometry*, Graduate Texts in Mathematics, vol. 150, Springer-Verlag, New York, 1995, DOI 10.1007/978-1-4612-5350-1. MR1322960

[L1] A. M. A. Licata, *Double covering of curves and duplication of semigroups and rings*, PhD thesis, Universitá degli srudi di Catania, (2020).

[L2] J. Lipman, *Stable ideals and Arf rings*, Amer. J. Math. **93** (1971), 649–685, DOI 10.2307/2373463. MR282969

UNIVERSITÁ DEGLI STUDI DI PALERMO
Email address: damiano.saccone@unipa.it

SELECTED PUBLISHED TITLES IN THIS SERIES

836 **Scott T. Chapman, Editor,** The Ideal Theory and Arithmetic of Rings, Monoids, and Semigroups, 2026

835 **Mee Seong Im and Tony Shaska, Editors,** Artificial Intelligence and Mathematics Research, 2026

834 **Mee Seong Im, Hanna E. Makaruk, Bach Nguyen, and Robert Owczarek, Editors,** Inverse Problems, Geometric and Topological Aspects of Representation Theory, and Low-Dimensional Topology, 2026

832 **Nils Bruin, David Kohel, and Chloe Martindale, Editors,** Arithmetic, Geometry, Cryptography and Coding Theory, 2026

831 **Irena Swanson and Kei-ichi Watanabe, Editors,** Lectures on Tight Closure and Its Applications, 2025

830 **O. Bezushchak, Yu. Drozd, P. Etingof, V. Futorny, R. Grigorchuk, and E. Zelmanov, Editors,** Modern Algebra Volume 2, 2025

829 **O. Bezushchak, Yu. Drozd, P. Etingof, V. Futorny, R. Grigorchuk, and E. Zelmanov, Editors,** Modern Algebra Volume 1, 2025

828 **Hafedh Herichi, Maria Rosaria Lancia, Therese Basa Landry, Anna Rozanova-Pierrat, and Steffen Winter, Editors,** Fractal Geometry in Pure and Applied Mathematics, 2025

827 **Carmen Caprau, J. Scott Carter, Neslihan Gügümcü, and Sam Nelson, Editors,** Algebraic Structures in Knot Theory, 2025

826 **Ahmed Laghribi and André Leroy, Editors,** Algebraic Structures and Applications, 2025

825 **Saurabh Verma, María A. Navascués, and Amit Priyadarshi, Editors,** Advances in Dimension Theory, Fractal Functions and Measures, 2025

824 **Hanna E. Makaruk and Robert M. Owczarek, Editors,** Inverse Problems: In Memory of Professor Zbigniew Oziewicz, 2025

823 **Mikhail Bershtein, Anton Dzhamay, and Andrei Okounkov, Editors,** The Versatility of Integrability, 2025

822 **Ahmad Barhoumi, Roozbeh Gharakhloo, and Andrei Martinez-Finkelshtein, Editors,** Recent Developments in Orthogonal Polynomials, 2025

821 **Bogdan D. Suceavă, Adara M. Blaga, Marian Ioan Munteanu, Cezar Oniciuc, Shoo Seto, Gabriel-Eduard Vîlcu, and Lihan Wang, Editors,** Recent Advances in Differential Geometry and Related Areas, 2025

820 **Sudeb Mitra, Rakesh Kumar Parmar, Saminathan Ponnusamy, and Toshiyuki Sugawa, Editors,** Geometric Function Theory and Related Topics, 2025

819 **Howard S. Cohl, Roberto S. Costas-Santos, and Robert S. Maier, Editors,** Applications and q-Extensions of Hypergeometric Functions, 2025

818 **Howard S. Cohl, Roberto S. Costas-Santos, and Robert S. Maier, Editors,** Classical Hypergeometric Functions and Generalizations, 2025

817 **Mikhail Khovanov, Joshua Sussan, and Anton Zeitlin, Editors,** From Representation Theory to Mathematical Physics and Back, 2025

816 **Luca F. Di Cerbo and Laurenţiu G. Maxim, Editors,** Geometry and Topology of Aspherical Manifolds, 2025

815 **Daniel Orr and Joshua Jeishing Wen, Editors,** Macdonald Theory and Beyond, 2025

814 **Susan Montgomery and Siu-Hung Ng, Editors,** Quantum Groups, Hopf Algebras, and Applications, 2025

813 **Thomas Creutzig, Julia Plavnik, and David Ridout, Editors,** Quantum Symmetries, 2025

For a complete list of titles in this series, visit the
AMS Bookstore at **www.ams.org/bookstore/conmseries/**.

FSC
www.fsc.org
MIX
Papier | Fördert
gute Waldnutzung
FSC® C083411

Kolibri 360 GmbH
Ferdinand-Jühlke-Straße 7
99095 Erfurt, Deutschland
produktsicherheit@kolibri360.de